Risk, Reliability and Safety Engineering

Series Editors

Prabhakar V. Varde, Reactor Group, Bhabha Atomic Research Centre, Mumbai, Maharashtra, India

Ajit Kumar Verma, Western Norway University of Applied Sciences, Faculty of Engineering and Natural Sciences, Haugesund, Norway

Uday Kumar, Luleå University of Technology, Luleå, Sweden

In this era of globalization and competitive scenario there is a conscious effort to ensure that while meeting the reliability targets the potential risk to society is minimal and meet the acceptability criteria towards achieving long term targets, including sustainability of a given technology. The objective of reliability is not only limited to customer satisfaction but also important for design, operating systems, products, and services, while complying risk metrics. Particularly when it comes to complex systems, such as, power generation systems, process systems, transport systems, space systems, large banking and financial systems, pharmaceutical systems, the risk metrics becomes an overriding factor while designing and operating engineering systems to ensure reliability not only for mission phase but also for complete life cycle of the entity to satisfy the criteria of sustainable systems.

This book series in Risk, Reliability and Safety Engineering covers topics that deal with reliability and risk in traditional sense, that is based on probabilistic notion, the science-based approaches like physics-of-failure (PoF), fracture mechanics, prognostics and health management (PHM), dynamic probabilistic risk assessment, risk-informed, risk-based, special considerations for human factor and uncertainty, common cause failure, AI based methods for design and operations, data driven or data mining approaches to the complex systems. Within the scope of the series are monographs, professional books or graduate textbooks and edited volumes on the following topics:

- Physics of Failure approach to Reliability for Electronics
- Mechanics of Failure approach to Mechanical Systems
- Fracture Risk Assessment
- Condition Monitoring
- Risk Based-In-service Inspection
- Common Cause Failure
- Risk-based audit
- Risk-informed operations management
- Reliability Cantered Maintenance
- Human and Institutional Factors in Operations
- Human Reliability
- Reliability Data Analysis
- Prognostics and Health Management
- Risk-informed approach
- Risk-based approach
- Digital System Reliability
- Power Electronics Reliability
- Artificial Intelligence in Operations and Maintenance
- Dynamic Probabilistic Risk Assessment
- Uncertainty
- Aging Assessment & Management
- Risk and Reliability standards and Codes
- Industrial Safety

Potential authors who wish to submit a book proposal should contact:

Priya Vyas, Editor, e-mail: Priya.vyas@springer.com

Prabhakar V. Varde

Risk-Conscious Operations Management

An Integrated Paradigm for Complex Engineering System

Springer

Prabhakar V. Varde
Reactor Group
Bhabha Atomic Research Centre
Mumbai, India

ISSN 2731-7811 ISSN 2731-782X (electronic)
Risk, Reliability and Safety Engineering
ISBN 978-981-19-9333-6 ISBN 978-981-19-9334-3 (eBook)
https://doi.org/10.1007/978-981-19-9334-3

This Springer imprint is published by the registered company Springer Nature Singapore Pte Ltd.
The registered company address is: 152 Beach Road, #21-01/04 Gateway East, Singapore 189721, Singapore

Dedicated to
Bhabha Atomic Research Reactor and my fellow team members for their cooperation, help and support all these years in Reactor Group;
My family: Wife Mrs. Meena Varde and daughter Ms. Saily Varde who stood by myside and supported in these challenging times, particularly for their understanding in this project

Preface

It was around 4 pm in the evening on June 2, 2019. I heard my daughter Saily's voice on my cell phone that her flight from Mumbai to Indore was diverted to Vadodara due to heavy air/weather turbulences, which she described as a frightening experience, and the flight had reached Vadodara. Later again when I called up 8:10 pm, she told me that now their flight was about to take off for Indore and finally she reached safely at Indore Airport. This story which lasted a good couple of hours and my experience in nuclear plant operational safety for over 30 years reaffirmed my motivation to write my next book on risk-conscious operations management. In this context, apart from engineering complexity, external factors put the whole operations management to new tests where decisions play the critical role. My daughter also informed that the condition of the emergency jackets, she was not sure, whether they have been checked for adequacy for use in postulated emergency scenario. More interesting observation was that, in her words, except the first-time fliers and one six-month-old baby, everyone was tense for the risk which a few hours ago was potential or passive now appears to be sort of real. This scenario certainly highlights postulation of role of risk management in operational environment, particularly for safety and mission critical systems. One more aspect that is vital in the context of risk or safety is that the attention for safety finds a paradigm shift. The responsible factors or causes for the undesired events suddenly catch attention. However, if the same aspects were argued before the accident or threat, it would have been challenging to put forward and accept for further actions.

Another motivation for me to work on this project was that, after publishing the book entitled *Risk-based Engineering* in May 2018, coauthored with Prof. Michael Pecht, it was natural for me to develop a subject 'risk-conscious operations management' where the focus is on management of assets such that risk is, not only acceptably low but further lower such that safety margin increases, while system performance and deliverables meet the set objectives, be it complex nuclear systems, process and chemical plants, space or aviation, etc. Even though the nuclear industry has always been relatively safe or conversely speaking risk-conscious society where safety is an overriding factor. After devoting my significant part of career on modeling for probabilistic risk assessment (PRA), I felt there is new approach required to understand and model not only human–machine interface, but also the organizational aspects with an aim to reduce the possibility of human error. Hence, I felt the need to have a single-point reference for operations risk-conscious management that deals with, apart from routine activities and anticipated transient, management actions to control deviations from normal operations or even abnormal/accident conditions. My experience in operations manning and later operations management and regulatory activities was the major driver for me to develop this subject not only for safety or conversely risk critical systems where risk-conscious approach can provide an improved framework even to meet the performance targets more effectively.

This book presents a new integrated paradigm for complex engineering systems. The subject also deals with a complex operational environment where the two otherwise contrary objectives, i.e., maintenance of high performance or deliverables and higher safety, i.e. ensuring lowest risk, are to be optimized such that plant meets all expectations of production while keeping the risk levels acceptably low such that operations of the system is justified in meeting societal objectives.

Even though the plant design procedures have human considerations as one of the major factors, the role of humans in operations management, at times become demanding. One of the major characteristics of the operations management is that even though enough automation, redundant provisions, application of diversity, availability of adequate thermal and nuclear margins, it is human interaction-intensive ecosystem particularly during emergency and the human factor plays a critical role for maintaining production objective while maintaining the lowest achievable risk levels. The history of accidents in operation of risk and mission critical systems, like energy-producing systems, e.g., nuclear, thermal, space, hydro systems, process, and chemical industry, and various modes of transport like aviation, railways, road transport, e-cybernetics, reveals that human error is one of the major contributors to accident. The message is clear and loud, operational aspects, particularly human factor improvements as also reducing dependence on human interaction particularly the one postulated for emergency scenario, requires special attention.

This book is based on my over 30 years' experience in operations and research on probabilistic risk assessment that includes modelling of human factor. A conscious attempt has been made to keep the narrative as generic such that specifics as also mathematics can be avoided, however, wherever required to communicate the concept, while avoiding the specificity a general approach has been used. In fact, this book is a sequel to the first book *Risk-based Engineering—An Integrated Approach to Complex Engineering Systems* published by Springer in 2018. This book has created a new subject 'Risk-based Engineering' where it takes the existing field of probabilistic risk assessment that led to development of risk-informed decisions and finally the risk-based engineering by providing an integrated approach where the deterministic and probabilistic approaches in risk assessment consolidate the subject into risk-based methodology. Risk-based engineering was published in 2018 and doing well among the researchers, academician and scholars, and students with ~ 20,000 downloads. In risk-conscious operations management, the idea has been further extended to operational environment keeping in typical characteristic of operations where apart from hardware and software system performance, human performance is critical particularly during deviation, off normal, and accident conditions.

The striking feature of this book is that it uses 'risk' as the central idea in place of 'safety' for one simple reason that risk has a mathematical connotation which provides opportunity for using parameters that can be measured unlike 'safety' which can be an ever-moving vague and imprecise target, where risk being a quantified entity tends to serve better that also include presenting the uncertainty. The second feature is, application of an approach that tends to be more holistic for treating human model not based on symptoms but based on the first principal treatment to human model for evaluation of human performance right going to deeper level of consciousness-the main driver of human system.

It may be noted that consciousness itself is a subject of research and there is no consensus among the spiritual/philosophical, neuroscience, quantum mechanics and chemical science experts as to what is consciousness. To address this challenge, it was required to have available references in the ancient book of records where there were strong references for understanding of not only consciousness but elegant input on human model. This book uses certain facts, like all agree that consciousness exists, consciousness can be related to a higher level to alertness and awareness and finally 'since we are conscious, we are existing'. This work is employing to the extent possible the applied and consensual concepts of consciousness, postulates and assumptions that are based on either philosophical or spiritual references such that evidence at experiential levels, if not at the measurement levels are available, to create a more robust human model.

Keeping in view the above requirements, the whole subject has been covered in ten chapters as follows: Chapter 1 provides the ***Introduction*** of the subject; Chap. 2 discusses one of the central theme, where a considerable research was required—the ***Consciousness***—this chapter reviews the state of the art in consciousness, identifies and formulates the applied attributes of consciousness and how it supports human cognition toward developing human factors for operational environment; Chapter 3 presents the reliability concept to support ***Dependency Modelling***, mainly includes hardware systems structures and components reliability improvement and risk reduction concepts; Chapter 4 develops and built attributes and model for '***risk-conscious culture***'—critical to, characterize operational approach to risk; Chapter 5 provides an overview of the integrated ***risk-based engineering*** that provides basic framework for risk-conscious operations management; further, Chapter 6 deals with ***risk simulation*** where an integrated approach using plant that deals with design and testing of simulator experiments along with the probabilistic risk assessment and deterministic safety assessment methods to predict the risk levels for postulated accident scenario; Chapter 7 presents ***human factor modeling***, where it works on developing an approach for human error precursor analysis such that human error can be eliminated or avoided at organizational level before it produces undesirable consequences, apart from this the second part of this chapter deals with human reliability modelling; Chapter 8 deals with implementation aspects of risk-conscious operations management and its applications. Chapter 9 investigates role of artificial intelligence/machine learning for development of operator support systems. This chapter presents the modelling of artificial intelligence and fuzzy logic system for development of an operator support system. This system has two major functions, viz. reactor transient identification employing artificial neural network and fuzzy rule-based system for performing the diagnostics. The objective here is to reduce the stress levels on the operator due to cognitive loads and thereby reduce chances of error during the emergency condition. Maintenance management is a major component complex engineering system in reactor operation. The experience suggests that human error contribution during maintenance activities contributes to not only system unavailability but also has potential risk implications. A new approach where the risk is the major driver called risk-conscious maintenance management system approach has been developed and presented in Chapter 10.

Since the human error is the major contributing factors to unavailability and risk, this book presents the risk-conscious operations management approach and expects that application of this framework has potential to risk reduction in complex engineering systems.

Mumbai, India

Prabhakar V. Varde

Disclaimer

The views expressed in this book are experience and opinion of author and not the view of any organization he is associated with. The book contains information derived from authentic and highly regarded sources. All the attempts have been made to ensure that the credit is given to the earlier work done by individual or organization by giving proper references or citations. Even though author's experience during his professional service has been reflected in this book, to the extent possible, care has been taken not to present any data and information of classified nature. Even though a formal permission was taken from the department, the author has been serving before his association with his employer as regular employee; further, conscious efforts were made to ensure that no information of classified nature forms part of this project. Reasonable efforts have been made to publish reliable data and information, but the author and publisher cannot assume responsibility for the validity, applicability or correctness of the material or the consequences of their use. Author and publishers have made their best efforts to ensure copyright is not violated and accordingly for the material used in this work where copyright permission was taken wherever required. However, if any copyright material has not been acknowledged, please do write and let us know so we may rectify in future edition of this work.

Contents

About the Author

Prof. Prabhakar V. Varde is a mechanical engineer by education, nuclear engineer by profession and an expert in the area of probabilistic risk assessment and its application as part of risk-based engineering. He started his career at the Bhabha Atomic Research Centre in 1983 as nuclear engineering trainee of BARC Training School. After completion of training, Dr Varde joined erstwhile Reactor Operations and Maintenance Group (now Reactor Group), served as operations engineering for Dhruva—a 100 MW research reactor at BARC and rose through the administrative ladder and retired in 2019 as Associate Director, Reactor Group. During his service, he completed his Ph.D. from Indian Institute of Technology (IIT) Bombay in 1996 and later focussed his research in nuclear safety in general and risk-based engineering, in particular, along with reactor-related services responsibilities. Dr Varde also served as Senior Professor and Member of the Board of Studies in Engineering Sciences of Homi Bhabha National Institute, Mumbai. He did his postdoctoral research at Korea Atomic Energy Research Institute, in 2002-2003. He is a visiting professor at Center of Advanced Life Cycle Engineering, University of Maryland, USA. He is also honorary professor at Amity Institute of Nuclear Science and Technology, Amity University, India. Dr Varde has over 250 research publications at national and international level. He has co-authored/edited more than 17 books/proceedings. In recent times, he received Homi Bhabha Group Achievement Awards in 2017 for 'Design and Development and Commissioning of Apsara reactor' as a team member and in 2018 as team leader for 'Development and Implementing Dhruva Nuclear Reactor Simulator', respectively. Dr Varde founded the Society of Reliability and Safety (SRESA) in 2010. He is also serving as President, Society for Reliability and Safety, India. He has also been serving as lead Chief-Editor for *SRESA International Journal of Life Cycle Engineering* published by Springer.

Chapter 1
Introduction

1.1 Background and Introduction

The nuclear industry is credited with maintaining the highest standard of safety; however, the three major accidents, viz. TMI (USA, 1978), Chernobyl (Russia 1986) and Fukushima (Japan, 2011), in a nuclear power plant provided learning lessons suggesting requirements for further safety improvement (International Atomic Energy Agency website 2021). The available literature on safety and records of accidents shows that human error is one of the major contributors to accidents (Swaton et al. 1987). Apart from the initiatives that mainly include use of innovative design, upgradations, use of simulator for training, deployment of digital technology, etc., the application of safety culture and convention on nuclear power plant safety were two major developments that can be attributed to improved safety records of the nuclear industry (International Atomic Energy Agency 1991a, 1994). Later, the development and application of safety guides and codes on the operational safety of nuclear power plants and research reactors were notable initiatives at IAEA to further strengthen safety (International Atomic Energy Agency 2008, 2006). After the 9/11 incidents, along with safety, the security aspects also form part of the safety culture in the nuclear industry (International Atomic Energy Agency 2004).

It is recognized that while there exists well-balanced automation in nuclear plants specifically considering the safety requirements, many operations activities are human interaction intensive, be it man–machine interactions, management, coordination of plant activities, communication, regulation; human factors are critical to operations, particularly during emergency scenario management, due to higher arguably considering cognitive stresses. Therefore, human factor considerations during the design as well as during operations management are a vital component to ensure plant safety while meeting operational performance goals and targets. Safety is an overriding consideration in nuclear and most of the safety critical systems; therefore, continuous improvement is part of 'safety culture'. This is the major driver for

P. V. Varde, *Risk-Conscious Operations Management*, Risk, Reliability and Safety Engineering, https://doi.org/10.1007/978-981-19-9334-3_1

improvement for human and organizational aspects that need to be further improved by employing more effective approaches such that there is higher resilience against human error.

Safety is basically notional and qualitative and bears a subjective connotation. As an ever 'moving up' target in perpetuity, it has since time memorial served well. However, in regard to rational-based systems, unlike the notion of safety, 'risk' has a mathematical basis and tends to provide a more effective and robust approach based on quantified and rational mechanisms for the identification, prioritization, planning, organization, maintenance, monitoring, characterization of uncertainty, diagnosis and management of deviations from normal operations and, more importantly, for accident mitigation and management of engineering systems. The insights on risk in risk-conscious culture are derived primarily from probabilistic risk assessment (PRA), the available reliability data with uncertainty characterization, deterministic documents and supported by experience and insights. The insight from PRA provides a vital and robust strategy for the identification, prioritization and management of safety issues in the plant. The root cause analysis in general and further down the human root provides fundamental causes that, if addressed, reduce or eliminate the chances of human error. This chapter provides an overview of salient features of a risk-conscious operations management (RCOM) approach envisaged in this book. Since the subject is operations management, there could not have been a better reference system than the nuclear industry, as the nuclear industry has demonstrated the highest standard in operations management and the same industry leaves no stone unturned to enhance operational safety. This book uses references from nuclear plants for communication of core ideas of risk-conscious operations management.

1.2 Salient Features of Nuclear Plant Operations

1.2.1 The System

The inherent characteristic features of nuclear plants that attract special safety considerations are (a) the ***chain reaction*** at a controlled rate produces heat and additional neutrons, (b) ***ionizing*** radiation that needs to be contained in a confined volume and monitored through surveillance in and around the plant, as this radiation is harmful to human health and (c) the ***decay heat*** that requires uninterrupted cooling even when the reactor is in a shutdown state (International Atomic Energy Agency 1999a). The reactor core is loaded with uranium fuel assemblies/bundles and moderators (thermal reactors) in a reactor vessel. The heat produced in the reactor core due to the chain reactor is utilized for power production in nuclear power plants, while in research reactors, neutrons are used for research, material irradiation and isotope production. The shutdown cooling system automatically cut in when the primary coolant system is not available and caters to the decay heat removal requirement from the core. Reactor systems are broadly divided into safety systems, safety support systems and process systems. Reliable operation process systems, such as reactor power regulating systems, reactor core cooling systems or primary cooling systems, electrical power supply systems and secondary and tertiary cooling systems, are required for meeting production objectives, including the electricity operation of the reactor, for example, the production of electricity and process systems.

The plant safety parameters are monitored by triplicate redundant instrumentation channels, and automatic action to shut down the reactor is based on the majority voting logic. The emergency core cooling system remains in a poised state and is activated when there is a breach in the cooling system boundary beyond a set limit that threatens the safety of the core. The containment is an engineered safety feature that isolates the reactor and associated systems directly connected to the reactor core from the outside environment and provides an additional barrier to radiation as part of the defense-in-depth approach in a highly unlikely scenario involving accidents. The control room is the hub for the control and coordination of operational activities and can be broadly divided into three parts: (a) the reactor console for reactor startup, shutdown and power maneuvering, (b) the hard or soft panels to display and control/alter the status of the systems and equipment remotely from the control room and (c) the reactor trip and alarm windows that attract operator attention through audio and visual alarm in case the parameter reaches its set limit. Of course, in recent times, there has been a move from analog to digital control rooms.

1.2.2 Safety Management of Operations Ecosystems—A Brief Overview

The safety analysis document along with design details that support the analysis of adequacy of the design against the national/international standards forms the major resource for safety management (International Atomic Energy Agency 2012). The overall operational safety functions are governed by the technical specification derived from the safety analysis report. The technical specification document stipulations mainly comprised the safety limits, limiting safety system settings and limiting conditions for operations. The maintenance and test intervals and organizational structures and functions also form part of the technical specifications. The supervisory activities of reactor and equipment status maneuvering, monitoring and surveillance on a 24 × 7 basis are performed by trained and qualified staff from a central control room, generally in the round-the-clock shifts. The technical specifications also stipulate operational staff requirements, positions, numbers and reporting hierarchies. The plant management with multilayer operating staff where the top management takes the decision making role, when required based on the input safety review input from regulatory agency. Operational management has a final responsibility for safety while ensuring reliable operation of the plant (International Atomic Energy Agency 1999b). In a broad sense, operational activities can be divided into four major categories: (a) operational activities involving man–machine interactions directly from control rooms and plant areas, (b) maintenance management, (c) safety services and (d) independent safety support reviews, including regulatory reviews, radiation protection, health monitoring and management. The major control room activities include supervisory tasks such as monitoring the status of the reactor and associated systems that include surveillance, authorization and coordination for activities with other agencies, such as maintenance staff, modification engineers, in-service inspection and radiation health physics staff. The plant operation shall follow stipulated conditions laid out in technical specifications. The identified activities are carried out using standard operating procedures. There is a separate set of procedures called emergency operating procedures. Radiation and health activities staff support the reactor operator in aspects related to radiation health and safety that include advice to operating staff.

There is a separate maintenance crew/staff that takes care of mechanical, electrical, electronics and instrumentation activities. All maintenance activities are performed through an authorization work-permit system. This essentially involves permission from the control room to initiate any maintenance or testing activity, and similarly, after completion of the maintenance activity, one more permission is required for putting the affected equipment back in service or in standby mode. Safety systems and safety support systems are tested periodically as per the surveillance schedule based on technical specification stipulations to ensure that they will cut in when needed. The reactor is shut down periodically for shutdown fueling and maintenance activities. The safety services activities include quality assurance, operational support reactor physics and reactor chemistry, training & qualification, documentation, regulatory interface, analysis & R&D support and inventory control.

1.2.3 Regulations

A regulatory agency oversees the safety activities of the plant and ensures compliance and operational enforcement. There is a three-tier review of regulatory activities and documents. The protocol for regulatory communication and review is governed by the stipulations made in the technical specification. Plants have a binding obligation to comply with the technical specification stipulations in total. Any waiver to the specifications shall be authorized in advance by the regulatory body. In case the plant's technical specification(s) are violated, a suitable regulatory review, as per the technical specification stipulations, is required to effect corrective action. Certain critical violations might require regulatory permission for the resumption of plant operations.

1.3 Consciousness—A Brief Overview

The objective of this review is to discuss the origin of state of the art in 'consciousness'—a key element of risk-conscious culture. The scope of the research covers the science and philosophy of two broad approaches. The oldest Vedic approach that has its origin approximately 1500–500 years BCE, also referred to as the spiritual approach (Nadar 2021). Even though the Vedic knowledge was passed on from generations to generations through oral recitation and memorization during the later period, it was compiled in the forms of Vedas and Upanishads. The philosophical and spiritual component of Vedic thoughts was later compiled into abstract form as dialog in Upanishads (Easwaran 2021) toward bringing out the salient concepts of '*nondual*' thoughts.

In a nondual system, consciousness is a pervading universal phenomenon. The root source of consciousness is outside of the human body, while it is all pervading and everywhere, including the individual human body. In fact, Vedas was the first to assert that consciousness is the life force in human beings. Everything in nature has consciousness as part of universal consciousness; only the quantum level differs. For example, humans have the highest level, and plants, animals and material objects have consciousness levels in this order. In humans, consciousness manifests as five levels: (a) physical, (b) vital or energy, (c) mental—subjective/emotional or qualia, (d) intellectual and (e) as self or pure consciousness—the highest level of consciousness (Muller 1962). Meditation and Yoga, along with Vedic insights into human constructs, are the major frontiers that provide direction for understanding the higher nature of consciousness. In fact, the Indian Hindu system there was a school of thought where the dualist philosophy was also actively persuaded. One of the major schools or case in point is *Sankhya Yoga*, which actively considered the mind–body paradigm in line with the modern Western approach to the science of consciousness.

In West philosophy, during the classical and medieval periods, the different aspects of mind took the center stage in philosophy toward enhancing the understanding of humans in general and intellect, in particular. It evolved into a 'dualist' approach in which the body and mind were distinguished as material and nonmaterial entities, respectively. This came to be referred to as mind–body dualism (Wikipedia 2022). Even though the Athenian philosopher Pluto, during approximately 500–400 BCE, postulated mind–body aspects and discussed entities such as 'Forms' and later Aristotle, the Greek philosopher provided the early philosophy of mind–body duality, it was French Philosopher in 1700 Century, Rene Descartes who discussed consciousness for arguably for the first time where he asserted that consciousness is fundamental to existence through his thought that '*I think therefore I am*' (The-Philosophy.Com 2008, Stanford Encyclopedia of Philosophy 2003). In line with the Vedic approach, the Panpsychism philosophy has a view that mind or mind-like aspects are ubiquitous and fundamental to reality and exist throughout the world (Wikipedia 2022b).

The available literature shows that in recent times, say for over 4 decades, there has been renewed interest in consciousness research and its applications; however, there are certain questions at the scientific level as well as at the philosophical level. There is a growing realization that the materialistic approach alone is not adequate to explain the subjective aspects related to consciousness. The famous 'hard problems of consciousness' raises questions about the explanation for subjective qualities such as the experience of emotional feeling by Charmer (2010). In recent times, Panpsychism Philips Goff has provided a very pragmatic approach to consciousness on the one hand while critically providing a perspective on the limitation of materialism due to approach (Goff 2019).

The major school of consciousness research in modern science can be divided into (a) neuroscience approaches, (b) quantum mechanics approaches and (c) computational approaches. The neuroscience approach considers that the brain is the source of consciousness. The neuroscience, keeping in view its objective problem, i.e., mapping/scanning the brain to look for physiological or psychological conditions works at the manifestation of physical symptoms to diagnose medical conditions, has explored the functioning of brain faculties. Here, the '*neural correlates of consciousness*' is one of the well-accepted concepts in neurological investigation (Baars and Gage 2010). Hamerrof and Penrose and provide the latest results that consciousness is produced not by the interconnection of neuronal connections but by the microtubules in neurons (Tononi et al. 2016). Tononi et al. postulate that consciousness is nothing but information or data in the brain, and they propose integrated information theory, where the Greek alphabet Φ (Phi) represents the measurable parameter of consciousness (Hameroff and Penrose 1996a).

The objective of this review is to use the state of the art in science and philosophy of consciousness to improve the understanding of human behavior in engineering operation management toward developing a risk-conscious culture and thereby work on risk reduction activities in normal as well as postulated abnormal plant scenarios. The scope of this research has been human consciousness for improving human factors and the wellbeing of the staff and the organization. Based on the understanding of the risk consciousness of the staff, work also extended to model factors on organizational risk consciousness. The above review, although not exhaustive as it does not discuss all ancient civilizations, reviews the ancient Vedic approach and Eastern approach, which essentially deal with ancient as well as with the progress made in modern physical or material aspects of consciousness, and provides the needed fundamentals and assumptions required to make foundations and structure of RCOM.

1.4 Risk-Conscious Operations management—The Framework

The four pillars of risk-conscious operations management, as shown in Fig. 1.1, are as follows: (a) the ***tenets of consciousness*** that provide the basis or guidelines for risk-conscious culture development as also the input for the risk-based engineering model,

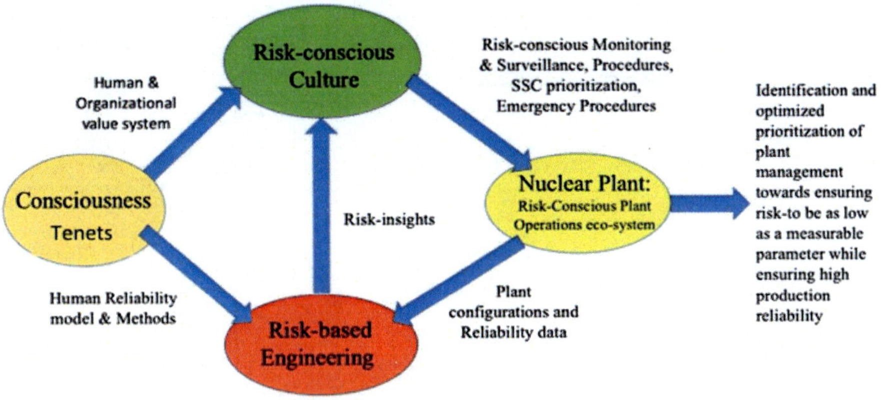

Fig. 1.1 Risk-conscious operation management model

(b) the second pillar is ***risk-consciousness culture*** development, (c) the next and most critical is the ***risk-based framework***, and (d) the third pillar refers to the ***plant configuration ad environment*** and addresses the development and implementation of plant-specific models and methods as part of risk consciousness culture.

The consciousness tenets in the RCOM provide input to the risk-conscious culture module. In the RCOM ecosystem, the risk-conscious culture module on human and organizational values support decision making along with input from the risk-based module for operational support. This covers normal operations, e.g., continuing plant operation based on risk-conscious criteria and guidelines. The experience and data generated in the plant eco system, not only on plant and human performance, become input to the risk-based module and further become feedback to the risk-culture module toward fine tuning of the system, as shown in Fig. 1.1.

Two major activities that can be targeted through the risk-conscious framework are, first, understanding and improving overall risk-conscious (rc) culture at the individual and organization levels; and second, the specific activities that include risk-conscious identification and prioritization of issues, (rc)-surveillance-test-interval estimation, rc-allowable outage time evaluation, rc-evaluation of normal and emergency operating procedures, and rc-identification of modifications, etc.

The following section provides a brief overview of the specific aspects/activities with respect to the reference human model that is central to the RCOM and the four modules of the RCOM.

1.4.1 Fundamentals RCOM

There are three major aspects that form the basis or fundamental that are at the core of the RCOM system. These are (a) reference human model, (b) tenets or assumptions of RCOM and (c) postulates of RCOM as follows:

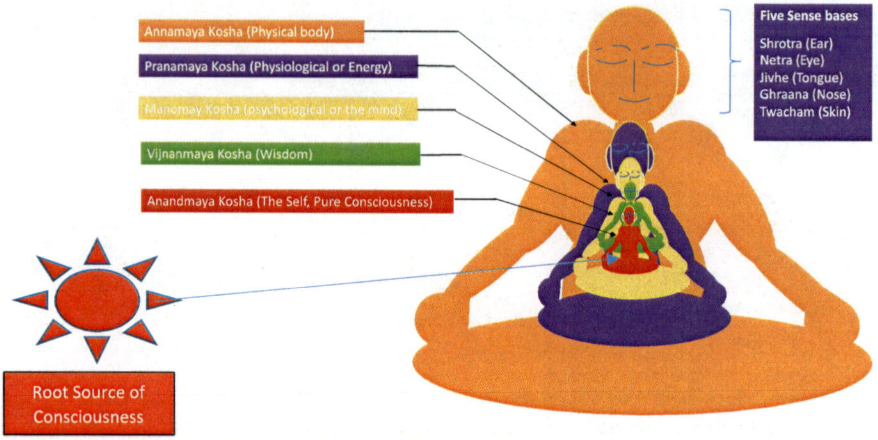

Fig. 1.2 The human model in the RCOM approach

1.4.1.1 The Human Model

Five manifestations of consciousness provide the fundamental for human model as (a) physical/body consciousness, (b) vital or energy consciousness, (c) mental—that includes thoughts, emotions, feeling consciousness, (d) wisdom—intellect memory consciousness and (e) self or pure consciousness. The first three levels have physical manifestations, e.g., the physical is gross, while the mental is sutler and energy is further, more sutler. The wisdom and self are nonmaterial forms of consciousness. Figure 1.2 shows the five sheaths of consciousness postulated by Rishi in their striving for self-realization—the goal of life in Eastern thought.

1.4.1.2 Consciousness Tenets/Postulates

Since the state of the art in modern science is still not mature, considering the requirements of RCOM, the knowledge base available in ancient philosophy has been employed where it is possible to fill the void.

(a) Consciousness is a fundamental and core entity of human beings and has the capability to reflect on the self as well as the external environment.
(b) Consciousness when considered has been considered the life force—as it relates to modern science as a sign of life.
(c) Consciousness is more than the physical symptoms at the physical level, as consciousness is all pervading and universal in nature; therefore, consciousness at various levels, viz. plant, organization, national and international level can be considered a reality with respect to the subject; in the RCOM case, it is risk consciousness.

(d) Consciousness can be measured at the physical level, and consciousness parameters should be mapped on a stress–strength model—a core requirement for risk-based engineering.
(e) Conscience in RCOM has been considered a subset of consciousness that addresses attitude, moral and value systems.
(f) The state of the art on consciousness in spiritual science, neuroscience, quantum theory and psychology complement many areas, and this knowledge can be integrated into a model of consciousness for the purpose of this research.
(g) Humans are basically spirits in nature; therefore, without the considerations of spirituality, human modeling cannot be considered complete.
(h) For a robust model of human reliability, consciousness should be the core consideration, apart from cognition, context, and many other stress-inducing factors.
(i) The state of the art in modern science does not yet convincingly explain the fundamental subjective human characteristics, such as emotions, feeling, attitude, and free will, at the experience level. However, these very subjective qualities often form the major driver and motivation for human performance.
(j) The spiritual process based on ancient learning provides robust and time-tested methods and approaches such as Yoga, meditation and Pranayama for understanding and managing subjective qualities or qualia.
(k) Awareness and attention are the major attributes of human consciousness. The awareness has two major attributes: human self-awareness (has spiritual connotation) and awareness of the environment around an individual. The collective consciousness of an individual in an engineering plant or setup is referred to as plant and system operational consciousness.
(l) The alertness quotient of an individual, which refers to the response efficiency to a situation along with action-taking capability, refers to cognitive as well as cognitive and consciousness qualities in humans.
(m) The potential role of the wisdom layer is critical to human performance, particularly in the control room emergency scenario, which requires interpretation and understanding of plant status, diagnosis of the situation and prognosis to postulate what is in the offing on time and spatial level in the plant, which also require heavy demand on memory to retrieve and use as appropriate input in support of decisions. These dynamics are studied on a plant simulator in RCOM.
(n) The human conscience aspects form a major input as part of modeling, and the evaluation of behavioral and ethical aspects, such as morality, commitment to the system, integrity and honesty, has also been considered under consciousness. For this study, the brain forms the physical link to understand the aspects related to conscience. However, the conscience in RCOM has been considered one of the subattributes of consciousness.
(o) Conscience has two major connotations: one during normal operations and plant emergency caused by equipment, human, external natural events, and the other deals with security aspects that deal with deliberate action performed by an informed mind.

(p) The term risk-based refers to an approach when the knowledge and insights from PRA and deterministic analysis are integrated along with robust considerations of human factors in support of risk-conscious operations management.
(q) Plant data, simulator sessions, monitoring of stress levels employing available instrumentations, such as EEG, fMRI, pulse, BP, diagnostic reports on morbid medical conditions, video recording to assess the physical motions and response forms the major monitoring and analysis methods in RCOM.
(r) In RCOM, the human root cause analysis framework provides a basic framework to understand the root human causes of failure.
(s) The subjective qualities have been adopted in the RCOM model employing a fuzzy logic approach where qualitative opinion or judgment—a real-time requirement in the operations environment—can be adopted in the risk model of the plant in support of analysis and decision making.

The insights and finding of consciousness protocol are mapped on the selection, training and qualifications, on-the-job monitoring, assessment of incident reports, through a feedback and corrective mechanism such that corrective program is initiated well in time as part not only human error reduction but also ensuring the wellbeing quotient in the plant.

1.4.2 Risk-Based Engineering

The risk-informed framework in the present form is more suitable in support of regulatory review and decision making. However, in regard to real-time operational scenario management and decision making, where the ecosystem is more dynamic and often more complex due to time constraints, complex analysis places a cognitive load on operations staff, especially during the management of emergency scenarios and decision making, to bring the plant to a safe state. The risk-based engineering approach has been adopted in risk-conscious operations management such that human factors can be integrated not only into the risk model of the plant employing the general characteristics of operations in general and the man–machine interface in particular.

Varde and Pecht developed a risk-based engineering (RBE) methodology that intuitively serves well the purpose of risk-consciousness operations management, as it considers monitoring and surveillance, equipment condition monitoring, prognostics and surveillance, and human interaction and shows capability to be dynamic in nature (Varde and Pecht 2018). The basic framework of RBE is shown in Fig. 1.3.

In the risk-based engineering framework, the first step is to identify the areas for improvement in support of plant operations management. In this approach, qualitative insights on risk from deterministic analysis are integrated with quantitative input from a qualified and approved PRA as part of integrated risk-based evaluation. Both approaches complemented each other to make a robust and holistic case. The results of the integrated assessment are validated against the available guidelines and criteria.

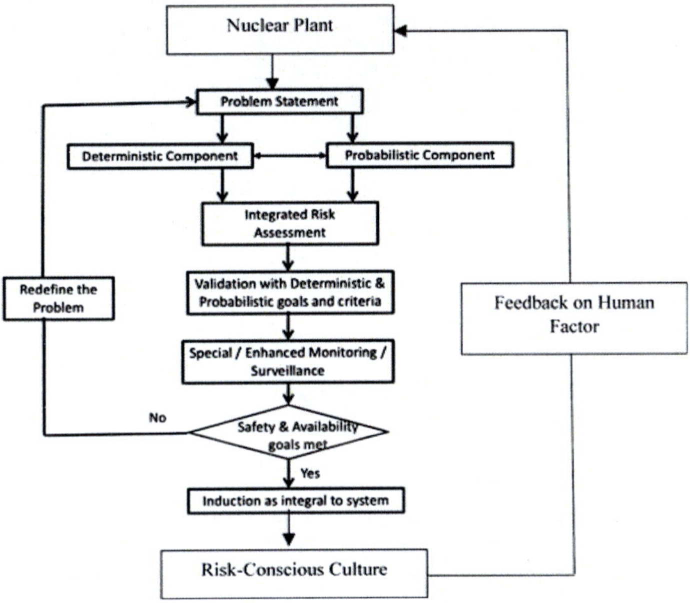

Fig. 1.3 Major elements of the risk-based engineering framework (Varde and Pecht 2018)

The complete operations management change program is supported by monitoring and surveillance based on the fundamental precursor parameters along with simulation support, if needed. Here, the condition monitoring or if required prognostics and health management program are designed and implemented to monitor the trend in parameter such that early indication of failure is available, such that corrective actions are implemented to avoid the consequences of failure. This approach is expected to ensure that key risk-reduction and performance criteria are adequately achieved to ensure regulatory compliance. The human considerations form an integral part of this framework, be it change/review of design, operation policy or authorization activities.

1.5 CQB Model

1.5.1 The Model

The human model proposed in the previous Sect. 1.4.1.1 has been extended in RCOM to include aspects related to interaction with the outside environment. The three Cs, i.e., consciousness as the fundamental component that supports the other two, i.e., 'cognition' and 'conscience' aspects of the human reliability model. It was referred to as the consciousness, cognition, conscience and brain (C^3B or CQB) model. In

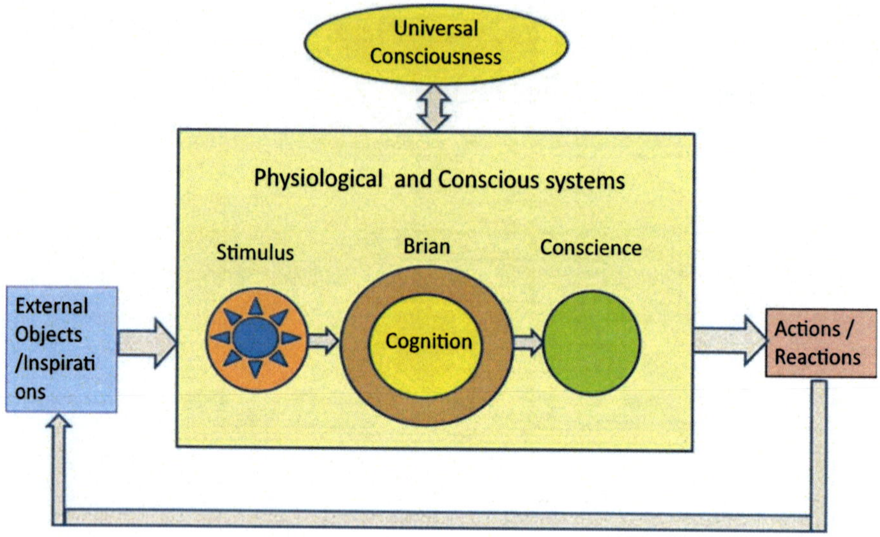

Fig. 1.4 Working human model in CQB

fact, consideration of consciousness and conscience exclusively is the new and unique feature of the CQB model. Figure 1.4 shows the salient features of the risk-conscious human model as applicable to operational management.

As the conventional practice in human reliability in CQB, it was asserted that awareness and alertness qualities alone are not adequate to capture human behavior. Consciousness, being the most fundamental attribute of human existence, has much deeper implications. The consciousness makes the human 'Aware' of self and surroundings. While neuroscience has developed a reasonable understanding of brain anatomy, physiology and the nervous system, there is very little understanding of how thoughts and emotions—such as love and anger are generated—often can be argued to be responsible for human behavior. Of course, there is reasonable understanding of brain faculties involved in processing consciousness and the effect of inadequate health of these basic brain faculties, amygdala, frontal, occipital, parietal lobes, sensory information processing regions, etc., on conscious behaviors such as reasoning, diagnosis, recognition, memory, vision and identification. However, very little knowledge of the processing of experiences such as emotions and attitudes is referred to as the 'hard problem of consciousness'. *It can be argued that the present notion of consciousness only deals with the manifestation of physical symptoms at the level of brain and not the consciousness per-say. This observation is in line with the human model proposed, where the physical manifestation of consciousness is just one of the five consciousness sheaths of the human model.*

Based on the data collected on the simulator experiment and experience with the control room scenario on incidents that involve mapping of human response to transients, body language, time taken to perform a task, communication aspects, success and failure of the task, the human model developed as part of CQB research was tested. It was realized that human reliability estimation employing only cognitive aspects and considerations of only human reactions is not adequate. The considerations of another two Cs, i.e., consciousness and conscience, tend to make the human model more robust and expected to reduce uncertainty in human reliability estimation. In the CQB model, in line with the structural reliability approach, the consideration of stress and strength provides the basis for the assessment of human reliability.

1.5.2 Human Performance Influencing Factors (HPIS)

Apart from the human model, the other factors that are vital to determining human performance are task characteristics, the system and the environment. The taxonomy proposed in CQB has stress–strength assessment at the core, i.e., stress generation potential and strength of availability of safety provision and human element should comply the applicable safety margin in terms of central value and associated uncertainty.

The term consciousness in the CQB approach has two connotations, viz. physiological, which considers the awareness levels of humans, and secondary, which connects people in the organization at the communication level. There are various methods for the assessment of parameters either qualitatively as subjective expert opinions or as measurable parameters. As mentioned in the previous section, electroencephalogram (EEG) can determine stress levels through the brainwave generated during awareness and activity levels of cortisol levels in saliva samples. Other measurements include pulse rate, blood pressure, blood oxygen level, monitoring and analysis of body language by video recording and during simulator sessions. Plant awareness is assessed during simulator training for emergency sessions and data on normal plant conditions, and the corresponding strength consideration comes from safety provision in the plant. In line with conventional approaches, the ratio of time required for any action to time available can also be used for stress quantification. The availability of a modern control room with a simulator facility and the checklist procedure are considered for assessing the strength parameters. Often, the use of fuzzy logic for processing expert opinions to assess the influencing factors forms part of the procedure in CQB.

The consideration of conscience has been for two reasons: one for staff attitude that affects performance adversely, such as commitment quotient, punctuality, etc., and the other is exclusively from security considerations that might include integrity, honesty and social conditions. It may be noted that the list of HPIF is not exhaustive and is indicative only. Each performance shaping factor has stress as well as strength consideration.

1.5.3 Human Reliability Model

Nonperformance probability is expressed using an exponential function. This is in line with the previously available models of human reliability. *Reference stress* refers to the stress acting on the operator in a normal working scenario, while the *actual stress* is the stress assessed during a specific emergency in simulator or control room conditions. S_R is the ratio of actual stress acting on the operator during the postulated initiating event condition or emergency condition in the simulator session or control room environment to the normal stress level during normal operations. The reference stress depends on the quality of the system design, man–machine interface, operator information and support systems, and other plant parameters. The nonperformance probability under stress ratio S_R is given as:

$$P(S_R) = 1 - \mathrm{C}_{\mathrm{sc}}(1 - \mathrm{P}_{\mathrm{o}})\exp\left(-\frac{S_R}{B}A\right) \tag{1.1}$$

$\boldsymbol{P_o}$ is the nonperformance probability under the reference stress condition. $S_R$isthestressratio. B is the stress normalizing parameter. It quantifies the operator training level and can be determined from the experimental data available on human reliability using the relation. Furthermore, the parameter A accounts for the nature of the task; for example, the value of A is lower for skilled-based jobs, higher for rule-based jobs and maximum for knowledge-based tasks. C_{SC} Conscience levels take a value of 1 for high integrity and 0 for the low conscience parameters, e.g., integrity, honesty or any parameter that challenges morality and ethics.

1.5.4 Human Root Cause Analysis

As seen in Fig. 1.5, in RCOM, human root cause analysis (HRCA) does not terminate at identifying human failures, such as slips, lapses, mistakes, etc., as these root types are not adequate to identify the corrective mechanism. It is a well-recognized fact that for a given failure, there could be broad categories, such as design, operations, maintenance, procedural or organizational or combination of any of these errors, but the common thread that runs through is human error. Therefore, human error is investigated further, down until the fundamental root(s) is reached. This provides an improved strategy that helps identify an appropriate corrective mechanism. The HRCS further investigates which sheath of consciousness is involved, whether it is physical, emotional, intellect, wisdom, etc. Once the sheath is identified, there is further downward investigation, as to whether the dominant nature is lack of consciousness, cognition or it is conscience (moral/attitude) caused the error. Like the hardware RCA, in HRCA, the physical mechanism for the failure is also established at the level of the brain/body. This helps in identifying the cause and level of stresses. These stresses, if found to be recurring (likelihood) or lead to serious consequences, enable decisions for the implementation of corrective action programs.

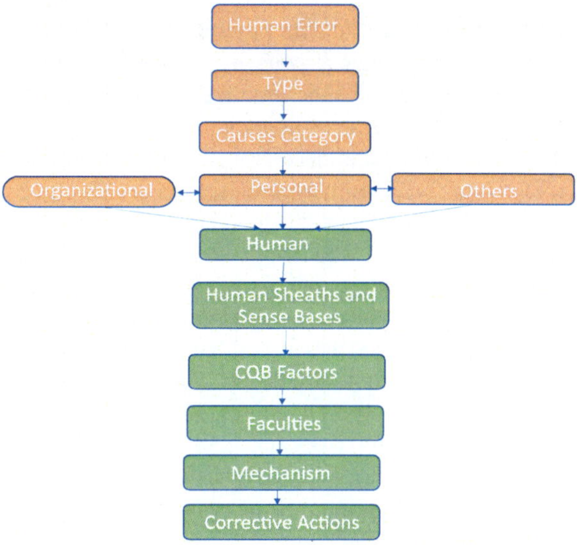

Fig. 1.5 Human root cause analysis flowchart

1.6 Risk-Conscious Culture

This section discusses the characteristics of risk-conscious ecosystems, the impact of risk-conscious approaches on safety management features, and the key requirements for the development and implementation of risk-conscious cultures. Before this, this paper proposes a definition of risk-conscious culture. For the purpose of safety critical systems, considering the operational and human aspects, the definition of risk-conscious culture has been given as follows:

> Risk-conscious culture in a risk-critical ecosystem which deals with the development of set of a human attitude, practices and characteristics that involves, value & learning-based, systematic and sustainable approach to assessment, implementation of corrective action program and management of risk; enabled through an integrated scientific and rational based risk model of the plant that integrates hardware and software failure with human-error-causative factors; toward ensuring risk reduction in terms of measurable parameter metrics.

The top management should be responsible and proactive in the implementation of risk-consciousness culture. Based on the administrative requirements derived from the protocol statements, the higher-level activities include determining the drafting of risk-consciousness policy and the scope and quality requirements for documentations, particularly PRA documents, broad guidelines/requirements for risk-conscious framework, etc., to initiating the development of risk-conscious culture. In fact, many inputs can be used from the available literature, such as safety culture, as these documents have been developed at the international level with the contribution of world renowned experts (International Atomic Energy Agency, Safety culture, Safety Series—A report by the International Nuclear Advisory Group 1991a).

Apart from the characteristic hazard to a given installation, such as nuclear and radiation safety for nuclear plants, other hazards, such as industrial risk, transport risk and chemical hazard, should be considered for the development of the risk-conscious culture and final characterization of safety. The risk-conscious culture, apart from the risk of accidents/incidents, should consider other risks, such as risk of fatality, injury, loss of reputation, risk of financial liability, loss of plant availability, reliability and frequency of interruptions and factors that adversely affect plant deliverables, etc.

The essential performance indicators of risk-conscious culture include perception and visibility of dynamic administration, ownership for and leadership at all the levels in the plant, for example, sharing of responsibility and liability for any task. The emphasis should be on innovative thinking and out-of-box solutions and an open appreciation of ideas and suggestions, whether big or small. An open atmosphere creative and periodic assessment of issues, where managements' role is engaging the staff in consulting, discussion. The perception of the staff, that the management cares for them and their happiness and wellbeing is one of the, is one of the major indicators of matured risk-conscious culture.

The PRA/risk-based insights need to be revisited to identify and prioritize areas for attention and actions. Root cause analysis (RCA) is an inherent part of risk-conscious culture. All the events cannot be subjected to detailed RCA; however, risk-sensitive cases should be analyzed until one reaches human error(s) responsible for the incidents. This analysis should be an additional input for the developing risk-conscious culture. The plant should have provisions in the organizational setup to compile the incident database that contains the type of human errors, e.g., error of omission, error of commission, lapse, mistake, communication error, etc. Furthermore, the role of attention, awareness, motivational and commitment level, physical and psychological wellbeing, etc., should form input to risk-conscious culture programs. Security aspects and security-sensitive information should be an input to risk-conscious activities (International Atomic Energy Agency and Culture 1991b).

Safety culture operates at various levels. First, the mechanism of developing insights from PRA and effectively communicating to staff the issues based on the risk ranking and the risk reduction associated with the subject is addressed; modification of training program for competency building based on risk insights developed and evaluating the feedback on improvement in attitude of staff; interaction with the management for the approach to be adopted to cater to regulatory and internal compliance requirements, a system of maintaining a dynamic database of incidents, learning/training. Organizing routine meetings for staff where all the plant safety-related subjects are discussed threadbare and collective brainstorming is employed for arriving at solution is one of the major drivers and a positive sign of a healthy ecosystem. The risk-culture improvement and its output are ongoing processes in plants.

1.7 Implementation

One of the major tasks is to create a system that starts with creating a risk-conscious operations management document. This document will provide guidance on implementing risk-conscious culture with respect to operations and maintenance activities. This document also provides the list of risk-significant systems and activities organized based on their priority based on the insights from the Level 1 PRA document.

The second important task is the database of human-induced or human error committed to their frequency and consequences. These insights will also form part of work permits or plant operations and maintenance procedure systems. Any computerized procedure, before its issue, will provide the past history of procedure activities performance, particularly human error, which suggests that these are the areas prone to error or failures and need focus & attention. Furthermore, the insights available on plant records on incident-associated investigations or root cause analysis will also form part of procedure authorization and issue systems such that the recurrence of incidents can be avoided. The formation of a system toward ensuring the wellbeing of the employ might pose a challenge. The idea is that a risk-conscious culture must ensure the mental wellbeing of members of the O&M group in plants. Group Yoga and meditation programs go a long way to enhance individual-level consciousness.

Routine technical seminars to spread awareness about the system operational status, modification, achievements, innovative applications, and expert lectures on the subject related to operational management should be organized. The scope of the risk-conscious culture should include risk aspects pertaining to (a) change in plant configurations management, (b) monitoring or surveillance or historical information and data, (c) procedure and its implementation, (c) planning & scheduling, (d) maintenance management, (e) training and qualification program, (f) reporting hierarchy, (g) gap in regulatory compliance implementation, (g) operating crew specific aspects, etc.

There are many factors responsible for the successful implementation of risk-conscious culture that include mandate authority from the management, availability of resources, a system for tracking the effectiveness of the program and staff attitude and wellbeing.

1.8 Remarks

The five guiding features of risk-conscious culture are (a) remaining perpetually conscious of plant health, related requirements and safety; (b) the availability of science and rational-based frameworks that include models, methods and data that provide a robust foundation for risk assessment and management; (c) professional competency building at the individual and organizational levels; (d) ensuring a conscious perception among the staff members that the organization cares for them, their wellbeing, and that the organization belongs to them; and (e) honesty, integrity and unquestionable commitment and dedication to the job and a systematic risk framework of the individual are the highest values that form. The availability of the PRA model of the plant provides an effective resource for developing tools and methods for creating a risk-conscious culture. For example, areas for improving human factors. The consciousness culture works at two levels, viz. it enables reducing the uncertainty in human reliability estimates by considering consciousness and conscience along with cognition by making the human reliability procedure more robust, and the risk-conscious culture developed based on the input on risk and human factor contribution to incident enable reduction of human-induced events. It is expected that the implementation of risk-conscious culture results in increasing the human happiness quotients through the Yoga and meditation techniques on the one hand and improving the safety level of the plant.

References

Baars BJ, Gage NM (2010) Cognition, brain, and consciousness—introduction to cognitive neuroscience. Elsevier

Chalmers DJ (2010) The character of consciousness. Oxford University Press, New York

Easwaran E (2021) (Introduced and translater), The Upnishads. Jaico Publishing, India

Goff P (2019) Galilio's error—foundation. For a new science of consciousness. Penguin Random House, UK

Hameroff SR, Penrose R (1996a) Orchestrated reduction of quantum coherence in brain microtubules: a model for consciousness. In: Hameroff SR, Kaszniak A, Scott AC (eds) Toward a science of consciousness. I. The first Tucson discussions and debates. MIT Press, Cambridge, pp 507–540

International Atomic Energy Agency (1991a) Safety culture, Safety Series—A report by the International Nuclear Advisory Group, SS No. 75 INSAG-4, Vienna

International Atomic Energy Agency (1991b) Safety Culture, INSAG Report Series 4. IAEA, Vienna

International Atomic Energy Agency (1994) Convention on nuclear safety, IAEA-INFCIR/449, IAEA, Vienna, 5th July 1994

International Atomic Energy Agency (1999a) Basic safety principles for nuclear power plants, 75-INSAG-3 Rev. 1. INSAG-12. IAEA, Vienna

International Atomic Energy Agency (1999b) Management of operational safety in nuclear power plants, INSAG-13 Report Series 4. IAEA, Vienna

International Atomic Energy Agency (2004) Code of conduct on the safety and security of radioactive sources. IAEA, Vienna

International Atomic Energy Agency (2006) Code of conduct on the safety of research reactors. IAEA, Vienna

International Atomic Energy Agency (2008) Conduct of operations at nuclear power plants, Safety Guide No. NS-G-2.14. IIAEA, Vienna

International Atomic Energy Agency (2010) The interface between safety and security at nuclear power plants, A report of the International Safety Advisory Group, INSAG-24. IAEA, Vienna

International Atomic Energy Agency (2012) Safety assessment for research reactors and preparation of safety guides, Specific Safety Guide No. SSG-20. IAEA, Vienna

International Atomic Energy Agency (2021) http://www.iaea.org. Accessed on 12 Apr 2021

Muller FM (Translater) (1962) The Upanishads Part II: Taittirya Upanishd. Dover Publications, New York

Nadar T (2021) The foundation of Vedas—science and technology of consciousness. http://www.youtube.com/user/DrTonyNadar. Accessed on 15th Apr 2021

Stanford Encyclopedia of Philosophy (2003). Dualism. https://plato.stanford.edu/entries/dualism

Swaton E, Neboyan, V, Lederman L (1987) Human factors in the operations of nuclear power plants-improving the way man and machine works, IAEA Bulletin, 4/87

The-Philosophy.Com (2008) French Philosophers, French Philosopher—Descartes: Philosophy. https://www.the-philosophy.com

Tononi G, Boly M, Massimini M, Koch C (2016) Integrated information theory: from consciousness to its physical substrate. Nature Rev Neurosci 17(7): 450–461

Varde PV, Pecht MG (2018) Risk-based engineering. Springer, Berlin

Wikipedia (2022a) mind-body dualism. http://en.wikipedia.org/wiki/mind-body-dualism

Wikipedia (2022b) Panpsychism. http://en.wiki.panpsychism/

Chapter 2
Consciousness

2.1 Introduction

The major premise of this chapter is that human factor considerations are critical operational ecosystems, as apart from human–machine interactions, organizational factors play a major role in achieving organizational safety goals and targets. Most of the existing human reliability modeling procedures have a major focus on cognitive modeling, while direct references to deeper aspects of consciousness are either superficial or not present. In the risk-conscious operations management (RCOM) approach, there is a special consideration for the development of a human model as well as the consideration of consciousness and conscience along with cognition. The premise of risk-conscious culture is that consciousness-associated deeper aspects and their attributes are critical to general human behavior that might lead to random (unintended) human error to address safety implications, while the conscience attribute provides an effective mechanism for intended or deliberate human failures that have security implications. For example, while consciousness attributes such as emotions, feeling of wellbeing, attention and awareness, and attitude affect human performance, moral, ethical, negatively conditioned/informed and motivated actions might pose security concerns.

Uncertainties associated with human reliability estimates adversely affect the final results in probabilistic risk analysis (PRA) (International Atomic Energy Agency 2010). There are inherent uncertainties associated with the existing reliability model due to the unavailability of a robust and explicit consideration of a human model. The consciousness, cognition and conscience and brain (CQB) human model developed as part of risk-based engineering by Varde and Pecht has been extended for RCOM application (Varde and Pecht 2018). However, the challenge in the context of operations management is to further work on modeling, keeping in view the available state of the art on consciousness. Even though there is a general consensus that consciousness considerations are integral to human behavior/health modeling and analysis, there are many questions for which researchers admit that there is a need for further work. For example, is it possible to comprehend the understanding

P. V. Varde, *Risk-Conscious Operations Management*, Risk, Reliability and Safety Engineering, https://doi.org/10.1007/978-981-19-9334-3_2

of consciousness employing a material science-based approach? What is the source of consciousness? Charmers "hard problems of consciousness" that deal with challenges in understanding qualia that explain emotions, feeling, etc. Is consciousness subject limited to human beings or extended to animals, materials and the universe as a whole? Is consciousness a measurable attribute? How can the ancient knowledge available on consciousness be adopted and integrated with modern material science to enhance the understanding of consciousness?

The authors research on consciousness pose one additional question; i.e., is consciousness the same as traditional Hindu Vedic philosophy and modern material science? The postulation is the Hindu philosophy, which postulates that consciousness is all pervading while material science looks for consciousness only at the level of the brain. Therefore, the next question is whether the symptoms available at the level of the brain are truly consciousness or its secondary effect that is a manifestation of consciousness at the material level, i.e., at the level of the brain.

The extensive review of various schools of consciousness right from the Vedic era to this modern material science age/time, understanding of the salient features of major available material science-based approaches and other related technological advances such as quantum theory, power of computational methods, information theory, neuroscience-based approaches are presented in this chapter. The objective of this chapter is not to discuss the R&D associated with consciousness but to use the available state of the art to develop an improved method for human reliability modeling to support risk modeling and analysis for risk-critical engineering systems.

2.2 Evolution of Science and Philosophy of Consciousness

Even though the Oxford Dictionary defines *consciousness as 1. The state of being aware of and responsive to one's surroundings, and 2. A person's awareness or perception of something* (Oxford Languages and Google 2022). However, there are contradictory views on the exact meaning of consciousness, the root source of the terms in various cultures, right from the ancient era to the modern period (https://plato.stanford.edu/entries/consciousness-17th/), its source, whether it is only a subject of human beings or universal entities, and many other gray areas that need exploration and discussion, such that the concept of consciousness and associated science and philosophy is interpreted and used in risk-conscious culture. As seen in Fig. 2.1, the evolution of the field of consciousness can be discussed in two major eras, i.e., during the period of Shrutis or Vedas on the one hand and later during the period of modern scientific and philosophical age started in the sixteenth century. However, the major scientific theories based on the material science approach have developed in the last 30–40 years. This period marks the active interest in the development and application of material science-based approaches to understanding, focused on human consciousness.

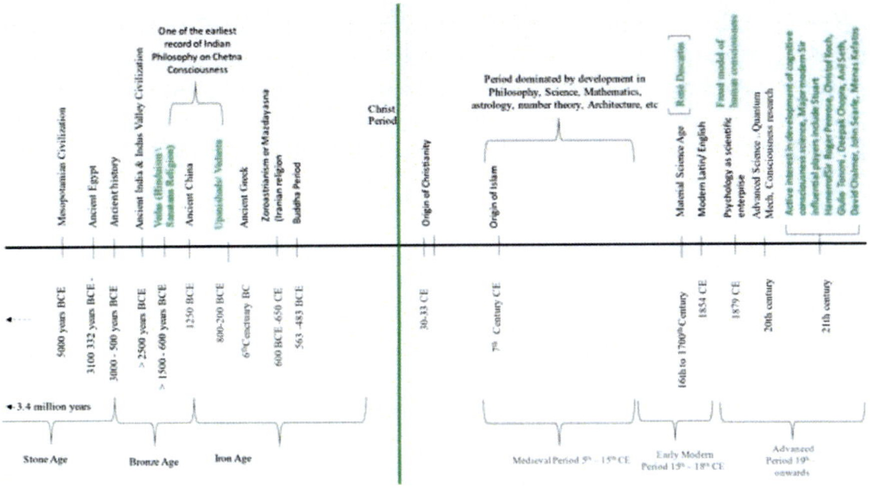

Fig. 2.1 Chronological representation of the development of philosophy and science of consciousness

Mesopotamian civilization has spread in southwestern Asia, particularly centered around what we call Iraq. The name translates to 'between two rivers', i.e., land areas between Tigris and Euphrates rivers. As one of the earliest religious systems in history to structure and be itself structured by, the complexities of a high civilization, Mesopotamian religions are of significant interest to historians, historians of religion and theologians (https://www.britannica.com/topic/Mesopotamian-religion/Religious-art-and-iconography#ref68281). Zoroastrianism, or Mazdayasna, is an Iranian religion and one of the world's oldest organized civilizations; it has unique features, such as dualistic thoughts (good and evil) and free will, and dates back to the 2nd millennium; however, it has a recorded history since 600 BCE and served for more than one millennium until 650 CE (https://en.wikipedia.org/wiki/Zoroastrianism).

> *"I regard consciousness as fundamental. I regard matter as derivative from consciousness. We cannot get behind consciousness. Everything that we talk about, everything that we regard as existing, postulates consciousness."*
>
> Max Planck
>
> The theoretical physicist who originated quantum theory, which won him the Noble Prize in Physics in 1918.

The era of Hindu Scriptures Vedas dates back to < 1500–600 BCE when Shrutis (by recitation and hearing); i.e., knowledge of the development of social life and practices, particularly spiritual components and philosophies that were passed on generations to generations 'orally', was documented during the Vedic period. On the one hand, the available literature could not succeed in dating the period of Shrutis, while on the other hand, the Hindu belief system ascertained that it was passed on to humans from a divine source—The God Almighty. Furthermore, the philosophical components of Vedas were compiled and explained in the form of Upanishads during 800–200, and some work continued until the modern period until 1600 AD. The translation of these thirteen major Upanishads from Sanskrit to English made this knowledge available to wider researchers and philosophers during the early nineteenth century (https://archive.org/details/thirteenprincipa028442mbp/pagen8/mode/1up?view=theater).

Arguably one of the first explicit and recorded references to consciousness can be traced back to the teachings of Adi Shankaracharya Upanishads as part of Santana Culture development during the early period of its development at ~ 800 BC (https://www.onelittleangle.com/wisdom/quores.saint.asp?mc=60). Since the Upanishads are derived from Vedas, the seed of science and philosophy of consciousness as an eternal phenomenon date back to a period much earlier than 1500 years BCE in Shrutis. Second, consciousness was considered fundamental to human existence, while it was considered all pervading. During the late 20th until recent times, a resurgence was seen in the development of cognitive consciousness, be it biological and neuroscience, psychology, computational and information sciences and finally quantum mechanics.

The period between 1600 and 2000 CE was considered the modern age and is now the classical age of science and philosophy.

This classical era has also been characterized as the material science era. Regarding consciousness philosophies, the thought of Descartes during the 16th–17th CE and later as part of a formal psychological science Sigmund Freud model of the 4 Level of Consciousness during the 18th CE dominated the concept of consciousness (Oliver 2017; Seventeenth-Century Theories of Consciousness 2020). In Fig. 2.1, the chronological representation covers only the major development and their sequence. The above is a sequence of developments and not an account of the exact gap and duration between each development.

Only major developments have been covered; for example, between the Upanishad and material science eras, there would have been some developments, but here, the objective is to discuss human reliability research; accordingly, the focus has been on related spiritual and cognitive-consciousness developments. There is no claim made out on accuracy on timeline and the aim is to present chronological aspects related to consciousness development therefore there can be some variation or uncertainty in epochs and duration as also there is no claim of completeness as also accuracy. The major idea is to capture major developments.

2.3 Consciousness and Human Reliability

In this section, the objective is to answer the question of why we need consciousness and conscience consideration and the role of these elements in making human reliability evaluation more effective. Let us first understand that it is not that human action/interactions pose adverse reliability implications. It is otherwise round that it is only human actions and interaction that keeps the world going be it transport, energy systems, aviation, space, nuclear, process and industrial systems. It is only testing conditions in relation to complex systems when human reliability becomes a critical aspect and requires continuous improvement. In normal situations, human reliability is comparable to hardware or even in some cases surpasses hardware and automated systems. The only issue is human or humans, who are adversely affected by context, fatigue, challenging situations, the paucity of time, and environmental and operational factors that lead to cognitive stresses. At times attributes such as overconfidence, casual approach, lack of focus and knowledge, and complacency led to random or systematic error. Similarly, security concerns call for factors associated with conscience, such as ethics and moral integrity, that pose security-related implications also require considerations of human reliability. One common denominator for safety critical systems is that modern systems are made robust and safe by incorporating the provisions for monitoring deviations from normal operations, automation where applicable for autocorrective features, redundancy, diversity and fail-safe and passive features such that human action is not expected during the initial period into the accident. However, other systems have become highly complex in terms of interconnection, dependability, and highly dynamic in terms of transients, and their hazard potential if they fail despite the best safety practices poses challenges, particularly during emergency conditions.

Major concern in safety critical is, even with the best design and operations features, if the accident occurs, the penalty could be very high due to their higher hazard potential. For example, the typical challenging scenario posing a threat of an accident during an aviation flight puts the pilot to test in respect of his knowledge and its timely retravel, comprehending the audio/visual and instrument and control signals, communication along with keeping an eye on track and operations puts the pilot in a highly complex situation that may not allow him to perform a proper diagnosis due to cognitive overload, lack of awareness, alertness that might make pilot susceptible to committing errors during emergency management.

The history of accidents in general shows that human error is one of the major factors contributing to accidents. This observation is generally acceptable, but it needs an all-encompassing factor consideration. For example, even the inadequate designs for accident scenarios or the improper operations of engineered safety features or safety systems, emergency aids, man–machine systems, data and information, and communication efficiency also make the tasks of operators challenging and often attributed to human error. It must be remembered that the operator's cognitive and conscious factors take a toll during emergency scenarios due to higher mental stresses. Therefore, even though as a direct cause operator error appears to be the contributing factor, a deeper analysis or to be specific human root cause analysis might reveal other factors that can again be attributed to human factors in design, ergonomics, communication, procedural, management, etc. The research on risk-conscious operations management revealed that the existing approach to root cause analysis as well as the unavailability of a human model may not allow the existing RCA approach to real-human roots responsible for failure. Therefore, a new paradigm, i.e., 'human root cause analyses', has been introduced to understand and identify the gap area as human attributes, anomalies and limitations that could pose potential for human error in emergency conditions. Considerations of consciousness and human precursor analysis are additional elements of the RCOM approach.

We conclude here that consciousness and conscience factors, along with cognitive factors, are essential factors to create a risk-conscious culture as follows:

- Unlike cognition, which is an individual characteristic, consciousness is an inclusive attribute that facilitates the creation of a plant culture in this case risk-conscious culture.
- The human attributes of consciousness, viz. awareness, alertness or remaining vigilant, are directly relevant to plant operations in general and emergency conditions in particular.
- Considerations of consciousness have the potential to reduce uncertainty in predicting human performance by making the human reliability model completer and more holistic. However, the challenge is that the state of the art in consciousness has many gray areas and lacks general consensus on many aspects.
- Conscience, which is considered a supplementary function of consciousness, is a critical human attribute that works at the level of hard work, dedication, sincerity, working in a team and is directly responsible for human performance under normal conditions. However, for security considerations, additional factors such as morality, ethics, integrity transparency, etc., come into play.
- Consideration of consciousness (along with conscience) is vital for creating risk-conscious culture such that various facets of operations can formally adopt this culture in surveillance and monitoring, reporting, communication, training and qualification, hardware quality assurance, engineering and safety analysis, and regulatory compliance.
- Considering the above, the RCOM approach adopts the CQB human reliability model for the development and implementation of risk-conscious culture not only at the individual and organizational levels but also with minor modification at the national level.

2.4 The Consciousness Framework to Support RCOM

The major issue in consciousness research is to create a framework that integrates philosophy and science available from ancient knowledge sources with the state of the art in modern sciences (sixteenth century onward) toward the development

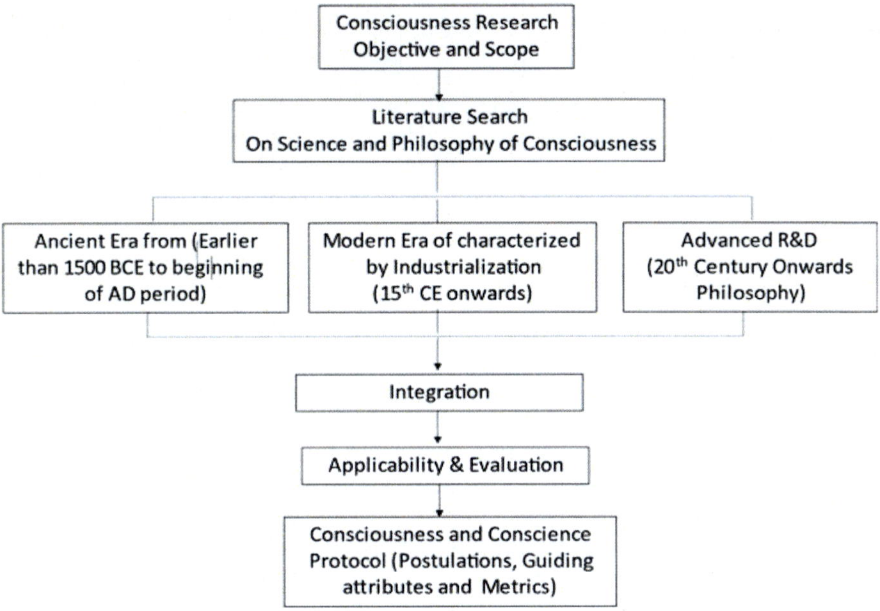

Fig. 2.2 Framework for consciousness consideration for RCOM

of a credible consciousness protocol and postulates for risk-conscious culture—central to the development of the RCOM system. Figure 2.2 shows the framework for development of consciousness protocol in RCOM. This calls for a systematic consideration of ancient wisdom/knowledge available in spiritual or philosophical books as well as the achievement of modern sciences. The following section addresses a review of relevant knowledge sources from the BCE era to the 20th CE and the early period of the 21st CE.

The objective of this literature is to develop a knowledge base on (a) understanding the fundamental nature, source, form/levels of consciousness, (b) applicability to human reliability in an operational ecosystem, (c) development of insight into organizational consciousness, (d) development of attributes of risk-conscious culture and finally (e) development of performance influencing factors and human reliability models to support risk assessment.

For the most part, the scope of review is to adopt the knowledge and insights for human reliability and organizational aspects or organizational factors. The aim of this work is to cover the full spectrum right from ancient age to the latest R&D on consciousness, such that apart from scientific, philosophical the higher-level spiritual knowledge such that the claim of in-depth human reliability modeling is reasonably met.

2.5 Literature Review

The literature review focuses on three major divisions, viz. Ancient Era, After Ancient Era and before modern industrial age and later classical science and industrial age and finally the advanced science or quantum age, as follows:

2.5.1 Ancient Era

During ancient times, i.e., the BCE era back to more than 5000–3000 BCE until 5th CE, one common observation is that all or most of the art, techniques, science of nature and philosophies were developed around first for human survival and second for understanding nature mainly human beings. One can only imagine now that after having adequate protection from natural threats, shelters, forming, food, communication and dependent social system that would have taken over a millennium year, the next or even a parallel development would have been 'developing a culture' set of or norms, practices, rituals of social living may be which would have taken over another > 1000 years that the societies in various geographical locations would have gone through.

Therefore, each cultural system would have seen this chronological order of events. Some culture would have advanced in one area, while the other would have excelled in another area, with many developments along the side of the major focus. For example, one culture would have advanced in warfare development, another in agriculture, another in art, and some other in the philosophy of human life, including spirituality. As discussed in the previous section. Of course, it is not possible to review all the systems.

> *You are the self, the infinite Being, the pure unchanging Consciousness, which pervades everything. Your nature is bliss and your glory is without stains. Because you identify yourself with false ego, you are tied birth and death. Your bondage has no other cause.*
>
> *- Adi Shankaracharya Hinduism Quote No. 3704 World Spiritual Heritage*

It is left to one's imagination how much time and effort and the development of knowledge, particularly the evolution of langue, facilitated the recording of ethos, script, grammar and critical aspects, i.e., the cultural development of society and poetic expression that we see in this recorded expression came into existence. In short, if we focus on Hindu books of cultural and philosophical developments, the Vedas, where the Shrutis were recorded by the great souls of that time, on subjects of highly cultural, philosophical and scientific in Sanskrit, etc., to produce this voluminous literature. The Hindu system asserts that the almighty godly energy divine intervention must have been there when these Shrutis were developed and passed on generations to generations 'with perfection'. We have no basis to either accept or reject this postulation. However, we must acknowledge that these were monumental efforts that span many eras. It takes imagination to realize that these great books of mankind were recorded by Veda Vyasa (https://www.hinduamerican.org/blog/veda-vyasa-the-sage-who-compiled-the-vedas) as a single system for writing four parts of Vedas, Viz, Rig Veda, Yajur Veda, Sam Veda and Athar Veda, with translation from Sanskrit to Hindi or English or for that matter any other language. The broad features of Vedas can be divided into three parts: rituals, religious practices for various social and family functions, second philosophy of life and third cultural aspects such as societal, musical and economics. All these aspects are woven with a common thread of spirituality and principles of nonduality—which is at the core of Hinduism. In this era, the development of Dhyana, meditation and Yoga forms part of human development to higher levels.

Keeping in mind the focus of this chapter, a decision was made to include in the scope of this book the review of Vedic Hindu culture. It can be argued that many other cultures would have focused on humanity, religious systems, etc., so why only the Hindu system. The simple answer is that spirituality has been at the core of the Hindu system since the ancient age, which is relevant to human reliability modeling, as humans are basically a spirit and consciousness is fundamental to human existence. The Vedas came into recorded form in 1500 BC; however, the striking observation is that before this period, this knowledge was passed on from generation to generation orally in the Sanskrit language 'Shrutis'. Now let join and imagine what we see in Vedas is even today the complex science and philosophies of rituals on one hand and the science and philosophies of life along with advanced knowledge of cosmology, mathematics, medicine, space and architecture, etc.

> *The only reason why you experience life and aliveness it is because you are conscious. We call this intelligence that make life happen consciousness*
>
> *- Sadguru*

That goes beyond human and living beings to include universal objects and phenomena and further deeper considerations (Vyasa and Presenter 2018). Veda in Sanskrit means knowledge. 'Chaitanya' means consciousness and means spirit or intelligence or to be more precise pure intelligence. It is a science and technology of consciousness, of awareness. Veda is the dynamics of this infinite field within, the science of consciousness, of the reality of one's own being. It is subjective science, with a technology of consciousness that allows us to dive deep within the Transcendental Technique (https://www.youtube.com/user/Dr.TonyNadar). The Upanishads are supportive books that explain the foundation and principles concepts laid out in Vedas. There is a general consensus in science and philosophy that consciousness is required for human existence (https://isha.sadguru.org/us/en/wisdom/whatis-consciousness-states-myths#1point; The Science of Consciousness 2022). In fact, cognitive consciousness is associated with awareness and alertness levels that arguably tend to influence the internal and external quality of human knowledge, feelings and emotions (Tyang et al. 2017).

> *Mind no longer appears to be accidental intruder into the realm of matter. We ought rather hail as the creator and governor of the realm of matter. Get over it and accept the inarguable conclusion. "The universe is immaterial-mental and spiritual." There are factors associated with consciousness that affect reality.*
>
> Pioneering Physicist Sir James Jeans

Human consciousness at the fundamental level has been defined as the *awareness of the self in space and time*, i.e., human awareness of both internal and external stimuli. Researchers study states of human consciousness and differences in perception to understand how the body works to produce conscious awareness. Consciousness varies in both arousal and content, and there are two types of conscious experience: **phenomenal**, or *in the moment*, and **access**, which *recalls experiences from memory* (https://opentextbc.ca/introductiontopsychology/chapter/2-2-psychodynamic-and-behavioural-psychology/#:~:text=and%20external%20stimuli.-,Sigmund%20Freud%20divided%20human%20consciousness%20into%20three%20levels%20of%20awareness,id%2C%20ego%2C%20and%20super). There is a general agreement that consciousness is subjective in nature; however, there are debates and discussions about whether it is one of the fundamental or more fundamental properties than the existing fundamental properties, such as four types of forces and time–space (http://www.sci-news.com/othersciences/psyscology/consciousness-fundamental). Even though the Vedic science considers that consciousness is not merely remain limited to humans, it has much wider and higher dimensions to such an extent that consciousness is a fundamental attribute of all existence and the source of consciousness is outside the human body (Sarvapriyanand 2018). However, in the state of the art in science and philosophy, there is no general consensus with respect to questions of whether consciousness is a local or universal phenomenon. Again, there is no general consensus on whether human consciousness works at the level of the whole body or only at the level of the brain. David Charmers notes that there is nothing that we know more intimately than conscious experience, but there is nothing that is harder to explain, and this is where his proposition 'hard problems of consciousness' argues that a reductionist approach that works so well for material science may not be the right framework for approaching consciousness and developing a new paradigm is required (Chalmers 1995). We can reiterate here that humans are spirits in nature. Understanding the relevant concepts, nature, mechanism of human consciousness and how it affects human attitude and behavior is critical and challenging to the subject of this book, i.e., risk consciousness. Furthermore, apart from human consciousness at the individual level, the organizational, national and international levels of risk consciousness are also important to realize full potential collective risk consciousness. Sigmund Freud divided human consciousness into three levels of awareness: conscious, preconscious and unconscious. Each of these levels corresponds and overlaps with his ideas of the id, ego and superego.

At this point, it will be an appropriate to use the knowledge available in the spiritual domain to create a human model where consciousness has been fundamental to existence. Furthermore, elaborates on consciousness that there are seven reflections, forms or manifestations of the self, i.e., an individual being as (i) physical consciousness—the sensation at body level, (ii) vital consciousness—energy levels (iii) mental consciousness—that contributes to mind, (iv) supraintellectual consciousness—the ability for questioning, imagination, reasoning, (v) consciousness proper to the universal beauty or blissful consciousness—natural component that supports the idea that we are part of nature and have the potential to realize bliss on the path of 'Anandamayi (Quantum Awareness where science and the Buddha meet 2005)', (vi) consciousness proper to infinite divine self-awareness—the component that defines self-realization I am the reflection of ultimate truth, (vii) consciousness proper to the state of pure divine existence—the essential nature of Brahman is pure consciousness (https://en.wikipedia/Chaitanya(consciousness)). Vedanta also identifies other forms of consciousness that are associated with worldly affairs. Often, particularly in this era of modern science, this material aspect of consciousness is being explored to support health care, develop ethical and moral behavior models, etc. (Introducer 2010).

In the last couple of decades, there has been an active interest in modern science for defining, exploring and understanding the science of consciousness. Here, there is a consensus that the science of consciousness is still in evolving stages, as the subject is complex and possibly difficult to capture not only with the traditional approach to science, i.e., Newtonian but also a challenge to even comprehend by advanced approaches such as quantum mechanics and computational models (Goff 2019a).

In this chapter, we will discuss various efforts toward understanding consciousness in various fields, viz. spirituality, philosophy, quantum mechanics, biological science, neuroscience, cognitive psychology and information science. There are some hard problems of consciousness, such as 'emotion, feeling, perception of color' and so on, that can be captured by the word qualia, which poses a challenge to understanding consciousness while these very attributes characterize the role of human consciousness (Chalmers 2006). There is some observation and evidence that spiritual philosophers and material researchers in the West are now breaking their traditional barriers toward understanding consciousness and are receptive to each other's idea of consciousness (Jyoti 2015).

The subject from a modern material science perspective is considered complex, as there are many questions that need to be answered, as discussed above. Therefore, there is renewed interest in exploring various aspects of consciousness that include dealing with questions such as whether consciousness is fundamental, whether consciousness is only human-specific or all creatures and material have consciousness, the acceptability of the universal view of consciousness, whether the materialistic approach is adequate to produce the required insight, and whether advanced sciences such as quantum mechanics can bridge the gap between existing spiritual knowledge and science-based approaches.

There is a general consensus that modern science's contribution to humanity is beyond comprehension. It has delivered all that was outwardly thought to be required for human comfort, happiness, welfare, security, safety, peace and harmony and so on. However, the next question is 'What kind of society we are now?' Are we happy, at peace with ourselves and with the surroundings? arguably the answer will be perhaps a big No. By exploiting the natural resources employing the material approach we could achieve short-lived happiness when we acquired some gadget, property, success, but we know it was short-lived and we were again back to the same situation almost.

> *A fundamental conclusion of the new physics also acknowledges that the observer creates reality. ....The stream of knowledge is heading towards a non-mechanical reality; the universe begins to look more like great a great thought than like great machine.*
>
> RC Henry, a professor of physics and astronomy at Johns Hopkins University, 2005

Why it is so? First and foremost, we had an inadequate model of true or inner happiness. We thought that happiness comes from outside of and we created the material world around us. The material world made, no doubt, our life easy and gave some temporary moments of happiness, often when we compared with others and found we are superior to a few but inferior to others around us and felt depressed when the situation was otherwise. This worked as perpetual trigger for the so-called rate race ...so-called competitive attitude ... further material success or failure ...

and finally the human felt often lonely, depressed and unwell physically and psychologically and medically. Many of these aspects affected the general human performance and behavior but more particularly the behavior, attitude and performance at workplace—the subject we target to address, here, in this book.

Therefore, what went wrong? We thought we worked very hard, did our duty with utmost sincerity and so on. We forgot that we are HUMAN only in part material entity, but, primarily the spirit or to be precise spiritual in nature. If this is not so, then why man were treated like machine or more appropriately like a robot. Accordingly, the traditional management system often uses the term manpower and now human resource in line with machine or money or any other physical entity. We treated man/women in an organizational structure as a machine to deliver the expected performance. Therefore, the HUMAN became MAN or WOMEN.

The biggest flaw in our model's existing model was that we primarily focused on the outside to be a precise external material world as a source of happiness and wellbeing. It is human fundamental nature to be self-aware or realized or conscious of our own being. The realization that the external world is within you as a reflection of the external world that you perceive as your understanding of the external environment. We are required to understand that everything is inside you, the source of our happiness, satisfaction, health, peace, love and so on, and there is a need to go ahead and look inside real peace, happiness or any other inspiration. The Eastern—spiritual and Western—material science have come to an understanding that all the realities of outside are within you—how you perceive the external world is your take and how to accept and fix the external world issues and finally accept and reject is also rests with you. Therefore, it is for our considerations that human wellbeing and human behavior and health can be ensured by a framework that, apart from physical, physiological and anatomical features, integrates spiritual practices, such as Yoga, meditation and faith.

The above arguments took cognitive and neuroscience to look deeper into the true nature of human beings. This aspect may have brought into the focus the science of consciousness. It can be recognized that cognitive activities, such as learning, memory, vision, auditory, touch and smell, are played out at the level of different faculties or part of the brain. Similarly, the brain activities related to 'thinking and memory' about the internal and external world at subtler forms what is called mind. For all this subjective phenomenon, consciousness becomes a fundamental requirement. Second, Eastern science and philosophy asserts that the mind is not localized at the level of the brain but operates from the cells all over the body, and in this sense, it is nonlocal.

Finally, we answer why we brought in consciousness as a subject while we are exploring the subject of human reliability. Human reliability is the probability that human action will be accomplished successfully in a given time under a given context. It is well-recognized in the area of risk and reliability modeling that (a) humans contribute significantly to accidents and (b) modeling humans is a complex task and contributes to a very high level of uncertainty such that the results of the analysis often become meaningless or inadequate for any serious considerations. It is also recognized that there is a need to modify the current human reliability techniques, which to a large extent are based on insights based on the field response or simulator data modified by the performance shaping factor again based on data collected from lab experiments or field data. However, this makes these models inadequate for the future prediction of human performance. Here, we argue that the human reliability model should be based on three Cs, i.e., apart from cognition, the consciousness and conscience playing at the level of the human body and brain.

The state of the art in human reliability does include considerations of cognitive sciences, but it fails to recognize that consciousness is more fundamental to humans, and without the explicit considerations of consciousness, it may not produce the desired results. Of course, the considerations of the third *C*, i.e., conscience, acquire added dimensions from security considerations for human factor modeling. This calls for considerations of apart from life quotient, i.e., attention, awareness, alertness, etc., and that accounts for energy levels, morality, ethics, etc.

Accordingly, the objective of this chapter is to understand the state of the art available in the area of philosophy and science of consciousness and develop a protocol comprising axioms/assumptions/tenets, guidelines, and update the existing framework of risk-based engineering and a broad category of performance shaping factors that can be employed to risk-consciousness operations management for complex engineering systems. The above work is realized through the updating of available procedures involving considerations of consciousness, cognition, conscience and brain (CQB) as part of risk-based engineering to model human and organizational factors in a complex engineering system. By doing so, the modified procedure will, on the one hand, cater to engineering aspects and operational aspects that are more intense in terms of human interaction in a given operational scenario during normal operations as well as in emergency conditions. As we will see, much of the complexity of consciousness becomes manageable as the end application, i.e., operations management with certain assumptions and boundaries.

The objective and scope of this chapter are to understand the state of the art in consciousness and explore the possibility of this knowledge to develop and incorporate human consciousness attributes into the development of attributes and elements for risk-conscious culture. Accordingly, this review covers the following aspects: (a) considerations of major categories/approaches toward creating a holistic view of consciousness; (b) in the conventional sense, major application areas of consciousness to societal and engineering and management systems; (c) state of the art in the source of consciousness and the role of brain faculties; (d) the capability of artificial intelligence to model consciousness; and (e) the potential of risk-based approaches to support risk-conscious applications.

2.5.2 *Major Evolutionary/Developmental History of Consciousness*

The understanding of consciousness can be divided into four broad approaches, viz. (a) based on traditional eastern science and philosophy based on the central theme of *nonduality*, also referred to as ideal monism, or mystic school of thoughts, (b) *dualistic* approach that considers mind and body as separate identity and (c) *panpsychism* approach and (d) material science-based approach, e.g., developed around neuroscience, and the most modern the quantum mechanics, approaches, etc.

The Vedic approach that is the foundation of Sanatana Principles of life is at the core of Hindu Science and Philosophy. Sanatanic philosophy of life is based on nondualistic principles. In the context of the present discussions, the mind and body are part of one unifying energy and not two separate entities such as material and nonmaterial, such as the idea followed in the dualistic approach. The Vedas and Upanishads advocate consider the consciousness all pervading and very well endorses the unifying forces and in line with the meaning of Sanatana that is 'no beginning or end'. In this context, the source of consciousness is outside the physical body and nothing but reflection or part of that only single source. In fact, the eastern literature believes that every entity poses consciousness only the level or consciousness quotient varies from infinite applicable to universal, through human, animal and material spread over a range in each case. Figure 2.3 shows the postulation of the all-pervading phenomenon of consciousness. This presents one question or clarification with respect to 'What we consider consciousness in modern science'. Is it that what we refer to consciousness as nothing but the manifestation of the primary source of consciousness at the physical level due to activation at the level of brain faculties?

There were three ancient major philosophies as part of ancient Chinese culture that still influenced the Chinese cultural system, viz. Confucianism, Taoism and Buddhism. Confucianism, the oldest culture, emphasized rituals, family bonds, worship of ancestors, etc. While spirituality, humanity and the soul were the major elements in Taoism, Buddhism shared many common grounds with Indian systems (https://www.nationalgeographic.org/article/chinese-religions-and-philosophies/). The origin of Buddhism was from India, and later, it spread to China. This will be discussed in the following paragraphs along with the Indian Vedic system. Human consciousness in Chinese thought is perceived at three levels: cosmological consciousness, which captures the objective world of being and becoming; second, the self that deals with the world of human reflection of self in which a human can distinctly be identified in this world of humanity and the third called political consciousness where the self for practical purpose in terms of desires, power creativity and freedom of expression (Consciousness: Chinese Thoughts 2019). In recent years, consciousness research has been actively pursued in line with the work performed in modern psychological sciences; however, the new theories proposed have some similarities as well as differences with classical theories, and there is consensus that there is considerable scope for theoretical and experimental development in consciousness science (Liu and Huo 2020).

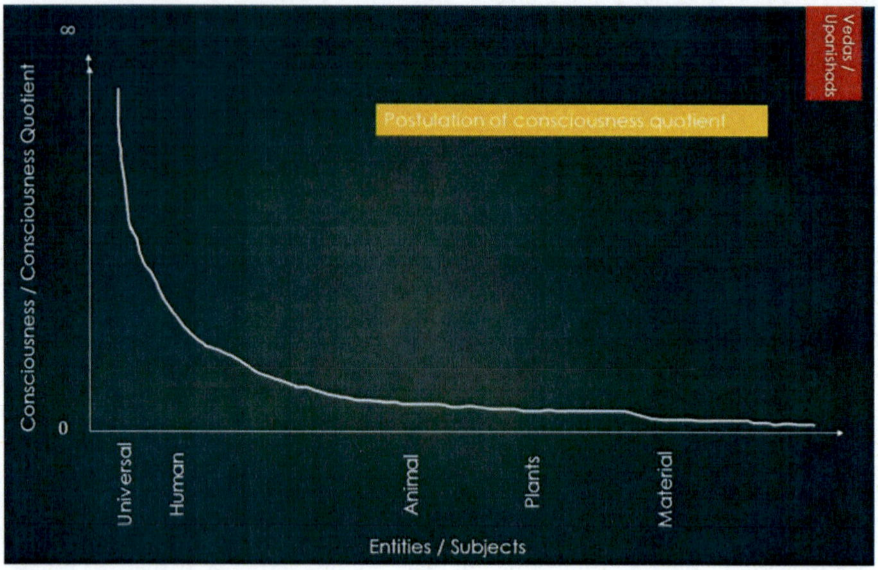

Fig. 2.3 Consciousness is all pervading and everywhere is one of the fundamental postulates of Vedic philosophy

Ru Xin presented a historical survey on 'Mind and Consciousness in Chinese Philosophy'. This work discusses Husserl's work on 'Philosophy as Rigorous Science dedicated to western philosophy'. In fact, now time has come to recognize the science and philosophy of ancient times, and in this context, the assertions are relevant. However, Husserl further insists that 'only with Greek we have do we find universal vital interest in the theoretical attitude and only Greek thought can be truly called philosophical. He thus refuses to place Oriental philosophy, mainly Indian and Chinese philosophy on the same level with Greek Philosophy, he regards the former as representing a mythico-practical attitude'. These views should although have universal appeal and appear attractive, however, when it sidelines the other great ancient philosophies like Vedic science and philosophies which deals apart from rituals, has at the core the science of medicine, mathematics, astronomy, music, human wellbeing and many more need considerations to further human and societal interests, which is the purpose of these philosophies and sciences (Ru 1984). It is vital to have an inclusive, integrated and serious approach to understand the true universal potential for the development of human and societal systems.

In the eastern approach, consciousness has been considered evolutionary. The unit of consciousness within the bodies of all species is indestructible. These individual units are qualitatively identical with each other but display a certain range of powers and abilities based upon the particular characteristics of the physical form they inhibit. During the evolutionary process, the imperishable conscious units transmigrate from lower to higher species, e.g., from apes to man. Thus, the Vedic literature describes the evolution of forms descending from higher to lower and the evolution of consciousness ascending from lower to higher. Vedas asserts that our universe is approximately 155.52 trillion human years old, and its total life span is 311.04 trillion human years (which is equivalent to 100 years of Brahma) (https://www.cs.ubc.ca/~goyal/creation.php#:~:text=Any%20theory%20of%20creation%20is,to%20100%20years%20of%20Brahma). Hindu Dharma believes in the evolution and involution of forms of life according to the Karma of individuals. It further asserts that Pure/higher Consciousness, which is the embodiment of '**Sat Chit Ananda**, i.e., Eternal Truth—Eternal Knowledge—Eternal Bliss' (Chandra Mauli 2018; Vedic theories of the universe 2022). As seen, the Chinese knowledge of cosmic consciousness and the principle of nonduality appear to be in line with the Vedic knowledge.

The Buddhist approach has liberation from miseries in human life as the major motivation and while asserting the principle of Anichha, i.e., 'principle of change in perpetuality' or 'change is as the lay of nature'; with this law Buddhism rejects the concept of 'Atman', which asserts the philosophy of rebirth, laws of 'Karma' (Indeed) that is at the core of Vedanta. From this, it can be inferred that Buddhist philosophy rejects the fundamentals of nonduality. Even though there are similarities in Hindu and Buddhist philosophy, as both were born in Indian ancient eras, one fundamental difference is that many schools of Buddhist Philosophy assert that there is no concept of Atman. The second difference is that the consciousness referred to in Buddhism as 'Vinnana' in the Pali language is not directly related to 'self', which is not in line with Vedic philosophy (Krishan 1984). It further differs from Vedanta in the sense that consciousness even though it exists, it is associated with the birth and death of humans, i.e., consciousness ceases after death.

However, the practice of meditation and religious practices are shared, although with slight differences. Therefore, it can be concluded that even though, in a way, the Buddhism approach is a subset of traditional Vedic science, there are differences, mainly that the Vedic approach considers that human beings have Atman or Soul, and it is eternal, while Buddhism has no concept of Atman. However, both approaches involve enquiry of the phenomenology 'self' by looking inside toward understanding human nature to achieve self-realization as the ultimate goal.

One of the major schools that still maintains the Buddha teaching and practice of meditation in its purest form is Vipassana (Hart 2012). The Vipassana technique, while presenting the science of real internal happiness and enlightenment, provides a training framework that involves guidance on moral conduct, improvement of concentration, awareness and equanimity and understanding the root cause of immediate and root causes of problems. This approach, when we look at it, appears to be meeting the scientific procedure to raise human quality. For example, for understanding the causes of misery and unhappiness, it is an established cause–effect paradigm that is tested and approved over generations. The algorithm for arriving at the root cause of suffering is as follows (Hart 2012):

If ignorance arises, reaction occurs
If a reaction arises, consciousness occurs
If consciousness arises, mind-and-matter occurs;

If mind-and-matter arise, the six senses occur
If six senses occur, contact occurs
If contact arises, sensation occurs
If sensation arises, craving and aversion occur.
If craving and aversion arise, attachment occurs
If attachment arises, the processes of becoming occur
In the process of becoming arises, birth occurs
If birth arises, decay and death occur, together with sorrow, lamentation, physical and
mental suffering and tribulations,
Thus, this entire mass of suffering arises.

Thus, ignorance has been identified as an immediate cause of suffering that leads to a chain of processes through consciousness, as above, and creates a vicious circle for misery and unhappiness.

The Greek philosophy arose in the sixth century BC, initiating the post-Socrates era, and had a western influence on its culture and scientific thoughts. Many philosophers continued through the Hellenistic period and contributed to the development of modern scientific culture. Some major players who shaped the Western philosophy on consciousness include Socrates, Pluto, Aristotle and later Descartes. The dualistic principles were at the core of Greek philosophy. In the context of humans, the mind and body are considered two separate entities. The Greek philosophies were essentially dualist in nature. Plotinus was the first Greek philosopher to hold a systematic theory of consciousness (Hutchinson 2018). His theory involves multiple layers of experience. Higher modes of consciousness provide human beings with a rich experience of the world with the subjective notion of world views on the true self and nature of reality. Furthermore, Descartes' principle of '*I exist as I think*' are some profound thoughts of dualism. Dualism is closely related to the thought of the philosopher Rene Descartes, native to the Kingdom of France (fifteenth–sixteenth century AD) (Seventeenth-Century Theories of Consciousness 2020). In dualism, the mind is nonphysical and works at the level of the brain—a material entity. Descartes asserted that mind and consciousness are sources of self-awareness and distinguished it from the brain—the source of intelligence. In fact, Descartes's theory known as Cartesian dualism (or mind–body dualism) on the separation between the mind and the body went on to greatly influence subsequent Western philosophies. In meditation on first philosophy, Descartes attempted to demonstrate the existence of God and the distinction between the human soul and the body (Meditations of René Descartes 2022).

As the philosophies and science of evidence and psychology on consciousness are evolving, a sort of convergence is emerging that integrates traditional/ancient knowledge with modern findings/observation. In this context, there is growing interest in *panpsychism—a relatively new approach* popular among researchers and philosophers (Goff 2019b). The major feature of this approach includes the view that mind and mind-like aspects are fundamental and ubiquitous features of reality and extend this argument to a universal phenomenon (Panpsychiasm 2021).

The last 30–40 years have seen a renowned interest in the development of science of consciousness. In the context of this book, the efforts are dedicated to understanding consciousness employing mainly a materialistic approach. Material science has made not only beginning but also developed the advanced science of cognitive consciousness or to be explicit—the physical manifestations at the level of brain toward understanding of human cognitive consciousness. In fact, these advancements have shown evidence of potential promising ideas or postulations that might take the science of consciousness toward creating support for ancient or previous knowledge for validating the concepts. This science addresses investigation or probing questions such as what constitutes cognitive consciousness, the source/sit of consciousness, whether it explains emotions or qualia, applications in diagnostics, understanding the subconscious mind and its potential to evaluate human behavior. Additionally, whether other than human entities, such as animals, insects, microbes, plants and materials, have consciousness is also being extensively investigated and debated.

The major schools of modern science-based approaches can be categorized as (a) bioscience and neuroscience, (b) quantum mechanics, (c) information theory and (d) computer-based approaches. Along with philosophies of the East, the above science-based approach to consciousness is discussed in the next section. The idea is to understand the major features of these approaches such that there is an input for developing an integrated framework for risk consciousness.

2.5.3 The State of the Art in Risk-Consciousness—Applications

The available literature shows that there are many risk-conscious real-time applications. It ranges right from some primitive or qualitative notion of risk to the applications based on formal risk studies. The 'conscious' component in these applications invariably demands a focus on human effort either in support of decisions, management or technical quality achievement. In short, the common thread through most of the applications discussed is 'make or include risk as factor' in support of activities or subject. In short, the level of depth and detailing varies from application to application, but the attempt has been to make 'risk' the central guiding parameter to address the requirements.

One of the major components for compliance to the sustainably by incorporating the zero-energy concept in the house design, however, due to design and loading variability (due to external conditions) or uncertainty there is risk factor that should form part of risk-conscious design. In this context, the risk-conscious design of off-grid solar energy houses was developed by Hu. H. as part of his doctoral research (Hu 2009). The timely response of an operator when required to actuate a safety function is a vital safety critical system to avoid failure or undesired consequences, which might include fatalities or serious losses during the operation of a system such as a process plant or a transport system. Further information flooding in the man–machine interface or network system during an emergency creates special challenges for the avoidance of consequences or risk. Taleb et al. discussed their work on the development of a risk-conscious vehicle collision avoidance system that addresses a risk-aware medium access control protocol to increase the responsiveness of their scheme along with associated developments (Taleb et al. 2010). McWilliam and Singh assume that all organizations, be it academic, social and so on, are basically risk organizations, as the organization is at risk if it is not able to deliver according to set performances and goals. In their research, they investigate the role of teacher collegiality and its effectiveness while keeping a check on ethics and morals in an academic environment and how to minimize the risk of isolation or otherwise so that set standards and targets are achieved (McWilliam 2003).

Oytam et al. researched the development of a technique to avoid the loss of information in a genome laboratory environment where the background batch noise tends to reduce the power of statistical tests that might produce uncertainty in the results of the tests, analysis and prediction, and they also discussed the role of risk-conscious correction of batch effects (Oytam et al. 2016). Lee et al. discussed the limitation of the traditional performance-based approach to the design of structures and buildings and proposed a framework employing a vector autoregressive process for capturing meteorological data uncertainty as part of risk-conscious design. This technique enables the generation of meteorological year data that provide probabilistic input for risk-conscious design by incorporating uncertainty into the model (Lee et al. 2012). Uncertainty is generally part of planning and requires a robust algorithm for the optimization of schedules and activities. Tometzki and Engell proposed a risk-conscious framework that accounts for uncertainty while employing two-stage stochastic mixed-integer models in which decisions must often be made under uncertainty (Tometzki and Engell 2010).

Sand and Engell worked on a real-time scheduling problem in a laboratory environment on batch processes under uncertainties. Every decision is associated with a certain level of risk, as it affects the future evolution and therefore is not predictable. This challenge is addressed by a moving horizontal approach with frequent optimization through stochastic programming to reflect the uncertainty. A real-time problem was modeled in this framework to validate the applicability of this approach (Sand and Engell 2003). Pegg explores the role of organizational cultures in workplace learning for those learning to be educational leaders and demonstrates the applicability of this risk-conscious approach through a case study performed on five school leaders (Pegg 2010). Tilman applies risk-conscious investment to maximize shareholder value in a financial institution and presents the case of large-scale financial losses and how they provide a unique capacity for improving investment through this risk-conscious approach (Tilman 2003). McWilliam discussed the role of accountability in an academic environment, specifically the factors such as value systems, which include moral and ethics played out through a risk-conscious approach to guard against declining standards and student performance (McWilliam 2004). Wilkinson, in this book entitled 'Anxiety in a risk society', investigates the sociological environment to answer the question while attempting to understand the hypothesis that we are more anxious as we are very risk-conscious or in other words a risk-averse society (Wilkinson 2001).

Guzowski et al., while working on energy efficiency projects, employ a risk-conscious framework to overcome the traditional barriers to widespread adoption of building energy efficiency measures in support of decision making. The work essentially deals with the development of a normative energy modeling approach to inform energy efficiency investments. The model enables sensitivity and comparative analysis with traditional approaches, such as audit and modeling (Guzowski et al. 2012). Terwiesch et al., while discussing risk-conscious operations of batch processes, emphasize the role considerations of dynamic optimization that takes into account model uncertainty and interrun variations as opposed to traditional nominal optimization toward reducing uncertainty in decision making (Terwiesch et al. 1995). Project financing and risk management often pose challenges in the space industry due to uncertainty in project schedules and other administrative and technical factors. Gerose and Nasini assess the application and role of risk-conscious attitudes in space project evaluation and management. It is stated that risk and opportunity management strategies are gaining ground in the space industry (Gerosa and Nasini 2001).

Lawrence et al. presented an approach to develop an experience-based approach to correct risk perception for natural events such as floods by integrating the effect of floods such that the approach to risk management is optimal, which enables a gap in understanding the risk associated with the events between the public, government bodies and other stakeholders and helps in developing a risk-conscious society (Lawrence et al. 2014). Situation awareness is critical to plant operations, particularly during emergencies. Mankowitz et al., while discussing the role of traditional hierarchical re-enforced learning, emphasize the significance of risk-conscious skills in improving situation awareness skills (Mankowitz et al. 2016). Lanne et al. discussed the role played by risk management and value creation in a typical business environment; however, the interrelationship between these two is not that well-applied. It is emphasized that the risk-conscious value creation approach can benefit the overall business environment as well as performance (Lanne et al. 2014).

2.5.4 Source of Consciousness and Role of Brain

Regarding the source of consciousness, there are two major views. Vedanta philosophy asserts that the principal source of consciousness is within and without, i.e., from the outside of the physical body (Durga et al. 2018). Therefore, humans share this principal source of consciousness, or life source, and this shared consciousness is fundamental to human existence and the energy for all mental and physical activities. According to Taittiriya Upanishad, consciousness is present in the fifth sheath called *Anandamaya kosha* within the human body, which is nearest to the blissful self (Sharvananda et al. 1989). This shared consciousness manifests at 7 levels, of which 3 levels that are directly relevant to humans are physical or physiological manifestations: vital (prana or energy) and mental consciousness. Other levels, as mentioned, include the superintellect and other cosmic aspects, as mentioned earlier in Sect. 2.1. Finally, the realities of the source of consciousness are outside, and the fifth kosha is the seat of consciousness, while at five levels, we have this manifestation of consciousness. Therefore, consciousness is a personal (internal) as well as an outside (external) subject. Internally, the manifestations are material phenomena. Consciousness fundamentally is universal and pervasive. *Therefore, every individual human may have his own subjective realities for the same objective entity be it a phenomenon, a natural scenario, a discussion in a team, and so on.* As per the ancient Hindu mythology, the discussion on the 'third eye' has reference to the pineal gland, which is considered the seat of enlightenment and higher consciousness. The concept of the third eye was part of Sanatanic philosophy, where the Shiva, apart from two sensory eyes for physical vision, had a third eye of wisdom. As per Vedic scriptures, the third eye is designated to facilitate communication with the divine powers and self-realization (Sanchetee and Sanchetee 2018). In ancient Egypt, the pineal gland was known as the seat of the spirit or the soul. They used the third eye as a route to higher awareness and consciousness. The Buddhists related it to spiritual awakening (https://www.thesleepguru.co.uk/meditation/the-magical-pineal-gland/). The pineal gland is responsible for sleep quality and the release of happy harmonic melatonin (Purucker 2016). A press article based on neurological research claims new evidence of the connection between the brain and higher consciousness. Here, the scientist relates it to the ancient Hindu philosophy of the third eye (of Shiva) related to mystical awakening. Dr. Panagaria postulates that the pineal gland, a single structure in the brain (between two halves), might be responsible for the physical body and intellect (The Hindu 2013). Descartes also regarded the small gland as the *conarion* or pineal gland as the principal seat of the rational soul, where it is discussed that my view is that this gland is the principal seat of the soul and the place in which all our thoughts are formed. The reason I believe this is that I cannot find any part of the brain, except this, which is not double (Descartes and the Pineal Gland 2013). Furthermore, the article entitled 'Humans: Harvard Scientists think they've pinpointed the physical source of consciousness' (Harvard 2022)'.

The modern sciences assert/argue that consciousness arises at the level of the brain. Neuroscientists, bioscientific theories and other computational schools have their research and arguments that attribute consciousness to brain-level phenomena. Observation of perception-related neurons in the prefrontal cortex has been found to be consistent with the theory of Christof Koch and Francis Crick, who postulated that neural correlates of consciousness reside in the prefrontal cortex (https://en.wikipedia.org/wiki/Neural_correlates_of_consciousness#History). Seth a neuroscientist, after his extensive research wonders on the capability of reductionist approach that worked excellently in material domain may not be adequate for consciousness research as 'consciousness is intrinsically private' (Falk 2021; Seth and Hohwy 2019). Hameroff and Penrose proposed a model involving orchestrated reduction where the quantum vibrations in 'microtubules' inside brain neurons are linked as a source of cognitive consciousness (Hemeroff and Penrose 1996). What we are discussing here are models of consciousness at the level of brain. Tononi's model has informational or quantum data, and knowledge is at the core of his integrated information theory (IIT) and is characterized by a '*phi*' (Tononi 2008*)*. The 'phi' metrics may not be directly related to source/sit of consciousness but has a model that refers to the quantum of data and information that it considers implicitly or explicitly as the basis of knowledge.

Modern science asserts that consciousness arises at the level of the brain and is physically reflected in various brain faculties, mainly the cortex area, while generally rejecting that the source of consciousness is outside of the human body and in fact outside of the brain (Dehane et al. 2017). This argument can make sense when we refer to the consciousness component as a physical symptom in various brain faculties, as it also supports the energy input to the brain to ensure proper functioning of various cognitive faculties of the brain and associated nervous system. This component of consciousness can be referred to as 'cognitive consciousness'. This postulation appears to adequately address the problem of source and 'operational regime of consciousness' while enabling the adoption of human consciousness to support the investigation and modeling of human behavior, particularly during stressed conditions. Second, with this distinction or categorization, the argument of source/sit of consciousness still remains the subject of research, and the associated complexity can be handled independently without greatly affecting human behavior aspect modeling and physically reflected in various brain faculties, mainly the cortex area, while generally rejecting that the source of consciousness is outside of the human body (universal consciousness) and further appearing to disregard the concept of the pineal gland as the seat/source of consciousness. However, the quantum mechanics approach in general and the model proposed by Pribram–Bohm through the holoflux theory of consciousness are inclusive in the sense that below the planks constant threshold, there might exist scope for cosmic considerations of consciousness (Joye 2016).

Lekov's author for the article 'The formula of the "Giving of the heart" in Ancient Egyptian texts' writes that in the ancient Egyptian era, the heart was considered the seat of consciousness, intelligence and emotions (Lekov 2022). In fact, in many cultures and traditions, including Hindu books religious book The *Ramcharitmanas* written in sixteenth century is an Awadhi version of The great Hindu Epic *Ramayana* written by Maharishi Valmiki during the period 7th to fourth centuries BCE (Goswamy 1633; Valmiki and Ramayana), considers heart the sit or source of emotions. However, Hinduism does not explicitly consider the heart to be the source or site of consciousness.

Chalmers has never been tempted by mysterianism. Chalmers has always believed that 'there's a solution out there somewhere and we ought to be able to find it. Or we ought to try. We're not going to know if there's no solution there until we try and try and try.' (Scientific American 2017). The above discussion of the source/locus of operation of consciousness remains focused on the faculties of the brain involved in conscious processing. Modern science has been actively trying to create theories and science about the source/siting consciousness, with an interest in modeling and primarily understanding human consciousness. Second, there is also an active interest in learning from ancient science and the philosophy of the sit of consciousness in the human body that needs to be explored.

The encouraging outcome about the state of the art in consciousness is that cognitive consciousness is central to modeling human behavior; therefore, for all practical purposes, we can say that subjective experiences and thoughts are communicated to the brain through sense organs and that brain faculties perceive the individual's perception of reality. When this perception is shared among a major part of the human resource right from top management to the grass root level, it is called organizational consciousness. This is where consciousness distinguishes itself from cognition; i.e., it has a way to become part of organizational metrics. This insight is a vital input to develop a risk-conscious culture for engineering systems.

2.5.5 *Consciousness and Artificial Intelligence*

The role of artificial intelligence and its application in machine learning has been growing rapidly in automation and decision making. The question is whether AI or computer-based operator aids have the capability to exhibit the attributes of consciousness. Furthermore, AI aids in operational ecosystems can support decision-making factors as part of conscience, such that it maintains the business 'value systems', viz. ethics, morality, attitude and integrity, so that apart from safety, the security aspects are not compromised. Dahene et al., while discussing the question 'What is consciousness, and could machine have it?', argue that the state of the art of present systems is only capable of implementation of computation processing that is akin to unconscious processing in the human brain (Dehane et al. 2017). While considering the information processing model for consciousness in their analysis, it is inferred that the information approach alone may not be adequate to reflect consciousness in machines unless the analysis in depth and nature is captured to address the complex issue. If we have simplified characterization of consciousness, as thoughts and emotions, we all experience then it is a challenge to establish whether machines have these. However, when researchers consider consciousness based on physical laws, then the consciousness level C0—a problem-solving computer can do this, C1—refers to the relation between the cognition system and the thought in question—can be evaluated, and the C2 category has a sense of knowing what we do. However, the overall conclusion is that we need a machine that can support these features. The best we can come up with is to say that consciousness is the thoughts and sensations we all experience personally which means we do not yet have a way of establishing whether it exists in something else, such as a computer (Mcrae 2017).

It appears that although there is a growing interest to make the machine conscious like humans, further work is required to make machines that can process the information with consciousness algorithms and technology. While discussing the potential of AI, it is vital to understand the behavior of AI systems as to how conscious AI systems respond to situations where common sense is critical to decision making. Nature briefing, May 5, 2020: Features and Opinion on 'How to give a computer common sense', comments: Computer scientists have built a system that can make simple guesses—for example, what happens after 'Gary stacks kindling and logs and drops some matches'. Machine learning techniques, such as neural networks, can make statistical Gausses after observing large amounts of data, but they are notoriously lacking 'common sense'. Other attempts relied on knowledge bases containing millions of rules and sentences handpicked by humans. COMET combines the two approaches, using a knowledge base to train a neural network. Lead researcher Yegin Choi says she was surprised that no one had tried this approach before. It's almost as if nobody bothered because they were so sure this would never work. COMET's Gausses are still statistically based, but at least they are often correct (Choi 2020). However, it appears that Gary Marcus and Yegin Choi had some encouraging results featured on the Quanta Magazine website, with the title 'common-sense-comes-to-computers' (Pavlus 2020).

The above discussion shows that there are active interests among the researchers to impart attributes that exhibit conscious behaviors in general, particularly the aspects related to conscience—which has a much larger dimension. Hence, it can be concluded that even for risk-critical systems, dependency on computers or operator aids should be addressed with extra care. Rather, it is required that robust technical and administrative provisions are made such that these systems do not induce a new element of security-related issues.

2.6 Major Philosophy and Science of Consciousness

In the preceding sections, a review of various schools of consciousness, along with a few important aspects for risk-conscious applications relevant to human factor modeling and analysis relevant to safety critical systems, was discussed. This section addresses thoughts and sciences more relevant to the development of human reliability to support the postulations, assumptions and enhancement of the CQB human reliability model as part of risk-consciousness culture in further detail. The idea is to create/upgrade the existing framework for risk-conscious operations management. The following section presents the state of the art of various techniques/models that include nondual, dualism, panpsychism, psychological, neurological, computational and quantum mechanism-based schools of consciousness.

2.6.1 Nondualism or Advaita

A sage, scholar, philosopher, poet, dialectician and religious reformer, Adi Shankara (circa 788–820 CE), was a foremost exponent and propagator of Advaita (nondual) Vedanta or absolute monism (Kapoor 2022). Vedanta is the asserts nondual approach to the existence—A Hindu spiritual approach to science and philosophy of consciousness (or Chetna in Sanskrit). The Vedas are the Vedanta (Upanishads derived from the philosophical component of Veda). The Vedas are a large body of religious texts written in Ancient India between approximately 1500 and 1000 BCE in Sanskrit. There are four Vedas (Rigveda, Yajurveda, Atharvaveda and Samaveda) (Chandra Mauli 2018). The Vedas are the source of collective wisdom, science of life and nature, tradition, and records of the evolving Hindu culture of a remarkable civilization. Vedanta philosophy asserts that the source of higher consciousness is outside and fundamental to human existence. Human consciousness is nothing but an extension of a higher level of consciousness, and mind and matter are created by the same source. Therefore, everything, be it matter or living being are all interconnected, and the source of human consciousness is the higher consciousness.

> Utterly destroy the false ego. Control the many waves of distraction. Discern the reality and realize "I am that". You are pure consciousness, the witness of all experiences. Your real nature is Joy. Cease this very moment to identify yourself with the false ego.
>
> Adi Shankaracharya
> Hinduism

Consciousness is responsible for all our experiences, including thoughts, emotions, feelings, love, empathy and behavior, for our inner growth, personal relationships and social interactions with the environments we live in. In other words, consciousness is the basis for different experiences that we consider our reality, including the observer, the object and our observation (Chopra 2019). The Vedanta philosophy asserts that all the reality is within you and the outside world is a reflection of what you have inside. The power of inward looking enables realization of a unified field through internal processes as well as including a cosmic unifying field of consciousness.

We normally have three states of consciousness: waking, dreaming and deep sleep. Samadhi is sometimes referred to as the fourth state of the ground state of consciousness. A primordial awareness that can become present continuously and in parallel with the other consciousness states. In Vedanta, it is called TURIYA (Samveda: Chandogya Upanishad, India Hindu Civilization: Indian Scripture, 6th–8th BCE; Chandyogya Upanishad 2022).

> Traditional science assumes, for the most part, that an objective, observer-independent reality exists; the universe, the stars, the galaxies, the sun, moon and earth, would still be there even if no one was looking.
>
> Deepak Chopra

There is a growing or enhanced understanding of consciousness that comes from quantum mechanics. This understanding shows a shift from the old paradigm that matters in reality and consciousness. The insights developed from the modern science on quantum field theory, where the existence of matter unlike classical physics is a probability of particle and field, will be discussed later in the section on quantum mechanics. However, it can be noted here that the nonduality has qualified as a concept with probability as there is probability when either a matter or field but both are related to energy as a common antecedent. Furthermore, the nonduality argues that we are not in the physical world instead the physical world in within us. This is related to the observer effect as the energy as matter becomes reality when we observe and in quantum mechanics relates the existence to the observer effect. We create the physical world when we perceive it consciously and that is why we observe it and, we create our experiences and our imaginations. When I say we, I do not mean the physical body I mean is consciousness that conceives, constructs and governs everything that we call physical reality. This model has been explored by cutting-edge scientists in the field of neuroscience and quantum physics. This is also the model that was explored by Sages and Seers in Bhagavad Gita (Chopra 2021).

John Searle asserts that consciousness is a very important subject in our life because it is a very necessary condition of life. It is the most neglected subject in our modern scientific and philosophical culture and notes that as there is curious reluctance due to (a) a traditional culture of dualism where consciousness is part of the soul; (b) we are a heavy-duty scientific world, science is objective while consciousness is subjective; finally, Searle asserts consciousness is a biological phenomenon and therefore a hard problem goes and postulates consciousness consists of all feeling, consciousness is real and irreducible and concludes it comes from one conscious field (Searle 2022). Here, it can be observed that there are two points: (a) Searle considers consciousness as a biological phenomenon; i.e., it might be more appropriate that he is referring to cognitive part consciousness, and (b) consciousness is a field keeping in the first observation, and he is probably referring to either quantum fields that relate to higher levels of consciousness or fields at the level of brain faculties. However, both observations can be related to the knowledge of consciousness as follows.

Even though, generally human consciousness is related to alertness, awareness (both inner and outer), feeling, emotions, etc., without its objective or target or level of realization. The *Upanishads* also referred to concluding or interpreting the philosophical component of Vedas, whereas in *Taittirya, Upanishad* (Upanishads (Taittirya) 2023) discusses seven folds/layers of consciousness. These seven layers of consciousness are (1) physical consciousness (cognitive consciousness), (2) vital consciousness, (3) mental consciousness, (4) supraintellectual consciousness, (5) blissful consciousness or consciousness proper to the universal beatitudes, (6) consciousness of self, or consciousness proper to the infinite divine self and (7) consciousness of Brahman or consciousness proper to the state of pure divine existence. These manifestations can be seen as the levels of evolution. The first one can be related to cognitive consciousness, which provides the link between ancient and modern science.

We must consider this argument that, generally, our focus or attention is on the external world or physical objects, and we perceive it as the ultimate reality. However, the deeper investigation through meditation reveals that the physical world we feel through our five senses, eye (vision), ear (hearing), nostril (smell), skin (touch) and mouth (taste), show only reality at physical levels. However, the modern physical sciences and spiritual practices have shown that the truth or realities are not the one that we perceived through our physical senses but subtler than we conceived in our physical consciousness. For example, the quantum mechanics and theory of relativity are showing us new facts and phenomena, earlier considered to be the ultimate truth of nature by classical physical science. For example, the probability associated with an electron in a given location through the principle of superposition, the role of observer and its limitation, the concept of time–space (classical physics has time and space) and many more findings confirm that consciousness is often built on a very superficial outer layer that can be related to the physical world and classical models. Consciousness can be related to the energy or vitality of our human faculties. For example, it can be argued that the neural correlates of consciousness (NCC) may represent the consequences of consciousness and may not be the consciousness itself. Even though the advance science is arguing that the microtubules—a microscopic tubular structure and not the neurons themselves or alone—are responsible for these sensitizations in various faculties involved in producing the consciousness or represent the sensitization levels.

The brain and mind represent two distinct components in the science of duality. For example, the focus of neuroscience is physical studies on NCC—a material entity along with its sensitization or energy levels, at the level of brain. However, there is a generally broad consensus that mind exists and is a nonmaterial entity. The thought process or physical activities might give rise to brain or mental activities, perhaps appearing to be studied by EEG studies performed in humans in neuroscience or psychology to understand mental consciousness. There are studies that characterize various wave functions, such as gamma, beta and alpha waves. For example, gamma waves are associated with near still brain states, such as meditation. When the brain is involved in the investigation of deep and complex studies, it might be characterized by EEG as a superintellectual consciousness where brain activities are focused, intense, complex in the sense of having heavy demand on memory and memory access/retravel and dumping processes, interpretation from many inputs/data/information, etc.

Furthermore, a higher level of consciousness represents the universal phenomenon that assumes that everything is conscious, right from human, animals, insects, viruses, amoeba have consciousness to material, i.e., nonliving entity (in traditional sense). Only the quantum of consciousness levels might be different. For example, humans in general have the highest quantum of consciousness elements, while animals have fewer. The concept of universal consciousness defines the concept of individualism and separateness in humans, which further evolves into a higher level of consciousness. Two human beings might share the same source of consciousness as if consciousness reflects higher levels of consciousness. However, the ability to receive higher levels of consciousness depends upon body specifics. This level of consciousness takes the human to a level of bliss where he can see that the physical body is not the ultimate reality, the reality is beyond this bodily confinement and there is a connection to the higher forces.

In the next level, there is a realization of real SELF, and this self is the truth—the infinite and limitless dimension and the awareness of SELF as not only a physical entity but also a higher divine force (Easwaran 2021). The last stage of realization is the establishment of the self in pure divine self—the ultimate truth. This is what is called mysticism the highest goal of human life—we are born with only one and only one purpose or the purpose of life is self-realization and experience of unification with the ultimate nondual entity.

From the above discussions, the following observations can be made:

1. Advaita's philosophy asserts the concept of nondualism; i.e., there is only one source of existence.
2. Vedic science and philosophy assert that the source of consciousness is outside of the physical body and that there are seven folds or manifestations of this consciousness.
3. The cosmic consciousness is all pervasive in nature, be it still or live entities.
4. Consciousness is most fundamental to human existence.
5. Consciousness is the basis for cognition and conscience, that is, neuroscience functions, such as perceiving or reflecting upon the information obtained from the 5 stimuli and the moral and ethical components, respectively.
6. The level of consciousness largely determines intelligence, knowledge, intellect, etc., and ethical aspects of life.
7. Consciousness is responsible for internal or spiritual awareness or growth for self-realization.
8. Consciousness is also responsible for interaction with the external world. Consciousness in humans is also responsible for qualities such as awareness, alertness, emotions, empathy, feelings, love and hurt/pain.

The Vedic approach is based on the human experience on different aspects of life and explores the mysterious phenomenon universe. Therefore, consciousness has been considered subjective in nature. To be precise, it is not objective and reductionist like the modern science approach.

2.6.2 Dualism

In the context of the philosophy of development, human beings, the term dualism generally refers to mind and body identity. In fact, there are schools of dualism in Western as well as Eastern philosophy. The Western approach to existence, particularly during the classical or mediaeval period, was based on mind–body dualism in the sense of material and metamaterial or even nonmaterial concepts. The dualist philosophy considers the mind as meta or nonmaterial, where the mind and body interact with each other and have their mutual dependence considered separate and distinct (https://en.wikipedia.org/wiki/Mind%E2%80%93body_dualism). This concept appears to reject or contradict nondualism or Hindu nondualism school assertions. The famous philosopher of classical times, Plato, proposed that true substances are not material but eternal forms. For an extended period in terms of a couple of centuries, this model prevailed in the west until the middle of the fourth century, when Aristotle disagreed in the Platonic Forms and asserted that forms are the nature and properties of things and exist as part of the things. With this argument, Aristotle could assert the body and soul relationship, i.e., sole as part of the body (https://en.wikipedia.org/wiki/Mind%E2%80%93body_dualism). As seen all through the mind and intellect was the main stumbling block that made mind–body dualism an accepted model.

Even though Arviat was a major school of nondualism, there were schools of dualism or Dvaita. Some of the descendent of Veda's thoughts, i.e., Vedanta, had the concept of dualism called Dvaita. Here, dualism refers to the 'reality'. In Vedanta schools, reality is fundamentally composed of two parts. This mainly takes the form of either mind–matter dualism or awareness-'nature' dualism in the Samkhya and Yoga schools of Hindu philosophy. These can be contrasted with mind–body dualism in Western philosophy of mind, but they also have similarities with it. The other schools of Vedanta, such as Mimansa, did not have any explicit thoughts on consciousness.

The thoughts of Plato and Aristotle found a basis of mind–body dualism and were widely accepted in the West. However, in the seventeenth century, French philosopher René Descartes extended mind–body dualism and identified consciousness with mind and distinguished it from the brain—the seat of intelligence. Descartes has been credited to provide the basis and foundation of modern Western philosophy with his thought regarding mind and conscious captures by the quote '*I think, therefore I am*' (Descartes-I-Think-Therefore-I-Am 2022). As it appears, this quote creates a foundation for consciousness, perhaps for the first time in the West, considered one of the dimensions of consciousness, i.e., 'intellect'. Keeping in mind the current lifestyle where the human is becoming full of logic and intellect, the Indian mystic Sadguru asserts that the life component of existence is a vital dimension of consciousness; therefore, he asserts that: it is time to restate the fundamental fact that '*you are, therefore you may think*' (Sadguru 2016).

There were questions regarding the mind–body problem, e.g., (a) related to their interaction, how nonmaterial mind interacts with physical body or brain, (b) the causal and mechanism aspects, i.e., how mind affects body and body affects mind [Dualism, Stanford encyclopedia of philosophy (Standard Encyclopedia of Philosophy—Plato 2022). The critics of Descartes' concepts of dualism could not find a robust sense or felt a lack of clarity of the model in substance. However, it can be argued that the major contribution of Descartes was that, arguably, it was for the first time that in the West, the concept of consciousness was introduced and of course considered part of the mind.

Hart W.D. in his chapter on 'Dualism' in a publication on 'A Companion to the Philosophy' discussed three major types (ontological aspects) of dualism that is related to mind and matter as follows (Hart and Dualism 1994):

(a) Substance dualism proposes that mind and matter are fundamentally distinct,
(b) Property dualism asserts another distinction that difference in properties of mind and matter,
(c) Predicate dualism has its root in the proposition that mind predicate to matter predicate is irreducible.

These first propositions have the assumption that the mind is not a nonphysical entity and that connects with the third preposition that the traditional material approach may not be capable of understanding the mind and its characterization. The second, i.e., property dualism, refers to the characterization based on physical property as a basis for distinguishing the mind from matter.

2.6.3 Panpsychism

It originated as part of Greek philosophy of mind. Panpsychism became the default philosophy of the mind until the nineteenth century in the West. However, it declined in the twentieth century, when the logic and overwhelming scientific temper captured the foundation of research and innovative thinking (Panpsychiasm 2021). However, there has been a renewed interest in Panpsychism in recent times toward understanding and exploring the hard problem of consciousness (Goff 2019b). The term panpsychism, as seen, has two terms: 'pan' and 'psyche'. The term 'pan' has usual meaning and refers to 'everything' or 'everywhere' or 'all', while 'psyche' refers to aspects of life 'soul', 'spirit', etc., and other subjective qualities. The panpsychism considers the mind or mind-like entities to be a fundamental and asserts in its original form to be present everywhere and in everything in nature or beyond. In short, the academic proponent of panpsychism ascribes the attribute of life or psyche to all entities in nature. However, many philosophers of the modern school of panpsychism view that quantum consideration of mind is the most predominant attribute of the sentient being, while for other less sentient objects to gross objects, such as rocks, the mind quotient decreases and becomes nonexistent as the objects become increasingly gross in nature.

Many panpsychism theories evolved over a period of time; for example, Cosmopsychism argues that cosmos is a unified object, and Panexperientialism argues for different fundamental elements, such as occasional experience, pancreativism and their ubiquitous nature.

There is resurgence of panpsychism thought. Of course, there are modifications to the original tenets of panpsychism, and the whole concept is being seen in the new light. This has been possible because there is a realization that the classical approach of material science that is suited for objective and reductionist barriers needs to evolve considering the requirements of subjective treatment and the recent research involving quantum mechanisms, which tends to question the classical approach. Advanced science, particularly quantum mechanics, has amply shown that classical ideas may not work when we move from gross material to micro and nanophenomena, the phenomenon at the wave–particle level.

Here, the most recent one is the arguments of Philip Goff in his publication 'Galileo's Error—foundations for a new science of consciousness' (Goff 2019b). The advance ideas proposed by Goff while reviewing the existing state of the art in consciousness and questioning some of the prohibitive elements of the classical framework certainly put panpsychism as one of the promising approaches to address the problem areas in consciousness research. In fact, David Charmer, who pioneered the 'problem-of-consciousness' research particularly related to the subject nature and qualia, often finds some of the arguments of modified panpsychism relevant.

2.6.4 *Scientific Background*

Based on the available literature, there appears to be consensus that human consciousness is a reality, and it exists as consciousness is critical to define existence, e.g., 'we are alive' (Consciousness 2021). Many eminent philosophers (such as David Chalmers and Thomas Nagel) and scientists such as Christof Koch and Tononi have rejected the idea that consciousness is directly produced by brain processes, and they have turned to the alternative view that it is a fundamental quality of the Universe (SCI-NEWS: Is Consciousness a Fundamental Quality of the Universe 2019). This paradigm is in line with the Vedanta Philosophy, which asserts that consciousness is all pervading and that human consciousness is concerned that its source is outside with a nodal or cognitive relay faculty that creates what we refer to in this book as cognitive consciousness a life force for 'living being' at the level of the brain. However, there appear to be varying views in modern science on the precise definition of consciousness (The Economists—What is consciousness 2015; Durham University 2019). In fact, the modern science approach is objective and reductionist in nature, while consciousness is a subjective phenomenon. There are strong opinions that there cannot be a science, in the traditional sense, of consciousness (Manaf

2021). This argument is based on the premise that unlike mind consciousness is a nonmaterial phenomenon. However, as discussed in Sect. 2.6.1, the Upanishads postulates that among the seven manifestations/components of consciousness, the first is physical consciousness; therefore, it can be asserted that the material science approach primarily focuses on the physical manifestations of consciousness at the human body level to be more precise at the level of Brain. As neuroscience asserts that the brain is the source of consciousness and the study of 'neural correlates of consciousness' (NCC) is the central piece of research on consciousness which is correlation in the brain areas. It can be assumed that neuroscience focuses on the physical manifestation of consciousness. Here, the feelings manifest as physical symptoms, and these symptoms further explain chemistry, and this in-tern explains the final effects at the biological/neuronal and psychological levels (Chalmers 1996).

In fact, renaissance or re-emergence of Western modern science-based approaches to understanding consciousness initiated and actively being persuaded for a couple of decades or may be over 40 years. The progress made in neuroscience toward understanding response at the level of brain of consciousness is one of the major areas of consciousness research.

There are challenges such as 'hard problem of consciousness' and 'mind–matter interactions' that are being addressed by employing mainly science-based approaches to the results, as we will discuss, are promising to the extent that the focus is on human consciousness (Chalmers 1996; Descartes and the Pineal Gland 2013). In fact, the scientific community to a large extent is of the view that consciousness is produced at the brain level and generally the investigation of a higher level of consciousness in recent times, even though it has not spread widely, but some philosophers are also scientists who are open to discussing this subject, such as David Chalmers, Koch and Tononi, as discussed earlier. Here, the quantum mechanics approach also appears to be at work as quantum consciousness, as the quantum physics community is investigating potential applications involving research into plank-level dimensions to address higher-level and human-level consciousness (Lite 2020).

Even though the Eastern philosophies assert that mind does not remain limited to brain human brain levels but across each cell of body, for the purpose of this study we will focus on the brain as the idea is to explore one of the important dimensions of the mind that is intellect and how it is affected due to qualities like various emotions that affect the performance in terms of awareness, alertness and other qualities related to human consciousness.

Therefore, before we start the discussion of individual approaches to understanding consciousness, we will discuss the brain and its constituent faculties and its connection to the human body to have a link for understanding the source, role of consciousness and its relation to cognitive functions. It will be more proper to discuss brain, cognition and consciousness in the following section, as the engineering community will appreciate the discussion on brain as mentioned above as a prerequisite for discussing the neuroscience and bioscience of consciousness.

2.6.4.1 Neuroscience/Biological Science

Figure 2.4 shows the nervous system of the human body, comprising two major parts, the central and peripheral nervous systems. The central nervous system comprises the brain and the spinal cord, while the ganglion and nervous system in body parts that communicate with the brain through the spinal cord form the peripheral nervous system. The brain in turn comprised the cerebrum, the brainstem and the cerebellum. Figure 2.5 shows the basic building block of the brain—the neuron, mainly responsible for cognitive functions of the brain. It has three major features, the nucleus at the center and axons that extend the central body connection to reach the dendrites. The nucleus is an oval-shaped structure in the body of the neuron. It contains the nucleolus and chromosomes necessary for the coded production of proteins within the cell.

The dendrites interface with the surrounding neurons and provide sites for synapses, i.e., sites for communication through chemical or biological reactions. The adult brain is composed of over 100 billion neurons. The central nervous system also has, as mentioned above, the neurons and neuroglia that support and protect the neurons. The corpus callosum is a strip of white matter that connects the two hemispheres of the cerebrum (Human Brain 2020).

In this section, we will discuss the role of various faculties involved in cognitive as well as consciousness processing faculties of the brain. This is a brief overview toward understanding the functioning of brain and associated faculties as well as the techniques and their purpose to characterize cognition and consciousness. First, we will discuss the superficial brain structure. Before this, we should become acquainted

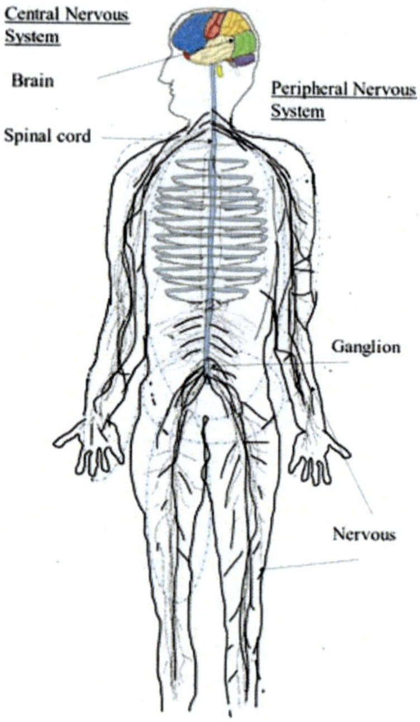

Fig. 2.4 Major features of nervous system

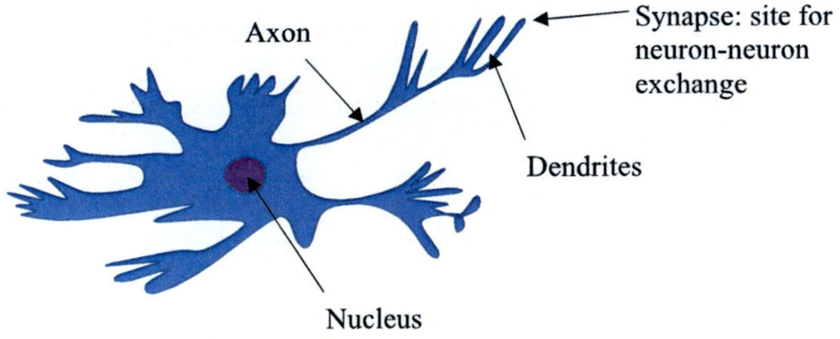

Fig. 2.5 Typical neuron—one axon and multiple dendrites

with a few terms relevant to understanding the functional faculties of the brain, as shown in Fig. 2.6. The major features are (a) gray and white matter, (b) convolution—a wrinkle on the surface, (c) a Gyrus—a bulge on the surface, (d) Lob—a region of the cerebrum, e.g., frontal lobe, parietal lobe, etc., (e) Sulcus—a shallow groove in the cerebrum, (f) fissure—a deep groove in the cerebrum and (g) cerebral cortex—the outer few mm of cerebrum.

The internal brain structure and functions will be discussed through the three brain images, as shown in Figs. 2.6 and 2.7. The first major part of the brain is the cerebrum, which is divided into four lobes, the frontal lobe (1), parietal lobe (5, 6), temporal lobe (8) and occipital lobe, as shown in Fig. 2.4. The cerebrum—the outer or anterior principal part of the brain in the frontal area of the skull—contains two hemispheres separated by longitudinal fissures and connected by the corpus callosum, as shown in Fig. 2.6.

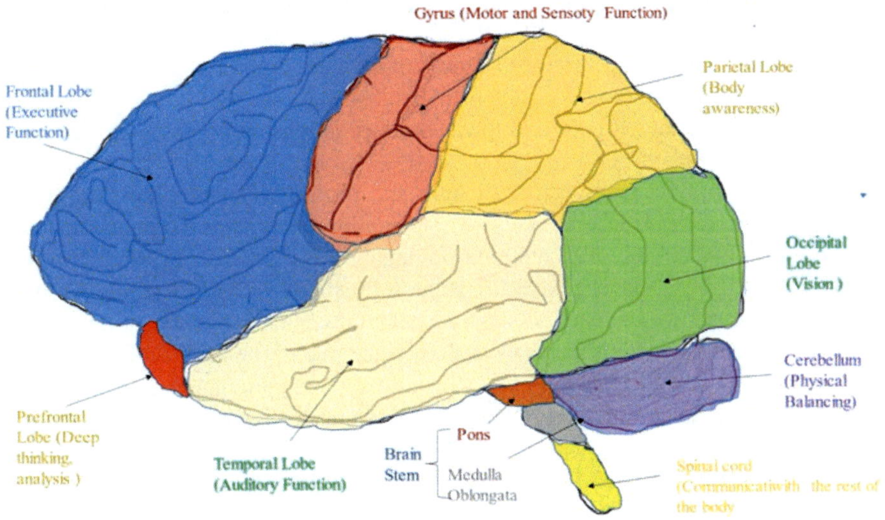

Fig. 2.6 Brain functional areas

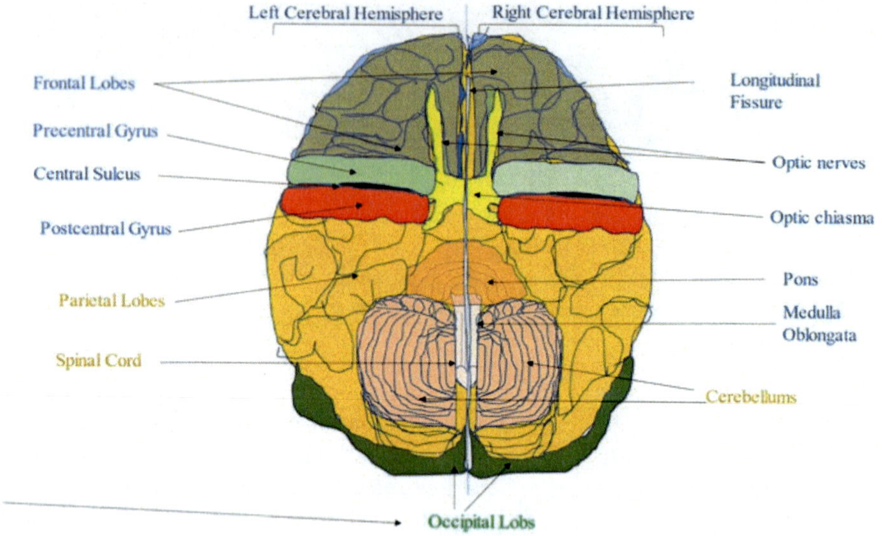

Fig. 2.7 Plan view of the brain showing the left and right halves and the location of the other sections in the two halves

The frontal lobe is involved in intelligent thoughts, planning, sense of consequences and rationalization—all directly relevant to the processing of data and information, particularly in the control room environment, in support of actions and decisions. The area between the frontal lobe and parietal lobe is shown as the precentral gyrus (2), central sulcus (3) and postcentral gyrus (4) in Fig. 2.6. The primary somatosensory cortex on the postcentral gyrus and Wernicke area is a general interpretive area. These lobs are responsible for imagination and give sense of time and space. The limbic system is comprised mainly of the amygdala, cingulate gyrus, hindbrain, etc., of the brain and addresses establishing emotions, as shown in Fig. 2.8. The limbic system links higher and lower brain functions and helps with memory storage.

Here, the amygdala plays a role in fight and flight, while the cingulate gyrus assigns specific steps to discern emotions. The hindbrain contains the pons, medulla oblongata and cerebellum. The medulla oblongata transmits information from the spinal cord to the brain and regulates life support functions such as respiration, blood pressure and heart rate. The pons acts as a neural relay center, facilitating information flow between the left side of the body and the right side of the brain and vice versa. Figure 2.9: Vertical plane cut section of the brain showing vital parts. The pons also facilitates the balancing function of the body, sleep and arousal and in the processing of visual and auditory information. The coordination of muscular activities, body balance and motor functions is performed by the cerebellum. The thalamus relays information to the cerebral cortex, the outer layer of the brain, while the hypothalamus controls the pituitary glands that secrete hormones and chemicals that regulate other glands in the body.

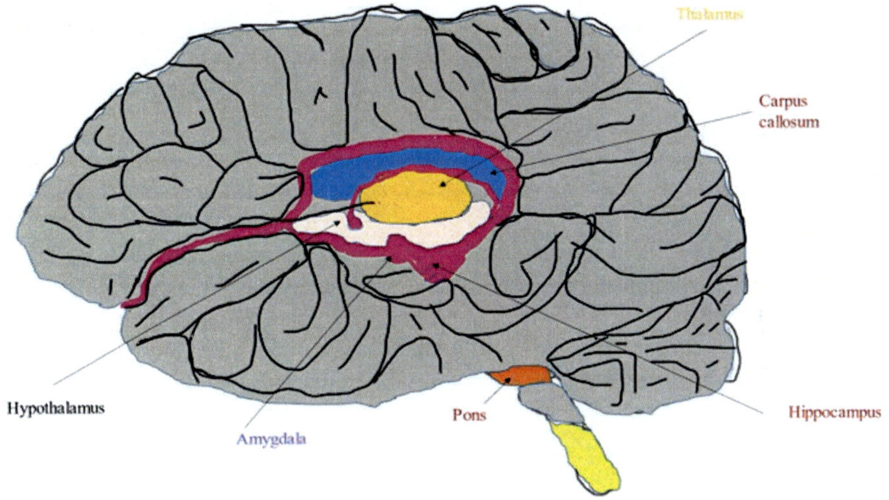

Fig. 2.8 Limbic system in the brain

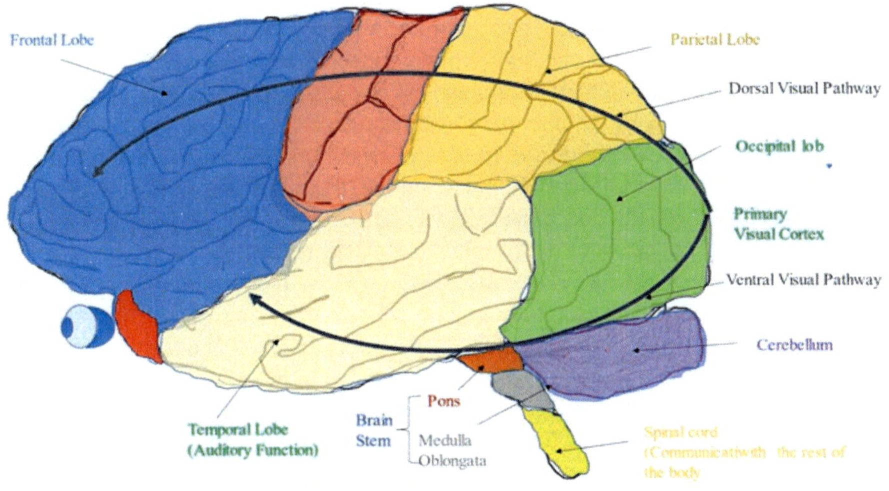

Fig. 2.9 Dorsal and ventral stream in relation to primary visual cortex in occipital lobe (region involved in visual processing)

There are three major scientific procedures in neuroscience to study consciousness: (a) experiments/studies on neural correlates of consciousness, (b) stimulation of electrical signals and (c) ablation. In neuroscience laboratories and clinics, experts have been applying these procedures to understand symptoms at the level of the brain by subjecting the patients to generating stimuli of various thoughts and events and trying to understand the response by employing functional magnetic resonance imaging (fMRI) or positron emission tomography (PETSCAN) images of local changes observed during the experiment. However, the study of neural correlates of consciousness has been used and reported widely (Imants 2019).

Neural Correlates of Consciousness (*NCC*) is defined as the '*minimal neuronal conditions jointly necessary for any one specific consciousness sensation*' (Crick and Koch 1990; Neural Collerates of Consciousness 2021). NCC constitutes the minimal set of neuronal events and mechanisms sufficient for a specific conscious percept. Neuroscientists use empirical approaches to discover neural correlates of subjective phenomenon, that is, neural changes that necessitate and regularly correlate with specific experience. The set should be minimal because, under the assumption that the brain is sufficient to give rise to any given conscious experience, the question is which of its components is necessary to produce it. In neuroscience or biological science and further in chemical science, the challenge is that a science of consciousness must explain the exact relationship between subjective mental states and brain states and the nature of the relationship between the conscious mind and the electrochemical interactions in the body (mind–body problem). Progress in nuclear psychology and neurophilosophy has come from focusing on the symptom rather than the mind. In this context, the neural correlates of consciousness (NCC) may be viewed as its causes, and consciousness may be thought of as a state-dependent property of some undefined complex, adoptive and highly interconnected biological system.

Although the measurement of consciousness or conscious fields is a question that is still a question as 'How to measure consciousness?', in neuroscience, there is a measurement system for monitoring the 'awareness' level or cognitive–consciousness level employing the Glasgow Coma Scale (GCS). It is a neurological scale that aims to provide a reliable and objective way of recording the state of a person's awareness state or cognitive consciousness. The GCS is the most common scoring system used to describe the level of consciousness in a person following a traumatic brain injury. Basically, it is used to help gauge the severity of an acute brain injury (What Is the Glasgow Coma Scale 2022). A person is assessed against the resulting points to give a person's score (Level of arousal in a patient).

There are many jobs that require attention rather perpetually, such as a machine operator, where most of the functions are manual, plant supervision and surveillance, assembly line of plants, vehicle driving, students in class and many more areas where attention plays a critical role. Researchers are actively pursuing the development of tools and methods that can track the alertness and mental focus of the human while performing a job. The most promising application for alertness monitoring is the EEG-based approach (Rahman et al. 2019). This involves monitoring the EEG signal from the brain, performing statistical analysis and intelligent processing of the data for comparison with the reference vectors to provide the level of alertness. This requires online monitoring/tracking of the data employing a head gadget and a system for integrated data processing and presentation and recording of results. In fact, typical instruments or techniques employed in the healthcare system to evaluate the consciousness level after administering anesthesia to the subject or headbands to monitor sleeping patterns or consciousness levels are some of the examples where along with EEG, other techniques are used to evaluate the subject alertness/sleep pattern/awareness or even focus or concentration levels of the subject (www.anandic.com).

2.6.4.2 Brain Areas Involved in Consciousness

It must be noted that consciousness is subjective in nature; therefore, the brain has no direct role in processing consciousness (Kotchoubey 2018). It is only cognition that is processed in the brain. Therefore, consciousness level is controlled by the brain. Furthermore, consciousness manifests at the level of the brain due to more human attitudes and behaviors and that in turn is controlled by the brain.

Human consciousness emerges on the interface between three components of human behavior, viz. communication, action (mental or physical) and decision making.

The following section will discuss the role of brain faculties in the formation of human attributes. The major human attributes concerned are interaction with the internal self as well as the external world and thereby produce formations, such as emotions, decision-making learning, ethical attributes, decision making, mental actions such as imagination, memory and general reactions to the internal and external world. As we have postulated, the quality of cognitive or physical consciousness depends on the fundamental external source of consciousness. This provides the necessary conscious energy for awareness, alertness, focus, attention and quality of attitude for external response for a reference cognitive system. The health or quality of cognitive faculties also determines the quality of behavior to external stimuli for formation.

> Consciousness is not a process in the brain but a kind of behaviour that, of course, is controlled by the brain like any other behaviour. Human consciousness emerges on the interface between three components of animal behaviour: communication, play, and the use of tools.
>
> \- Boris Kotchoubey

The following section presents the role of cognitive consciousness that is critical to human attitude, behavior and response or reaction to the external world. It may be noted that since we are talking about cognitive–consciousness attitudes, there can be an overlap in cognitive function in the brain and consciousness. It can be expected that there can be an overlapping or gray area where the cognitive and consciousness interface. Table 2.1 shows the mapping of cognitive-conscious mechanisms for various functions and stimuli that determine human behavior or attitude.

Table 2.1 Role of cognitive consciousness in the brain that supports potential human formation or response to external or internal stimuli

S. No.	Stimuli	Major brain faculties/area responsible for quality control	Reference quality attribute/index	Conscious stimuli	Human attitude	Failure/anomaly
	Vision and visual perception	• Eye/sight • Brain faculties: occipital lob orbitofrontal, • Somatosensory area • Dorsal stream: • Ventral Stream • Interior temporal cortex (It takes 300–500 ms to develop this consciousness) • Cones and genetic	• Vision = 6/6 • Perception of 'where': desired reflexes • Perception of 'what': desired reflexes • Comparison with past perception: desired retrieval reflexes • Color and contrast distinguishing capability	– Conscious vision: conscious observation, comprehension and perception – Subconscious visual perception and its correlation with the real-time requirements	• Vision awareness, alertness, for creating a perception about 'what, where and when and how much', • Judgment based on apriori visual images information supported by faster reflexes • Mental or physical action or decision making • Visual monitoring and actions	• Compromised vision • Deficit in movement and action/judgment • Deficit in/identification of faces, equipment landmarks, settings • Threshold of neural activity from unconscious to conscious processing • Error of judgment due to inadequate visual stimuli consciousness processing • Aging-related deficiencies • Color blindness • Judgmental error

(continued)

Table 2.1 (continued)

S. No.	Stimuli	Major brain faculties/area responsible for quality control	Reference quality attribute/index	Conscious stimuli	Human attitude	Failure/anomaly
	Hearing and auditory perception	• Ear/hearing Auditory pathway: ear canal, tympanic membrane, Vestibulocochlear nerve • Brain faculties: cochlear nerves, auditory cortex in temporal lobe	• Hearing level ranging from 0 to 120 dB HL (from normal (blue), …profound (red) for a frequency range from ~ 150 to 8000 • Articulation index and speech intelligibility index. Or audibility index (Amlani et al. 2002) • Perception of distance • Perception of subject • Perception of 'where' • Perception of 'what' • Comprehension	Conscious listening Grasping, conscious inferencing or judging capability of physical object and parameters	• Listening and grasping • Inferencing • Parallel processing of audible input and inferencing • Filtering of input for discerning and decoding • Cognitive processing for developing formation and perception	• Hearing-impairment beyond acceptable level • Deficits in the ability to detect changes in pitch, localize sounds in space, or understand speech • Damage to the auditory cortex can disrupt various facets of auditory perception • Aging-related deficiencies • Failure to sense an abnormal sound or vibration in time due to lack of attention or alertness is one of the major contributing causes to failure • Judgmental error

(continued)

Table 2.1 (continued)

S. No.	Stimuli	Major brain faculties/area responsible for quality control	Reference quality attribute/index	Conscious stimuli	Human attitude	Failure/anomaly
	Executive functions	• Cognitive processing: intelligence quotient 100 < IQ ≤ 120; • Decision-making capability. mental stability (particularly resilience to stressful conditions • Brain faculties: prefrontal lobe which includes orbitofrontal cortex (for sensory communication and limbic system, cerebrum, long-term memory and short-term memory, and area involved into reward-punish function, somatosensory area, anterior cingulate, Basel ganglia	• Intelligence quotient: 100 < IQ ≤ 120 • Reasoning, • Rationalizing, • Evaluation of options, Inferencing and • Decision making, etc. • Emotional response • Production of dopamine	• Conscious-thinking and performance of executive with conscience considerations, moral, • Ethics and best interest of stakeholders including public	• Guidance, leadership, • Management functions • Planning, reasoning • Decision making, • Technical and Administrative competency • Knowledge-based behavior and actions, • Holistic view of the situational • Value addition potential organizational consciousness • Behavior control, emotions, • Impulsive behavior • Memory and learning	• Error of judgment • Lack of leadership and management quality • Not taking responsibility for outcomes • Lacking team spirit, unethical or immoral conduct • Lacking commitment to organizational consciousness and conscience and culture

(continued)

Table 2.1 (continued)

S. No.	Stimuli	Major brain faculties/area responsible for quality control	Reference quality attribute/index	Conscious stimuli	Human attitude	Failure/anomaly
	Touch somatosensorily sensation	Sensitivity to physical touch, pain Other physical sensations • Brain faculties: somatosensory cortex, nerves, ganglia • Peripheral: the spinal cord, skin and body parts	• Touch sensation • Communication of emotions, empathy, love, assurance, sharing	Conscious-touch, emotion perception, feeling	• Sharing humanly attitude • Development of team spirit • Handshake and embracing	• Emotional bond is critical to organizational consciousness. Lack of emotional bonding or feeling of lack of emotional bonding has been suspected to be seen as major root cause for organizational failure
	Smell	• Nasal cavity: smell receptors in nasal passage, combination of olfactory receptors somatosensory cortex	• The sense of smell is closely connected with memory reinforces the perception and thereby uses smell as tool during plant walk downs and supervision If it's a perfume or typical smell, then it creates corresponding emotions	Conscious-smell power is key to identifying fouling phenomenon		• Failure to sense an abnormal smell due to lack of sensation or poor attention or awareness that leads sensory to failure involving foul smell like burning smell from electrical circuit, foul smell due to dead rodent in the plant areas that could be symptom of incipient failures

There are certain principles relating to the neural events that lead to perceptions of the various sensory experiences. Touch and vision have been most thoroughly investigated, but there is good reason to believe that all other sensory experiences are dependent upon similar neuronal mechanisms (McGeer 1978). Nevertheless, Table. 2.1 presents visual consciousness as well as specific features associated with other stimuli, and faculties have also been discussed. Here, our special interest is focused on the neural events that are necessary for giving a conscious experience.

If we understand the role of brain faculties involved in consciousness considering visual stimuli, we see that there are three major areas that include the dorsal stream, ventral stream and inferior temporal cortex. The dorsal stream projects the signal to the parietal cortex responsible for movement and relative spatial assessment of object that deals with 'where is it'. The ventral stream addresses the question 'what it is' in terms of the identification of new objects.

Here, the ventral stream evaluates the relative size and orientation of the object that is basically at qualitative evaluation. The primary visual cortex addresses *orientation, edges and boundaries.* In the secondary visual cortex, there are processes dealing with the *separation of objects from their backgrounds* and *object features of intermediate complexity*, such as *geometric shapes.* This processing is vital for the development of conscious perception of external world objects. The inferior temporal cortex addresses feedforward and backpropagation from-and-to the original neuronal signals and is responsible for developing fine-tuning of the formation of perception with the original image (Brain areas involved with consciousness, 2022). Damage to the dorsal stream is shown to lead to deficits in movements and actions rather than problems with visual perceptions. On the other hand, damage to the ventral stream results in difficulties in recognizing objects rather than actions such as grasping objects even if they are not recognized (Fig. 2.9).

Within the later ventral stream, there are specialized visual modules, most of which are clustered in the inferior temporal region. Different features are processed in these different modules within the ventral stream and may take different lengths of time to be processed. *The ventral stream calculates the size and orientation of objects in relation to one another rather than against any absolute measure.* In contrast, *the dorsal stream has an absolute scale.* Perception relates to reflectance properties that show the difference between the surface of an object and surrounding surfaces. Within the outline contour, *reflectance properties* are also important for determining color and texture. The *ventral stream addresses color, texture, form, brightness, size, shape, orientation and perception of motion.* Consciousness involves the processing of images, and clusters of neurons in the visual cortex form a pattern to represent images, established through the analysis of neural correlates of consciousness.

Damage related to any one module in the ventral stream can result in localized deficits, such as not being able to recognize faces or landmarks. All this processing, however, happens at a still unconscious stage of the ventral stream. Although the ventral stream is referred to as conscious, consciousness is only involved with the end product of a long line of ventral processing.

The limbic system is responsible for human behavioral and emotional responses and is located deep down or in the center of the cortex above the brain stem, as shown

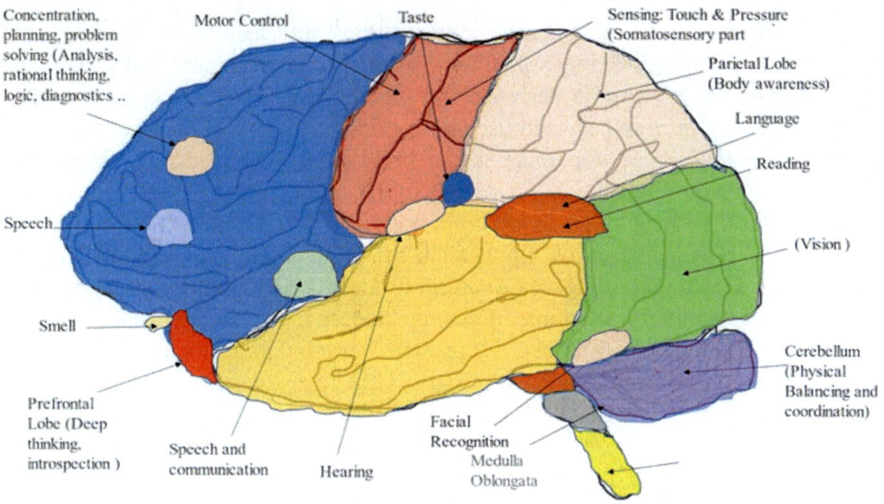

Fig. 2.10 Brain faculties/cortical areas involved in consciousness in various lobes

in Fig. 2.8, comprising three major parts, the hippocampus and amygdala, as well as the thalamus and hypothalamus (System 2022). The hippocampus in the brain is (a) a memory center of the brain responsible for formation of episodic memories and are coded and stored, (b) produces of new neutrons and provides basis for plasticity, and therefore responsible for learning faculty of brain. The amygdala is responsible for emotions or response to emotions and thereby provides emotional content to memory as well as the formation of new memory. The thalamus and hypothalamus deal with the production of important hormones and the regulation of thirst, hunger, mood, etc. The functions of the basal ganglia are reward processing, habit formation, movement and learning. The orbitofrontal also helps to signal the significance of events to the hippocampus memory-forming region.

The final stage for all this processing is the basal ganglia, the processing of which is immediately upstream of action and behavior and governs the release of dopamine that is crucial to learning and memory.

Generally, emotions can be divided into positive, neutral and negative. Happiness, deep satisfaction and empathy are some of the examples of positive emotions, while fear, frustration, anger, sadness and disgusted are negative emotions. Neutral emotions are an expression of equanimity and are considered vital for spiritual growth. This is a stable condition of mind or consciousness where there is no craving or aversion. Negative emotions are mostly detrimental to concentration, focus and stability in human behavior and adversely affect human performance. However, at times the challenging component of negative emotions could be a motivating factor for improving performance. Controlled positive emotions are an asset for human performance and a catalyst for positive motivation; however, too much excitement or overconfidence could reduce concentration or focus, which is not good for performance.

Dopamine, an organic chemical, is a neural transmitter and plays multiple roles in the human brain and body. Dopamine is critical to neurological and physiological functions. Its structure is derived from amino acids. The human body makes dopamine, and the nervous system uses it for communication among neurons. The largest concentrations of dopamine in the brain are found in the basal ganglia, amygdala and prefrontal regions, particularly the orbitofrontal region. Dopamine-producing neurons located in the midbrain appear to be influenced by the size and probability of rewards based on inputs from the orbitofrontal and amygdala to the basal ganglia. Dopamine acting on the ventral striatum reduces inhibition and releases the output of behavior. Furthermore, dopamine release is also an important influence on learning and memory. Figure 2.10 presents various area of brain involved in processing of consciousness.

2.6.5 *Modern Scientific Approaches and State of the Art*

2.6.5.1 Integrated Information Theory of Consciousness

The basic motivation of the integrated information theory (IIT) of consciousness proposed by Tononi is to understand the relationship between brain and consciousness (Tononi 2008). The integrated information theory of consciousness presents a theory about what consciousness is and how it can be measured. It has its premises on two phenomenal properties of consciousness. These are differentiation—the availability of a very large number of consciousness experiences and integration—the unity of such experiences (Tononi 2004).

The basic assumption is that consciousness exists and it is there. The IIT is built around a basis framework comprising four axioms. The axioms of IIT state that every experience (a) exists intrinsically, (b) has composition and structure, (c) the information states that the experience is specific and (d) is integrated, i.e., unitary and definite. Furthermore, there are postulates in IIT corresponding to these axioms. These postulates provide a physical description of these axioms. The theory lays down the framework that tracks the procedure from phenomenology, i.e., experience to physics, involving neural mapping in brain toward offering the physical substrate to that can link for every experience a set of neural correlates.

The postulate for the first Axiom 'intrinsic nature of experience' states that the Physical Subtract of Consciousness (PSC) also exists intrinsically and then in turn for something to exist intrinsically it must have a cause–effect power. The axiom of composition states that the experience is structured, i.e., composed of many several phenomena. The corresponding postulate of the composition state that the PCS also follows a cause–effect phenomenon. The third postulate for the axiom of information states that the PSC must specify the cause–effect structure or cause–effect repertoire of all mechanisms of a system. The postulate of integration states that the cause–effect structure of a PSC must be unitary; i.e., it should be irreducible. The irreducibility of a conceptual structure is measured as integrated information represented by Φ

(Tononi et al. 2016). The notable feature of IIT is that it has demonstrated the identification of physical substrate of consciousness by maxima of intrinsic cause–effect by characterizing the major complexes with high firing, low firing and burst firing in cortical areas of the brain.

There are some aspects that need further research to bring IIT from the laboratory environment to real-time application. The major procedural elements of the IIT involved in this clinical research are characterizing the experience, formulation of conceptual structure and finally mapping of PSC on cortical area—and this is a complex recourse consuming and demanding task.

There are some critics of IIT regarding the definition of consciousness, even though IIT has made huge mark in the consciousness research but looking at the approach it might be more suitable for clinical applications, while the larger area of consciousness research which deals with human behavior during engineering application may be a little farfetched dream. However, the important aspect of IIT is that the clinical component of this research can be explored for application in engineering environments when the state of the art of IIT is integrated with other approaches, such as spiritual and quantum approaches, to develop models for individual and organizational consciousness.

Finally, IIT concludes that consciousness is a fundamental quantity, that it is graded, and that it is present in infants and animals. Finally, the conclusion is that it should be possible to build conscious artifacts.

2.6.5.2 Global Workspace Theory

The global workspace (GWS) theory of consciousness was proposed by Bernard J. Baars in 1988 to explore a model of cognitive consciousness (Baars 2005). Interestingly, a theater setting has been used to explain the components of consciousness not only at the level of the brain but also at other internal faculties, sensory systems and the use of resources, as shown in Fig. 2.11. Given the context that is self, that could be internal or external, the subject of thought, i.e., intentions, expectations or formulation of perception after a rigorous cognitive process, viz. diagnosis, analysis or interpretation of signals or even a transient phenomenon, etc. In fact, GWS provides an elegant framework to define the role of consciousness and cognition, which is a critical component RCOM where the role of the connection or overlapping function needs to be understood between consciousness and cognition or simply the gray line between consciousness and cognitive functions.

GWA models output, i.e., action, based on sensory input. The supporting elements for cognitive-conscious processing require training on interpretation and language automatisms, i.e., the state of consciousness levels or low or impaired consciousness for an action or response.

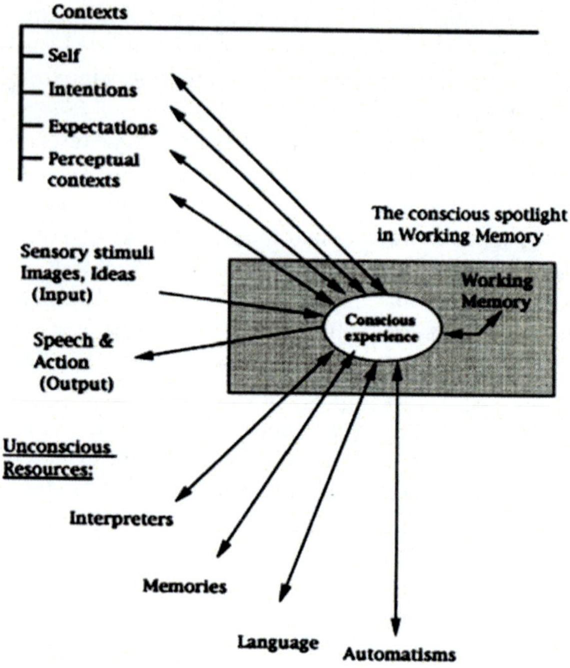

Fig. 2.11 Model of GW theory of consciousness (Baars 2005)

2.6.5.3 Quantum Mechanics Approach to Consciousness

There are two major procedures in quantum mechanics, the Schrodinger equation (Landaue and Lifshitz 1974) and the making of measurements that involve testing or manipulation of a physical system to yield a numerical result (Measurement in Quantum Mechanics 2022). The quantum mechanics approach enables modeling of quantum development in our system, which takes beyond present technology, possibly that restriction to remain in the material domain, which many feel, make any approach restrictive to model a phenomenon-like consciousness. Furthermore, it is noted by experts that there are two major developments in this century: the first is Einstein's theory of GR, and the second is quantum mechanics. There is a consideration among the experts that a combination of these two great theories might enable seeing the gap in existing state of the art that could be relevant to understanding the phenomenon at the level of the brain.

The quantum mechanics approach enables modeling of quantum development in our system which takes beyond present technology, possibly that restriction to remain in material domain which many feel, make any approach restrictive to model a phenomenon-like consciousness. Furthermore, it is noted by experts that there are two major developments in this century: the first is Einstein's theory of GR, and the second is quantum mechanics. There is a consideration among the experts that a combination of these two great theories might enable seeing the gap in existing state of the art that could be relevant to understanding the phenomenon at the level of the brain.

Orchestrated—Objective Reduction (Orch-OR) Theory

This theory was proposed by Penrose and Hamerrof and has its premise in the biological philosophy of mind that postulates that consciousness originates at the quantum level inside neurons rather than at the connection between neurons (Penrose and Hameroff 1995). The moment of conscious awareness is orchestrated by microtubules in the neurons in the brain. In Orch-OR, the mechanism is held to be a quantum process called objective reduction that is orchestrated by cellular structures—microtubules. Orch-OR is a theory that attempts to explain how quantum mechanical processes in the brain result in consciousness in the human brain (Collins 2015).

Orch-OR has been considered because objective reduction is noncomputable for physical sources of consciousness. Furthermore, objective reduction occurs when the eigenstate state of quantum superposition differs by the miniscule space–time factor (Lucas 2014). The consideration is that the Orch-ORR is that superposition must collapse within a specific time to avoid losing its quantum nature. This time is referred to as the 'reduction time'. In the Orch-OR model, the reduction time is a function of gravitational self-energy $E_{G.}$ The self-energy is in turn a function of mass distribution between states. Accordingly, the reduction time τ is given as:

$$\tau = \frac{h}{E_G} \tag{2.1}$$

where

$$E_G = \frac{1}{G} \int (\nabla\varphi_2 - \nabla\varphi_1)^2 \mathrm{d}x^3$$

E_G in turn is a function of the gradient of the eigenstate, h is Plank's and G is the gravitational constant. The major conclusions of this research are that (a) consciousness is a process in the structure of the universe connected to the brain via quantum computation in microtubules ('Orch OR'), (b) OR-based primitive feelings in the universe may have prompted the origin of life and driven its evolution and (c) resonating brain microtubules with transcranial ultrasound (TUS) can safely and painlessly improve mood and offers promise for Alzheimer's disease, traumatic brain injury and other disorders (www.quantumconsciousness.org).

While being critical of the computational theories, the ORCH-OR states that there is an existing idea that consciousness arises from some complicated computation; however, the developer of ORCH-OR has some reservation on this line of approach. Consciousness seems to be not computing, as there are laws of physics, and understanding quantum mechanics might provide a more holistic or integrated approach to understanding consciousness.

Qualia Research Institute Approach

Research and development on mathematical structure of consciousness has been the core theme toward understanding the complexity of qualia (Discovering the Mathematical Structure of Consciousness 2020). Human attributes such as feeling, sensation, perceptions and emotions that are experiential and subjective are collectively referred to as qualia. The understanding of effect and associated modeling of qualia still pose a challenge for consciousness research. QRIs have been actively involved in developing the science of consciousness and the aspects related to qualia or subjective experiences. At the outset, this approach has two core philosophical premises for its research and development: (a) qualia formalism, which asserts that subjective human experiences have a mathematical framework, and (b) the second one addresses 'valence realism', which asserts that the valence of conscious experience can be measured objectively. The two premises form the basic fundamental of the QRA approach. The QRA approach proposes that if subjective experience can be represented as mathematical objects, then the symmetry of the objects can be an indicator of feeling or qualia, e.g., level of pleasant experience.

Symmetry is defined as the quantum of variance under any physical transformation. Furthermore, the symmetry theory of valance, claimed, yields some concrete, testable predictions. Stimulating the brain at harmonious frequencies via transcranial magnetic pulses through the skull should induce states of higher emotional valance. Similarly, the stimulation of the vagus nerve, which relays information between the brain and the rest of the body, should feel more pleasant when it is synchronized with harmonious music. Recently, Selen Atasoy, a neuroscientist now at the University of Oxford, developed a paradigm that may turn out to be incredibly useful for testing the theory, as well as for quantifying emotional valance in general. Atasoy's method hinges on the notion that the connectome—a system of neural pathways in the brain or the structure of all connections of the brain—resonates at certain natural frequencies (Shinozuka and Why-We-Need-to-Study-Consciousness 2019).

QRI claims that emotional valance corresponds to the weighted sum of the consonance, dissonance and noise in the harmonics of a given brain state. We calculate the dissonance between CSHWs in a way that is similar to computing the dissonance of a combination of musical notes. Like sound, the brain harmonizes with alike frequencies (i.e., frequencies falling within a critical bandwidth/and higher amplitudes will cause mutual dissonance, and the total dissonance is equivalent to the sum of the dissonance between all possible pairs of harmonics) (Shinozuka 2019).

Pribram–Bohm Holoflux Model

The Pribram and Bohm model has the concept of implicate order as the basis for modeling phenomena at the quantum level and provides a more fundamental approach to the subject—consciousness, particularly the Charmers' Hard Problems of Consciousness (Joye 2021). The Implicate and explicate orders are ontological concepts for quantum theory coined by theoretical physicist David Bohm during the

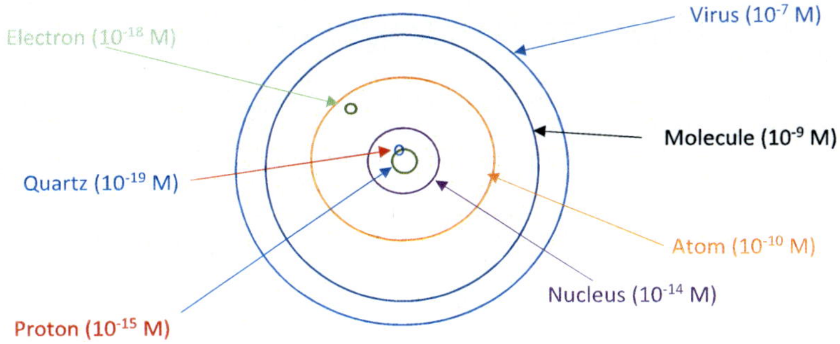

Fig. 2.12 Visualizing explicate order of center, everywhere

early 1980s (Bohm 1980). The implicate order is seen as a deeper and more fundamental order of reality. In contrast, the explicate order includes the abstractions that humans normally perceive (Implicate and Explicate order 2022). Figure 2.12 shows a typical visualization of the explicate order domain that also depicts atomic descriptions of viruses. The Pribram–Bohm Holoflux theory of consciousness proposes that consciousness manifests as energy and consciousness resonating between two domains: (a) the space–time explicate and (b) the spectral implicate. Consciousness has been formed by quantum effects in or between brain cells (unlike neurons, as discussed in the previous sections) in the holographic brain theory of neuroscience. The traditional approach to neuroscience assumes that quantum effects will not be significant in brain cells at the plank scale. We know Planck's constant as follows:

- Plank length: 1.616199×10^{-35} m; and
- Time: 5.39106×10^{-44} s

The implicate order is at the center everywhere at the bottom of space at the plank's length of 1.6×10^{-35} m, as shown in Fig. 2.12. In particular, the concepts were developed to explain the bizarre behavior of subatomic particles, which quantum physics struggles to explain. As shown in Fig. 2.13, the scale depicts the implicate and explicate order frameworks for understanding the same phenomenon or aspect of reality. In the holoflux theory of consciousness, energy is hypothesized and shown to support both local and nonlocal properties, as shown in Fig. 2.14. This thesis emerges from an integral evaluation of evidence drawn from three sources: (1) the holonomic mind/brain theories of Karl Pribram, (2) the ontological interpretation of quantum theory by David Bohm and (3) the hyperphysics of consciousness developed by Pierre Teilhard de Chardin. Applying an integrated methodology to superimpose and correlate seemingly disparate (Shelli 2016).

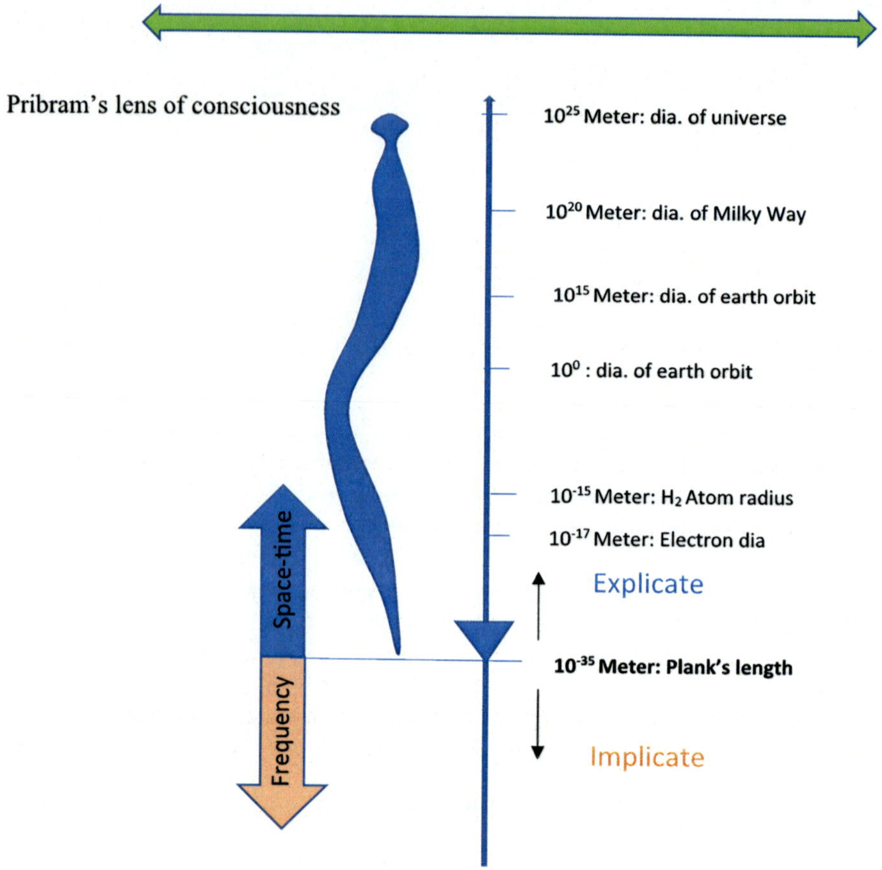

Fig. 2.13 Scale of scale: explicate and implicate order

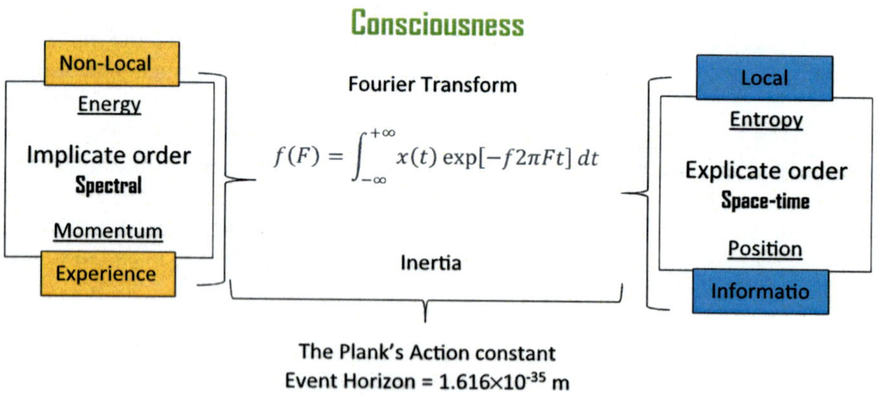

Fig. 2.14 Pribram–Bohm Holoflux model (Joye 2021)

> "Ultimately the entire universe (with all its particles', including those constituting human beings, their laboratories, observing instruments, etc.) has to be understood whole, in which analysis into separately and independent existent parts has no fundamental status"
>
> David Bohm
> (https://samim.io/p/implicate-and-explicate-order-are-ontological-concept/)

David Bohm presents his thesis on undivided wholeness and of implicate and explicate orders. He used these notions to describe how the appearance of such phenomena might appear differently or might be characterized by varying principal factors, depending on contexts such as scales. The implicate (also referred to as the 'enfolded') order is seen as a deeper and more fundamental order of reality. In contrast, the explicate or 'unfolded' order includes the abstractions that humans normally perceive, as shown in Figs. 2.12 and 2.13. In the enfolded [or implicate] order, space and time are no longer the dominant factors determining the relationships of dependence or independence of different elements. Rather, an entirely different sort of basic connection of elements is possible, from which our ordinary notions of space and time, along with those of separately existent material particles, are abstracted as forms derived from the deeper order. These ordinary notions in fact appear in what is called the 'explicate' or 'unfolded' order, which is a special and distinguished form contained within the general totality of all the implicate orders (Bohm 1980).

Unified Field Theory of Consciousness

In classical physics, the strong nuclear force, electromagnetic force, weak nuclear force and gravitational force are considered to be fundamental forces. The unified field theory allows all that was thought of as fundamental forces and elementary particles to be written in terms of physical and virtual fields. Therefore, in modern physics, all four fundamental forces are mediated by fields, and the standard model of particle physics, as shown in Fig. 2.15, results from exchange gauge bosons. The exchange particle gluon mediates the strong nuclear force between neutrons and protons that keeps the atomic nucleus in a stable state, while in electromagnetic interactions, the photon acts as the exchange particle. The weak force responsible for the radioactivity mechanism is mediated by the W and Z bosons. Similarly, there is a postulation of the exchange particle graviton that facilitates the gravitational force. In the standard model, all the fundamental particles (fermions) can be seen on the left metrics, while the exchange particles that facilitate forces are shown on the right, referred to as group of Gauge Bosons and Tensor Bosons.

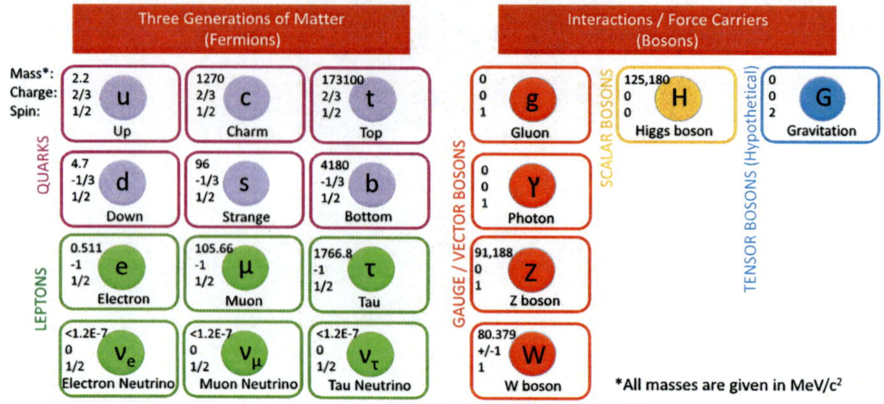

Fig. 2.15 Standard model of elementary particles (Unified Field theory 2022)

The standard model provides the foundation of unified field theory in quantum mechanics. Even though the situation is like 'work in progress' toward establishing a universally accepted model of unified field theory, for the want of combining the general relativity and quantum mechanics to offer 'theory of everything'.

Hagelin postulates an analogy between unified field theory and the state of self (unified existence) and asserts that this state of consciousness benefits human attitude through experiments in a controlled laboratory environment. The 4th stage of consciousness is achieved by a Yogy (spiritual Yoga practitioner working for self-realization) through transcendental meditation and is characterized by reaching a state where there is only 'self' by losing the duality of inside and outside—Hagelin wonders 'Is Consciousness the unified field'.

Hagelin (2017) further discusses whether consciousness-like unified fields in physics are the ultimate reality with the considerations of string theory. While pointing to the limitation of classical physics, where the limit to explore the unified field is limited by microscopic dimensions, the quantum mechanism through the arguments of the quantum field provides an elegant mechanism to model the unified field. In the consciousness model, the unified field is characterized by an ocean of intelligence, dynamism and infinite silence, and there is no material at all; only vibration waves and energies can be depicted in Fig. 2.16.

It is interesting to note that the Vedic/Upanishad's approach to consciousness is that the 4th stage of consciousness is achieved by a transcendental meditation program that enables the conscious mind to identify itself with the unified fields or union with the higher self. In the CQB model, consciousness has been considered a fundamental attribute of humans, and Hagelin's model has direct relevance to human factor modeling in a risk-conscious ecosystem. In fact, the limitation of quantum unified field theory does not directly affect the applicability to human factor improvement.

Hagelin presented the results of some experiments employing the electroencephalography (EEG) technique. EEG is an established method of evaluating cortical

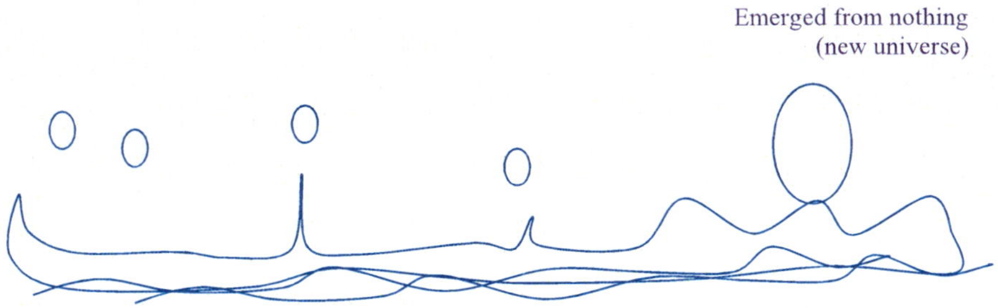

Fig. 2.16 Representation of unified field of consciousness (Chen et al. 2015)

activities. There are two parameters modeled in EEG using the data collected from the experiment. The probability (*p*)- and correlates of EEG coherence (v). The *p* probability parameter provides the result of the hypothesis, where the null hypothesis provides the value of the experimental objective. EEG correlates of EEG coherence are defined as the normalized crosspower spectrum per frequency of the two signals recorded simultaneously at different sites of the scalp and are a sensitive method for assessing the integrity of the structural connection between brain areas (Chen et al. 2015).

The results of the experiment were presented and showed that the subjects (candidates involved in the studies) who took up transcendental meditation showed remarkable improvements in terms of correlates of EEG coherence that include human attributes, creativity, learning, grade-point average, H-reflex (a reflectory reaction of muscles after electrical stimulation of sensory fibers), moral and ethical reasoning (attribute specific to conscience) and transcendental consciousness (that addresses awareness of with and without). The effective size for attributes that include creativity ~ 0.7, field independence ~ 0.6, practical intelligence ~ 0.5, mental efficiency ~ 0.45 and fluid intelligence ~ 0.4 with corresponding probability p was shown to be < 0.001, < 0.005, < 0.006, < 0.003 and < 0.001, respectively, were *p*-parameters for improving human attributes and performance, as shown by the coherence index obtained (Hegalin 2001).

In another experiment, EEG alpha coherence and slow alpha recording from frontal and occipital derivations during relaxation or Transcendental Meditation (TM) and performance of controlled experiments involving a group of subjects indicated a decrease in occipital power, and more intense TM sessions showed a significant increase in frontal coherence above 0.95. Frontal alpha coherence has been found to be a more sensitive discriminator of the TM technique than alpha power (Dillbeck and Bronson 2009). While discussing the results of experiments where EEG correlates of subjects in meditative condition were developed to understand the sensitivity of meditation to assess change in cognitive capability of the subject.

EEG coherence mapping shows that the creativity and H-reflex improve significantly, while concept learning Hagelin discusses experiments performed by EEG monitoring of brain waves with various conditions, including meditative states (Is Consciousness the Unified Field 2014). The EEG characterization shows that the brain wave pattern remains in the alpha (8–10 Hz) range and resembles a unified model typical of quantum models with various levels of cognitive control of the meditators.

Hagelin concludes on the question 'Is (pure) consciousness the unified field' and puts forth four arguments in favor of a unified field and consciousness based on the analogy he discussed and the evidence obtained in his experiment as follows:

1. There is a quantitative correspondence between pure consciousness and the unified field.
2. Experimentally observed data show the long-range field effects of consciousness.
3. Certain classical siddhis (supernatural abilities) unambiguously demonstrated place consciousness at the level of a unified field.
4. Philosophical and neuroscientific challenges have their premise in the derivative of matter.

2.7 Summary and Conclusions

Considering the available state of the art on science and philosophy of consciousness in general and human consciousness in particular, keeping in view the subject, i.e., requirements of risk-conscious cultural development, this section identifies the state of the art at the philosophical and scientific levels as also major assumptions, postulates and considerations that lay the foundation for the development of a risk-conscious approach for engineering operations management:

1. Consciousness is fundamental to human existence.
2. Consciousness manifests at 7 levels. Apart from the cognitive consciousness (physical component or physical manifestation of higher-level consciousness), it manifests at another six levels. The manifestation at the 4th level is referred to as transcendental consciousness—a state of bliss.
3. Ideally, a state of bliss (Anandmayi), peace and happiness are intrinsic and innate human qualities. However, generally, the human strives to attain these qualities through material's approach, while the truth is spirituality is the only path to attain these qualities.
4. An integrated approach is envisaged that employs the best of the insights and knowledge available from traditional/ancient philosophies as well as modern material science.
5. Self-realization is the ultimate goal of human existence, and this goal can be achieved through the practice of meditation.
6. The major attributes of consciousness that form directly part of risk-conscious culture are awareness (internal and external), alertness, focus, concentration, attitude and conscience (moral and ethics). Consciousness qualities called qualia, which refers to intuition, feeling and emotions, are considered to have

a physical basis for the purpose of RCOM. The higher attributes of consciousness that need to be developed are cultivated through spiritual processes and require special attitudes and environments. For example, Yoga and meditation are integral parts of life.

7. The 6 (5 sense-bases + mind) stimuli of seeing, listening, smelling, speech and taste, sensation, and thinking are the sources of external and internal awareness.
8. Human constitutions have voluntary (autonomous) and nonvoluntary functions. Involuntary activities are breathing, cardiovascular, pulmonary, nervous and muscle control and are intrinsic in nature, and voluntary activities involve physical or motor movements of organs and enable physical actions in maneuvering the external environment, such as walking, talking and writing.
9. Human brain faculties give rise to cognitive function or activities, learning, sensation, listening, speech, memory, processing logic and information, rationalizing, perceptions, etc.
10. The human mind is a product of the brain even though the brain is a material entity comprised of thought and information.
11. Humans interact with the outside world through sensory, cognitive and conscious powers.
12. Humans share consciousness at the family, organizational, city, country and world levels.
13. The Prana is essential to enable life forces in humans, but acquired consciousness is required for cognitive and motor function.

The source of human consciousness is the pure consciousness that is fundamental to all material and nonmaterial world. The state and level of consciousness provide the energy for cognitive functions in the brain. Right consciousness along with spiritual attitude are the conditions for human conscience that drive the value system for human society. While on one hand the philosophy provided the right insight into the understanding of qualia and the neuroscience has been researching to understand the same on other, as such the 'hard problem of consciousness' in particular, does not pose much gap in using qualia for applying consciousness to address engineering problems. Without the considerations of consciousness, it is not possible to provide a robust model of humans for behavioral or performance-related aspects. All humans are connected at the plant, organization, country and global levels. A human being is a consciousness entity. Hence, apart from individual consciousness employees, peers, citizens and international community consciousness considerations should be reflected in modeling risk scenarios. Even though a given group or population is exposed to the same set of situations, the reality formed in the individual mind might very and might be different, as reality is an internal attribute. The aim of human life is self-realization, and achieving the improvement of performance attributes in an engineering environment is an obvious and natural benefit that is available to the individual and organizations collectively. There are techniques and methods such as Yoga and meditation that lead to the development of conscious human beings with the right performance (cognitive functions) and value (conscience) systems that lead to success, which includes stability, peace and happiness.

References

Anandi-.thesleepguru.co.uk: The magical pineal gland. [Online]. Available: https://www.thesleepguru.co.uk/meditation/the-magical-pineal-gland/. Accessed 20 May 2022

Amlani AM, Punch JL, Ching YT (2002) Methods and applications of the audibility index in hearing aid selection and fitting. Trends Amplif 6(3):81–129. https://doi.org/10.1177/108471380200600302

Archive.org (1921) Oxford University Press. [Online]. Available: https://archive.org/details/thirteenprincipa028442mbp/page/n8/mode/1up?view=theater. Accessed 11 May 2022

Areas of brain involved in speech production and comprehension. Psychology-info.com. [Online]. Available: https://psychology-info.com/areas-of-the-brain-involved-in-speech-production-and-comprehension. Accessed 27 May 2022

Baars B (2005) Global workspace theory of consciousness: towards a cognitive neuroscience of human experience. Prog Brain Res 150:45–53

BC Campus. Psychodynamic psychology. [Online]. Available: https://opentextbc.ca/introductiontopsychology/chapter/2-2-psychodynamic-and-behavioural-psychology/#:~:text=and%20external%20stimuli.-,Sigmund%20Freud%20divided%20human%20consciousness%20into%20three%20levels%20of%20awareness,id%2C%20ego%2C%20and%20super

Bohm D (1980) Wholeness and implicate order. Routledge, UK

Brain areas involved with consciousness. Quantum mind: exploring quantum theories of consciousness. [Online]. Available: http://quantum-mind.co.uk/brain-areas-involved-consciousness/. Accessed 26 May 2022

Chaitanya (Conscioiusness). Wikipedia, [Online]. Available: https://en.wikipedia/Chaitanya(consciousness). Accessed 21 April 2021

Chandyogya Upanishad. Wikipedia.org. [Online]. Available: https://en.wikipedia.org/wiki/Chandogya_Upanishad. Accessed 22 May 2022

Chalmers D (1995) Facing up to the problem of consciousness. J Conscious Stud 2(3):200–219

Chalmers DJ (1996) The conscious mind—in search of a fundamental theory. Oxford University Press, New York

Chalmers DJ (2006) Phenomenal concepts and the explanatory gap: new essays on consciousness and physicalism. Oxford University Press

Chandra Mauli MS (2018) Vedic evolution: its philosophy and science

Chen CC, Hsu CY, Chiu HW, Hu CJ, Lee TC (2015) Frequency power and coherence of electroencephalography are correlated with severity of Alzheimer's disease: a multicenter analysis in Taiwan. J Formos Med Assoc 114:729–7.35

Chinese Religions and Philosophies. [Online]. Available: https://www.nationalgeographic.org/article/chinese-religions-and-philosophies/. Accessed 15 May 2022

Choi Y (2020) Nature briefing—how to give computer common sense. Nature

Chopara D, quotefancy. Deepak Chopra Quotes, [Online]. Available: https://quotefancy.com/quote/793290/Deepak-Chopra. Accessed 20 Sept 2022

Chopra D (2019) Discovering the reality of consciousness. [Online]. Available: https://deepakchopra.medium.com/discovering-the-reality-of-consciousness-a7107a9b992b. Accessed 2022 May 2022

Chopra D. Quantum physics and consciousness. YouTube. [Online]. Available: https://www.youtube.com/watch?v=X8l9AprftVw. Accessed 4 April 2021

Crick F, Koch C (1990) Towards a neurobiological theory of consciousness. Semin Neurosci 2:263–275

Collins AL (2015) Orchestrated objective reduction: quantum physics and its implications in human consciousness. College of William and Mary—Degree Program thesis

Consciousness. Wikipedia. [Online]. Available: https://en.wikipedia.org/wiki/Consciousness. Accessed 21 April 2021

Consciousness: Chinese Thoughts (2019) Encyclopedia.com. [Online]. Available: https://www.encyclopedia.com/history/dictionaries-thesauruses-pictures-and-press-releases/consciousness-chinese-thought. Accessed 15 May 2022

David Chalmers Thinks the Hard Problem Is Really Hard—Consciousness will still mystify us even if we scientifically solve it, philosopher predicts. Scientific American, 10 April By John Horgan on April 10, 2017. [Online]. Available: https://blogs.scientificamerican.com/cross-check/david-chalmers-thinks-the-hard-problem-is-really-hard/. Accessed 22 May 2022

Descartes and the Pineal Gland. Stanford Encyclopedia of Philosophy, 18 Sept 2013. [Online]. Available: https://plato.stanford.edu/entries/pineal-gland/. Accessed 20 May 2022

Descartes-I-Think-Therefore-I-Am (2022) Newlearningonline.com. [Online]. Available: https://newlearningonline.com/new-learning/chapter-7/descartes-i-think-therefore-i-am. Accessed 25 May 2022

Dehane S, Lan H, Kouider S (2017) What is consciousness and could machine have it? Science 358:486–492

Dillbeck MC, Bronson EC (2009) Short-term longitudinal effects of the transcendental meditation technique on EEG power and coherence. Int J Neurosci 14(3–4):147–151

Durga TK, Sridhar MK, Nagendra HR (2018) Consciousness in Upanishads. Int J Sanskrit Res Ananta 4(5):66–69

Discovering the Mathematical Structure of Consciousness. Qualia Research Institute—QRI. [Online]. Available: www.qualiaresearchinstitute.org. Accessed 30 May 2020

Durham University, 11 Nov 2019. [Online]. Available: https://www.dur.ac.uk/news/allnews/thoughtleadership/?itemno=40191. Accessed 25 May 2022

Easwaran E (2021) The Upanishads. Jaico Publishing, Mumbai

Falk D (2021) Anil Seth finds consciousness in life's push against entropy. Quanta Magazine, 30 Sept 2021. [Online]. Available: https://www.quantamagazine.org/anil-seth-finds-consciousness-in-lifes-push-against-entropy-20210930/. Accessed 22 May 2022

Gerosa S, Nasini F (2001) Project financing and risk management: a new challenge for program management in the space industry of the third millennium. In: 4th European project management conference, 2001. London

Goff P (2019a) Galilio's error: foundations for a new science of consciousness. Penguin Random House UK, London

Goff P (2019b) Galilio's error—foundation for a new science of consciousness. Rider/Ebury Publishing, London

Goswamy T (1633) ShriRamcharitmanas, Gorakhpur (India). Geeta Press

Guzowski LB, Graziano DJ, Heo Y, Muehleisen R (2012) Testing a streamlined project evaluation tool or risk-conscious decision making: the Chicago loop energy efficiency retrofit initiative. In: IEEE summer study on energy efficiency in buildings

Hagelin J (2017) Is consciousness the unified field? A field theorist's perspective. Create Space Independent Publishing Platform

Hamroff SR A brief history of the consciousness. In: Science of non-duality (SAND). www.quantumconsciousness.org

Hart W (2012) Vipassana meditation—the art of living: as taught by Acharya S.N. Goenka, Igatpuri (Maharashtra): Vipassana Research Institute

Hart WD (1994) Dualism, in Oxford, Blackwell, Wikipedia. In: Guttenplan S (ed) A companion to the philosophy of mind. Oxford, Blackwell, pp 265–267

Hegalin J (2001) Intelligence 29:419

Hemeroff SR, Penrose R (1996) Orchestrated reduction of quantum coherence in brain microtubules: a model for consciousness. In: Hamerrof SR, Kasznaik AW, Scott AC (eds) Towards a science of consciousness; the first Tucson discussion and debate. Cambridge Press, MA

Hollnagel E (1998) Cognitive reliability and error analysis method. Elsevier, Oxford

Hu H (2009) Risk-conscious design of off-grid solar energy houses—PhD thesis. Georgia Institute of Technology, Georgia

Hutchinson DM (2018) Plotinus on consciousness. Cambridge University Press

HAF-Oum. HinduAmerical. [Online]. Available: https://www.hinduamerican.org/blog/veda-vyasa-the-sage-who-compiled-the-vedas. Accessed 9 May 2022

Hinduism. One Little Angel.Com. [Online]. Available: http://www.onelittleangle.com/wisdom/quotes/saint.asp?mc=60. Accessed 21 May 2022

Human Brain. Wikipedia, [Online]. Available: https://en.wikipedia.org/wiki/Human_brain. Accessed 21 April 2020

Implicate and Explicate order. [Online]. Available: https://en.wikipedia.org/wiki/Implicate_and_explicate_order. Accessed 30 May 2022

Is Consciousness the Unified Field (2014). Science and non-duality.com. [Online]. Available: https://www.youtube.com/watch?v=LPhgDfT4Zpc. Accessed 31 May 2022

Imants B (2019) California institute for integral studies. Categorical modeling of consciousness states—consciousness: ideas and research for twenty-first century 7(7), Article-1. Wikipedia [Online]. Available: https://digitalcommons.ciis/Vol7/issue7/1. Accessed 21 April 2021

In 2022 the science of consciosuness, Arizona (2022)

International Atomic Energy Agency (2010) Development and application of level 1 probabilistic safety assessment for nuclear power plants, SSG-3. IAEA, Vienna

Introducer E, Translator: Easwaran, The Upanishads. Jaico Publishing, New Delhi

Isha.Sadguru—What is consciousness—6yths Busted and 4 States Explained. Isha Foundation, [Online]. Available: https://isha.sadguru.org/us/en/wisdom/whatis-consciousness-states-myths#1point

Joye SR (2016) THE Pribram–Bohm hypothesis a topology of consciousness. J Nat Soc Philos 12(2):114–136

Joye SR (2021) Topological approach to the hard problem of consciousness: The Pribram–Bohm Holoflux theory. [Online]. Available: file:///C:/Users/pvvar/Downloads/THE_PRIBRAM_BOHM_HOLOFLUX_THEORY_OF_CONS.pdf. Accessed 30 May 2022

Jyoti A (2015) Consciousness: where science and spirituality meet. Anand Niketan Trust, Pune

Kapoor SK (2022) Adi Shankara exponent of Advaita Vedanta. Tribune India, 23 May 2022. [Online]. Available: https://www.tribuneindia.com/news/archive/lifestyle/adi-shankara-exponent-of-advaita-vedanta. Accessed 23 May 2022

Kelley SD. Monitoring consciousness Using the bispectral index™ during Anesthesia—a pocket guide for clinicians. Anandic Medical Systems/www.anandic.com

Kotchoubey B (2018) Human consciousness: where is it from and what is it for. Frontiers in Psychology: Theoretical and Philosophical Psychology. [Online]. Available: https://doi.org/10.3389/fpsyg.2018.00567/full#:~:text=Consciousness%20is%20not%20a%20process,and%20the%20use%20of%20tools. Accessed 25 May 2022

Krishan Y (1984) Buddhism and belief in Atma. J Int Assoc Buddhist Stud 7(2):117–135

Landaue LD, Lifshitz M (1974) Chapter 3—Schrödinger's equation. In: A short course on quantum mechanics, pp 67–115

Lanne M, Markus J, Mervi M (2014) Risk-conscious value creation. In: Annual RESER conference—2014. Finland

Lawrence J, Quad D, Becker J (2014) Integrating the effect of flood experience on risk perception with responses to changing climate risk. Nat Hazards 74:1773–1794

Lee BD, Sun Y, Hu H, Augendroe G, Paredis CJ (2012) A framework for generating stochastic meteorological years for risk-conscious design of buildings. In: IBPSA. USA. (www.ibpsa-use.org) / http://www.researchgate/publication/161913987

Lekov T. Academia.edu: the formula of the giving of the heart in ancient Egyptian texts. [Online]. Available: https://www.academia.edu/9011055/The_formula_of_the_Giving_of_the_Heart_in_Ancient_Egyptian_Texts. Accessed 25 May 2022

Lite G (2020) Consciousness creates reality—quantum physics explains, 12 January 2020. [Online]. Available: https://369news.net/2018/05/22/consciousness-creates-reality-quantum-physics-explains/. Accessed 25 May 2022

Liu J, Huo Y (2020) Recent advances in research on the association between consciousness and unconsciousness in China. Theory Psychol 30(1):99–120

Lucas J (2014) The face of freedom. Comment on consciousness in the universe: a review of the Orch-OR theory by Hameroff and Penrose. Phys Life Rev 11(1):87–88

Measurement in quantum mechanics. Wikipedia. [Online]. Available: https://en.wikipedia.org/wiki/Measurement_in_quantum_mechanics. Accessed 29 May 2022

Manaf AK (2021) YouTube: Vedic theories of universe—ancient Indian scriptures, excerpts from comparative cosmology by AKIF Manaf J. [Online]. Available: https://www.youtube.com/watch?v=wTjx8kiJsNg. Accessed 25 May 2022

Mankowitz DJ, Tamar A, Mannor S (2016) Situational awareness by risk-conscious skills. [Online]. Available: https://arxiv.org/pdf/1610.02847.pdf. Accessed 20 May 2022

McGeer PLE (1978) Perception, speech, and consciousness. In: Molecular neurobiology of the mammalian brain. Plenum Press, New York

Mcrae M (2017) There are three kinds of consciousness's and computers have mastered one, says study. Sciencealert.com. [Online]. Available: https://www.sciencealert.com/artificial-intelligence-three-types-of-consciousness. Accessed 4 April 2021

McWilliam E, Singh P (2003) Safety in numbers? Teacher collegiality in the risk-conscious school. In: AARE/NZARE conference. Auckland

McWilliam E (2004) On being accountable: risk-consciousness and doctoral supervisior. In: AANE annual conference, Melbourn, educational research conference

Meditations of René Descartes. Britannica.com. [Online]. Available: https://www.britannica.com/biography/Rene-Descartes/Meditations. Accessed 18 May 2022

Mesopotamian-religion. [Online]. Available: https://www.britannica.com/topic/Mesopotamian-religion/Religious-art-and-iconography#ref68281

Mind-body-dualism. Wikipedia.org. [Online]. Available: https://en.wikipedia.org/wiki/Mind%E2%80%93body_dualism. Accessed 21 April 2021

Nadar T, Vedas. Youtube. [Online]. Available: https://www.youtube.com/user/Dr.TonyNadar

Neural Correlates of Consciousness. Wikipedia. [Online]. Available: www.en.m.wikipedia.org/wiki. Accessed 21 April 2021

O (1921) The thirteen principal Upanishads translated from the Sanskrit with an outline of the philosophy of the Upanishads and an annoted biography. [Online]. Available: https://archive.org/details/thirteenprincipa028442mbp/pagen8/mode/1up?view=theater. Accessed 11 May 2022

Oliver S (2017) The river of consciousness. Picador, U.K.

Oytam Y, Sobhanmanesh F, Duesing K, Bowden JC et al (2016) Risk-conscious correction of batch effects: maximizing information extraction from high throughput genomics datasets. BMC Bioinform 17(32)

Oxford Languages and Google (2022) Google, [Online]. Available: oxfordlanguages.oup.com. Accessed 15 May 2022

Pavlus J (2020) Artificial intelligence: common-sense-comes-to-computers-20200430. Quantum-Magazine.Com, 30 April 2020. [Online]. Available: https://www.quantamagazine.org/. Accessed 21 May 2022

Panpsychiasm. Wikipedia. [Online]. Available: https://en.wikipedia/panpsychism. Accessed 25 February 2021

Pegg AE (2010) Learning to lead the risk-conscious organization: an empirical study of five English primary school leaders. Leadersh Educ 13(2)

Penrose R, Hameroff SR (1995) Ochestrated objective reduction. J Conscious 99(2)

Purucker GD (2016) Man in evolution. Theosophical University Press (2016)

Quantum Awareness where science and the Buddha meet (2005) Tag archives. [Online]. Available: https://quantumawareness.net/tag/r-c-henry/. Accessed 20 Sept 2022

Rahman AM, Rashid MM, Khanam F, Khurshid M (2019) EEG based brain alertness monitoring statistical and artificial neural network approach. Int J Adv Comput Sci Appl 10(1):431–442

Ru A (1984) Mind and consciousness in Chinese philosophy: a historical survey. In: Tymieniecka AT (ed) Phenomenology of life in a dialogue between Chinese and occidental philosophy. Part on Analecta Husserliana Book Series, vol 17. Springer, Dordrecht

Sadguru J (2016) The inner engineering—a Yogi's guide to Joy. Penguin-Ananda

Samveda: Chandogya Upanishad, India Hindu Civilization: Indian Scripture, 6th–8th BCE

Sanchetee P, Sanchetee SC (2018) Relevance of pineal gland: science versus religion. Yoga Mimansa 50:41–48

Sand G, Engell S (2003) Risk-conscious scheduling of batch processes. Comput Aided Chem Eng 15:588–593

Sarvapriyanand S (2018) Consciousness explained in 200 s Talk at IIT Kanpur. Kanpur

Searle J. On shared condition—consciousness. YouTube. [Online]. Available: https://www.youtube.com/watch?v=eqDgt12m26c. Accessed 21 May 2022

Seth AK, Hohwy J (2019) Open science in consciousness research (Editorial). Neurosci Conscious 2019(1). 10.1093.nc/niz018

Sharvananda S (1989) Isha, Kena, Katha, Prasna, Mundaka, Mandukya, Aitareya and Taittiriya, Upanishads. Text and translation with notes. Mu. Up.; III.1.3. SriRamakrishna Math, Madras

Shelli SR (2016) The Pribram–Bohm Holoflux theory of consciousness: an integral interpretation of the theories of Karl Pribram, David Bohm, and Pierre Teilhard De Chardin. California Institute of Integral Studies, San Francisco, CA. https://www.heartmath.org/assets/uploads/2016/05/the-pribram-bohm-holoflux-theory-of-consciousness.pdf

Shinozuka K (2019) Discovering the science of consciousness. In: STEM SUMMIT 7 Macmillan learning Scientific American. Austin, Texas

Shinozuka K (2019) Why-we-need-to-study-consciousness. Scientific Americans, 12 Sept 2019. [Online]. Available: https://blogs.scientificamerican.com/observations/why-we-need-to-study-consciousness/. Accessed 30 May 2022

System L. Limbic System. Queensland brain institute. [Online]. Available: https://qbi.uq.edu.au/brain/brain-anatomy/limbic-system. Accessed 29 May 2022

Science-Alert.com. Harvard. [Online]. Available: https://www.sciencealert.com. Accessed 20 May 2022

Sci-News.com. [Online]. Available: http://www.sci-news.com/othersciences/psyscology/consciousness-fundamental. Accessed 4 April 2021

SCI-NEWS: Is Consciousness a Fundamental Quality of the Universe? 14 January 2019. [Online]. Available: http://www.sci-news.com/othersciences/psychology/consciousness-fundamental-quality-universe-07291.html. Accessed 25 May 2022

Seventeenth-Century Theories of Consciousness (2020) Stanford encyclopedia of philosophy. [Online]. Available: https://plato.stanford.edu/entries/consciousness-17th/. Accessed 11 May 2022

Seventeenth-Century Theories of Consciousness (2020) Stanford encyclopedia of philosophy. [Online]. Available: https://plato.stanford.edu/entries/consciousness-17th/. Accessed 15 May 2022

Standard Encyclopedia of Philosophy—Plato (2022) Stanford.edu. [Online]. Available: https://plato.stanford.edu/entries/plato/. Accessed 25 May 2022

Taleb T, Benslimane A, Ben K (2010) Towards an effective risk-consciousness and collaborative vehicular collision avoidance system. IEEE Trans Veh Technol 59(3)

Terwiesch P, Ravemark D, Rippin DT (1995) Risk-conscious operation of batch processes. In: A postprint volume from the 4th IFAC symposium, Helsingør, Denmark, 7–9 June 1995, Helsingør, Denmark

The Economists—what is consciousness, 10 Sept 2015. [Online]. Available: https://www.youtube.com/watch?v=ir8XITVmeY4. [Accessed 21 April 2021].

The 'third eye' connection. The Hindu, 21 July 2013. [Online]. Available: https://www.thehindu.com/. Accessed 20 May 2022

Tilman LM (2003) Asset/liability management of financial institutions: maximizing shareholder value through risk-conscious investment. Euromoney Books, 1st edn.

Tometzki T, Engell S (2010) Risk-conscious solution of planning problems under uncertainty by hybrid multi-objective evolutionary algorithms. Comput Chem Eng 35:2521–2539

Tononi G (2004) An integrated information theory of consciousness. BMC Neurosci 42(5)

Tononi G (2008) Consciousness as integrated information theory (IIT): a provisional manifesto. Biol Bull 215:216–242

Tononi G, Boly M, Massimini M, Koch K (2016) Integrated information theory: consciousness to its physical states. Nature Revies-Neurosci Macmillen

Tononi G (2008) Consciousness as integrated information: a provisional manifesto. Biol Bull 251:216–242

Tyang CMT, Amin HU, Saad NM, Malik AS (2017) Review article: The influences of emotion on learning and memory. [Online]. Available: https://doi.org/10.3389/fpsyg.2017.01454/full. Accessed 12 May 2022

Upanishads (Taittirya). Self-Realisation. [Online]. Available: https://upanishads.org.in/upanishads/sa/ishaupanishad/self-realisation. Accessed 23 May 2022

Unified Field theory. Wikipedia. [Online]. Available: https://en.wikipedia.org/wiki/Unified_field_theory. Accessed 31 May 2022

Valmiki M Ramayana, India, 7th to 4th centuries BCE

Varde PV, Pecht MG (2018) Risk-based engineering. Springer, Singapore

Vyasa V, Presenter AV (2018) Tirth, Four Vedas—Rig, Yajur, Sam and Athar, India. Manoj Publication (Recorded History 1500–1600 BCE)

Vedic (Hindu) Theory of Creation. UBC. [Online]. Available: https://www.cs.ubc.ca/~goyal/creation.php#:~:text=Any%20theory%20of%20creation%20is,to%20100%20years%20of%20Brahma. Accessed 15 May 2022

Vedic Theories of the Universe (2022) YouTube, [Online]. Available: Hinducosmology #vedictheoryofuniverse #vedictheory. Accessed 15 May 2022

Wikipedia. Neural correlates of consciousness. Wikipedia, [Online]. Available: https://en.wikipedia.org/wiki/Neural_correlates_of_consciousness#History. Accessed 21 May 2022

Wilkinson I (2001) Anxiety in 'risk' society. Routledge-Tailor & Francis Group

What Is the Glasgow Coma Scale? Brainline. [Online]. Available: https://www.brainline.org/article/what-glasgow-coma-scale#:~:text=The%20Glasgow%20Coma%20Scale%20(GCS,of%20an%20acute%20brain%20injury. Accessed 26 May 2022

Wikipedia. Wikipedia, [Online]. Available: https://en.wikipedia.org/wiki/Zoroastrianism

World Spiritual Heritage—Sankara. OneLittelAngle.com—Hinduism-Adi Shankaracharya. [Online]. Available: https://www.onelittleangle.com/wisdom/quores.saint.asp?mc=60. Accessed 10 May 2022

World Spiritual Heritage—Spiritual quotes of Shankara. One Little Angel.com. [Online]. Available: https://www.onelittleengel.com/wisdom/quotes.saint.asp?mc=60. Accessed 5 May 2022

Chapter 3
Dependability Engineering

3.1 Introduction

Dependability is the system property that integrates attributes such as reliability, availability, safety, security, survivability and maintainability (Avizienis et al. 2001). Alongside industrial evolution, particularly safety critical systems such as nuclear plants, space, aviation, industrial processes and production systems, the approach to perceiving risk, reliability and safety as part of dependability modeling and management has also evolved over the years. Here, the present study is looking at the past for learning lessons toward understanding, at least partly, the future requirements (Lannoy 1990). There are four major or vital aspects directly relevant for risk modeling and management or derisking activities in operating plants: first, the hardware and software; second, the human factors; common cause, and fourth, external events and associated threats. The science of control and protection has moved in the last 50 years from valve-based design through solid-state logic to digital systems, and further efforts are on hand to use the intelligent or machine learning approaches in control and protection judiciously. The software components are integral parts of digital control and protection systems along with application aids such as computer-based maintenance management systems and operator information/advisory systems.

The basic question that arises with respect to any asset or service or software system is 'How dependable is it?' for any stakeholder and, more importantly, for the operating organization itself. For operating organizations, this factor provides a strategy for the optimization of availability to identify and put resources primarily for risk reduction and performance improvement (Rubino and Sericola 2014). For stakeholders, this consideration relates to the timely delivery of products or services. For regulators, dependability indicators can provide an added or even the needed oversight of a plant's risk levels and the delivery potential of the plant in support of just decisions.

P. V. Varde, *Risk-Conscious Operations Management*, Risk, Reliability and Safety Engineering, https://doi.org/10.1007/978-981-19-9334-3_3

For safety critical systems, dependability is all the more relevant, as along with the traditional life cycle parameters such as reliability, availability and maintainability, the deterministic as well as probabilistic notion of risk assessment and management forms additional and important dependability metrics. Here, apart from the core hazard risk, other aspects, such as security and financial risk components, often form part of the holistic risk assessment and management approach. In all these cases, the dependability evaluation with respect to human factors and common cause failure forms the integral element of risk assessment and management.

In systems engineering, dependability is a measure of a system's availability, reliability, maintainability, and, in some cases, other characteristics, such as durability, safety and security (International Electrotechnica lCommission (IEC) 2022). For safety critical systems, it can be argued that dependability has two connotations, the traditional deterministic and conventional or the relatively recent one, the probabilistic that seemingly operates under the reliability, availability, maintainability and safety (RAMS) program. The conventional approach to 'dependability' has probabilistic notion at its root. The traditional approach to design, operation, maintenance and regulation is deterministic in nature, which is proven and time tested. In the deterministic approach, the principles of defense in depth that include the philosophy of conservatism, quality assurance, redundancy, diversity, fail-safe design and single failure criteria are central to life cycle management. However, over the years, design, operations and maintenance have evolved, and even though safety remains an overriding factor, the need for improving plant deliverables has also drawn even more attention.

One of the major areas that work for both derisking on the one hand and reliability improvements on the other is data collection and analysis that facilitate the development of probabilistic metrics for dependability. All the indicators of dependability require performance data of the plant, which include records of individual systems, structures and components and human, software and service failures from plant records.

Dependability and associated attributes, such as reliability of process operations, maintainability of safety and process equipment, risk assessment and management functions, are directly relevant to risk-conscious operations management. There are two reasons, the plant SSC reliability is not only critical to cater to derisking but also important for process systems, as in many accident sequences, the process system failure contributes to the initiating event, and these initiating events are generated due to the poor reliability of the process system. Second, the plant shutdown or trip introduces delivery risk; for example, a power supply to grid will fail due to tripping/shutdown of a nuclear power plant, interruption of production in a process system will adversely affect delivery of the product, and failure of equipment due to poor reliability will adversely affect the service potential. Of course, human error contribution to accident risk is an important aspect; nevertheless, this chapter is dedicated to hardware dependability, while the human factor forms a common thread across most of the chapters in this book.

The scope of this chapter includes the introduction to statistics and probability systems reliability, availability and maintainability—the attributes of operations management. A brief discussion on security risk to physical assets has been provided. The whole discussion has been oriented to application to safety and mission critical systems with reference to nuclear plants.

3.2 Risk-Conscious Approach

Even though the risk-conscious approach that provides the fundamental to risk-conscious culture and that in turn makes up as risk-conscious operations management is the theme of this book, it is vital to understand the role of dependability in RCOM. There are two components to be considered: risk and consciousness. First, we will discuss risk. It was considered appropriate that risk and not safety should form one of the parameters of dependability for obvious reasons. Decision making whether at the individual level or at the management level is governed by risk perception. Why risk and not the safety? First, risk has a mathematical connotation; i.e., risk can be estimated. In fact, risk is a quantitative attribute and provides a better assessment of the asset. Furthermore, along with risk, mathematics allows the evaluation of uncertainty associated with the statement of risk. Therefore, risk along with uncertainty provides a more complete statement. The quantification algorithm enables assessment of increase or decrease whenever there is a change or modification to system, policy and practices. All the arguments discussed with respect to risk and uncertainty as an attribute do not reflect well for 'Safety' as an engineering attribute. Safety does not have a mathematical connotation; uncertainty is represented by the factor of safety, which tends to be subjective and based on a highly conservative, not so sound, argument in general and accounting of variability in particular.

The second attribute or RCOM is 'conscious' or 'consciousness'. It is generally recognized, be it philosophy or neuroscience or spirituality, that consciousness is fundamental to human existence. We think therefore we are. In fact, the consciousness that modern science talks about is cognitive consciousness, and neuronal correlates of consciousness (NCC) are the subject being used for research as well for monitoring the physical and mental health of an individual. Traditional human reliability techniques do have direct considerations of cognition, but consciousness per se does not form part of either human behavior or human reliability directly. In fact, there is no systematic and in-depth reference or consideration of the 'human model' in human reliability evaluation. For argument's sake, even the direct parameters related to human performance, such as awareness and alertness, and the ethical parameters related to security risk have not been holistically considered.

There is a consensus that the human factor is important and carries a very high dependability component, as the record of accidents shows that human performance or nonperformance has been the major contributing factor to the accidents. The second observation has been that human error data generally carry higher uncertainty than hardware components. It is also recognized that human modeling is a complex phenomenon. In this direction, the RCOM approach has two major considerations, dealing with 'risk' while 'safety' has been considered a subjective upward-moving target that every ecosystem should try to achieve and application of the consciousness, cognition, conscience and brain (CQB)-based approach such that the risk estimates are more holistic and dependable. This has direct reference to risk-conscious culture implementation in the plant.

3.3 Role of Dependability Modeling

In the risk-conscious approach, the role of dependability starts with the assessment of component reliability and unavailability as well as the human error rate toward providing the risk statement at the plant level while providing the unavailability statement at the system level. Let us understand with the example of initiating event 'loss of off-site power'. Here, the plant performance is directly dependent on the reliability of off-site power as the failure of off-site power results in a reactor trip (in case there is no alternate source that makes up for grid power failure, it is compensated by the alternate source). Simultaneously, or in a short interval, other process systems, including the coolant system, also trip, and demand is generated for safety systems toward ensuring safe shutdown and shutdown cooling, as nuclear reactors, even after shutdown, keep producing decay heat, which is a small fraction of normal power. This decay heat removal is a critical function as decay heat, if not extracted continuously, has the potential for serious fuel damage. Furthermore, the unavailability of off-site power for extended duration could be cause of concern as the plant risk margin increases and may lead to station blackout conditions. As already discussed, plant trips adversely affect delivery reliability at the same time plant unavailability adversely affects the quality of supply or unavailability of services at the client's end. In a similar manner, the security-related event also has potential adverse consequences, from writing off the plant to the long shutdown required to fix the problem.

The design and operation of the plant are based on ensuring the required redundancy, diversity and safe failure approach such that one failure at the component level or train level does not adversely affect plant availability and capacity for risk aversion. In fact, routine maintenance activities are performed by removing one component or train from service for a relatively short duration without adversely affecting the plant deliverables. Similarly, for many system configurations, the plant technical specifications allow plant operation to be uninterrupted for a specified time to restore the failed component; if not, then only the plant is shut down.

The above discussions show that design efforts are performed for reducing the dependability from component to system to plant and between plant to cater to the same services by implementing the principle of redundancy, diversity and fail-safe operation. The probabilistic evaluation provides insights into adequacy and level of dependency in terms of reliability, unavailability and finally in the form of a statement of risk.

3.4 Dependability Assessment

3.4.1 Interpretation of Dependability

The standard definition of dependability in systems engineering is '*It is a measure of a system's reliability, availability and maintainability' and might include other characteristics, such as durability, safety and security.* Although the basic definition extends the scope to include durability, safety and security, keeping in mind the requirements of safety and mission critical systems, changes are proposed as follows. While safety levels can be considered as a moving target in an operating organization, 'Risk' for the reasons mentioned in the introductory chapter is favored, as 'risk' is a quantifiable and thereby a measurable parameter forms an effective achievable goal.

Therefore, the proposed dependability definition will use risk as one of the parameters along with reliability, availability and maintainability. Let us examine the existing taxonomy of dependability to effectively use the dependability framework for safety or mission critical systems.

The reliability, availability and maintainability, as indicated in Fig. 3.1, often have probabilistic notions, as they are based on historical or, in some cases, data available from testing or generic data and can at most be an assessment function and lack prediction capability. Of course, it does indicate a trend, but the major limitation is reliability and availability attributes in probabilistic notion estimate the probability and do not provide any information about the future trend or instant of failure, of course with associated uncertainty, a critical requirement for operations management. Furthermore, the notion of safety, confidentiality and integrity tends to be qualitative and often vague and imprecise, which introduces relatively large uncertainty and constrains their application in support of management. The fuzzy logic that has the capability to capture linguistic variables for estimating integrity and confidentiality is expected to be more effective. Of course, in risk-conscious management instead of safety, the risk, unavailability and failure probability metrics have been used to capture the risk potential of events.

The system design addresses the aspects indicated for 'mean'—Fault prevention, tolerance, removal and forecasting. For example, in the nuclear industry, these functions are achieved by implementing national and international approaches/practices, for example, defense in depth a conservative approach factor safety application, and the principles of redundancy, diversity, fail-safe design and quality assurance in construction such that the possibility of unsafe events can be reduced and meet the compliance criteria while ensuring the reliability, availability and maintainability functions. With respect to operations, parameter trend monitoring from triplicate voting logic, e.g., the 2-out-of-3 or 2-out-of-4 logic, for each signal coming to the control room of the plant and automatic actuation of the safety shutdown system ensures unsafe consequences of failure. The monitoring system is fault-tolerant, as the signal component/channel fault does not have any consequences unless two channels out of the three confirm the existence of fault deviation. Of course, the following maintenance corrective action can easily be addressed without adversely affecting

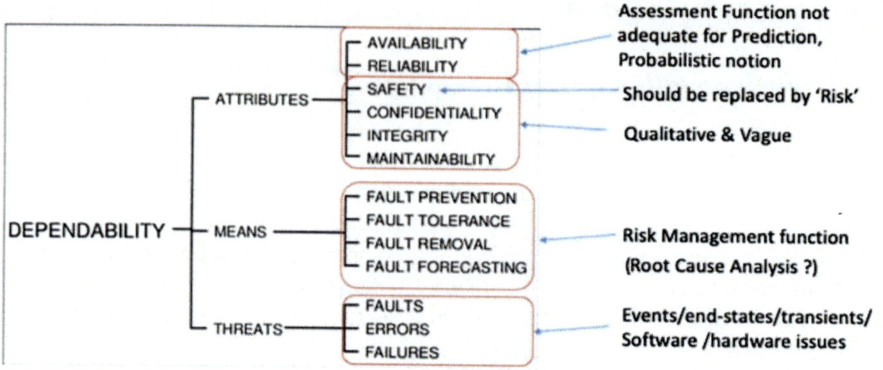

Fig. 3.1 Comments on the traditional dependability metrics and its limitation to application of risk-conscious applications

the dependability parameters. It may be noted that the current systems design ensures fault or parameter trending as forecasting measures; however, the systems capability for transients or in some instances of external events need to be verified.

The third branch, i.e., the 'threat', constitutes failures, faults and errors or, more specifically, human error that determines the dependability attributes. It has been mentioned earlier that whether it is a fault, error or failure, the common thread is human error, be it during design, construction or operation affects the system dependability potential. However, the important point is that it is not the independent failure that affects safety critical systems but the 'common cause failure (CCF)', as it produces undesirable consequences despite redundancy, physical separation, etc., measures. With this, the human error should be an exclusive consideration for dependability metrics for safety critical systems. Hence, the proposed metrics for dependability to support the case of risk-conscious operations management are as Fig. 3.2.

3.4.2 Deterministic Approach

The safety critical systems are designed and operated by primarily employing the deterministic approach, employing multiple barriers and multiple protections employing conservative approach redundancy and diversity, and failing-safe criteria to make the system resilient against a single failure. The supplication of codes and standards renders the design application of consolidated knowledge, experience and requirement of factor of safety to address uncertainty in site specific aspects, material, design process, operational challenges including human factor and common cause failure aspects to a great extent. The deterministic safety analysis was performed by analyzing the plant response against postulating the set of internal as well as external initiating events to demonstrate the safety of the plant. The safety insights gained

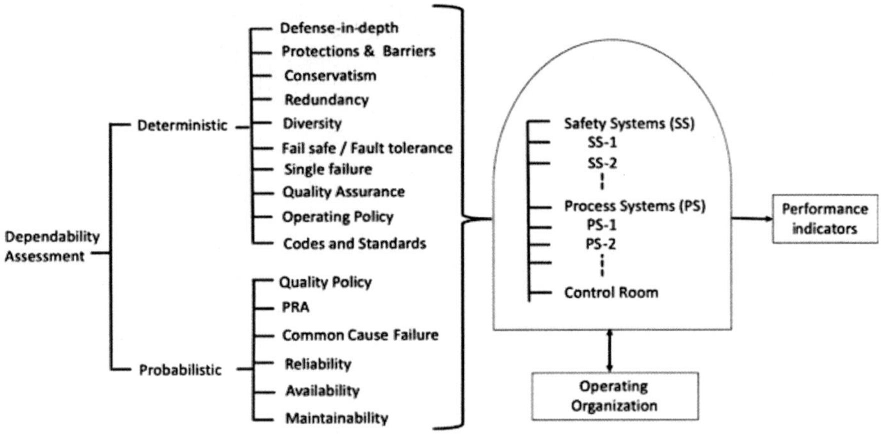

Fig. 3.2 Dependency model for complex engineering systems

from the safety analysis form a major input to draft station policy documents or technical specifications documents. The design safety case is subjected to a multitier regulatory review and approval for acquiring consent for operation.

Systems Structures and components and services of nuclear plants can be divided into two categories: process systems, which require operating reliably to cater to predetermined/target/deliverables, and safety systems, which remain dormant during normal operation of the plant and become activated when required to change the status from operation to shut down or control a deviation or terminate the accident sequence or mitigate the consequences of an accident. The engineered safety features of the plant, such as reactor containment, are designed to confine the radiation in containment during any accident scenario or even in extreme conditions; if it is needed, discharge the release at a control rate. The exclusion zone around the nuclear plant and emergency preparedness plans are designed to eliminate or reduce the radiation consequences in the public domain.

The station operations safety management should comply with the technical stipulation. Any change or incident departing or potential for departing the technical policy is referred to as a safety-significant event and should be subjected to regulatory review. Operations management deals with day-to-day operations of the plant generally round-the-clock manned by the trained and qualified crew while complying with stipulations of technical operations policy. These include safety limits, maintaining margins with limiting system settings, and limiting conditions for operations. The maintenance wing performs maintenance activities periodically as stipulated in the technical operational policy of carrying out preventive maintenance and support in system testing in coordination with the operating crew.

Major operational supervision activities include reactor start-up, power maneuvering and shutdown, start/shutdown of plant equipment as part of configuration management, plant parameter monitoring and surveillance, configuration management, coordination with maintenance and other agencies. All tests and maintenance, including modifications, were carried out through approved procedures. The plant is shut down periodically for fueling and shutdown maintenance activities.

The major performance indicators of the plant are (a) number of safety and operational events, (b) plant and system availability, (c) plant capacity factor, (d) number of plant trips/shutdowns, (e) trend monitoring of important parameters, (f) exposure to hazards and other hazard measurement data, (g) industrial events, etc.

Safety- or safety-related events are reviewed by the plant safety committee and further subjected to regulatory review as per the station policy document.

The defense-in-depth principle and adherence to the stipulated maintenance of safety and integrity of barriers and level of protection, redundancy, diversity, single failure criteria, fail-safe operations, adherence to schedules and quality assurance programs are hallmarks of operations management. Ensuring the implementation of an organization chart approved by the regulator, the maintenance of and qualified trained staff, and other surveillance and core management activities are also the responsibility of operations management.

3.4.2.1 Enhanced Defense in Depth

The defense-in-depth concept, although originally started from defense establishment, was adopted by many sectors, including the nucleus. Defense-in-depth philosophy is generally credited to a large extent for excellent safety records or the success story of the nuclear industry and is considered a proven strategy. In fact, in-depth defense has been extensively employed in security systems of nuclear, space and many other systems, particularly with respect to physical protection, information security and cyber systems, where multiple layers of protection are ensured assuming that the treat is a reality.

However, over a period based on the experience in last over 40 years, there have been two developments. On the one hand, there is recognition or weakness of the defense-in-depth approach, such as overconservatism and the subjective notion of safety that is qualitative and at times might tend to be arbitrary. Second, related to improved technological development and understanding that requires reconsideration as there are many advanced scientific and engineering developments, particularly in material science, understanding of degradation mechanisms, computational capabilities, information technology, availability of performance data on SSCs, other energy systems that can coexist with nuclear and relatively better framework for understanding of human behavior, etc., on the other.

One of the major developments is that the nuclear industry has been actively engaged in developing advanced reactor systems such as modular reactors, which are inherently safe systems that involve the use of features such as negative voids and temperature coefficients of reactivity, as well as the application of passive thermal hydraulic and mechanical systems for ensuring core cooling even without any motive or mechanical or electrical power. Consequently, many advanced reactor systems with advanced technology and improved confidence are being developed and require a safety philosophy that retains the fundamental requirement that safety remains as the overriding factor, and plant deliverables can be improved significantly.

Furthermore, one of the major developments is that in relatively recent times, probabilistic risk assessment (PRA) techniques have matured and are being used as part of regulatory review and operational applications either as complementary and or supplementary tools to existing deterministic approaches, essentially based on defense-in-depth philosophy.

There have been extensive efforts toward reviewing the traditional defense approach concepts. For example, The USNRC recognized that the term (defense in depth) has been in use for many years; it has not always been used or defined consistently. Some observations have been made based on historical records that identify the consistencies as inconsistencies in the use of the defense-in-depth philosophy by the USNRC (United States Nuclear Regulatory Commission 2016). There is a recognition that the defense-in-depth principle is partly reflected in a probabilistic risk assessment (PRA), but not all of the defenses-in-depth levels are included in the models. In addition, events included in PSA are not typically labeled with defense-in-depth information. With this background, SKI has developed a methodology for risk-informed assessment of defense in depth considering LOCA scenario evaluation. A general insight is that PSA could, however, be a powerful tool to assess the status of various defenses at depth levels in a nuclear plant (Holmberg and Nirmark 2008). Furthermore, there is an increasing interest in the use of best-estimate computer code toward having a realistic safety margin and understanding and evaluation of uncertainty (International Atomic Energy Agency 2008a). The capability of the traditional defense-in-depth approach must be upgraded or enhanced for new generation or advanced plant designs.

The IAEA sponsored/spearheaded project on Innovative Nuclear Reactors and Fuel Cycle (INPRO) presents a framework that tends to enhance defense in an in-depth approach. The INPRO strategy has at its core the sustainable reactor system with a scope ranging right from mining, fuel production, irradiation in reactor systems, transportation for reprocessing and waste management. The salient features of the enhanced defense in depth are the enhancement of all five levels of traditional defense in depth by the use of probabilistic goals and criteria and the use of advanced technology to consolidate the five levels. In this framework, a holistic safety assessment strategy comprised of a deterministic and probabilistic approach to support all five levels, viz. prevention, control of abnormal operations, controlling the design basis regime, ensuring low core damage probability and finally containing the released radioactive material. This strategy envisages increased emphasis on inherent passive features. This strategy seeks to achieve a low core damage frequency along with other probabilistic criteria for advanced systems (International Atomic Energy Agency 2020). There is guidance available that proposes a framework and detailed approach and methodology for safety assessment for innovative and advanced nuclear energy systems (International Atomic Energy Agency 2004a, 2008b). The end point of the enhanced defense-in-depth strategy is that, even in the case of accidents with severe core damage, emergency radioactivity releases from the plant large enough to require evacuation of the population must be made 'very unlikely'.

3.4.2.2 Quality Assurance

Quality assurance is an integral part of the defense in the depth approach. The quality of the hardware, material, software, environment and human resource management system is fundamental to the reliability, availability and maintainability and further to the assessment of risk. In fact, to become a candidate for dependability assessment of the system, structure and system, human resources should conform to quality specifications.

In fact, there is an implicit assumption in dependability assessment that the entity in question has stipulated 'quality', as quality conformance is the first condition; for any system, structure, component, software, service, or human resource; for considerations of dependability. One of the major reasons for quality considerations in dependability analysis is that quality-related issues, be it design, manufacturing, environment, etc., have been found to be one of the dominant root causes for independent in general and common cause failure (CCF) in particular during the operational phase of the plant.

3.4.2.3 Human Resource Qualification

It is recognized that the human factor is the major contributor to accidents. In this context, the training and qualification program in the operational ecosystem deserves attention, focus and resources in the organization. The training and qualification program prepares the staff in foreseeable activities in plant operations and maintenance that include supervisory activities viz. monitoring and surveillance, coordination of maintenance, test, modification and ensuring compliance with technical specification stipulations are some of the activities. One of the major areas where the staff is trained is operational vigilance not only for normal operations but also for emergency conditions. The staff is trained through a comprehensive procedure comprised of the completion of individual systems checklists, followed by a successful written test, walking down, and finally an interview by the qualification and authorization committee to judge the competency of knowledge of plant systems, procedures for normal and emergency scenarios and technical specifications and their implementation. The attitude toward safety, response capability for a given scenario, reflexes, interpersonal skill, leadership quality to handle a complex scenario and finally understanding of the responsibilities that come with manning the plant are also judged by the committee comprised of operations, maintenance, design, radiation protection, physics and chemistry, and regulatory experts.

Generally, the qualification license is given for a predetermined period of 3 years. Then, again, the staff must undergo the relicensing procedure to obtain authorization for a specific role/duty.

3.4.2.4 Technical Documents

The plant safety reports and policy document derived from the safety report form the major document that stipulates plant operating policy and conditions, including the safety system settings, requirements of staffing, responsibilities of staff, and surveillance test and maintenance policies. On plant management requests, these documents are approved by the regulatory agency periodically.

The technical manuals, which include design, operation manuals and emergency procedures, all go through a quality assurance procedure such that the approved procedures are available for the trainees at all levels. These documents are prepared by experts in the field based on plant administrative and management policy. To ensure that the documents serve the intended purpose, i.e., support effective and efficient training of staff, the documents are peer-reviewed to check for objectivity, facts check, accuracy of information, correctness of data, legibility, completeness, and finally for errors so that these documents reflect the as-built and up-to-date status of the plant, procedure and policies. Finally, these procedures are compiled, and for each system, separate documents with required instructions for its use are issued for use by the trainee and plant staff.

Some of the records, such as plant summaries, incident reports and various best practices, also form the part of achieves and are updated periodically. Apart from this, the approved engineering drawings that include process and instrumentation diagrams and engineering drawings are also maintained and updated by the engineering group of the plant.

To ensure that systems and policies are reflected/updated time to time, these documents are reviewed periodically, say after a period of 5 years.

3.4.2.5 Domain Specific Scientific and Engineering Services

The QA activities also include review and reporting of implementation of regulatory compliance. Engineering modifications at the plant, system and component levels are ongoing activities in plants and parts of engineering service activities. The review of the modification proposals is coordinated by the engineering services staff, while the review is performed by various maintenance and operations agencies, such as the requirements, feasibility, testing requirements and schedules. Safety benefits and scheduling requirements for these proposals are defined at the outset. The modification proposals are then subjected to a plant level safety committee, and if the proposal is safety significant, then it requires regulatory review. Quality assurance in design and manufacturing and implementation of the modification proposals. The quality assurance (QA) activity in an operating environment generally ensures the material, manufacturing, audit and regulatory review, including internal inspection, in-service inspection and condition monitoring aspects of the plant.

Once implemented, the modification is again subjected to qualification and testing to meet the set operations and regulatory conditions before putting the modified system for regular use.

3.4.2.6 Maintenance Management

Maintenance management, which is primarily performed on a preventive basis, is a vital activity, and the intent is to ensure the required availability of safety and reliability of the safety and process systems. The activities include servicing, overhaul, checking, repair, removal from service and bringing back the affected equipment to operational state process systems or poised or standby states for safety systems. Maintenance management is an organized technical activity, and the role and responsibilities are defined in the operations policy document. Depending on the SSC maintenance requirements, there are four maintenance wings: mechanical, electrical, electronics and instrumentation. The periodicity of the test and maintenance for safety significance and equipment of higher operational importance are performed, mostly as given in the plant policy document. Any deviation from the policy document attracts regulatory review and is reported along with the justification and proposed next scheduling.

3.4.2.7 In-Service Inspection

To ensure the integrity of SSCs in the plant, in-service inspection of components is performed with a periodicity say of 5–15 years depending on the safety significance of the SSC. The program is essentially based on likelihood and consequences due to loss of integrity of the component—more generally associated with passive components, such as piping, supports and structural components. The inspection schedule and methods should comply with the recommended codes and standards of the ISI. In case there are findings that indicate deviation from the expected integrity, then either it is corrected such that it meets the acceptance level, or if the observation is not significant that warrants repair/correction, then a decision is made either to enhance additional surveillance or examination to verify that degradation is well within the acceptable level. A separate ISI document that provides the complete inspection program, duly approved by the regulatory agency, is the mandatory requirement for consenting and further reauthorization for a plant.

For rotating machines, based on the safety significance and equipment required directly for process system operations, such as pumps, motors and alternators-generator sets, a condition monitoring program is implemented and maintained. Machine online/offline temperature, vibration, alignment, offset and oil quality, in some cases wear monitoring, are performed to track the degradation in machine performance. The instrumentation often provides alarm when the parameter reaches the set limit such that corrective action is initiated to eliminate or reduce the chances of failure and thereby avert or reduce the consequences of failure.

3.4.2.8 Performance Evaluation

This activity addresses integrated performance evaluation of plant systems, generally performed annually. The major performance indicators and their trend monitoring are (a) operational availability, (b) capacity factor, (c) number of plant trips and shutdowns, (d) safety-significant event, if any, (e) industrial safety event, if any, (f) radiation dose consumption and other observation on radiation-related activities (nuclear plants), (g) overview of fueling operations (nuclear plants), (h) status of regulatory compliance and (i) availability of safety systems. Apart from this, the review of the incidents, its root causes, performance of process and safety systems also forms the part of performance review. The performance review essentially provides the quality of operations management over a period. Any significant observation that needs attention for corrective action forms the recommendations of the performance review.

The performance data collected from the records of performance review reports of certain periods form the input information for generating a reliability database for generating plant-specific data to be used for the evaluation of dependability attributes—reliability of plant systems, unavailability of safety systems, maintainability of components and equipment and risk estimates of plants.

3.4.2.9 Security Risk

Apart from safety, the considerations have been becoming part of asset protection as well as the staff and public from undesired and deliberate attempts to adversely affect asset operation. The major difference between safety and security is the consideration of undesired events or component failures caused by random events. In the case of security, the major consideration involves deliberate attempts to induce failure to create undesired consequences. As seen here, in the case of security evaluation, the scope of modeling involves the probability of mala-fide or malicious intention to induce undesired events.

Even though security aspects have played a major role in ensuring aversion to threats from internal and external elements or human-induced informed events, it was after September 9, 2011, an event involving the use of aviation systems to create catastrophes, that security risk was considered the core part of asset management practices. This event is a typical case of attacking vulnerable and high consequence objects involving deliberate mala-fide intention causing the loss of public life and property and meeting other objectives. Safety critical systems as usual took the lead in enhancing security aspects of the assets. The reference to security aspects will be made considering one of the safety critical systems, viz. nuclear plants.

Along with safety, security should form part of the plant policy document to ensure the protection of the public, property, occupational workers and the environment from potential hazards. The technical specification document, which traditionally has safety as the fundamental driver for deriving safety limits and limiting safety system settings and surveillance requirements, should also be restructured on security requirements. The plant organizational chart should be revised such that security aspects are addressed in not only surveillance but also online reflection of authority and line of communication. The overlap of safety and security should be identified and addressed suitably to avoid any contradiction. The general experience is that for a larger part, safety and security requirements might have a similar approach for implementation; however, formal evaluation to identify the areas requiring an approach that meets safety and security objectives should be implemented to the area of overlaps. The training and qualification program should implement a risk-conscious culture that includes both safety and security requirements. The staff should be trained in the areas where safety and security aspects contradict and therefore the provision of address these complex aspects that have been provided in the plant. All the plant emergency operating procedures should be revised based on the insights available from security risk assessment. The vulnerability of the plant to design basis threats or initiating events should be identified, and accordingly, emergency procedures should be developed for the identification protection, control, mitigation and management of accident scenarios.

There are two major aspects of plant security: physical protection and cyber systems. The threat could be insider and outsider threats. The security assessment requires identification of the design basis threat, vulnerability and associated consequences for the corresponding scenario. Plant physical protection should comply with the elements of the defense in depth toward mitigating or neutralizing potential threats. Typical provisions include zoning, fencing, distance, vigilance, authorization and permissions. The major operational provisions are surveillance that includes periodic activities, viz. area visits for identification/detection of deviations in the area, testing, simulation and maintenance, performance records of physical systems, adequately trained staff and monitoring of ethical behaviors that include integrity, dedication, etc., for timely detection of threats.

With advances in technology, while on the one hand benefiting technical and societal systems, the same technology can also become a potential threat. Therefore, the assets should be protected from this advanced technology that the insider presence may not require. Some of these threats are the impact of external objects (aircraft, etc.), electromagnetic pulse (EMP) devises or even nuclear electromagnetic pulse (NEMP). For example, electronic hardening is a well-known technology for EMP threats.

The traditional deterministic approach, even though well-proven, tends to be subjective, particularly in regard to decision making, as the use of qualitative attributes requires interpretation. For example, the qualitative notion of safety, reliability and redundancy tends to be subjective. The probabilistic approach provides an opportunity to use the performance data of hardware and humans and to some extent on software when evaluated employing probabilistic models to produce quantified estimates of reliability, availability, maintainability and risk. Furthermore, the protection of assets against cyberattacks requires a commensurate technology and a management system.

3.4.3 *Probabilistic Approach*

Traditionally, the major attributes of dependability reliability, availability, maintainability and safety and even security had a qualitative notion, as the performance data of SSCs and including human error data were not adequate to apply mathematical treatments to these attributes. However, over a period of time, performance data on engineering systems started pouring in, and industries and organizations developed a system of data collection and analysis toward providing the estimates of the above terms under a single umbrella reliability, availability, maintainability and safety (RAMS), which along with a defined framework found application as attributes of dependability found later applications for dependability modeling. This section will present how these quantified attributes are considered part of safety critical systems.

3.4.3.1 Reliability Fundamentals

Reliability is defined as the *probability that a component or a service or a product will survive or function without interruption for a given mission or a period of time (referred to as mission time) under stated conditions.* Mathematical reliability is expressed as:

$$R(t) = \Pr(T \geq t/(c1, c2, c3 \ldots)) \tag{3.1}$$

where

$R(t)$ Mission statement of reliability for time 't'
t Mission time (unit of time)
T Random variable

It may be noted that in the current context and in future systems, software will play a key role, and the above equation cannot be applied to systems having software components on it, e.g., digital systems. The reason is that the above equation is meant for components or systems that age with time (t) due to mechanisms such as wear and corrosion.

Nevertheless, this definition of reliability is the standard definition and has wider acceptability for hardware systems and service applications. This model is valid at all the levels, be it plant, system, structure and finally at the component level. For example, if we consider a complex mega plant like that delivers either power, medicine or sample for diagnostic or therapeutic application, viz. a research reactor, then the reliability of delivery can be estimated. We need the historical records as to how many times this delivery failed and second a mission time, i.e., the total time of continuous operation for qualified irradiation such that the sample has the required activity levels in (curies). These two data, generally referred to as failure rate λ and mission time for operation t. If we assume that the failure data follow an exponential distribution, then the equation for component reliability for a given mission time t and failure rate λ per unit time, is written as follows:

$$R(t) = \exp(-\lambda t) \tag{3.2}$$

This equation provides an estimate of the reliability of service from the subject reactor. For argument's sake, the data showed a failure rate of 5×10^{-3}/h and mission time (time required for achieving the target irradiation level) is say one week, i.e., 7 days (168 h).

The reliability of sample delivery services can be computed as follows:

$$R(168\ \text{h}) = \exp\left(-5 \times 10^{-5} * 168\right)$$

The reliability of the sample delivery services has been estimated as 0.992. This estimate is a key parameter for the service provider as well as healthcare services about the quality of the product in terms of required curie (sample strength) as well as the probability of successful delivery. This information enables an end user of this service or product to make decisions regarding the dependability of the supply from a given reactor. Apart from reliability, other attributes, such as availability in support of informed decisions.

This is the equation for reliability at the component level. There is an assumption that the data follow an exponential distribution. The relation between reliability and failure probability is as follows:

$$R(t) + F(t) = 1 \tag{3.3}$$

$F(t)$ is the statement of mission failure probability. Accordingly, the failure probability of the component is given as

$$F(t) = 1 - R(t) \tag{3.4}$$

Further for exponential distribution

$$F(t) = 1 - \exp(-\lambda t) \tag{3.5}$$

For the case considered above, the failure probability of delivery from a plant is 1 – 0.992 = 0.008 or 8×10^{-3}/demand. Conservatively, this can be expressed as 8 failures in 1000 demand (given the historical data and the mission time for sample production).

The above numerical illustration shows how plant performance data can be used to estimate the service/delivery reliability for any plant. This works even better for client as to whether the service is dependable or not.

There are two major procedures for reliability evaluation, either using mission mode operation or load-based evaluation, i.e., stress–strength-based modeling. For structural systems, the stress–strength model where apart from initial evaluation during design, the aging management requires the reliability of structures considering remaining strength and stress data. This information is vital for scheduling aging management programs for old plants. The following section discusses the concept of design reliability, i.e., the design of systems employing the stress–strength approach. The procedure is used after obtaining information on the loss of strength due to corrosion, cyclic loading, erosion, etc., after completion of the design life of SSC in support of life extensions and aging management.

3.4.3.2 Design Reliability or Inherent Reliability

The expectation of reliability of a system being designed is called design reliability. To obtain assurance of reliability considerations of material properties and degradation potential for postulated operational and environmental conditions, lifetime loads, variability in operational conditions, human factor interactions, on the one hand, and meeting the established code/standard, quality assurance procedures and maintainability consideration in the operational phase, on the other. For any product, component, system, plant and service to qualify for the operational phase, it has to be designed such that the design meets reliability expectations. The design for the reliability approach follows the stress and strength considerations shown in Fig. 3.3.

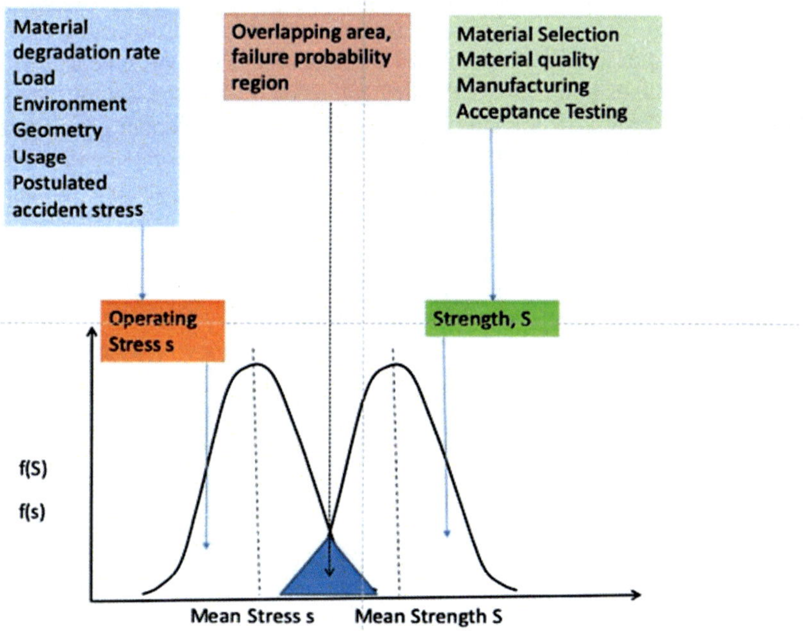

Fig. 3.3 Stress–strength interference to capture variability in design for reliability

The above discussion is for the same of completion of operational management as there are, in a limited sense, design activities in support of modifications and retrofits as part of engineering services. Varde and Pecht present the details on the subject and establish the relevance of this aspect as part of risk-based design (Varde and Pecht 2018). The existing approach to design is based on the traditional deterministic procedure. Of course, extensive work is being done to employ the stress–strength model-based approach to complex systems.

3.4.3.3 Operational Reliability

Operational reliability is the reliability of the system realized during the operational phase of the system. However, during the design phase, there are established methods to obtain high-reliability levels, such as stringent quality control and quality assurance procedures and testing of individual components and, to a great extent, full-scope or limited-scope integrated prototype testing by creating real-time tests and the environment as much as possible. Finally, the residual integrated test operations are targeted to ensure that while design intents are met, the system operations are as intended.

However, failures are encountered when the plant enters the operational phase, particularly during the initial operational phase, which is generally attributed to quality issues. However, deeper insight reveals that the reason includes the learning phase related to operational and maintenance, design issues, dependability, or cross connection between two or more systems, manufacturing errors, human–machine interface issues, deviation in operations and maintenance environments and associated errors, changes in operational policy and so on. Often, the lack of communication on proper design specifications or postulated operating conditions between the design/project team and manufacturer, particularly critical components, becomes a source of loss of reliability of components and systems during the operational phase.

Before we address the reliability risk associated with complex systems, we will briefly discuss the different models when we approach from component- to system-level reliability. There are many approaches to reliability modeling, (a) reliability block diagrams, (b) fault trees or success trees, (c) Markov models and (d) petri nets.

The following section will discuss reliability modeling employing the reliability block diagram approach.

Series System Reliability

An engineering system where the minimum required components are connected accordingly to their function chronological order, such that the first component is required to initiate the process taking input from a source, the output of the first forms input for the second and the output of the second becomes input for third and so on. This is an example of a dependent system where the successful operation of a given component in a series of connected components depends on the preceding component. Therefore, the series system configuration shows dependency among the

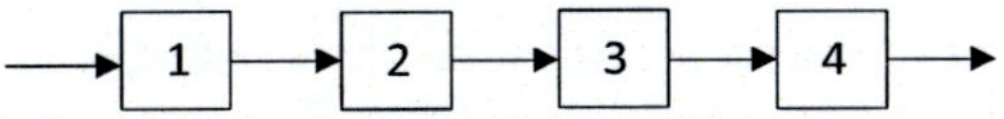

Fig. 3.4 Series system configuration comprising four components

components for successful operation of the system. In this series, even if one component fails, then the system fails. That is, for successful operation of the system, all the components should provide failure-free operation. The reliability block diagram of a system where four components are connected in series to meet a given function is shown in Fig. 3.4. The arrow at left-most indicates 'input', and the right-most arrow is the 'output' of the system comprised of 4 components connected in series configuration.

We know that, for reliable, i.e., uninterrupted operation of the system, all 4 components should function, and any one of the component failures disrupts the path from input to output, and there for even one component failure takes the system to a failed state. Finally, we say that for System's success or Reliability for a given mission time t, Components 1 AND 2 AND 3 AND 4 should function for the success of the mission time t. We express the system reliability for a mission time t as $R_S(t)$ as:

$$R_s(t) = R_1(t)(\text{AND})R_2(t)(\text{AND})R_3(t)(\text{AND})R_4(t)$$

In Boolean mathematics, the operator AND represents intersection mathematically expressed as $\cap$ or product or multiplication ($\times$). The operator OR represents Union, mathematically expressed as $\cup$ or $\vee$ and operates for summation/addition (+). Accordingly, the mathematical expression for series system reliability

$$R_s(t) = R_1(t)R_2(t)R_3(t)R_4(t)$$

$$R_s(t) = \prod_{i=1}^{4} R(t)_i \tag{3.6}$$

We can write the general expression for the series system reliability $R_s(t)$ of n components as follows:

$$R_s(t) = \prod_{i=1}^{n} R(t)_i \tag{3.7}$$

Assuming that the failure rate of the components in the system is constant; i.e., it follows an exponential distribution, the reliability expression, as a function of failure rate λ_i and mission time t, is given as follows:

$$R_s(t) = \exp\left(-\sum_{i}^{n}(i)t\right) \tag{3.8}$$

This implies that the system failure rate is the summation of all the individual component failure rates. In the practical interpretation of this argument, only those components that significantly contribute to the mission of the system should be part of the system. The reliability of the series system increases with the component individual failure rate and part count rate. In case a situation is encountered in which the component function is needed in the system but the individual component failure rate is high, in such a situation, a parallel redundancy is used in place of a single (relatively high failure rate component, which is necessary for the system function), i.e., in place of a single component, two components are connected in parallel such that in the case of unavailability of one of the two components, the second component function meets the functional requirements. In the next section, we will discuss parallel system reliability evaluation.

Illustration 3.1

We have a small segment in a process pumping system comprising a suction valve, a pump and a discharge valve. The function of this system is to take water from an underground sump and supply the water to an overhead tank. To keep the illustration simple, we ignore the pipeline, which is otherwise an integral part of the system.

Assumptions

1. Since the tanks and pipelines are passive structural components, they have very low failure rates compared to the pump, valves and filter; hence, they are not considered in this analysis and are not considered.
2. The failure rate data from a generic reputed source adequately meet the purpose of this analysis, and the same has been considered for the analysis.
3. Only the identified failure modes given in the table have been considered adequate for the analysis.
4. All four components in the system are subjected to maintenance on a monthly basis to minimize the failure.
5. The failure rate of the component is constant; therefore, an exponential distribution was used for the reliability evaluation.

System and component boundary identification

In reliability analysis, the system boundary must be defined to determine the scope and objective of the analysis. The system boundary is indicated in Fig. 3.5. The system boundary includes filter F, valve V-1, pump P and valve V-2. Similarly, for components, the boundary must also be defined to make it clear what constitutes the component. For example, the pump boundary includes the mechanical pump, motor, power supply, bearings and, most importantly, human error in operation and maintenance that causes the pump to fail. The same argument is applicable to valves V-1 and V-2 and the filter. Only the physical boundary of these components defines these components.

Objective and scope of the analysis: To estimate the reliability of the pumping system for a mission time t of 150 h. considering data obtained from a generic source.

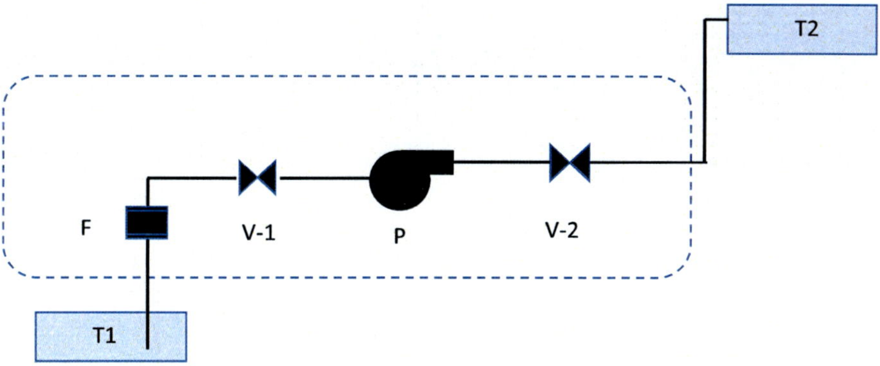

Fig. 3.5 Pumping system that pumps water from an underground tank T1 and to an overhead tank—a common example in residential as well as industrial/process industry

Table 3.1 Input data for reliability analysis

S. No	Component	Component code	Failure mode	Failure rate/h	Reliability (for mission time of 100 h)
1	Filter	*F*	Choke	2×10^{-5}	0.998
2	Suction valve	*V*-1	Fail to open and fail to remain open	3×10^{-4}	0.970
3	Pump	*P*	Fail to start and fail to run	5×10^{-4}	0.951
4	Discharge valve	*V*-2	Fail to open and fail to remain open	3×10^{-4}	0.970

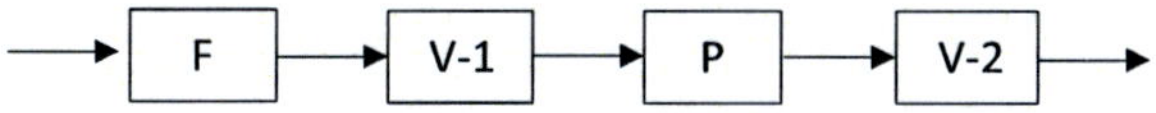

Fig. 3.6 Reliability block diagram for the pumping system

There are four components in the system considered here, viz. the filter, suction valve, pump and discharge valve. All four components are connected in series. There is a source of water from the underground tank, and the output, i.e., discharge from the system, is received in an overhead tank.

Having made assumptions of the analysis and defined the scope, objective and boundary of the components, the next task is to collect the information that includes the mode of component failure and failure rate such that reliability computation can be performed. Table 3.1 provides the input for the analysis.

The reliability block diagram for the system is as Fig. 3.6.

The system failure rate in a series system is given as:

$$\lambda_s = \sum_{i=1}^{n} \lambda_i \tag{3.9}$$

$\lambda_s = \lambda_F + \lambda_{V-1} + \lambda_P + \lambda_{V-1}$
$\lambda_s = 2 \times 10^{-5} + 3 \times 10^{-4} + 5 \times 10^{-4} + 3 \times 10^{-4}$
$\lambda_s = 1.12 \times 10^{-3}$
For a mission time of 150 h, the system reliability is:

$$R_s(t) = \exp(-\lambda_s t)$$
$$R_s(t) = \exp\left(-1.12 \times 10^{-3} \times 100\right)$$
$$= 0.894.$$

If we use the equation

$$R_s(t) = R_F(t) R_{V-1}(t) R_P(t) R_{V-2}(t)$$

Input reliability values from table as follows

$$R_s(t) = 0.998 \times 0.970 \times 0.951 \times 0.997 = 0.893 \text{(higher - order terms were ignored)}$$

The system reliability considering the failure data given in the table for the components for a mission time of 100 h is 0.894.

The reliability estimates are basically a function of the failure rate for the components in the system. However, the failure rate of components given similar configurations and applications, even though they are procured by employing the same quality acceptance norms, design margins for operation, and from the same supplier, might vary in different plants. For example, there are *n* power plants, and all the plants using the same configuration, components, design and quality, vendor, might experience different failures. If the experience goes, then yes, the failure rates of these components for similar applications vary from plant to plant. One of the major reasons is the operations and maintenance culture of these organizations, particularly the awareness for consequences of failure and how close monitoring of these systems employs online/offline, periodic surveillance to eliminate the incipient failure or reduce those consequences of failure. The level of awareness of the staff for the importance that is attached to quality policy and standards in operation and maintenance practices, which include policies, training, qualification dedication, commitment of organization and particularly people, and procedure, material used, maintenance of records, feedback or learning from failures—that sums risk-conscious culture of the organization. We call risk culture, as in an operational scenario, we have a system for which designers have incorporated the provisions to eliminate or mitigate the consequences of the risk of unreliable operation. In the operation ecosystem, we are dealing with residual risk. The domain of residual risk for unreliability operation is very small. May of the order of is very small, as seen in the above example. The reason for

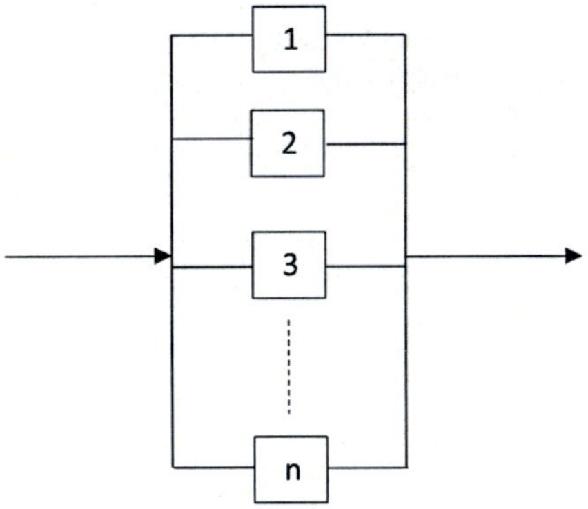

Fig. 3.7 Reliability block diagram for parallel configuration

the new or remaining or residual failures that precipitate during operation will be revealed only when the system operates. With the close monitoring and analysis of the precursor signature, such as an increase in bearing vibration and temperature for pumps, pressure drop across the filter and strainer, rattling noise from the valve disk, or vibration or leakage or even choking of valve passage with time by observing the drop in pressure across the valve, can be detected in time before the system collapses. The high alertness and awareness of the staff make it easier to understand the degradation before a complete failure. Similarly, a policy in an organizational setup where failures and their causes are discussed periodically such that corrective mechanisms include root cause analysis can provide insight into eliminating or reducing these failures.

Hence, reliability is not the typical characteristic of system configuration and components but also the reliability or risk consciousness at the plant, individual and organizational levels. The heightened awareness about what is expected in terms of quantities targets which can be, if required translated into understandable and appreciable quality attributes as linguistic variables such that it becomes plants' language of maintaining high O&M standards.

Parallel System

When systems require higher reliability that is not possible to achieve with series configuration alone, the parallel configuration or standby configurations are deployed. For example, assume a reliability target of a safety system to be 0.9999. In such cases, parallel system configurations are used. In parallel configurations, more than one component is placed such that if one component fails, then the other parallel can cater to the safety function. The system operation will be considered to fail if the last operating component in the system fails before meeting the target mission time.

Figure 3.7 shows a reliability block diagram (RBD) for the parallel system having n units in parallel as follows:

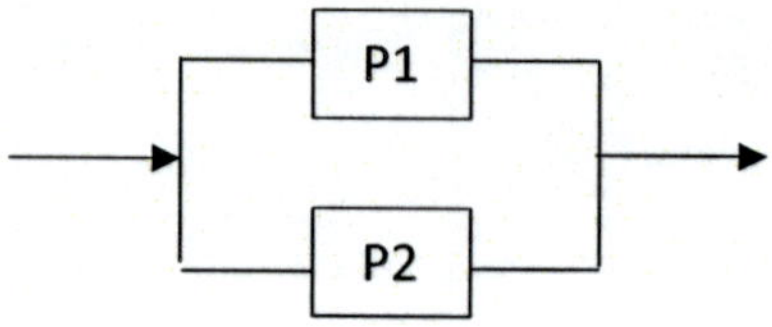

Fig. 3.8 Reliability block diagram for a two-unit parallel system

Similar to the series system, the RBD parallel system also has an input and output to represent the source and deliverable of the system and the block that represents the components. In the parallel configuration shown in Fig. 3.7.

If $R_1(t)$, $R_2(t)$, $R_3(t)$, … $R_n(t)$, is the reliability for the 1st, 2nd, 3rd … Nth block, respectively. We know that the failure probability $F(t) = 1 - R(t)$; therefore, the failure probability of the component can be given by $F_1(t)$, $F_2(t)$, $F_3(t)$, … $F_n(t)$, for the 1st, 2nd, 3rd … and nth components in the system.

The parallel system fails only when the last; i.e., nth component fails; therefore, the system reliability for a parallel configuration for a given mission time of t can be mathematically expressed as

$$R_p(t) = 1 - \{F_1(t) \cdot F_2(t) \cdot F_3(t) \ldots F_n(t)\}$$
$$R_p(t) = 1 - \{[1 - R_1(t)] \cdot [1 - R_2(t)] \cdot [1 - R_3(t)] \ldots [1 - R_n(t)]\}$$

The generalized equation for a parallel system can be given as:

$$R_P(t) = 1 - \prod_{i=1}^{n}[1 - R_i(t)] \tag{3.10}$$

Let us consider an example of 2 pumps connected in a parallel configuration, and we need to find the reliability of this system for a mission time of 100 h (Fig. 3.8).

We assume that the failure rates for the two pumps are the same at 5×10^{-4}/yr;

The reliability of $P1$ and $P2$ considering an exponential distribution can be given as

$R(t) = 0.950$

From Eq. (3.10).

$$R_P(t) = 1 - [1 - R(t)]^2$$
$$R_P(t) = 1 - [1 - 0.950]^2$$
$$R_P(t) = 0.9975$$

This shows that when two components are connected in parallel, the reliability increases; as seen in this example, the reliability of one pump is 0.950, while for the two-pump configuration, it is 0.9975, i.e., two times with two pumps.

In the previous example, one should note that there is an assumption that all the components in the parallel system are independent. This is the reason that with each

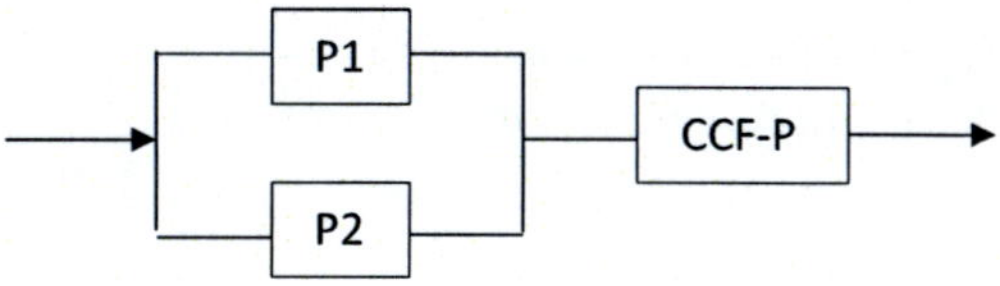

Fig. 3.9 Common cause failure consideration for parallel system reliability

additional component in a parallel system, the reliability increases by that many folds, i.e., twice for two components, three times with three components and so on. In actual practice, the reliability does not increase with each additional component, as there are many common factors that cause more than one component or all the components in parallel configuration. For example, the common source of power supply, the same source of input, the same make of the component, the same location that affects more than one component to fail, etc. In fact, human error has the potential for common cause failure, e.g., error in lubrication parallel unit, procedural error causing adversely affecting all the units in parallel system, similar maintenance error committed on all the units and so on. These factors result in the common cause failure (CCF) of more than one component in a parallel configuration. The common cause database shows that this common factor adversely affects system reliability. For example, if the common cause factor is 0.1, then the common cause failure rate will become 10% of the independent failure rate. The independent failure rate for the pump is 5 $\times$ 10^{-4}/yr; accordingly, the CCF rate for the pump will be 0.1 times (5 $\times$ 10^{-4}/yr) = 5 $\times$ 10^{-5}/yr.

If we redraw the reliability block diagram for the parallel system with CCF, we will see (Fig. 3.9).

This is a combination of series and parallel systems often referred to as complex systems. For the mission time of 100 h. The CCF probability is

$$R_{\text{CCF}} = 1 - \left[1 - \exp(-\lambda_{\text{CCF}} t)\right]$$
$$R_{\text{CCF}} = 1 - \left[1 - \exp\left[-\left(5 \times 10^{-5}\right) \times 100\right]\right]$$
$$= 1 - 4.98 \times 10^{-3} = 0.995$$

It follows that the parallel system reliability $R_P(t) = 0.9975$ is modified by a CCF-P of 0.995.

The reliability of the modified two-pump system becomes 0.9975 * 0.995 = 0.9925, which is less due to CCF. The CCF and other aspects will be discussed later in detail. In reliability analysis, CCF consideration is inevitable; otherwise, we obtain very optimistic results that could be misleading.

Standby System

In the parallel configuration, all the units were operating in parallel, and the system reliability was estimated with the failure of the last operation. The question of why to operate all the units as it is recourse consuming also results in aging of all the units.

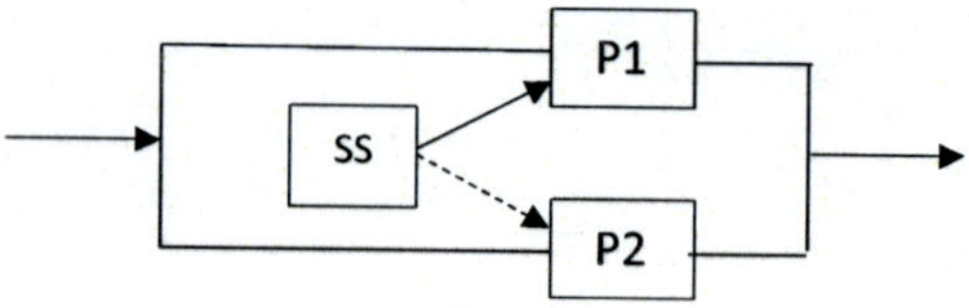

Fig. 3.10 Two-unit standby system with a switching unit

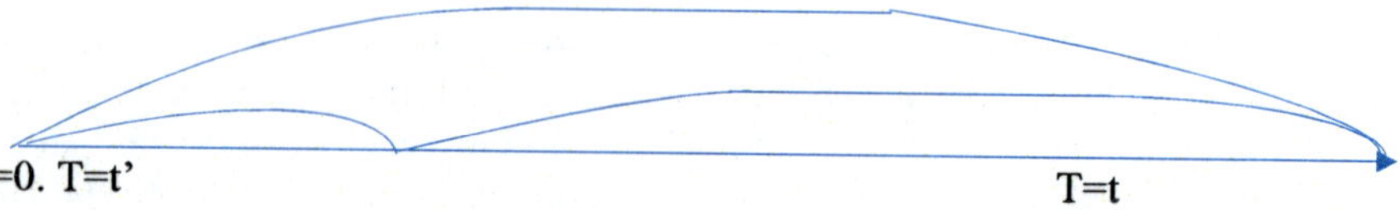

Fig. 3.11 Operational configuration of the standby system

Therefore, the standby configuration where only the required number of units are operated and if the operating unit fails, then the ideal or standby units are pressed in service by a sensing and switching system, which is similar to a watchdog monitoring the performance of the operating units, and in the case of failure of one or more than one unit, presses other standby units into operation such that the required number of unit conditions is satisfied.

Figure 3.10 illustrates a standby system of a pumping system configuration comprising two pumps, one operating and the other standby poised to be started in case of failure of the operating pump, by a sensing switching system.

The reliability of the two-unit standby system can be comprehended as follows: Either the operating unit successfully operates for the given mission time t OR the operating unit fails at time t' and the switching system senses the failure of the operating unit and sends commands to the standby unit and the standby unit has not failed in standby mode and starts successfully and meets the mission (Modarres et al. 2010) (Fig. 3.11).

For the purpose of writing a mathematical expression for the above condition, we assume that pump-1 was operating and pump-2 was operating in standby mode.

We create the following notations:

$R_1(t)$ = Reliability of pump-1 for a mission t;

λ_1 = Failure rate for pump-1;

$f_1(t)$ = failure probability density function for unit-1;

$R_{ss}(t')$ = *Reliability of switching system up to time* t';

$R_{2s}(t')$ = *Reliability of the 2nd unit in standby mode;*

$R_2(t-t')$ = *Reliability of unit 2 for* $(t-t')$;

The reliability equation for the above system is written as

$$R(t) = R_1(t) + \int_0^t f_1(t')dt' \cdot R_{ss}(t') \cdot R_{2s}(t') \cdot R_2(t-t')dt \tag{3.11}$$

$$R(t) = \exp(-\lambda_1 t) + \int_0^t \lambda_1 \cdot \exp(-\lambda_1 t')\cdot \exp(-\lambda_{ss} t') \cdot \exp$$
$$(-\lambda_{2s} t') \cdot \exp[-\lambda_2 (t - t')]\mathrm{d}t'$$

Performing the integration and rearranging the equation, we obtain

$$R(t) = \exp(-\lambda_1 t) + \frac{\lambda_1 \exp(-\lambda_2 t)}{\lambda_1 + \lambda_{ss} + \lambda_{2s} - \lambda_2}\{1 - \exp[-(\lambda_1 + \lambda_{ss} + \lambda_{2s} + \lambda_2)t]\}$$

To simplify the above equation. If we assume that the failure rate of the switching system and standby unit is 0, i.e., $\lambda_{ss} = \lambda_{2s} = 0$;

$$R(t) = \exp(-\lambda_1 t) + \frac{\lambda_1}{\lambda_1 - \lambda_2}[\exp(-\lambda_2 t) - \exp(-\lambda_1 t)] \quad (3.12)$$

If we assume that the two units have same failure rate, i.e., $\lambda_1 = \lambda_2 = \lambda$;

$$R(t) = \exp(-\lambda t) + \lambda t \exp(-\lambda t)$$

Or

$$R(t) = (1 + \mathrm{t}) \exp(-t) \quad (3.13)$$

This is the equation for standby system reliability considering that the failure rate of the units is the same and the failure rate for the switching system and standby unit in standby mode is zero.

3.4.3.4 Availability

Availability is the probability that the system will be available on demand to cater to the intended function. Availability is a function of the length of time the system remains down for repair and maintenance; hence, availability is the characteristic of a maintained system. Plant deliverables are a direct function of plant factors (Electric Power Research Institute 2004; Garrick et al. 2004). Often, for complex engineering systems where scheduled periodic maintenance or unwarranted repair is an integral part of the system operation, the availability factor provides important performance indicators of the dependability metrics. For a system, the availability is often defined as (Fig. 3.12):

$$\text{Availability}(t) = \frac{\text{Up - time}}{\text{Up - time} + \text{down - time}} \quad (3.14)$$

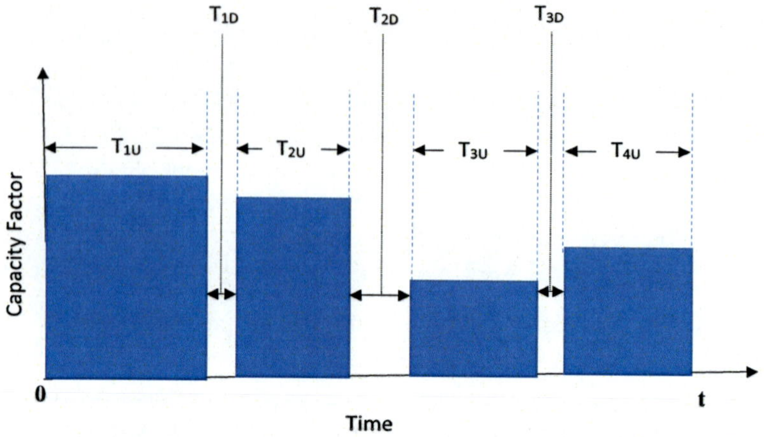

Fig. 3.12 Availability representation through the uptime and downtime of plants

The availability factor for the above operating and shutdown modes of the plant can be given by:

$$A(t)\frac{T_{1U} + T_{2U} + T_{3U} + T_{4U}}{(T_{1U} + T_{2U} + T_{3U} + T_{4U}) + (T_{1D} + T_{2D}T_{3D})}$$

Or

$$A(t)\frac{T_{1U} + T_{2U} + T_{3U} + T_{4U}}{(\text{Total Time})}$$

where T_{nU} and T_{nD} are the times when the system was up- or operating and down- or shutdown, respectively. The availability factor does not capture plant capacity factor information and only shows the ratio of the total time when the plant was in operation and the total duration under consideration. However, the availability factor shows that a series of shutdowns essentially includes the maintenance profile of the plant.

There are situations when a system availability assessment is required for a system having enhanced complexity; i.e., the system can have more than two states, the effect of repair and needs to be accounted for. In such situations, which require the modeling of dynamic states, state-space modeling employing Markov analysis is needed.

For the purpose of illustration, let us consider triple modular redundancy or two-out-of-three voting logic in a power station. The system is considered good if at least two out-of-three monitoring and protection channels are healthy. Therefore, the system is healthy if all three channels are healthy; if one channel fails or is removed for maintenance for a short time, the protection system permits this maintenance; however, the plant policy is to put back the channel in service after completing maintenance after a stipulated time as permitted by the plant policy.

There are the following states of the system:

State-1: **Healthy state**; all three channels are operating successfully.

State-2: **Degraded state**: System is operating with only two channels. One channel is out for repair/maintenance.

State-3: **Failed state**: Two or three channels failed.

Conditions

(1) The condition is that maintenance is permitted only for one channel at a time.
(2) Maintenance cannot bring the control logic to a healthy state from failed state.
(3) From state 1 to state 2, transition takes place by failure of any one of the three channels (channel *A*, or Channel *B* or Channel *C*: Failure).
(4) State 2 to state 3 transition takes place by failure of any of the two operating channels; Say if Channel *A*; has failed for transition from 1–2, the failure of Channel *B* or Channel *C* takes the system from the 2nd to 3rd state.
(5) State 1 to State 3 transition takes place by common cause failure of all the three channels.

The Markov model for the two-out-of-three voting logic can be developed as Fig. 3.13.

To write the differential equation for this system, we define

$$\frac{dP_1}{dt} = -3\lambda P_1(t) - \lambda_c P_1(t) + \mu P_2(t) \tag{3.15}$$

$$\frac{dP_2}{dt} = 3\lambda P_1(t) - \mu P_2(t) - 2P_2(t) \tag{3.16}$$

$$\frac{dP_3}{dt} = 2\lambda P_2(t) + \lambda_c P_1(t) \tag{3.17}$$

Solving the above equation for determining the probability of the system in States 1, 2 and 3;

We will use the Laplace transform approach to solve the above equation:

The initial condition at $t = 0$ probability of being in state 1 is; and at $t =$ infinity $P_1(t) = 0$

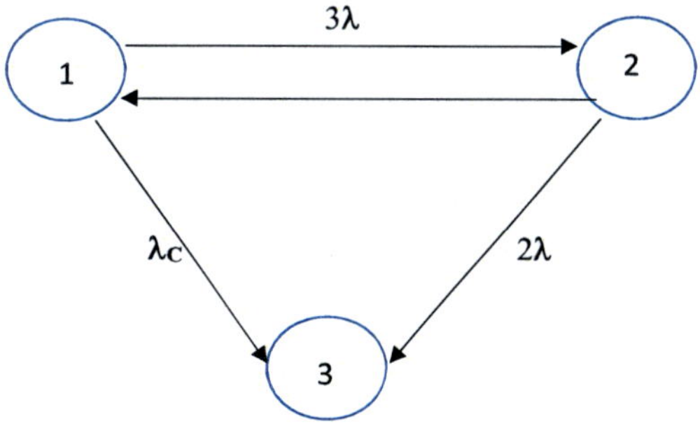

Fig. 3.13 Markov state-space modeling of 2-out-of-3 (*g*) modular redundancy

$$1 - sP_1(s) = -3P_1(s) - \lambda_c P_1(s) + \mu P_2(s) \quad (3.18)$$

$$sP_2(s) = 3P_1(s) - \mu P_2(s) - 2P_2(s) \quad (3.19)$$

$$sP_3(s) = 2\lambda P_2(s) + \lambda_c P_1(s) \quad (3.20)$$

Rearranging Eq. (3.18) as follows:

$$\begin{aligned} 1 &= sP_1(s) - 3P_1(s) - \lambda_c P_1(s) + \mu P_2(s) \\ 1 &= [s - 3 - \lambda_c]P_1(s) + \mu P_2(s) \end{aligned} \quad (3.21)$$

Rearranging Eq. (3.19) as follows:

$$\begin{aligned} 3P_1(s) - sP_2(s) - \mu P_2(s) - 2P_2(s) &= 0 \\ 3P_1(s) - [s + \mu + 2]P_2(s) &= 0 \end{aligned} \quad (3.22)$$

Rearranging Eq. (3.20) as follows:

$$\lambda_c P_1(s) + 2\lambda P_2(s) - sP_3(s) = 0 \quad (3.23)$$

The above, (3.21)–(3.23), are now athematic equations with three variables that can be solved for evaluation by arriving at the solution of the Laplace equation and then taking the inverse-Laplace to evaluate the probability of the system being in states 1, 2 and 3. Here, the probability for State 1 and 2, i.e., $P_1(t) + P_2(t)$ will give us system Availability of the 2-out-of-3 (G) voting logic (with an assumption that plant policy permits one channel maintenance for a preset time and this consideration has been factored in here for considering state 2 for defining availability) and the probability of the system reaching state 3; i.e., $P_3(t)$ will give an estimate of system unavailability.

3.4.3.5 Maintainability

Maintainability has both deterministic and probabilistic components. The considerations of maintainability form part of the design right from the conceptual stage through preliminary design, final design and later as part of construction, commissioning and later operations. Maintainability consideration during initial stages provides maximum returns, and later, any new considerations of maintainability pose challenges ranging from being prohibitive to very expensive in terms of large investment of resources and at times compromising plant performance indicators such as safety, capacity factor, availability factor, deliverables and so on.

One of the major issues in design for maintainability is lack of experience of the design team in operation and maintenance aspects of the plant. This suggests that experienced O&M should be involved in the design, construction and commissioning stages of the plant. Another challenge that the design team faces is postulation of the actual operating condition down the line couple of years to when the plant is facing an aging problem. For example, locating a heat exchange in a location where in situ servicing poses a challenge, locating a valve with a spindle location such that operation of the valve poses a perineal challenge, locating underground drain lines without much consideration of future access to these locations, locating a piping system in such locations where in-service inspection becomes either prohibitive or poses challenges of developing costly and complex technology that often becomes prohibitive. Maintainability also means making all the repair and overhauling support resources, including maintenance stations with adequate space, finances, adequate inventory, and trained manpower resources and procedures.

At the outset, it can be appreciated that the maintainability is central to good operability and finally achieving high-performance indicators, such as availability, reliable operation, safety and mainly plant ability to deliver the intended products in a timely manner.

Considering the above, maintainability can be defined as the capability of a maintenance and operating environment to facilitate the high-quality maintenance function in an efficient and effective manner such that equipment and services are restored to operational readiness in a timely manner. In the above definition, the term 'capability' is replaced by 'probability', and minor rephrasing can be performed to mathematically facilitate the formulation of maintainability. Accordingly, maintainability is defined as the *probability that a maintenance and operating environment will facilitate the implementation of a high-quality maintenance function in an efficient and effective manner such that equipment and services are restored to operational readiness in a timely manner.*

The above conditions are a prerequisite to define maintainability in a probabilistic sense; therefore, the maintainability $M(t)$ as a function of time, referred to as repair time, can be defined as:

$$M(t) = P(T \leq t \| C1, C2 \ldots) \tag{3.24}$$

P is the probability, and T = a random variable representing repair times.

The repair, overhaul time of equipment has been observed to be random due to reasons that can be attributed to maintenance conditions, level of damage, number and expertise of manpower, and tools and fixture availability, to name a few.

This led us to consider the probability density function of the repair time $g(t)$. The repair rate μ/hr, performed on the equipment, provides us with another parameter that is required to define the maintainability.

$$M(t) = \int_0^t g(t)\mathrm{d}t \tag{3.25}$$

Equation (3.25) is the probability that a repair will be performed within time t and is basically the definition of maintainability. Furthermore, if the repair rate μ is exponentially distributed, then the pdf of repair can be given as

$$g(t) = \mu \exp(-\mu t) \tag{3.26}$$

$$M(t) = \int_0^t \mu \cdot \exp(-\mu t)\mathrm{d}t \tag{3.27}$$

Integrating Eq. (3.27) yields the expression for maintainability for constant repair rate, i.e., exponential distribution as follows:

$$M(t) = 1 - \exp(-\mu t) \tag{3.28}$$

Furthermore, to evaluate the expression for Mean Time to Repair (MTTR), we proceed

$$\mathrm{MTTR} = \int_0^\infty t \cdot g(t)\mathrm{d}t$$

$$\mathrm{MTTR} = \int_0^\infty t \cdot \mu \cdot \exp(-t)\mathrm{d}t$$

The above equation is the product of two parts: t (first terms and exponential part the second term, so we do integration by part as follows:

$$\mathrm{MTTR} = \left\{ t \cdot \frac{\exp(-\mu t)}{-\mu} - \cdot \frac{\exp(-\mu t)}{-\mu} \right\}$$

Putting the limit values ∞ and 0, we obtain

$$\mathrm{MTTR} = \frac{1}{\mu}$$

Discussion: The above discussion on maintainability brings out the following points: (a) Maintenance management is essentially considered an organizational feature more than location-specific or equipment repairs/overhauls being performed by maintenance teams. There is an important observation made that often the errors are committed while reassembling back the equipment under maintenance. The role of human factors and supervision is critical, particularly when maintenance activities are coming toward completion. The implicit assumption in the above model is that there are established procedures for each of the maintenance activities that need to be followed. Apart from this, there should be a database of maintenance activities where data logging should be performed for the repair time for various equipment. This activity can be done as part of reliability centered such that we can qualify, or benchmark the maintainability as one of the performance indicators of the plant. Periodic review of the data on unduly trending resource requirements and equipment performance based on the given maintenance strategy can provide invaluable insight into the strength and weaknesses of the maintenance management program.

3.4.3.6 Probabilistic Risk Assessment

Probabilistic risk assessment (PRA) is an analytical approach to estimate the risk from a complex engineering system by postulating potential accident scenarios and assessing the response of the plant to provide a quantitative and integrated statement of risk. This approach is now part of residual risk assessment for mission critical, safety critical, process and chemical plants. The PRA technology is considered sufficiently mature to be used as risk-informed/risk-based applications. The major steps involved in the performance of PRA are:

1. Plant familiarization and reliability data collection
2. Formulation of list of applicable initiating events
3. Accident sequence modeling
4. System analysis
5. Data collection and analysis
6. Human reliability analysis
7. Common Cause Failure Analysis
8. Quantification of the plant model
9. Uncertainty and sensitivity Analysis
10. Interpretation of the results of the analysis and documentation.

The PRA project starts with submission of a project report by a PRA team member comprising reliability/PRA experts, data analysis, design or operation team members led by a project leader who has good experience with PRA modeling-associated requirements. The performance of PRA is a resource-intensive activity; therefore, the support, involvement and approval of the leadership in an organization are a key element for the success of the project. The PRA team, along with PRA modeling, should also extend the scope to include the development of risk-informed/risk-based applications. This serves two purposes: (a) management will always look forward to some applications, such as risk-informed In-service Inspection, that might bring down inspection efforts, and (b) through the application, the model tends to be validated, reflecting the as-built features of the plant.

The drafting of the project report and plant familiarization are the two tasks that can go simultaneously. Presentations on plant siting, design and operational, plant walk down, study of safety analysis report, design documents, operational documents, such as technical specifications, plant procedures, emergency operating procedures,

Event Sequence Modeling

The objective of formulating a list of initiating events is to make the list as complete as possible by postulating all the internal as well as external transients and conditions. A

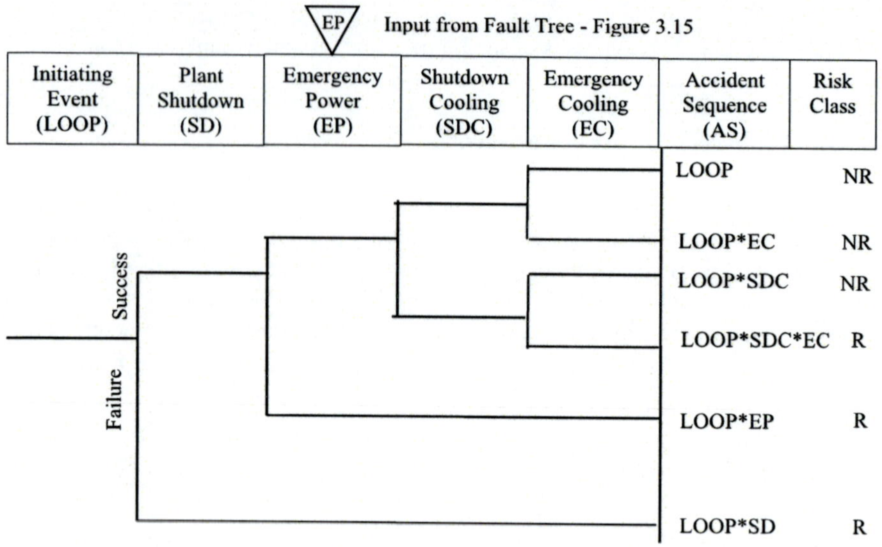

Fig. 3.14 Accident sequence modeling employing the event tree approach

preliminary hazard analysis of the plant provides, as a first step to kick-start the work on development of list of initiating events. There are two major steps, first to make the list based on the information available from all the sources. The plant operating experience, the list of initiating events from generic sources, and the experience of similar plants. Of course, if the PRA is being performed for a first of a kind design, then the job becomes slightly more challenging; however, the logic model of the plant and other insights along with the information available from generic sources will help in drafting a list of initiating events. The initial list is subjected to screening analysis to ensure which event should be considered for detailed analysis using the screening criteria. And what are those events they do not need detailed modeling consideration and can be closed by deterministic or probabilistic considerations. A final list of initiating events and the information on the safety function of the plant that includes safety systems and engineering safety features and human actions along with success criteria or failure criteria enable the PRA team to start the work on accident sequence modeling. Event tree modeling is one of the most elegant and well-accepted graphic frameworks to propagate plant responses to an initiating event. Event trees are developed for all the initiating events. The output of the accident sequence modeling is a list of accident sequences. For example, let us see an event tree for an initiating event, loss of off-site power (LOOP), as follows (Fig. 3.14).

The event tree is an inductive graphical technique to propagate the accident scenario based on the plant response to an initiating event. The safety functions are arranged in a chronological sequence, and the success/failure of these safety functions determines the stages of the plant state. This event tree has been developed to demonstrate the accident sequence development for an initiating event—loss of off-site power (LOOP) to the plant. The top three accident sequences (LOOP, LOOP*EC, LOOP*SDC) indicate that there is no risk (NR) to the plant as shutdown and cooling functions are met successfully. The last three accident sequences (LOOP*SDC*EC, LOOP*EP, LOOP*SD) have risk implications, as in two accident sequences, cooling of the plant is adversely affected due to unavailability of electric power, and in the last accident sequence, the shutdown function failed and poses a risk to the plant. The instantaneous plant response is to shut down the reactor. For the quantification of accident sequences, please refer to the Event Tree/Accident Sequence section in Chap. 5.

System Analysis

System analysis addresses the unavailability modeling of engineering systems, particularly safety systems, engineering safety features and safety support systems. The role of these systems is to remain in standby mode and actuated by normally automatic command-generated control and protection system-based sensing of failure or inadequate operation of the process system. As seen, in the event tree shown in Fig. 3.15, the safety systems are arranged, as seen in the header boxes on top in an order they are required to meet the safety function. The plant shutdown system (SD) and shutdown cooling system (SDC) are safety systems and are required whenever even in the normal operational phase of the plant, they are required to shut down and maintain in the shutdown state, while emergency power (EP) comprising emergency diesel generator sets is a safety support system. The emergency cooling system (EC) is an engineering safety feature in the plant and is required only during emergency conditions when the normal process cooling system and shutdown cooling system fail to function, which is a low probability scenario.

As part of the system modeling program, the unavailability analysis is performed, normally employing fault tree analysis. Therefore, as part of the PRA program, system analysis is performed for the safety systems shown as the header row of the event tree, system modeling. In this section, a simplified fault tree is shown for the emergency power system of the plant.

The emergency power (EP) system comprises three Diesel Generator (DG) sets in parallel configuration. The individual DG set has capacity such that even one DG set operation is adequate to cater to the shutdown loads of the plant. The definition of failure criteria is required for system modeling. The system failure criteria is 3-out-of-3 DG failure or, conversely, the system success criteria is at least 1-out-of-3 operation such that all shutdown shut down load requirements are met. The second requirement is defining the TOP event in the fault tree. The fault tree for the EP system is shown in Fig. 3.16.

The fault tree is a top-down approach; therefore, we start from the top event 'Emergency Power' (EP). In this modeling, the emergency power failure can occur due to (a) DG-Independent failure as shown in the left intermediate event OR (OR-gate) (b) due to CCF of all the DGs simultaneously.

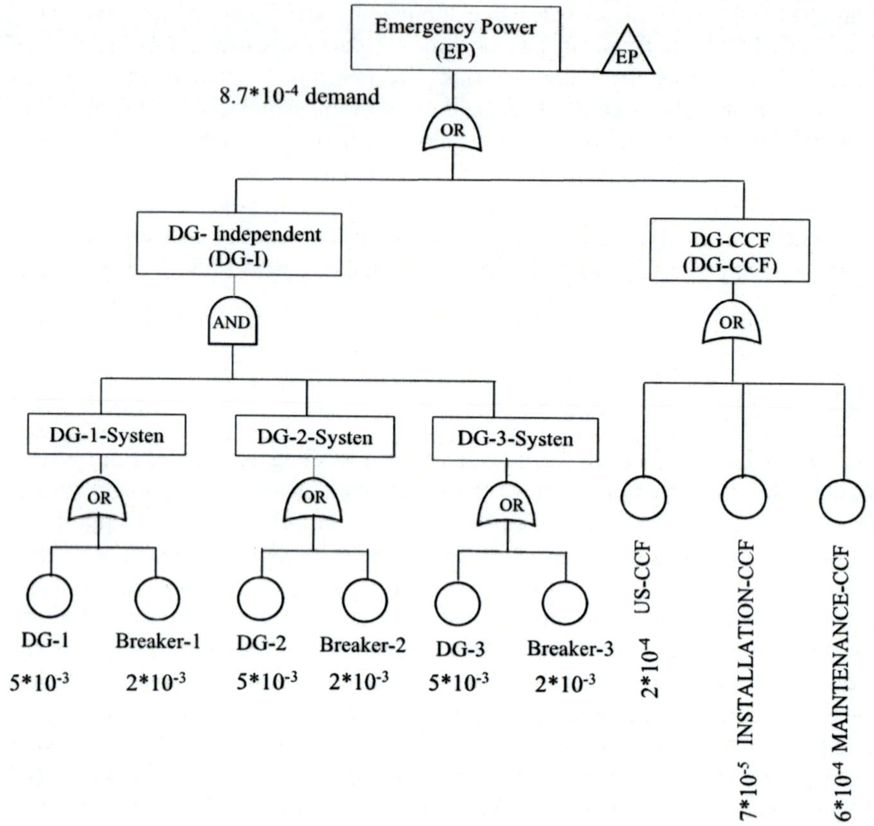

Fig. 3.15 Fault tree for the emergency power (EP) system

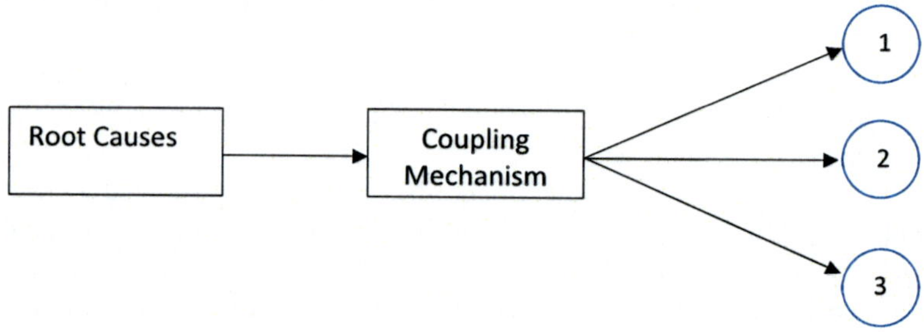

Fig. 3.16 Role of root cause and coupling mechanism in CCF events (International Atomic Energy Agency 1992)

Common Cause Failure CCF Modeling PSA. There are two types of CCF modeling: (a) explicit CCF and (b) implicit CCF. The explicit CCF events are those that are modeled as dependent events, e.g., power supply to redundant pumps; these are called systematic events and are epistemic events. These events occur basically due to lack of knowledge. The other, i.e., implicit CCF, falls under the category that is due to random or unforeseen phenomena, and in the analysis, it is accounted for separately. Here, the adequacy of the defense provision for CCF is checked.

In fault tree analysis, we stop going down when we arrive at the basic event shown by the circle.

We write the equation of the Top event as follows:

$$
\begin{aligned}
T(\text{EP}) &= \text{DG} - \text{I} + \text{DG} - \text{CCF} \\
&= \left[(\text{DG} - 1 - \text{System}) \cdot (\text{DG} - 2 - \text{System}) \cdot (\text{DG} - 3 - \text{System})\right] + 1 \\
&[\text{BUS} - \text{CCF} + \text{INSTALLTION} - \text{CCF} + \text{MAINTENANCE} - \text{CCF}] \\
&= [(\text{DG} - 1 + \text{Breaker} - 1) \cdot (\text{DG} - 2 + \text{Breaker} - 2) \cdot (\text{DG} - 3 + \text{Breaker} - 3)] \\
&+ [\text{BUS} - \text{CCF} + \text{INSTALLTION} - \text{CCF} + \text{MAINTENANCE} - \text{CCF}); \\
&= (\text{DG} - 1 \cdot \text{DG} - 2 + \text{DG} - 1 \cdot \text{Breaker} - 2 + \text{Breaker} - 1 \cdot \text{DG} - 2 \\
&+ \text{Breaker} - 1 \cdot \text{Breaker} - 2). \\
&(\text{DG} - 3 + \text{Breaker} - 3) + [\text{BUS} - \text{CCF} + \text{INSTALLTION} - \text{CCF} \\
&+ \text{MAINTENANCE} - \text{CCF}] \\
&= [\text{DG} - 1 \cdot \text{DG} - 2 \cdot \text{DG} - 3 + \text{DG} - 1 \cdot \text{Breaker} - 2 \cdot \text{DG} - 3 + \text{Breaker} - 1 \\
&\cdot \text{DG} - 2 \cdot \text{DG} - 3 + \text{Breaker} - 1 \cdot \text{Breaker} - 2 \cdot \text{DG} - 3] + [\text{DG} - 1 \cdot \text{DG} - 2 \\
&\cdot \text{Breaker} - 2 + \text{DG} - 1 \cdot \text{Breaker} - 2 \cdot \text{Breaker} - 3 + \text{Breaker} - 1 + \text{DG} - 2 \\
&\cdot \text{Breaker} - 3 + \text{Breaker} - 1 \cdot \text{Breaker} - 2 \cdot \text{Breaker} - 3] + \text{BUS} - \text{CCF} \\
&+ \text{INSTALLTION} - \text{CCF} + \text{MAINTENANCE} - \text{CCF};
\end{aligned}
$$

The next step is data collection for the basic events shown in the fault tree by circles, such as DG-1, Breaker-1, DG-2, Breaker-2 and BUS-CCF. For the present case, let us assume that the analysts arrived at failure probability from some identified generic source. Additionally, we assume that the DG-1, DG-2, DG-3 failure probability is the same = 5×10^{-3}/demand, and the Breaker-1, Breaker-2 and Breaker-3 failure probability is also the same = 2×10^{-3}/demand.

CCF for BUS-CCF, INSTALLATION-CCF and MAINTENANCE-CCF arrived from a generic data source as 2×10^{-4}, 7×10^{-5}, 6×10^{-4};

Using these data, we arrived at the top event, i.e., the EP system failure probability was 8.7×10^{-4}/demand. The CCF factor is a dominant contributor to the top event. The essential aspects related to CCF modeling are discussed in the net section as special areas in risk and reliability modeling.

The failure probability of the top event is plugged into the event tree top header event, EP. Along similar lines, the system failure probability for other events, SD, SDC and EC, also plugged into the event tree. Furthermore, it was found from the plant record that the failure frequency of off-site power is 1.0/yr. This value is used for the LOOP in the event tree.

Impact of Data and Interpretation of Results

This section has been included to highlight certain associated aspects that are vital for appreciating and interpreting the results, which is an important consideration in PRA modeling. First, data collection and analysis are resource-intensive tasks, and the quality of data impacts the results of the analysis. Second, the lack of modeling and analysis of common cause failure and further in-depth coupled with quality of data directly impact the results of the analysis. The experience suggests that in safety critical systems, the CCF contribution, as we have observed in the given fault tree, dominates the results. Furthermore, the human error probability forms a critical component of the analysis, as many human errors could have a CCF component. Next, the output of the fault tree forms input into the event tree analysis for one of the header events. This, along with quantified input for initiating events as well as other safety systems or recovery failure probabilities, enables the evaluation of the accident frequency of various accident sequences. Finally, we have not considered uncertainty considerations in this section, and it has been assumed that the point estimates (median/mean) are acceptable as the final result. However, uncertainty and sensitivity analysis also form part of the PRA modeling.

3.4.3.7 Probabilistic Security Risk Assessment

Probabilistic risk assessment is an established safety assessment tool in the nuclear industry. Interest in performing security risk assessment is also growing in the nuclear industry. Particularly, after the 9/11 terrorist attack on the world trade center, the safety and security community and the governments and international organizations, such as the IAEA, revisited their models, procedures and compliance guidelines, such that the probability of security risk was reduced as much as possible. The available literature shows this reaction to the evolving security scenario has proven to be a catalyst for fine-tuning or even developing advanced strategies and methodologies for security risk assessment. Some of these models use probabilistic metrics in support of creating the risk-informed framework, and there are concerns raised about adopting the PRA framework for security assessment (https://www.ncbi.nlm.nih.gov/books/NBK373719/). One of the major concerns is that modeling for threat or design basis threat (DBT) is not exactly as straightforward as the consideration of accident sequence PRA, where the response of a safety system in response to an initiating event is central to providing the risk scenario in terms of a combination of events. The reason is subjectivity, as security events are subjective matter where considerations of human factors for performing an intended malicious act pose a challenge for modeling threat perception.

A detailed study to adopt the PRA framework for probabilistic security risk assessment (PSRA) is a viable option. There appears to be a consensus that security modeling, even while using probability metrics in a limited sense, for example probability of detection, probability of success of stealth acts, probability of crossing manned barriers, etc., probability of employing forged entry documents or even defeating information security barriers have been used to model scenario. Furthermore, the probability for modeling the success/failure of countermeasures on a given path or multiple routes requires data and tests, and expert opinions have been used.

It is recognized that in PSRA, there are certain tasks/analyses that require attention and might involve special tools and software. For example, developing the approach routes to identified facilities on a campus and associated success/failure based on expert opinion but simulation software is required to validate the model. However, these tasks are performed in security risk passement using either deterministic or probabilistic approaches, and the procedures are available in the open literature. Here, we identify the major challenges vis-à-vis the approach adopted to address the challenges for major steps of probabilistic risk assessment as discussed in the preceding section.

3.5 Special Areas

There are some aspects in dependability analysis that need to be discussed in this chapter, as the quality of analysis depends on these factors. For example, the appropriate factoring in of common cause failure may produce overly optimistic results that defeat the purpose of the analysis. Even though root cause analysis is not performed for all failures, it is only performed for potential risk-significant events, and adequately performed RCS is critical to plant operations derisking. The quality of data is vital for the analysis, as the insights developed from the study are used for evaluating the risk ranking of the SSCs for developing the surveillance program in general and maintenance and in-service inspection program in particular. In fact, these insights into identification and prioritization provide required input for aging management and life extension programs. The discussion will not be complete without exploring the role of intelligent or machine learning approaches in dependability analysis.

3.5.1 Common Cause Failure Analysis

The term common cause failure (CCF) as defined in IAEA Glossary as '*Failures of two or more structures, systems or components due to a single specific event or cause*' (International Atomic Energy Agency 2018). However, from risk considerations, the instant of failure is also relevant. Therefore, for the purpose of the RCOM approach, the CCF definition has been formulated as '*failures of two or more structures, systems or components due to a single specific event or cause during a short interval such that it has risk implications'. The definition of duration is specific to the situation or threat for which the CCF is being considered.* CCF analysis is a key exercise that brings out the dependency in the system (Table 3.2).

In fact, without the considerations of CCF, the reliability analysis might provide very optimistic results and could be misleading. It would not be an exaggeration to state that without CCF considerations no results should be accepted. Therefore, considerations of common cause failure (CCF) are an integral part of the design process, particularly for safety systems that include safety support systems (International Atomic Energy Agency 2004b).

The importance of CCF is highlighted by the fact that apart from safety analysis, an independent CCF analysis is often performed to ensure potential degradation in redundancy, diversity and fail-safe provision in the plant by simulating all postulated accident scenarios. As seen, this is not a safety or specifically a regulatory requirement but a risk-conscious culture where a separate document, such as a safety analysis report, is developed based on CCF analysis modeling and insights toward ensuring that CCF does not form a major contributor to the accidents. CCFs are those single failures or events that can instantaneously fail redundant provisions and defeat the very purpose of redundant systems in the plant. The design of the system configurations, layout, physical separation and so on are the deterministic defenses against the CCF events. For example, if there are three redundant pumps in an emergency train, then the source of power is from three different sources, e.g., emergency diesel generator (DG) sets, so that there is an independence of the power supply for the redundant. Unit. Similarly, the diesel generators are located at different locations so that local events do not affect all the DGs. For example, out-of-three DGs, two DGs are kept away in a separate building at higher elevations so that flooding events do not affect the DGs kept at higher elevations. Furthermore, the two DGs located in the same building are physically separated such that any fire event does not affect another DG. The two injection trains connected to the cooling systems for injection as well as the compensating function of the coolant system during a breach in the coolant system are laid out through two different routes such that any physical problem in one location will not affect either train. The four protection system channels are placed out such that there is a sufficient gap between the four, protection cabinet and provision of fire protection for the postulated fire category. Human error could also be a source of CCF events, including calibration error that disables the redundant functions, leaving redundant pump breaker breakers control in the wrong position, say test position after maintenance, if not noticed in subsequent surveillance, disable the pumps on demand to switch on.

In safety analysis, the plant configurations are checked for potential common cause failure events, and corrective actions are initiated to correct the CCFs. Often, CCF factors are revealed during the operational state of the plant. These events are treated as significant safety events, and plant restartup requires regulatory clearance.

Table 3.2 Identification of challenges and feasibility of addressing the issues while implementing PRA procedures for PSRA

S. No.	Major procedural steps in PRA	Challenge	Possible solution or identifying advantage, compensating provisions, bottleneck areas
1	Plant familiarization	• Plant and site layout familiarization and identification of security risk-significant features • Safety and security overlap identification	The scope of walkdown and survey goes beyond plant to include site features, that include pathways, physical protection systems that include barrier and levels, adequacy, and availability of hardware-based, electronic, information management features as also location and adequacy of countermeasures provisions *Record recommendations for improvement based on the walkdown and surveillance. Organizational provisions to ensure implementation of security-related risk-conscious culture* *Identify the plant systems, features, actions where the safety and security conflict exist*
2	Formulation of list of applicable initiating events	Design basis threat formulation, that might include outside/inside threat, applicability of cyber security evaluation requirements, impact of object from outside aircraft crash, etc., requirements related to evaluation of electromagnetic pulse (EMP) from over, off-site, outside/inside, etc.	Wherever, it is not possible to evaluate the probability of postulation event a conditional probability of 1.0 should be assumed. Science-based approaches will be required to assess vulnerability of the plant to various structural and EMP attacks. The regular evaluation of surveillance and intelligence insight is often more critical than probability *The site should be demonstrated to be protected against the design basis threat scenario either by simulation or mockup and drills* *Generate recommendations based on the observations made during the step*
3	Accident sequence modeling	Evaluation of human error probability (random) and intended or deliberate human actions	Use of dynamic event tree provides an effective mechanism for security sequences modeling. Use of fuzzy approach for human error probability where event-specific data are not available, for intended or deliberate unethical actions to cause failure the probability of 1 should be used unless we have adequate data on how ill-intended action can fail *Again, these probabilities can be evaluated to be realistic based on expert opinion evaluation*

(continued)

Table 3.2 (continued)

S. No.	Major procedural steps in PRA	Challenge	Possible solution or identifying advantage, compensating provisions, bottleneck areas
4	Physical protection system analysis	There are two aspects physical barrier and protection, cyber security features and manning that include surveillance, supervision, intelligence gathering; Role of insider outside nexus could defeat the series of protection measures	Fault tree, failure mode and effect analysis, etc., like PRA are applicable to PSRA also. Literature shows the models and methods for defeating the barriers, breaking locks, force opening of doors, disabling audio–video protection measures. Monitoring of attitude, integrity levels, lapse in duty, behavioral problems often can be detected *Analysis of probability of failure for physical system is rather more in line with safety system analysis baring some specific aspect to security systems*
5	Data collection and analysis	What is relevant and what is not; what is adequate and what not. Data with relatively higher uncertainty	Involvement of defense and security expert along with plant and organizational team, involving utility service staff, administrative staff, intelligence data collection system, organizational and national conscience should be factored in while finalizing the data sets. Higher uncertainty is better than no data as at least qualitative input like accidence sequences can be used for system backfitting, retrofits and modifications *Data collection and analysis acquire consumes maximum resources particularly human-hours and testing even in PRA. In case of PSRA it further poses many constraints*
6	Human reliability analysis	Same as item 3 in respect of human random and intended actions	Fuzzy logic approach, expert opinion and data are available in security or safety domain from other systems and events can be used *Fuzzy logic provides an elegant mechanism to account for qualification based on qualitative input*

(continued)

Table 3.2 (continued)

S. No.	Major procedural steps in PRA	Challenge	Possible solution or identifying advantage, compensating provisions, bottleneck areas
7	Common cause failure analysis	Availability of data on CCF for random failure CCF is a term more relevant to random failure of multiple equipment due to a single cause in a relatively short span of time such that safety is defeated	International and national level awareness is there regarding inadequacy of data on CCF, and extensive efforts are on hand for over two decades targeted at safety (US Nuclear Regulatory Commission 2007) improvement. International Atomic Energy Agency (2009), Nuclear Energy Agency (2008) CCFs are more relevant to protection functions. For intended or deliberate actions, the organizational culture could be the CCF. That is typically, deficiency is there, but protection adequacy is less. CCF is applicable to analysis of multiple barriers and levels of protection measures, same team/staff performing same or similar wrong action in more for redundant features, e.g., wrong setting of alarms in more than one location of for redundant features *Often the CCF is termed as monsters as it eats away all the redundancy provided for assuring higher level of safety and security. The focus should be on defense against the CCF in the system—a deterministic view*
8	Quantification of the plant model	Even after all the efforts few data are not available	Assume the data and uncertainty bounds based on expert opinion. This step is related to generating quantified output of the model using the quantified data for hardware, software and human actions random and intended one, that not only presents to the team the point value but also uncertainty bound *Data collection is time-consuming but assuring quality of data is key to final results, to counter garbage-in-garbage out, involve domain as also statistical and reliability expert. To some extent the shortage of data can be compensated by applying Bayesian approach as matter of policy where the plant-specific data can be combined with generic data to implement the best-estimate approach*

(continued)

Table 3.2 (continued)

S. No.	Major procedural steps in PRA	Challenge	Possible solution or identifying advantage, compensating provisions, bottleneck areas
9	Sensitivity analysis	Requires performance of simulations considering the events with rather higher uncertainty bounds and assumptions or assumed data	This step helps in understanding the sensitivity of given assumptions and uncertainties. Certainly, there will be cases where they are high impact items. In such cases experiments involving life testing, creating demands on a mockup facility, simulations employing physics of failure studies. For example, effect of higher temperature due to fire on functional capability of a protection item can be better understood by file simulation or experiments at different temperature severity. Same argument applies to higher humidity, dust, etc. There are test chambers available for life testing in hostile environment *Use the new insight for fine-tuning the security risk model*
10	Interpretation of results of the analysis and documentation	Often it is not all that easy to interpret what the model is trying to communicate	Involvement of subject expert like security, defense and vigilance staff is vital for this step *Interpretation of result is key to utilization of the insights of the study for incorporating modifications, retrofits, performance of detailed aging studies. PSRA project and modifications can go hand-in-hand as the analysis proceeds. This is an efficient approach as many shortcomings are noted during plant walkdowns and initial analysis*

Dependent failures, and more specifically CCFs, can therefore be thought of as resulting from the concurrence of two factors, one that provides a susceptibility for components to fail or to be unavailable from a particular root cause of failure and a coupling mechanism that creates the conditions for multiple components to be affected by the same cause. This may be illustrated as follows: For dependent failures or more specifically cause failure, can be considered to occur because of two conditions or factors, (a) root cause of failure and (b) coupling mechanism. For example, failure of all the redundant trains of protection occurs due to an increase in humidity in the area. Here, the increase in humidity is the coupling mechanism, while the aging of the relay that led to the failure of the relay in stuck-closed conditions is the root cause. The root cause is the most basic reason for a component failure, which if corrected, could prevent recurrence of the fault/failure. There could be one or more root causes associated with the failure. The root causes also depend on the level of depth/details being investigated.

At the outset, all the root causes have their roots at human error, be it in design, fabrication, operations (actions and procedures) or maintenance practices. The coupling mechanism is the process that causes multiple degradation or impairments together and identifies the influences that created conditions for multiple components to be affected [NEA]. As seen in Fig. 3.16, the root cause along with the coupling mechanism, if it becomes reality, the results in otherwise redundant units failing simultaneously. Probabilistic risk assessment (PRA) modeling enables quantitative assessment and incorporation of the CCF contribution of the SSCs into the risk model of the plant. As mentioned earlier, generally, the experience is that CCF events are the major contributor to safety system unavailability and thereby become major contributors to the risk. Quantification of. CCF events enable the prioritization of the CCF list and accordingly help design and operating organizations provide insights into eliminating or reducing the potential of CCF weakness as part of safety management programs.

We will discuss the most conservative and used model, i.e., β-Factor model as follows. β-Factor is the ratio of failure rate due to CCF events and total failure rate. Accordingly,

$$\beta = \frac{\lambda_c}{\lambda}$$

λ_c: CCF failure rate of the component
$\boldsymbol{\lambda}$: Total failure rate,
Here $\lambda = \lambda_i + \lambda_c$.
λ_i: Independent failure rate of the component

$$\beta = \frac{\lambda_c}{\lambda_i + \lambda_c}$$

CCF modeling and analysis are performed as part of probabilistic risk assessment (PRA). The PRA insights show that CCF events are the dominant contributor to the risk, i.e., to the core damage frequency (main result of Level 1 PRA) of the plant.

3.5.2 Human Reliability

Assessment of human reliability or otherwise round the human error rate is more complex than the hardware systems. The reason is that humans are a spirit. Consciousness, cognition and conscience (ethics and morals) are complex factors that determine human response, unlike hardware, where the absence of this variability makes jobs relatively simple. As the literature shows, predicting the human response is not straightforward. As the human response can be influenced by complex psychological and phonological factors, level of consciousness and cognitive functions at the level of brain. Chapter 7 on human factors in operation is dedicated to human reliability in operations. Here, we will have a limited discussion as to how human error forms part of dependability. Let us consider an example of the human action associated with the manual 'opening' of a valve in a hydraulic line. The relatively straightforward procedure that is in line with the hardware system involves the estimation of human probability by the data collected from the field. Here, the probability of human error in opening a valve is computed by using the data on the total number of opportunities available for a given human action in opening the valve and the total number of failures committed in opening the valve. The human error probability in this case is estimated as follows:

$$\text{Human error probability (foraopening the valve)} = \frac{\text{Total number of failure}}{\text{Total number of opportunity}} \tag{3.29}$$

This estimation is applicable to reference conditions and based on the assumptions that all the *C*-factors (consciousness, cognition and conscience) have normal/reference states, the stress levels (normal operation, training levels (as the staff is qualified for the given activity) and learning factors (adequate experience is available) as well as operational (operator is not stressed) and environmental stresses are normal (no harsh environment to adversely affect human performance). However, for argument's sake, if the data collected implicitly reflects the applicable factors, for a given application, then this approach is preferred. For example, the data collected from the record on the opportunity for opening a valve and the total number of failures in similar conditions that are postulated for scenario modeling or factoring in risk assessment are best suited to the above model. However, in risk or reliability modeling, the data collected from the field might not reflect the emergency scenario being developed. In fact, data availability per say is often a challenge. Therefore, the above approach has limited application, and a general handbook approach is used. In this approach, the base human error probability is estimated based on the associated human stress level, quality of the interface, etc. However, at present, we will continue to focus on the fundamental definition of human reliability.

Here, there is the question 'Can we look at the "success" domain to define human reliability and then estimate the probability of human "failure"?'. For example, proposing human reliability 'Probability of completion of a given task in a given period of time under stated condition' in line with the definition of hardware systems. Even the consideration of the 'time' factor is reasonable, as the probability of human failure depends on the availability of a time window relative to the time required to act. However, one aspect separates the 'human' with 'hardware'—human has consciousness as the fundamental entity that works at the level of cognition brain or mind. That means the level of consciousness, here for argument we simplify interpret as 'awareness', i.e., awareness, of his motivation, information about the plant and working of valve as also the signal or feedback parameter he should look at like open/close indicator, reading of pressure, flow in the line, feel of the force he should have when opening of the valve and the most importance of the action and interconnection/interdependency of this action on other systems, if any. The above information is processed internally in a human being, unlike a hardware component.

The above discussion shows that there is a need to learn from the present human reliability models and complement these with consciousness and cognition and sensory information to give a model for human reliability. Accordingly, the reliability for human action is given as

$$R(t) = P(f(t, t_w)C_s, C_x, C_o) \tag{3.30}$$

Assume that the ratio of t and t_w follows an exponential distribution

$$R(t) = Exp\{(-[t, t_w])/C_s, C_x, C_o\} \tag{3.31}$$

$R(t)$: Reliability for a given human action in time t_a.

t: time required to complete the human action.

t_w: Time window available for completion of action.

$P(f(t_w, t_a))$: Probability of successful completion of action as a function of given action time and available time window t_w;

C_s: Consciousness human internal conditions:

C_x: External condition or context.

C_o: Organizational Consciousness.

This research assumes that consciousness is fundamental to defining human beings and further supports the functioning of cognition, mind and conscience quality. Consciousness also provides a reference to further define the subconsciousness, unconsciousness and other deeper realities of human existence. For the purpose of the above human reliability modeling, the term consciousness, C_s, might capture one or many of the factors, including the awareness of the staff about the responsibility, equipment, system, interaction levels, interdependences of the affected system with other systems, feedback/response of the plant in terms of direct indicators (meter showing status of valve), process parameter response, perception of the whole exercise and the plant, limiting conditions for operations, physiological and stimulus response capability, human conscience level (honesty, integrity, dedication and so on), attitude toward the plant and organizations (dedication level, proactive approach), and cognition levels (memory, reflexes, intelligence, prognostics capability and so on).

The external conditions or context might include one or combination of the following: man–machine interface, complexity of the environment defined by primary and supporting levels (skill-based, rule-based, knowledge-based and availability and adequacy of information and advisory system, stress levels, environment, ground benign, humid, high temperature, hostile; availability and adequacy of checklist, communication and monitoring level between control room/supervisor, availability of feedback mechanism and so on;

C_o—organizational consciousness level affects organizational factors that again affect human mental conditions, such as stress, plant culture including general human considerations, humility, human consideration (absence of indifference between plant management and staff), staff happiness levels, sensitivity of organization to opportunity and recognition to outstanding performance while sending the right signals among the staff.

The effectiveness and applicability of the human reliability improvement process can be significantly increased by developing a risk-conscious culture by performing the root cause analysis such that process of root cause analysis for a given failure should not stop at 'Human error' but further down to the level where we get the information on internal and external factors, as mentioned above. This is a resource-consuming activity, as it will require an integrated approach that requires the application of knowledge of psychology, physiology of the brain, neuroscience and spirituality.

3.5.3 Root Cause Analysis

Root cause analysis (RCA), even though it is performed more as part compliance procedures, is often recommended by the regulatory authority. Even though RCA has been discussed in this book at length, in this chapter, the aspects directly related to probabilistic and deterministic components of RCA have been discussed.

First and foremost, it must be recognized that RCA should not remain limited to brainstorming sessions. If the general objective of the RCA is to avoid recurrence or reduce the occurrence of the events significantly, then an in-depth analysis is necessary. Of course, all the events do not require in-depth and detailed analysis, but events having potential risk-significance require proportionate resources to do the justification and to meet the analysis objective that the recurrence of the event or the probability of recurrence is reduced significantly.

This book recommends the development of a strategy to address the requirements of RCA and a framework such that the probability or impact of risk-significant events is reduced, while the other events that do not appear to be risk-significant are evaluated to identify their potential to become precursors to risk-significant scenarios, if any. As these events may also not be important from risk consideration, but if the events cause adverse effects on plant availability and reliability, then appropriate measures should be allocated to reduce interruptions or unavailability components to the plant.

For a significant event, the RCA should go down to the level of identifying the human roots, and in human roots, the factors, be internal or external, contributing to the failures should be identified and dealt at the management level. It is the responsibility of the organization to make available all the resources available, which include expert human resources either inside the plant or external experts, lab facilities for testing and investigating the failure, a framework for surveying similar failures to obtain additional insights, and a mechanism to share these results with regulatory bodies. Provision should also be made regarding decisions related to the RCA program in an organization such that risk-significant and precursor to risk-significant failures are not missed. Furthermore, the RCA program should ensure that timely corrective action and monitoring program is implemented.

Details about the HRCA (Human RCA) have been discussed in the Chapter on Human Reliability analysis.

3.5.4 Uncertainty Analysis

There are four major factors in PRA that need focus to deliver a quality PRA: modeling and analysis CCF, human factor considerations, adequacy and accuracy of reliability data and characterization of uncertainty. One good aspect is that the probabilistic approaches provide an elegant framework for accounting for uncertainty. However, here the challenge is, if we are using 90% confidence bounds for characterizing epistemic component of uncertainty, then it is well known that the availability of data to characterize the 5% and 95%, i.e., lower and upper bound, respectively, are rare. Most of the modeling addresses mean or median estimates. Considerations of 95% bound in all our analysis become too conservative, but the question is, in the interest of the risk management will it not be appropriate to consider 90% confidence. This requires a cultural shift to name the target with an upper or 95% bound and not the median value in the probability distribution for unavailability for system and core damage frequency at the plant level.

The second and bigger challenge is that stipulations are there for acceptable uncertainty as to be 'as low as possible' be at SSCs or plant level, but there is no general guidance or formulation what is acceptable. One practical solution to address this issue is that plants should avoid, as for as much as possible, the use of plant-specific data, at least for the risk-significant component; i.e., where the risk importance is high. Furthermore, rigorous sensitivity analysis and the analysis in risk simulation mode is the only option such that the analysis represents use of the best of the state of the art available in risk and reliability analysis.

3.5.5 *Potential Role of Intelligent and Advanced Systems*

One of the solutions to reduce human error during normal operation as well as emergency conditions is to develop and deploy the operator and maintenance aids based on machine learning or intelligent algorithms. The deployment of smart technology is another approach to address issues such as calibration error or to address the instrumentation drift problem. With the use of digital technology, where software algorithms to a certain extent address the issue of drift, further conscious use of smart/intelligent technology on a case-to-case basis can be adopted. Every new technology, for example, the use of digital technology, even though it has brought many advancements in monitoring and surveillance, the regulatory body has its reservation due to software issues in the digital system.

The available literature shows that there is increasing interest in the deployment of intelligent operator support systems for monitoring, prevention and emergency support in the control room. Chapter 9 presents a model of the operator support system where the transient identification and following diagnostic requirements are performed by an artificial neural network (ANN) and a fuzzy rule-based system, respectively. The identification and diagnostic features of the operator system reduce the cognitive load that is the cause of stress, particularly while handling the emergency scenario, which is responsible for the error in decision making. The operator support station acquires online data and information in online mode and supports the operator in decision making. The system operates in an open loop, i.e., even though the data are coming to the operator support system and analysis is also performed, the decision making is totally subject to the operator's discretion.

The use of robots in areas not accessible during normal operations as well as during emergency operations due to either space or radiation level constraints provides an effective strategy to address operations, particularly during emergencies. The sensor data acquired from various public locations, related analysis and further decision making during off-site emergency management provide an effective alternative supported by adequate logistics.

References

Avizienis A, Laprie JC, Randell B (2001) Fundamental concepts of dependability, CS-TR-739/ULCA CSD report No. 010028/Lass Report No. 01-145. University of Newcastle. https://www.researchgate.net/publication/2408079_Fundamental_Concepts_of_Dependability

Electric Power Research Institute (2004) EPRI (Electric Power Research Institute). Probabilistic consequence analysis of security threats—a prototype vulnerability assessment process for nuclear power plants, technical report No 1007975. EPRI, Palo Alto, CA

Garrick J, Hall JE, Kilger M, O'Tool (2004) Confronting the risks of terrorism: making the right decisions. Reliab Eng Syst Saf 86(2):129

Holmberg J-E, Nirmark J (2008) Risk-informed assessment of defence in depth: LOCA example: phase 1: mapping of conditions and definition of quantitative measures for the defence in depth levels, SKI report 2008:33. SKI, Sweden

International Atomic Energy Agency (1992) Procedure for conducting probabilistic safety assessment of nuclear power plants, SS-50-P4. IAEA, Vienna

International Atomic Energy Agency (2004a) Methodology for the assessment of innovative nuclear reactors and fuel cycles, IAEA-TECDOC-1434. IAEA, Vienna

International Atomic Energy Agency (2004b) Procedure for conducting common cause failure analysis in probabilistic safety assessment. IAEA-TECDOC-648. IAEA, Vienna

International Atomic Energy Agency (2008a) Best estimate safety analysis for nuclear power plants: uncertainty evaluation. IAEA, Vienna

International Atomic Energy Agency (2008b) Guidance for the application of an assessment methodology for innovative nuclear energy systems—inpro manual—safety of nuclear reactors. IAEA, Vienna

International Atomic Energy Agency (2009) Protection against common cause failures in digital I&C systems of nuclear power plants. IAEA, Vienna

International Atomic Energy Agency (2020) INPRO methodology for sustainability assessment of nuclear energy systems: safety of nuclear reactors, IAEA-TECDOC-1902. IAEA, Vienna

International Atomic Energy Agency (2018) IAEA safety glossary terminology used in nuclear safety and radiation protection. IAEA, Vienna

International Electrotechnica lCommission (IEC). Electropedia: the world's online electrotechnical vocabulary. Electropedia, [Online]. Available: https://www.electropedia.org/. Accessed 16 June 2022

Lannoy (2017) Risk management, safety and dependability: looking back from 1990 to 2015: which future? ISTE Ltd. London, UK—openscience.fr, UK, France

Lessons learned from Fukushima Nuclear Accident for Improving Safety and Security of U.S. Nuclear Plants: Phase 2. National Library of Medicines, [Online]. Available: https://www.ncbi.nlm.nih.gov/books/NBK373719/. Accessed 15 June 2022

Modarres M, Kaminskiy M, Krivtsov V (2010) Reliability engineering and risk analysis—a practical guide. CRC-Press, FL

Nuclear Energy Agency (2008) ICDE project report: collection and analysis of common cause failures of switching devices and circuit breakers, nuclear safety, NEA/CSNI/R(2008)5. NEA, Paris, France

Rubino G, Sericola B (2014) Successive operational periods as measures of dependability. IRISA Campus de Beaulieu, Cedex, France

United States Nuclear Regulatory Commission (2016) Historical review and observation of defence in depth, NUREG/KM-0009. USNRC/BNL, USA

US Nuclear Regulatory Commission (2007) Common cause failure database and analysis system: event data collection, classification and coding, NUREG/CR-6268, Rev. 1. USNRC, Washington DC

Varde PV, Pecht MG (2018) Risk-based engineering—an integrated approach to complex systems—special reference to nuclear plants. Springer

Chapter 4
Risk-Conscious Culture

4.1 Introduction

Even though the nuclear industry deserves to be complimented to set very high standards of safety, when we look back and review the major contributors to the three accidents, namely, TMI in 1979 (USA), Chernobyl, 1986 (Erstwhile Russian Federation, now Russia and Chernobyl, 2011, Japan), it becomes apparent that human or organizational factors were some of the factors that either caused design issues to these accidents directly or indirectly (International Atomic Energy Agency 1991) (International Atomic Energy Agency 2020a). On the other hand, if we look at the existing knowledge-based tools, methods and practices, including the training approaches considering human factors in risk assessment, considerations of safety culture, etc., that exist today have improved, there is a consensus among industry experts to improve these in support of safety and reliability improvement of nuclear plants (IAEA 2022c; International Atomic Energy Agency 2019a).

For example, the major lessons learned from TMI are operator errors and organizational factors, improvement in the man–machine interface, design deficiencies and judgmental limitation with the regulatory body in the interpretation of the intensity of the situation and, of course, inadequate management of emergency conditions, particularly communication with the public (Elmallah 2014). Even though the earthquake-induced tsunami initiated the accident at Fukushima, the other factors responsible included design weakness, shortcomings in safety culture and gaps in safety systems, which were the major contributing causes to the accident (Gil 2021). One general observation could be that human and organizational factors were the root causes along with limitations in emergency management systems that led or aggravated this event to reach the level of disaster. Furthermore, the conclusion of the investigation report of the commission on the Fukushima Daiichi accident that was reported in the National Diet of Japan concluded that the accident at Chernobyl in Japan was termed a man-made disaster (Investigation Committee 2022).

Each of these accidents provided invaluable learning lessons at the national and international levels in terms of improving the understanding and safety philosophy.

P. V. Varde, *Risk-Conscious Operations Management*, Risk, Reliability and Safety Engineering, https://doi.org/10.1007/978-981-19-9334-3_4

For example, the lessons learned from Chernobyl, one of the major paradigms called safety culture, was introduced by the INSAG group (International Atomic Energy Agency 1986, 1992) of the International Atomic Energy Agency and further expanded on basic safety principles for nuclear power plants (International Atomic Energy Agency 1999). Keeping in mind the major recommendations with respect to the role of safety culture in nuclear plants, the IAEA published an INSAG-4 exclusively on 'safety culture' in 1991, which provides guidelines for establishing safety culture in operational nuclear power plants. Report was with respect to Chernobyl, the document on safety Furthermore, IAEA, with interest and support from member states, developed the IAEA-TECDOC-1329 on safety culture in 2002 (International Atomic Energy Agency 2002). The concept of safety culture has been increasingly employed and has become an important tool to improve safety. Later, IAEA spearheaded these efforts worked on the review of safety culture in over 30 years in the nuclear industry and its impact on nuclear safety improvement (International Atomic Energy Agency 1996). Some specific efforts from IAEA include developed documents that mainly include Safety Report Series 83 entitled performing safety culture self-assessment in 2020 and for regulatory agencies (International Atomic Energy Agency 2020a). These developments were testimony to the interest of member states in the application of safety culture for nuclear power plants.

Apart from safety culture, the IAEA launched another instrument to expand the scope of safety through the convention on nuclear safety by encompassing the role and responsibility of safety not only at the organizational but also at the national and international levels through administrative, legal and governmental mechanisms. It was recognized as a learning lesson in nuclear plant accidents that the accident in one country has the potential to adversely affect the societal mindset not only of the country where the accident has happened but also of other countries, particularly neighboring countries sharing borders with the country of origin of radiation. Hence, it was required to have a global instrument that ensured that every member country had safety infrastructure, legal provision and an effective mechanism to address safety at the national and international levels. To further strengthen the safety in nuclear power plants, IAEA developed a conventional on nuclear safety—an instrument of IAEA with the objective toward maintaining a high level of safety worldwide through the improvement of national measures and international cooperation (International Atomic Energy Agency 1996) and other conditions for ensuring the safety of nuclear plants world over, by laying down conditions that include provisions for legal, administrative and compliance conditions.

In the late nineties, the application of probabilistic risk assessment (PRA) and the evolution of the methodology to an extent to be considered sufficiently mature to complement the deterministic methods found increasing application risk-informed approaches in regulatory decision making (International Atomic Energy Agency 2005).

All these developments further consolidated the safety approach to nuclear plants and that is where the nuclear industry grew, and the IAEA data (2019) show over 443 nuclear plants are operating and generating ~ 10% share of electricity worldwide (International Atomic Energy Agency 2019c). *Let us face the fact right, all these success stories were possible as the human actions avoided many and, in some cases, mitigated many potential adverse scenarios to become reality or accident by controlling, preventing, correcting, and mitigating adverse conditions. This could be a hallmark of the highest standard of training and qualification programs in the nuclear industry. However, the accident history also shows that the human factor has been one of the major contributors to the chain of events that lead the path of accidents, either as an initiator or as a lack of action or error in managing the situation.* Two major conclusions can be drawn, i.e., (a) designing the tools and methods for incorporating inherent safety features such that human interactions during off-normal conditions can be reduced and (b) revamping the approach to considerations of human factors and their role in accident management.

In this chapter, the second aspect dealing with human factors will be discussed through the application of risk-conscious culture to complement the existing approaches discussed in the preceding paragraphs. The rationale for the application of risk-conscious culture is as follows:

(a) A robust and more complete human model is a fundamental requirement for understanding behavioral aspects in human reliability modeling and analysis.
(b) Consciousness and conscience are the two major factors apart from cognition operating at the level of the brain and should be part of developing a human model.
(c) The human factor aspects in risk-conscious culture are based on the fundamental of human existence—the 'consciousness' in the first place often referred to as mental consciousness or awareness, which tends to be a subjective aspect in traditional approaches to science.
(d) Safety improvement or 'risk reduction' on a perpetual basis is the overriding priority, and hence, it should manifest at the individual and organizational levels as culture collectively.
(e) Risk, as derived from the probabilistic methodology, deals with undesired outcomes of any real-time system and is defined as the likelihood of undesired consequences.
(f) A perpetual awareness or human quality that remains a focused risk element of the process or human behavior that becomes a thinking model of risk reduction.
(g) The risk-conscious culture is based on the integrated, structured, measurable and rational-based risk-based engineering framework.
(h) The operational environment dominates a significant component of human factors. Here, it is necessary to deal with, apart from objective criteria and guidelines, the subjective aspects, such as attitude, emotion, and feelings, and requires a cultural approach. This approach acts as a catalyst for development and growth that incorporates learning and adoption into new ideas for the improvement of safety.
(i) Risk-consciousness culture assumes that the level and intensity of connectedness and communication among individuals and in the organizational hierarchy is one of the important factors for risk reduction.

(j) The 'safety improvement' in terms of measurable parameters is an ever reaching and moving target in risk-conscious culture.

Before we go ahead, we need to define a few terms, such as residual risk and culture, oath, need to develop a risk-conscious culture.

4.2 Residual Risk and Risk Perception

The operations of complex engineering systems, such as nuclear, space, aviation, chemical and process plants, railways, essentially deal with ensuring higher dependability attributes of plant systems. The complex systems comprised broadly of process and safety systems. The dependability of these systems is ensured mainly by ensuring higher reliability of process systems and availability of safety systems through an effective maintainability program along with other supporting activities, such as high-quality risk reduction that also includes human resource qualification programs. The reliability and availability of process and safety systems are the two major parameters that make nuclear plants more dependable in the context of safety while catering to the energy need. The process and safety systems are designed to meet the safety and production objectives while giving overriding priority safety objectives. As a safety practice, the failure of process systems is postulated, and systems are provided to identify the deviation, prevention, control and mitigation of accident scenarios by incorporating safety systems. These provisions are expected to reduce the risk levels to an acceptable level. The probabilistic risk assessment risk approach creates a model of the plant by considering the likelihood of postulated initiating events and assessing the response of the plant (mainly safety functions that include safety systems and human actions) to provide the statement of residual risk.

Risk is a measurable parameter and mathematically expressed as

$$\text{Residual Risk} = \sum_i \text{likelihood}_i \times \text{consequences}_i \tag{4.1}$$

Suppose the area under the circle represents the desired safety level and the area shown in green shows the safety levels achieved by incorporating safety features and operational environment; then, the residual area reflects the residual risk from the operation of the plant (Fig. 4.1).

The residual risk from a system here has been expressed by summation of likelihood and consequences of an accident sequence for the *i*th postulated scenario. This expression forms the basis of the evaluation of risk employing probabilistic risk assessment.

Hence, for safety critical systems, the definition of residual risk is the remaining risk component after accounting for all the postulated initiating events and assessing the response of the plant in terms of performance of safety function for the respective initiating event(s).

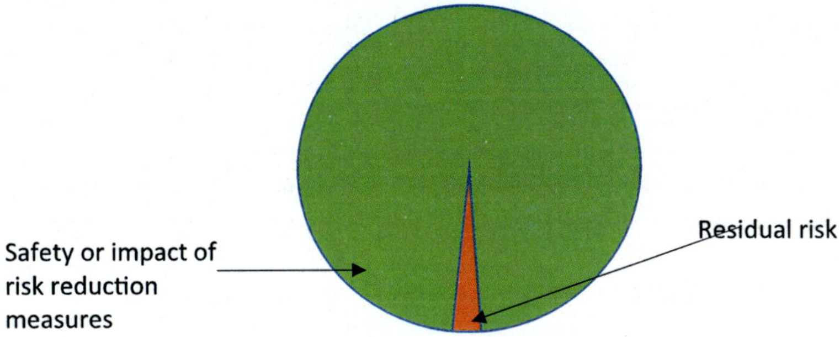

Fig. 4.1 Representation of the residual risk concept

Accordingly, the residual risk can also be expressed as

Residual Risk = Inherent risk−impact risk control measures in place.

4.3 Culture

The Wikipedia definition of 'culture', which is more related to the social system of development or evolution, is expressed as 'the way of life of groups of people, meaning the way they do things' (Wikipedia 2022). The more specific definition of culture, specific to organizations dealing with safety and security, is 'what is memory to an individual is culture to organization' (IAEA 2022b). Different groups, organizations, and habitants may have different cultures. A culture is passed on to the next generation by learning, culture is seen in people's writing, religion, music, clothes, cooking and that is what they do.

Another definition in the technical domain is company culture, how they start their day, their language of interaction, documentation, line of authority, way of paying respect among colleagues and to superior, management's approach and attitude to level of openness or independent thinking, questioning, their priority, frequency and mode of meeting and discussions, provision in the organization to ensure that staff has access to sports, music and social gathering as targeted toward increasing productivity by creating an atmosphere of wellbeing.

Some of the attributes of individuals and organizations ensure that the risk aspects are identified and assessed, and corrective actions are implemented in a timely manner toward making the system risk environment at an organizational quality level of employees toward the goal of a risk-free system.

4.4 Safety Culture

Since its inspection, the nuclear industry has been implicitly risk-conscious and has included the principles of monitoring and surveillance, controlling deviation from normal operation, protection of accidents, prevention of accidents, mitigation of accidents and emergency management. The application of defense in depth involves multiple levels of protection and barriers as part of a safety management program. The extension of the defense-in-depth philosophy in operations management is implemented employing a conservative and prescriptive approach to safety management to account for uncertainty while dealing with risk in the assessment and management of operations scenarios. For example, the availability and maintenance of containment integrity in TMI-2 kept the radiation confined to the containment volume and protected the public in the vicinity of the plant from severe radiation consequences. However, it is also argued that the weakness in emergency management in the public domain was lacking in many accidents. In fact, the human and organizational factors, particularly those strongly related to weakness in safety cultural aspects in Fukushima, made this accident referred to as the man-made disaster. It was evident that the lack of a strong nuclear safety culture was an important contributing factor to the Fukushima Daiichi accident (National Academic Press 2014).

In one of the leaflets of IAEA on 'safety culture', it is noted that '*In some circumstances when severe accident happens, analysis indicate that safety margins had been eroding steadily for years. This can result from people gradually accepting declining conditions in safe work practices and ignoring the brought on by this decline that may have unnoticeably drifted toward prioritizing other concerns over safety. RISKS MIGHT HAVE BEEN PLAYED DOWN, because "nothing has happened", which can eventually lead to a severe event occurring*' (Culture for Safety 2022a).

The international framework of IAEA recognizes safety as an overriding priority, leadership, and accountability for safety while assuming that safety is integrated into all activities and safety is learning driven. Furthermore, each of these elements has implementable attributes.

The IAEA glossary (International Atomic Energy Agency 2019b) defines safety culture as '*the assembly of characteristics and attitudes in organizational and individuals which establishes that, as an overriding priority, protection and safety issues receive the highest attention warranted by their significance*'.

4.5 Risk-Conscious Culture

There are five major elements of risk-conscious culture that provide the foundation for risk-conscious operations management for a complex system. This section discusses (a) governing principles, (b) risk and reliability elements, (c) human and organizational elements, (d) technical elements and (e) evaluation metrics and criteria. The aim is to present the salient features of the risk-conscious culture. As the topic is subjective, in the sense that different engineering, particularly complex systems, have their specific domain characteristics, the discussion is general and not specific to any particular complex safety critical system. Nevertheless, a framework can be adopted for a specific domain/system depending on the need and specific requirements.

4.5.1 Governing Principles

The discussion is on two specific topics, i.e., oath and then the fundamental principles of RCC. The Oath might be either part of organizational policy or could be a part of the overall policy that governs the plant system, human and associated activities.

4.5.1.1 The Oath

The organizations/services that directly deal in the public interest have a culture of oath taking by the individual employ. To implement this element, the organization should have a 'Oath' that an employee can take either at the time of joining, or at the initiating stage of an organization, the individual or group involved in operations management must take this oath. This oath should include the element of consciousness and conscience dedication to ensure that the employee's overriding priority is working to ensure that potential from risk from normal operation as well as accident conditions is minimal to the public and property in the vicinity of the plant in a manner that public confidence in nuclear energy remains high.

One of the examples is proposed as follows:

I am part of risk-conscious culture, and I pledge consciously that:

- I shall not be compromising on my duty and be aware, attentive, vigilant for my duty toward plant related to risk aversion matters.
- I shall work to the best of my capabilities toward discharging my duties toward achievement of plant reliability and availability targets.
- I shall be honest and transparent in dealing with matters related to reporting, communication, action in plants and organizational matters.
- I shall maintain high confidentiality on sensitive matters of the plant.
- I shall not take or do undue favor or benefit from my dealing with my colleagues or outside party or person.
- My colleagues in the plant are like family members to me; therefore, I will strive to maintain an all-inclusive attitude for wellbeing and happiness.
- I shall always work toward maintaining a team spirit while working as a group in plant matters.
- I shall not compromise my integrity
- That my duty toward identification, proactive action toward prevention, control, mitigation management.

The above one could be considered to develop and design the Pledge for the plant or organization and is administered to all the employees and associated persons before they enter into the plant or organizational business. Oath is a mandatory requirement for the plant business and is applicable to a person within as well as without plant areas.

4.5.1.2 RCC Principles

Risk-conscious culture has been developed based on some existing philosophical/scientific ideas and inputs, postulations and few assumptions that provide the basic framework for the three major elements of RCOM, i.e., risk, consciousness and culture as follows:

1. Safety and security are overriding and perpetual considerations and form part of important goals.
2. Human consciousness is fundamental to human existence and realization of internal and external reality.
3. Humans are basically a spirit, and the major idea of the RCOM is to further improve human performance and work quality, and therefore, spiritual practices, such as Yoga and meditation form an integral part of RCOM
4. A collective human behavior and attitude characterizes the organizational culture.
5. The risk-based approach is central to the implementation of a risk-conscious culture.
6. Achieving excellence in reliability, availability and quality should be followed at the individual and organizational levels.
7. RCOM culture.
8. The motivation for any culture flows from top, viz. government and organization policy through the plant environment to grass root levels.
9. Setting the highest standard for risk culture in the beginning is the initial condition for program success.
10. The target should be excellence in any respective area, and the rest will follow.
11. Apart from the happiness quotient of the employees, how much pride they have in identifying themselves with the organization are the two major factors that are the corner stone for safe and productive organizations.
12. High conscience levels determined by high value and ethical and moral behavior are the major attributes for improved safety, security, and productivity of the organization.
13. The monitoring or tracking of culture in measurable parameters is key to the sustenance and management of risk-conscious culture.
14. Training organizations in collaboration with other sections and groups plays a critical role in the implementation of risk-conscious culture.

However, the available literature shows that there is a consensus that consciousness is not fully understood. The major reason is that understanding consciousness requires stepping out of the established modern material science procedures, viz., reductionist framework, measurement logistics, etc. It appears only advances in quantum mechanics on one side, and the schools of spiritualism are inching toward making progress on consciousness. Nonetheless, there is good consensus on some aspects, viz., consciousness exists, consciousness is subjective in nature and a source of human emotions, feeling or more precisely qualia, etc. Notwithstanding this description, consciousness in RCC is a vital element of the human model, and the attributes of consciousness, such as alertness, awareness, empathy, happiness and the considerations of cognitive consciousness. It has been assumed that the quality of cognitive function of the brain, such as memory, perception, logic and analysis, is driven by the level of consciousness. Furthermore, consciousness is also the driving force for human consciousness that is responsible for ethical and moral behavior, which includes work attitude, integrity, commitment, honesty and devotion to work, transparency in official dealing with organizational aspects and dealing with agencies outside the organization.

Consciousness, cognition and conscience work at the level of the brain and determine human behavior, action, reaction and response at the physical and mental levels. Furthermore, the human interacts with the surroundings through the five sensory systems, and in risk-conscious culture, it is referred to as the *eye* for conscious seeing or visual interaction; similarly, the *nose* for conscious smell; *ear* for auditory functioning, i.e., conscious listening; *mouth* for conscious speech and taste; physical body for sensation; and brain for *cognition* and memory for analysis. Another strong assumption derived from spiritual science is that 'inner happiness' is an innate quality to human consciousness, and with spiritual practice and methods, one can regain it.

The RCC developed in this research has two major objectives. First, operations management as an operations ecosystem is a human interaction-intensive activity, and human performance has a considerable impact on safety. Second, this risk-based approach provides an elegant and effective framework for risk-conscious culture. In a risk-based approach as applied to risk-conscious culture, the quantified insights available from PRA study can be used along with deterministic insights and integrated with operational judgment to make it practical and integrate it with individual and organizational functions. Here, the integration happens among the probabilistic, deterministic, and operational experience to make the culture risk conscious. The risk-based approach developed by Varde and Pecht has been adopted for developing risk-conscious culture (Varde and Pecht 2018). The subject of human reliability and the fundamentals of risk-based culture will be discussed in the following chapter.

The risk-conscious culture approach assumes that every country and organization has collective risk consciousness, where risk perception plays a major role. The risk perception itself is driven by many factors that include the traditional values and existing socioeconomic and spiritual following and geography and demographic conditions of a nation. Finally, everything sums up into the political wishful thinking and policy toward risk aspects for industrial systems. In India, the major wakeup call was the Union Carbide accident in 1983 that shook the nation but not the international community. This event formed a backdrop for political bias in dealing with foreign business houses operating in India. The second major event is COVID-19, which established that risk from either accident or biosystems need not be an isolated phenomenon, and the message is WE ALL ARE CONNECTED. COVID-19 also brought about the fact that like every individual, a nation also has its own typical consciousness. This is reflected in the behavior and response to the COVID-19 situation. COVID-19 has made it earlier to characterize a country's consciousness and conscience, and this suggests that (a) a nation's policy is governed by the national consciousness and conscience and (b) a risk-conscious policy can be developed by a nation making the nation consciousness and conscience effectively.

To implement the risk-conscious policy and legislation at the national level, there is a need to have a liability-benefit analysis with respect to complex systems that will again be based on the insights into the technical factors risk factors interpreted and modified, if needed, employing expert elicitation or any other subjective tool. The national risk-conscious policy and legislation provide the major motivation for translating into organizational and plant risk-conscious policy. It may be noted that here, the available literature needs to be reviewed to consider the element and philosophy of safety culture. At the organizational level, the responsibility of developing and implementing the risk-conscious culture is the responsibility of top driven and operated by middle management such that the staff is sensitive and motivated to follow the risk-conscious culture.

One important factor of risk-conscious culture is identifying 'loss of reputation' as one additional factor that is driven by public opinion or society at large. If we learned, one lesson from the three major nuclear accidents is that the risk of accidents should be addressed at any cost. We cannot afford accidents, as accidents often have international repercussions. Therefore, risk-conscious culture should be developed and implemented and reviewed with full 'will' and motivation with regulatory or political oversight. This means that the implementation of risk-conscious culture and its periodic review should be one of the major elements of the regulatory framework. Hence, development, drafting, implementation, monitoring, regulatory oversight and governmental policy on risk culture for operating plants should follow the highest standard on safety.

Achieving excellence in risk-conscious culture should be an overriding factor over operational performance factors, such as availability and capacity. The culture demands that the organization takes pride in declaring that they have achieved excellence addressing risk-conscious factors. For example, the administration and the people of Indore take pride in claiming that their town has won the government recognition, placing as CLEANEST CITY IN INDIA in fourth successive terms. One of the factors, apart from administration motivation, is the cleanliness-consciousness of the citizens of Indore. Here, it does not matter that the literacy rates in state (MP), economic status, where Indore belongs is less than many states, in fact no body is bother. Because they know, when they can make cleanliness as reality, other smaller things will follow.

In risk-conscious culture, it is assumed that excellence achieved through maintaining the right consciousness, i.e., right awareness and thought, ethics, dedication and commitment, led to organizational growth as well as an individual's inner growth and happiness. Because at the core human being is happy in nature.

4.5.2 Risk Model

Keeping in mind the requirements of the operational environment in any complex safety critical system where on the one hand, ensuring risk reduction is an overriding priority, while ensuring intended deliverables, on the other hand, often poses challenges. This calls for an integrated approach where hardware, human, and software-based systems can be represented for providing a model that is as close to the real-time environment and is as complete as possible toward addressing the risk and reliability targets. Further, in the integrated approach facilitates considerations of probabilistic & deterministic approaches; use of qualitative and quantative data, success / failure criteria; and to a large extent complement and suppliment each other arrive at decisions as part of risk-informed approach.

4.5.2.1 An Integrated Approach

An effective and elegant way to model risk for a complex system is by integrating the deterministic and probabilistic approach in terms of methodology that involves exchange of insights between the two approaches toward effectively reducing the uncertainties in the results. Integrated risk-based engineering (IRBE) or simply risk-based engineering (RBE) employs an integrated approach involving the use of risk insights from deterministic and probabilistic risk assessment, the treatment of human error, uncertainty modeling and finally prognostics and surveillance for monitoring the change that has been affected. Therefore, the risk-conscious culture uses the IRBE framework for developing risk insights in support of operations management.

The PRA methodology provides a structured framework for risk assessment starting from qualitative modeling of accident sequences employing essentially the event tree approach; system analysis through fault tree methodology; quantification using performance data or generic data along with the uncertainty estimates; the Monte Carlo approach for generating the system level cut-sets and the accident sequences at the plant level; estimation of risk-importance measures; and finally, the presentation of integrated statement of risk in terms of core damage frequency along with the uncertainty band and sensitivity analysis for various assumptions made in the analysis. The graphical representation of the model, the provision for tracing of the data along with the documentation makes this procedure rational based, transparent and traceable for facts and figures used in the model.

4.5.2.2 Risk Modeling Prerequisite

The plant should have infrastructure for the development of a level 1 probabilistic risk model of the plant if the objective is to assess the risk in terms of reactor core damage frequency, i.e., plant internal accident scenario modeling. If the scope includes containing scenario modeling to assess the risk of various release categories from the containment, then level 2 PRA will form part of the scope. However, if the target includes accident/emergency management in the public domain, then level 3 PRA is required for the assessment of risk, i.e., likelihood and consequences in the public domain. However, the scope of this section does not include level 2 and 3 PRA. It has been observed that for any PRA to serve operational applications requires the active involvement of operations staff to capture the essence of operational events and occurrences. Therefore, it is good practice for the organization to invest in training and for plant staff in probabilistic risk assessment and reliability modeling. Before the PRA project is initiated, it is required that quality attributes/indicators be developed for various activities, such as initiating event selection, system modeling, accident sequence modeling, reliability database generation, quantification of the model, accident frequency assessment, uncertainty and sensitivity analysis, formulation of results and insight, documentation, and employing available international/national standards. The third major requirement is that only PRA with regulatory approval should be employed for the given application.

4.5.2.3 Data Collection and Analysis

Data collection and analysis are critical parts of risk modeling. Considerable efforts are required for transforming the raw data available from the plant records to reliability estimates of hardware, software and human error data. Again, the training of operational staff on data collection and analysis enables the development of a quality database. As part of conscious risk culture, it is required that an online data collection and analysis system be developed and updated to support the maintenance of the living PRA or risk monitor model. The system should have provisions and guidelines for the acceptance of data considering the level of uncertainty, procedure used for data collection, and periods selected for quantification. For example, for a diesel generator that has served for 40 years and recently showing higher failures, it is futile to consider 40 years of data, as during the useful life of the diesel generator, for over 33 years, except for the initial period, the performance of DG was good. These useful period data will mask the failure data experienced recently in five years and will not provide the true condition of the DG performance. Similarly, the same argument cannot be provided for a stainless-steel pipeline that has served for over 45 years. Here, the accumulated data collected over 45 years are an input for evolving database generation and updating.

Data collection and analysis is a subjective task and requires obtaining the right perspective for a target application.

4.5.2.4 Qualification of the Model

The ASME or IAEA procedures provide an effective reference for assessing the quality in terms of completeness and correctness of the model of the as built and as operated system, adequacy of data, interpretation of results, documentation for every major step involved in creating the PRA model and its application. For a given application it is required that the scope, objective and level of details as also use of plant-specific data forms some of the major considerations.

4.5.2.5 Accident Precursor Analysis

Even though it is not always possible to look at the potential benefits of a precursor analysis program, the accident precursor program should be part of a risk-conscious culture. The detailed and in-depth analysis of many accidents might reveal the symptoms or precursor events that have the potential to become an element of an accident sequence (Corcoran 2004). In this sense, the probabilistic risk assessment through the accident sequence analysis program may be able to provide quantified risk-significance of events and functions (Jang et al. 2017).

Accident precursors are those operational events, groups of events, or even near miss events, factors including human actions that have the potential to propagate to lead to unacceptable consequences. In the traditional sense, precursor analysis employing a deterministic approach forms an important element of operational feedback analysis. In regard to the analysis of precursor event analysis, which has potential implications for core damage, probabilistic risk assessment provides an elegant mechanism to map operational events. The formulation of a list of accident sequences, along with the likelihood and consequences associated with the respective accident sequence, is one of the objective functions of PRA IAEA (2004).

The risk-based engineering framework envisages assessment of precursor events employing deterministic screening and assessment along with PRA and finally assessment of conditional core damage frequency with respect to subject accident sequence. Furthermore, risk-importance measures, e.g., risk reduction and risk achievement, enable risk identification and prioritization programs and thereby the management of corrective actions or monitoring programs to track safety performance.

Identification and Evaluation Operational Precursor

Probabilistic risk assessment provides a systematic framework for the analysis of precursor events. However, the availability of PRA study is a prerequisite and might make precursor analysis prohibitive in the implementation of risk-conscious culture. In this section, a generic model has been developed where apart from frequency and consequences, factors, job complexity based on the nature of the job in terms of knowledge, rule or skill along with associated weightage factors. Even though these precursor events mostly belong to a normal operational environment, the stress factor has been accounted for by considering the available time window and the time required for the job. Five levels of consequence category form part of this modeling. The final output of the empirical model predicts the risk factor associated with the subject precursor event.

$$\text{Riskfactor, } R = f(\text{Human factor}(H) * \text{frequency of event} * \text{Time} - \text{window} - \text{Factor available} * \text{consequence}) \quad (4.2)$$

$$\text{Human factor} = f(\text{Job Complexity})$$

$$J_c = f(Q * w);$$

Where: $Q_i = 1\text{–}10$

$W_i = 0.1\text{–}1.0$

$Q =$ Quantum of a task

$W =$ Weightage of task

$Q_i =$ Quantum of a job type.

One of the considerations in RCOM for a typical job analysis is that a job or activity can have all three components, viz., knowledge, rule and skill based, in different proportions, which is different than the conventional procedures in human reliability analysis. RCOM the job complexity formulation is as follows*:*

$$J_c = \sum_i (Q_i * w_i)$$

$$J_c = (Q_1 w_1 + Q_2 w_2 + Q_3 w_3) \quad (4.3)$$

Q_1: Quantum knowledge-based of the job; $w_1 =$ Weightage for knowledge-based component

Q_2: Quantum rule-based of the job; $w_2 =$ Weightage for rule-based component

Q_3: Quantum skill-based of the job; $w_3 =$ Weightage for skill-based component.

Precursor risk factor

$$R = J_c * ER * TF * C$$

$\varepsilon_p =$ *Event rate or precursor event failure rate (per year)*

TF = Time window factor given by

TF = t/T

t = Median time for the task

T = Available time for the task

C = Consequences factor $1 \leq C \leq 10$.

The risk factor can be given by the following model:

$$R = J_c * \epsilon_p * [t/T] * (C/n)$$

n = a constant

Accordingly, the equation for risk factor for a job is given as follows:

$$R = J_c * \epsilon_p * [t/T] * (C/n) \tag{4.4}$$

The consequence level matrix is given in Table 4.1. There are five levels. Level 1 has the fewest consequences, while Level 5 has the most consequences. Equation 4.4 estimates the risk factor associated with a given event as a function of human factor, reflected in the job completion factor, precursor event rate ε_p/yr, ratio of actual time for completion of the job to available time, and potential consequence associated with the event. n is the normalizing factor.

Annexure C illustrates the estimation of risk factors for typical precursor events for identified consequences. There can be subjectivity in the chosen parameters in the model; however, the table shows a procedure for estimating the risk factor associated with a precursor event.

4.5.2.6 Domain Knowledge

One of the schools of consciousness developed a model of consciousness employing integrated information theory (Fallon 2022). At the core of this model is collection and sharing of information that becomes knowledge at the level of the brain through neuronal communication and sharing of memory at the level of the brain. However, this is one of the most widely considered models of consciousness operating at the level of the brain, where brain mapping has been shown to produce neural correlates of consciousness. For risk-conscious culture, it is vital to consider that the technical knowledge base of the plant acquired although study of documents, discussions, meetings, on the job training and learning on design and operational and safety aspects of the plant, an integral part of the acquiring knowledge of the plant. Furthermore, the risk-conscious culture formally inducts, the probabilistic risk assessment, emergency drill experiences, understanding of root causes of failures and potential threats posed by precursor events. To emphasize further, the root initiator for TMI-2 incident was a small malfunction in the feed water system that led to resin ingress into the system and choked one of the strainers (Rempe and Knudson 2013). This event led the operation crew to use compressed air to dislodge the resin to clear the chocked path. It must have been a routine practice to address these issues; however, the use of compression to operate the valves in the feed water system was used to clean the

Table 4.1 Typical precursor events (PE) and potential consequence factor

S. no.	PE category	Consequence level (CL) metrics	Examples of precursor events (PE) and associated consequences
1.	Level 1	$1.0 \leq CL \leq 1.5$	• Events ranging from no consequences to local consequences in terms of short-term unavailability of equipment • Inadequate industrial safety provisions • Inadequate training program • Deficit maintenance management • Inadequate operation procedure • Not able to maintain good operational practices • Inadequate safety culture • Perpetual noncompliance to regulatory recommendations
2.	Level 2	$1.6 \leq CL \leq 3.0$	• PE Events resulting in frequent anticipated operational occurrences or transients in the plant consequently causing plant delivery schedules adversely • PE having potential for unavailability of safety systems and engineered safety features • Industrial safety event having potential for minor to medium injuries or health effects or compromised industrial provision and practices • PE having potential for plant incidents or significant event • Discovery inadequate or flawed training policy or procedure; inadequate maintenance policy procedure

(continued)

Table 4.1 (continued)

S. no.	PE category	Consequence level (CL) metrics	Examples of precursor events (PE) and associated consequences
3.	Level 3	$3.1 \leq CL \leq 5.0$	Events/human error/software failure having potential for common cause failure Events having potential got involving technical Specification violation
4.	Level 4: PE involving supervisory, monitoring, data logging, diagnosis, prognosis inadequacies	$5.1 \leq CL \leq 7.0$	• Events involving anticipated operational occurrence (AOO) having potential for accident condition (Class IV power failure condition, if not recovered manifests into station blackout scenario • Industrial accident involving permanent disabilities • AOO events extended to station blackout due to failed recovery
5.	Level 5: Potential human error having potential to adversely affect availability of safety systems	$7.1 \leq CL \leq 10.0$	• PE having potential for design basis event • Industrial accident having potential for more than one fatality and sever injuries and disability • Events have potential to cause core damage on PRA model

strainer. This led to closure of the turbine inlet valve and tripping of the turbine. The rest is the history.

The point being discussed here is that the operating crew should be trained in systems interconnection and intercommunications. In normal operation, the actual work should start only by approved procedures from the control room. While performing an operational checklist, the possibility of deviation cannot be ruled out. In such situations, efforts should be made to revert toward system normalization. Once some unexpected symptoms are observed, then the right approach is to resolve through discussion at the team level as well as at a higher level.

Furthermore, right at the checklist level, the training program should extend the scope to include the training on risk model of the plant, which includes fault tree analysis of all the safety systems, accident sequence model of the plant, failure criteria at system and plant level, risk-significance of SSCs and human actions, potential modes of common cause failures, and accident management strategy. The demonstration of simulator experiments should be part of risk-conscious culture.

The domain knowledge of an individual is a key performance indicator judged through interviews, discussions, responses, participation in meetings, etc.

4.5.2.7 Human Root Cause Analysis

Human root cause analysis in RCC refers to performing a root cause analysis (RCA) for risk-significant events until the human root responsible for the human error or organizational failures are analyzed in depth. The special feature is not identifying the human factors responsible for the failure but deeper down the human limitations or deficiencies, in line with the hardware failure, that were responsible for these failures. For example, if the direct cause is distraction from the job that led to error of omission or commission than what are the factors that led to the subject failure. This aspect has been discussed in detail in Chap. 7 on Human Factors in Operations. The HRCA forms an integral part of the risk-conscious culture in an operating plant.

4.5.2.8 Interpretation of Objective to Subjective and Vise-Versa Information

One of the advantages of the PRA is that it provides numerical results of the safety significance of SSCs and human action in terms of conditional core damage probability or risk-importance measures. However, these estimates are vital, but often this input is just one input, whereas the operational environment requires considering certain other operational aspects in support of decision making. On the other hand, certain situations often do not allow the assessment of numerical estimates due to the inadequacy of data or new components. However, the operational expert can provide qualitative input on these aspects. For example, the rupture disk employed in an emergency core cooling system is a unique component. The experience of testing has shown that the reliability of these components is 'very good'. The question is how to convert this qualitative estimate into a quantitative estimate. Here, the fuzzy logic approach has been used in risk-conscious culture such that this expert opinion can be converted into quantified input. This transformation of information from quantified to qualitative and vice versa provides a bridge between PRA and the operational environment and facilitates decision making in a conscious manner.

4.5.3 Organizational and Human Elements

Given that human factors are one of the major contributing factors for accidents worldwide, human and organizational factors have become important considerations for risk-conscious cultural implementation programs. Often, it has been seen that even if the individual human is identified as a direct cause for the accident across the human–machine interface, the detailed and in-depth analysis points to factors such as deficiencies in the interface design, environmental factors, procedural deficiencies and lack or inadequate training. Keeping this in mind, the organizational factors become very important.

4.5.3.1 Organizational Factors

Background

A detailed root cause analysis of the records of incidents/accidents in the plant or similar plant or generic sources with applicable features reveals organizational factors responsible for undesired performance of the plant. The review of the PRA model of the plant is another source that provides the prioritized list of component, software, and human error where the analysis and correlation between other events reveals the organizational factors. For example, inadequacy of training could be a factor that is responsible for deficient training programs. Modifying the training checklist or as a built design and drawing might work to reduce the repetition of incidents.

Identification of Strength and Weakness

Every nuclear plant has its own safety culture or quality program. This culture might include a system of plant safety management that might include all or some of the elements of the attributes we have discussed. Here, it is necessary to critically understand and assess the strength and weakness of the present culture and how it will be possible to improve the safety of the plant employing risk-conscious culture. In fact, a comparison with the traditional culture provides an important matrix to assess the improvement.

RCOM Policy

A written and approved organizational policy at the highest organizational level is one of the important cornerstones of risk-conscious culture. The policy should be derived from governmental or legal documents where the mandate and the ultimate responsibility of ensuring the maintenance of risk-conscious culture should be part of the policy. It should be ensured that the policy and the intent reach every individual right from top management to grass root level workers, including contract workers and researchers, even though they may not be directly involved in the operational activities of the plant.

Collective Consciousness

At the operational level, the organization should be risk-conscious—the individual and collectively the thinking should be risk-conscious in terms of attitude that includes risk-based thinking, risk-awareness and attention, approach to physical and mental work, decisions, communication, and administration. Everyone should recognize in line with risk-conscious policy that risk-free operation is an overriding priority to operational performance excellence. In addition, true excellence is when the organization is demonstrating lowest achievable risk indicators like system unavailability and core damage frequency, incident free records while achieving the excellent operational indicators in terms of capacity and availability factors.

Review of Existing Policies and Practices

Before initiating activities for the implementation of risk-conscious culture, it is necessary to review the existing status and approaches for risk identification, risk assessment and management. This includes assessment of operation and maintenance management activities, infrastructure that includes level and effectiveness of training and qualification, engineering modification review provisions and approval scheme, effectiveness of quality assurance program, scientific adequacy that deals with core management and material integrity management program; provision for threat detection, identification of area location, means and methods for threat aversion; and most important—the attitude of employ to risk identification and management. This review can be completed more gainfully by reviewing incident reports, root cause analysis reports, near miss events and operational indicators. If the reliability studies or PRA of the plant is available, then these studies could be subjected to review for identifying and prioritizing the area for improvement.

This step provides the necessary impetus for the implementation of risk-conscious culture.

Safety Performance Review

Periodic safety performance review is key to implementing, monitoring or tracking, evaluating the effectiveness of derisking provisions and programs. The integrated safety performance review involves recording and tracking the deterministic and probabilistic high-level attributes as well as the low-level attributes. The major performance indicators, for example, availability and capacity factor, number of incidents and events, and number of unscheduled shutdowns, provide the risk and reliability indicators. Standard attribute tracking helps understand plant strength and weakness and thereby areas for improvement. These attributes are becoming an integral part of decision making on incentives at the individual and organizational levels. The insights from root cause analysis, waste management and fueling-related aspects help specific improvements in the plant. Furthermore, insights into regulatory compliance levels provide insights into the effectiveness of facilities catering to regulatory recommendations. Industrial safety, apart from radiation safety, is important, and the trends in events and accidents or near misses and their management provide information on improving industrial safety features, including regular drills and exercises.

Operational Elements

Even though risk assessment and modeling is one of the core elements in RCC, there are other elements, such as training and qualification, quality assurance, technical audit and documentation, that form part of RCC. The training and qualification provides and prepares the new and the existing staff through checklist procedures, hands-on training and simulator training sessions and exercises for postulated conditions and emergency scenarios with respect to the response capability required given the risk perceptions. For example, proper training based on knowledge and models built based on probabilistic risk assessment, incidents and events taking into account the defense in depth levels that are inherent to the plant enables them to require timely action/response. Management provisions such as attaching the risk-significance level to various plant SSCs, areas, actions and procedures are some of the factors that have the potential to reduce the chances of error. Similarly, the implementation of a quality assurance program complying with regulatory or available codes and standard guidelines along with periodic plant walkdowns and surveys for hardware and human factor aspects help assure response/action readiness. Implementation of the in-service inspection and condition monitoring approach along with review of these data periodically collected data/information at suitable intervals, employed as part of risk-conscious culture, has the potential to enable the building of the required resilience through additional prognostic capability into the passive and active SSCs. The role of implementing a risk-based audit program that ensures compliance requirements is key to increasing the efficiency of regulatory communication and compliance.

Furthermore, the documentation system designed and implemented employing the available standards and codes or based on plant-specific experiences goes a long way in tightening administrative, technical and regulatory response capabilities.

4.5.3.2 Human Factors

Once the major metrics of organizational consciousness are understood, the next step is to consolidate the gap area or requirements of human factors. In fact, here it is two pronged approach, i.e., to understand the general strength and weakness at the level of the individual and then develop a program for reference or desirable metrics for human factors, keeping in mind organizational policy and risk-significant requirements. In this sense, we take input from organizational factors and the individual staff level, the reference attributes and work toward achieving the required metrics.

General

Human modeling in risk-conscious culture is based on the CQB framework for human reliability modeling proposed in risk-based engineering (Varde and Pecht 2018) along with additional factors required to model risk-conscious culture, here referred to as enhanced-CQB (e-CQB). This section considers attributes other than those that support the case of the human reliability assessment approach as part of quantitative probabilistic risk assessment. However, there is one more aspect that is considered critical to address uncertainty in the human behavior model. This is what attributes do not directly support risk assessment but are crucial for improving human performance as precursors to human failure. These factors at times often overlap and may remain exclusive from the point of human behavior modeling. These attributes are considered to support modeling, a new term coined here as Potent Performance Indicators. For example, the attributes considered for assessing the human happiness quotient, wellbeing of employees, etc., may not directly, unlike performance shipping factors, human reliability is affected by certain happiness and wellbeing factors.

The enhanced-CQB framework comprised factors, viz. consciousness, cognition, consciousness and brain and physical body for interaction with the surroundings through six stimuli and physical actions. The physiological and psychological factors that determine human wellbeing also form part of enhanced-CQB.

Consciousness

The standard and more widely characterization of consciousness related consciousness to level of awareness, alertness, perception, and capability for cognition through the physical senses. Swami Sarvapriyananda, while discussing the Vedanta definition of consciousness, explains that 'Whatever you are aware of is not consciousness—we have a negative definition of exclusion, i.e., anything that you are aware of like, site, light, sound, smell; all these are objects of consciousness, however, *"that which is aware of them is consciousness" that means consciousness is awareness* of experience or awareness in which experience is possible'. For example, you are looking at a Glass Jar on the table, Jar is object, you are looking through the eyes—eyes are also object; you have observation in mind—the mind is also an object. We need to eliminate all the objects to get to the consciousness; therefore, whatever is left is consciousness—that is YOU and ONLY YOU and your experience is consciousness. Furthermore, the *Advait Vedanta* teaches nonduality, i.e., everything including the material world itself is also consciousness—referred to in Vada as Maya (Sarvapriyananda 2013).

In the spiritual sense, awareness primarily includes inner of self or inner reality perceived as also outer awareness of surroundings. For risk-conscious culture development, we have used an integrated approach in which characteristic elements of spiritual, neurological, information theory, and quantum mechanics have been employed to formulate an integrated concept of consciousness. Spiritual philosophy suggests that we all share the same nondual source of consciousness and consciousness is all pervasive, be it human, animal, material and finally nature at some level and measure. The principle of collective consciousness that transforms into organizational consciousness and national-level consciousness at a higher level has also been derived based on the existing models.

Second, we are connected, which means that if one individual or group's action mental or physical has consequences and the effect propagates to organization and at times at the national and world levels. On the other hand, the objective approach to consciousness employing integrated information theory proposes that information with integration levels in the brain through neuronal connections creates knowledge, feelings, and emotions that can be mapped at the level of the brain. In our approach, technical knowledge is acquired while training and qualification and while discharging the duties in real time, and behavioral aspects form fundamental elements of operations technical and management aspects. More information and knowledge about the subject means more consciousness of the subject—here, it is risk-conscious operations management.

The neurological school of consciousness proposes that consciousness is an aspect related to some brain faculty functioning or neuronal correlates of consciousness that enable understanding of consciousness. Often, this research generally focuses on the study of diseases, such as Alzheimer's disease, sleep disorders, and in some cases psychological conditions. The neurological and psychological science helps the organization to draw its health checkup metrics and its periodicity or at least can provide an input on adequacy of the current health checkup procedure using the advanced knowledge generated through this advanced research.

Extensive research is being done worldwide to understand consciousness through the application of the quantum mechanics approach. In fact, it will not be an exaggeration that quantum principles while considering wave–particle or quantum field theory provide valuable insights and converge the Western approach using a material science approach eastern spiritual philosophy that considers consciousness as all pervasive and not only related to the human brain. There are debates and discussions on the spiritual philosophy of nonduality. It can be argued and postulated that the framework thus far has shown promising potential. The current world largely operates in Newtonian science. Therefore, the essence of different approaches and the available state of the art will be suitably adopted in the risk-conscious approach.

Given this background the philosophy and science are concerned, there is a consensus that even though consciousness is has been accepted as critical attribute for human existence at a level where we agree that we are conscious and that is how we interact with our being and outer (inner as well as outer) and the surrounding. *We are conscious.* So WE ARE and that is the basic considerations in eCQB model. Therefore, we come across an understanding that consciousness is the fundamental attribute of humans, and we further assume that the cognition, conscience, and brain faculty functions are working as we are conscious. The next insight from philosophy is that humans are born happy at the core of their being. Therefore, happiness in the eCQB approach has been considered the integral attribute of human beings. The general attributes associated with consciousness are awareness, alertness, perception, intuition, emotion and feeling; as seen, they are critical in the operations management environment. The only bottleneck is that all these attributes are subjective in nature, and the traditional material science approach is objective and reductionist—and this aspect poses a challenge for using these attributes in modeling human behavior.

Nevertheless, if we consider the subject application, i.e., risk-conscious culture approach development, we have our ways. As mentioned, these consciousness attributes are subjective in nature, on the one hand, while what we need is identifying these factors absorbing its subjectivity but basing our decision on trends and the interrelationship of these attributes with potent performance-shaping factors (PPSFs), then a decision or action can be taken to improve, address, and correct the issue.

The eCQB approach also considers that the human mind operates at three levels: (a) conscious, (b) subconscious and (c) unconscious (Ramsay 2019). Conscious activities, such as concentration, alertness, awareness, are active in the current situations, reading, listening, and watching the traffic or supervising in a plants control room the status of various systems and equipment and updating your knowledge with new insights or even emotions associated with your interaction with surrounding activities. The subconscious state is a memory location that for recent experience, observation, image, or feeling, emotions and recalled or retrieved as reaction or response. The unconscious state is a deep state in which all our memories and experiences reside. According to Sigmund Freud's model, human behavior or attitude is governed by the unconscious mind and many vital attributes, such as emotions, that include feelings that provide a basis for many decisions based on the memories of past experiences (McLeod 2015). This experience or conditioning manifests as our belief, habit and behavior and often reflects on human personality. Often, the dream state is associated with the unconscious state of the mind. These three conscious states are critical to understanding and comprehending the human personality or trait and provide vital insights for understanding other consciousness attributes and human behavior.

Awareness Quotient

We further revisit the definition of the Vedanta definition of consciousness as the 'experience of awareness' (Sarvapriyananda 2013). Furthermore, we expand 'individual's experience of awareness of the object'. This phrase provides a fundamental and vital pillar of consciousness—that is what an individual in his experience has a perception through his awareness of physical world. My experience is very personal. When two persons see an object, their experience might be different. This experience becomes their knowledge and their perspective on the object. Hence, right awareness will generate the right experience and that in term will generate right knowledge.

Awareness quotient even in the objective sense, which is concerned with plant and system status, design and operations modes, significance of these systems and functions in terms of operational and risk indicators along with their limits of operations and safety forms the core system of the knowledge base of the plant. Furthermore, awareness about the system during incidents and plant response through the knowledge of PRA documents along with deterministic provisions that include barriers and levels and assessments of levels of redundancy, diversity and potential for CCF forms a core system for risk and liability factors in risk-consciousness culture. In risk culture, apart from the risk of incidents/accidents, the risk of losses due to interruption and shutdown, risk of not meeting deliverables and risk of security also result in a holistic awareness of risk.

Alertness Quotient

Plant supervision, primarily from the control room on the status of the reactor and support operations and safety system and the corrective action in case of deviation to ensure normal plant operations, forms the core component of operational management. When the plant is in a shutdown state, various test, surveillance and maintenance activities are coordinated. The instrumentation and control design and operations philosophy has requirements to alert the operator in the control room particularly and in the local area of the plant based on requirements, hardware and software provisions in the form of audio/visual indications alarm graded according to their safety significance such that the operator can be alerted in advance about the status of the plant. Operator alertness plays a critical role during operation and shutdown. The review of incidences from historical records and the study of incidences from similar plants from the information available at the national and international levels provides the role and potential of the alertness quotient in plant safety and the corrective mechanism.

Emotional Quotient

Often, in regard to teamwork, particularly when operations management activities are involved that require an attitude for communication, coordination with more than one department agency and at the same time working under a well-defined hierarchy or organizational setup, managers with a higher emotional quotient might prove more affective for the successful completion of their jobs. Psychological science provides an effective mechanism to evaluate the emotional quotient, and this procedure can be adopted while considering the operational organizational requirements.

Happiness Index

Only happy employees in an organization can make the organization not only productive but also work toward making the plant and system risk-free. One of the major attributes that can be considered here is how closely the staff takes pride in his association with the organization. If the discussion in the, not meeting room but also canteen, or at home with family members, with colleagues in social gathering is about the risk reduction, then we say that we have risk-conscious thinking individuals. Furthermore, the organization is risk-conscious if it creates the right atmosphere for periodic meetings, group exchanges, and enthusiastic individuals who want to identify, assess, and take time-bound corrective actions to address the issues posing threats or potential threats of incidents. (Musikanski et al. 2017).

The happiness index of the employee has been considered in the Development of Happiness Index, as discussed here is to some extent in line with the approach adopted for assessing the Gross National Happiness Index (GNHI) (Ura et al. 2021). At the plant level, the happiness index supports production and risk reduction objectives, as both require the staff to be aware and alert about plant operations and corrective actions.

These subattributes can be divided into material factors and spiritual components for happiness as follows:

A. **Material Components**

Material components are those attributes that are derived from the physical and at a certain level of mental wellbeing. Material components are those attributes that seemingly work to make an individual confident that the organization and colleagues recognize her/his talent, services and technical ability and might also make him think that he is better or above average, particularly compared to his peers. The individual as well as the family's wellbeing in terms of physiological and psychological health goes long way for assuring individual happiness. There is a feeling of comfort with the kind of family acquisition, such as home, property, children's education and other social and economic factors that go toward ensuring material happiness. The following are some of the factors that can be indicative of the nature of the attributes of the employee happiness index:

A.1 **Remuneration**

Remunerations, allowances and performance-related rewards commensurate employee's contribution forms the umbilical cord between the employee and the organization that ensures employee and his family's nurturing, care and energy required to lead a life of dignity and honor in given social surroundings. The thinking that my organization is giving me such and taking care of me and my family not only needs but also luxury makes him pride and feel connected with the organization. This pride and association with the organization goes long way in reflecting higher performance and dedication of an employee.

A.2 **Professional Contributions**

A system of reward that recognizes excellence in the organization through individual and teamwork. The technical and professional contribution forms the hall mark of an organization in terms of sending the right signal down the line to propagate a culture of excellence, thereby increasingly encouraging emulating and motivating others to set new records of higher excellence. This system should be balanced and have punishment as a balancing factor, which means that the organization should also punish inadequate performance in a graded manner and send the right signal to employees in the organization that the organization is not tolerant of bad or inadequate performance. For a sustained growth of organization, the system of award/punishment should be implemented, without any bias and ensuring that complete transparency is maintained in the administration and management system.

A.3 **Carrier record**

The periodic performance review of the employee is a critical and vital instrument to assess the contribution of an employee. A transparent, rational-based system where employees should feel happy about their assessment is one of the factors for a sustained happiness quotient in an organization. It is a complex task for the organization and poses a challenge to achieve this, however, not impossible, if the robust, trusted and balanced communication system is in place between employee and the management.

A.4 **Physiological and psychological wellbeing**

Not only disease free but also overall physiological and psychological wellbeing is a basic requirement for human happiness. In the eventuality of health issues, the availability of a reliable health care system that assures timely and quality services for employees and their families is not an option but a must to maintain the happiness quotient of employees.

A.5 **Family wellbeing**

The remuneration, as discussed above, should be adequate to ensure that the employee meets not only his basic requirements but is also able to afford the luxury that is commensurate to his position and accepted in the social setup. The employee and his family should be able to afford their children's education without stressing his normal lifestyle.

B. **Spiritual Component**

Humans are basically spirits, and considerations of spiritual factors, along with material factors, make this human model more complete. These factors work at the core of human existence or human consciousness and provide an effective way of modeling human behavior, mental as well as physical. There are many spiritual schools based on different philosophies that provide ways and means to work on spiritual or inner or deeper levels of mind that eventually raise factors such as alertness and awareness along with ethical attributes that provide inner stability as well as the human qualities that work at the individual, interpersonal or organizational levels to raise human performance qualities and efficiency. These schools include Vedanta (Sarvapriyananda 2013), Buddhist (2022) and one of the special techniques called the Vipassana school of meditation (Hart 2020), which claims to preserve Buddhist culture and the technic of meditation. In risk-conscious culture, these paramitas are proposed to enhance inner human qualities such that higher human abilities are realized in different conditions of reactor operations management. The Paramita are Yoga, meditation and pranayama techniques have been embraced by the spiritual tradition of east right since the Vedic age—thousands of years before Buddha achieved Nirvana and it contributes to be the core technique for improving upon and self-realization. We do not need to prove the power of these techniques. Just sit in meditation posture for 10 min and focus your attention on your breath and see for yourself how you feel. Similarly, mankind has seen the benefits of the Yoga and pranayama techniques. The United Nations in 2014 recognized the Yoga, and now this technique is being performed worldwide to improve human wellbeing.

Mental Resilience/Stability

All weather condition mental stability or resilience is a key metric that needs to be considered right since the selection and training period of the staff. Response capability assessment during simulator sessions along with hands-on training is one of the most effective approaches for mental stability assessments. Here, the objective is that the individual demonstrates a composed and stable behavior, particularly for corrective emergency actions. Mental stability is the key for cognitive stability and is critical for required human response in given time and other constraints.

Cognition

The dictionary meaning of cognition is the process of knowing in the broadest sense, including the perception, memory, judgment. Accordingly, the cognitive capabilities include recall or retrieve information from memory, analysis of the situation with given input, arriving at inference, using intuition and finally decision making. These activities, depending on the situational requirements, work independently or collectively. The major task involves generating the right perception applying the rationales or logic. The one attribute that captures the essence of cognitive capability is 'intelligence'.

Intelligence

As per Wikipedia Intelligence in many ways: the capacity for logic, understanding, self-awareness, learning, emotional knowledge, reasoning, planning, creativity, critical thinking and problem solving (Wikipedia, Online). In the context of complex system operations, intelligence as an attribute can be defined as 'It is the capability of human beings to see reality through information, data and logic processing and its interpretation in the optimum time in support of decision making'. Human cognitive faculties, like processing of sensory information, memory locations, neural information processing, toward getting the right perspective of the situation in support of decision making. In psychology, there are established procedures for assessing the intelligence quotient. These assessment procedures can be modified suitably to reflect operational aspects, particularly the assessment of behavioral attitude during normal as well as emergency conditions. From the above discussion, the major attributes used to evaluate intelligence are as follows.

Agility of Memory

The ability to access the stored data and technical details from memory efficiently retrieve the required details to support the knowledge base along with other mental factors. It helps to have a good knowledge of system parameter settings, location of equipment, and many more plant data and information in support of operational jobs. Generally, it is not difficult to identify individuals with good memory.

Perception

The ability of abstraction from the data and information collected from the sensory stimulus and stored in the memory as a pattern for operational activities, such as comparison of systems performance while relating to their distinguishing features, forming perception about employees' attitude toward work without going into many details. Often, perception as part of intuition also supports the function of decision making.

Logic Processing

Logic processing essentially means dealing with if-and-then constructs as antecedents and outcomes as consequences. The knowledge created can be seen as the development of production rules that provide a way to create and store knowledge about plant behavior in terms of binary response. For example, the straightforward and critical plant instrumentation and control functioning is designed around logic processing either employing the old relay-based logic or in modern time employing software logic as part of digital systems. In the operations environment, the operating configuration and status of systems and equipment depending on requirements are integrated through logic processing, and the staff is supposed to have knowledge of the processing logic. Furthermore, the technical specification prescribes the limits and conditions that are configured around the implementation of safety logic.

Interpretation

Interpretation is the processing of available data, audio and visual signals to draw inferences or insights into reality. In the context of the control room environment in a plant, it is often required to interpret the visual and audio signal information available from the mimic panel on the visual display units or hard mimic panels to understand the plant and system in status and determine the course of action or corrective actions. Plant design and operational knowledge along with plant dynamics and correlating the available information to arrive at the inference form the core mental/cognitive activity in support of decisions.

Analysis

Analysis is a systematic process to model and understand system behavior with some assumptions and constraints as the prerequisite while working with given input parameters processed through a model of the system or a transfer function and conditions to generate the desired output. Analysis as an activity can range from simple tasks that involve correlating a couple of input signals to understand system configurations to complex tasks, such as the analysis of the loss of coolant accident (LOCA) scenario and its consequences. However, in the context of operational cases, the availability of complex and dynamic analyses, such as LOCA analysis, loss of off-site power (LOOP) or loss of regulation incidents (LORI), are ensured in safety analysis reports, and the staff is trained in emergency operating procedures (EOPs) with respect to these accident sequences. In operational environments, analysis or diagnosis is routine task to understand plant configuration and system behavior.

Learning Capability and Decision Making Quotient

Even though the learning capability of the staff can be assessed during the training, the decision making quotient can be tracked only during real-time control room- or plant area-specific metrics. Learning capability is an attribute that enables the accumulation of knowledge during training as well as while discharging duties during real-time conditions. The new insights from incidents and events, change in policy and practices, and upgrading of hardware, software and regulatory observations are some of the few opportunities that enhance learning and keep an individual updated.

Decision making capability plays a major role in normal operations, deviations from normal operations, emergency response and further accident management. Improved decision making based on diagnosing and comprehending the condition for controlling deviation and bringing back the plant to normal operation. If the plant could not be brought back to normal operation and automatic design basis options keep the plant under safe conditions. However, even though adequate automation exists, the history of accidents world over shows that operator decision making is vital for plant safety. Furthermore, the last defense on emergency management might involve planned actions. In these situations, the operator or emergency control room computerized aids provide advice for emergency response based on diagnostics based on the sensor inputs and site to control room communications.

Requirements of learning and diagnostics should be identified based on safety analysis based on the insights from accident analysis.

Conscience

Major Attributes of Conscience

Ethical behavior or moral principles/attitudes are generally implicit but can be taught or inculcated as part of culture based on the value system of the organization. Ethical and moral behavior is an important cornerstone of safety in general and security of the organization be it internal or external threats. Therefore, risk consciousness plays a critical role in overall risk reduction. At the entry level of an employee, the major parameters assessed through written tests and interviews are formulated in such a way that in principle focusses on the evaluation of individual capability, expected performance and the ability to work in a team through parameters such as educational levels, earlier experience and records, recommendations and experience. However, there is an implicit attempt to understand conscience levels that is determined by his expected ethical and moral behavior. These inputs form part of decision making regarding his/her entry into the organization. The periodic personal evaluation during the service, through parameters such as integrity, dedication and commitments to work also provides a system for tracking individual conscience levels.

Often, it has been seen that this system needs a review such that a more meaningful and useful system can be created. For example, personal integrity is assessed by binary logic, i.e., a person's integrity is 0 or 1. The second limitation is that the metrics need to be more elaborated and need guidelines for marking. Without guidelines and organizational moral principles, the metrics might lose their meaning. Furthermore, how to evaluate its impact also need to be provided through a metrics. Table 4.2 provides the major attributes of the conscience metrics.

As mentioned earlier, the next and most important job at the organizational and individual levels is developing a grading scale (very high, high, medium and low) instead of binary levels (e.g., yes or no). The grading system helps the evaluators make appropriate decisions based on the observed data or incidents.

Other Consciousness Factors

Human ability to remain equanimous is critical to human attitude and behavior. Equanimity can be achieved by self-observation in different life conditions. The bottom line is one should avoid the two opposites, i.e., craving and aversion. Craving for the things we like, i.e., higher position or promotion, good office, a car and desire for higher things in this material world, on the other hand aversion for unpleasant situation or conditions, like even a although of 'not getting promotion', or children not doing well in academics, or could not manage to buy a car this year, and son. One should have tolerance for certain things not going the way we want or other's uncomfortable behavior toward us or our family or by office colleague. In fact, the attitude of compassion toward fellow beings in office and society should be the hallmark of our personality. In a safety critical system operational environment, it is required that the employee has the capability to be equanimous not only during normal operation but also during treat or emergency conditions. As during emergencies, plant staff, particularly control room staff, need to maintain the right balance, as the science of human error suggests that during stressful conditions in the control room, the cognitive capacity for understanding, diagnostics and analysis is adversely affected. That is, when the human needs this capability most, it is degraded. Therefore, the training, qualification and selection should ensure that the staff expected to handle treats and emergency scenarios have robust cognitive ability/stability.

Table 4.2 Conscience attributes in RCOM

S. no.	Attribute	Description	Some of the possible mechanisms to achieve the attributes
1	Ethos	It can be defined as a collection of qualities or values like ensuring high-quality products and services keeping customer satisfaction as in perpetuality. Based on the value systems, it is possible to distinguish one organization from other	Identify one area or business houses among the available in the open domain, chose more than one business houses compare their bottom line and spending. The organization spending more on social corporate responsibilities meeting the highest standards in customer satisfaction in production and after sales and services
2	Moral Conduct	Concerns with the principles of right and wrong behavior. Acting with the best interest of organization while being conscious of the public safety, security and service. The organizational value system based on the high moral values is the hall mark of any quality organization	Organization should have a written policy defining the attributes of moral conducts right from top management to grass root level
3	Integrity	Strength of moral human character that are unshakable during trying conditions while living and seemingly following the high principles of human conduct	The major attribute of attribute is dependability. Some of the attributes of integrity are loyalty, uprightness and sincerity. Monitoring and assessing these attributes might ensure higher integrity levels
4	Honesty	Human quality principles of unquestionable truthful behavior in the best interest of nation and organization	Can be judged through exchanges that includes at individual, meeting, and group discussion level
5	Transparency	It is an attribute that defines 'visibility of our actions, relations within and outside of organizations' respecting the best interest, particularly safety and security principles of conduct of the organization	This attribute even though implicit in behavior it requires organized efforts to establish itself. For example, copying an email to superior for official matters, involving apart from stack holders, other concerned in discussion and decision making
6	Commitment	Ability to meet and respect the organizational project/work deadlines	If we observe an individual for a modest period say couple of years and it will be possible to evaluate their commitment levels

(continued)

Table 4.2 (continued)

S. no.	Attribute	Description	Some of the possible mechanisms to achieve the attributes
7	Dedication	There is thin line between dedication and commitment; in the context of RCC, dedication is a perpetual attribute that combines concentrated sincere efforts for achieving desired quality and organizational objective and goals. It works at individual as also at team level	Some of the characteristics like giving their 100% at physical and intellectual level with utmost sincerity while meeting the quality and organizational goals or assigned jobs. In the context of RCOM, it means ensuring and remaining committed to risk reduction efforts on continuous basis

Hard work, persistent or painstaking efforts, dedication to the duty and sincerity are the needed qualities that a human should have to achieve excellence. The area of operations management requires one to gain insight into the design and operational aspects of mega systems and to understand the micro- to nanolevel aspects, such as neutronics and nuclear aspects, to understand the reactor core kinetics and safety aspects. This requires highly energetic aspects, and training and qualification programs teach reactor-related aspects, but individual human factors are at the core of human performance.

Wisdom means performing your Dharma; here, Dharma means right action and right thinking. How one knows what is right and what is wrong. The Vipassana approach is a very simple and elegant way to approach the definition of Dharma. Any action which is in the interest of all and does good to all and society at large is right Dharma, on the other hand any action that harms to others is Adharma (opposite of Dharma). In the context of organizational behavior, we further define the Dharma of an individual as the performance of official duties by maintaining high morality and value such that the organization benefits. Generally, we acquire knowledge by reading, observing and listening to others; however, true knowledge comes from direct experience. Hence, experiments, feelings and questioning attitudes have been considered higher levels of wisdom. The action performed in accordance with Dharma principles, keeping in mind the plant society at large, is the key attribute of wisdom.

Emotional quotient is as important or often more important in the management of technical functions. This attribute is vital as part of the management and coordination function toward creating a positive environment and keeping the individual and organization connected. A member of the organization that needs support for his or his family-related issue goes down a long way when all the employees become a family to the member passing through a rough phase. The organizational colleagues are ‘first friends and then colleagues’—this approach, can temper down or help is addressing the negative side of the feeling of jealously, unhealthy competition and grid, which often hinders effort to create a healthy environment.

4.5.4 *Technical Elements*

4.5.4.1 Salient Features—An Overview

There is a need for a more structured, measurable, and individual or organizational conscious-driven environment that is interpretable, integrable by operational staff and management to address management such that risk levels are well within acceptable guidelines.

Figure 4.2 shows the technical framework for the implementation of a change management and corrective action program that is based on periodic safety performance, where the focus of feedback is the identification of human and organizational factors (as it is assumed that sustainable improvements are only possible to improve conscious factors) to meet the objective of a perpetual system of derisking, i.e., working for achieving the 'near zero risk target', e.g., CDF $< 10^{-6}$/yr.

However, the open literature defines risk-conscious culture as an integrated and conscious operational management environment of organizational and individual values, attitude, behavior, connection, communication driven by overriding priority to risk assessment, monitoring, management and mitigation through the application

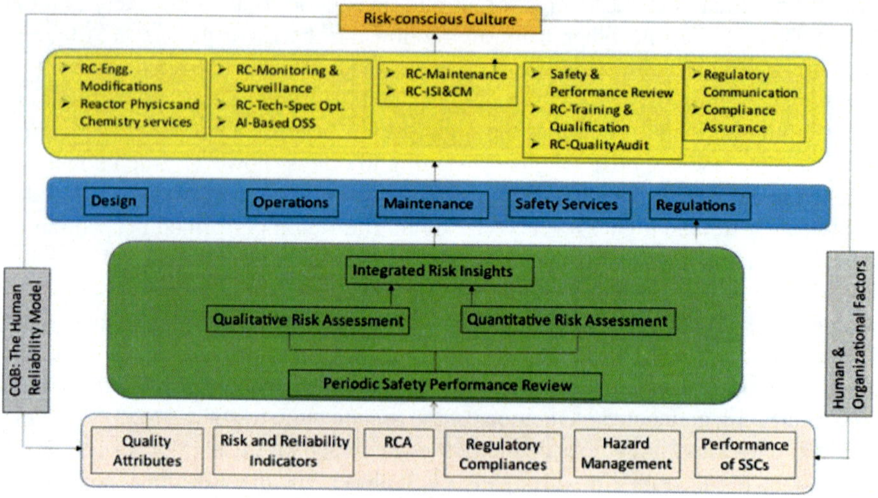

Fig. 4.2 Technical framework of risk-conscious culture

of a risk-based engineering framework toward providing holistic, systematic and rational-based actions and decisions.

Taking a cue from the above discussion, the definition of risk-conscious culture is proposed as follows: *A system that represents a collective human consciousness and associated attributes for a proactive approach to predicting, monitoring, assessing, and implementing the corrective action and feedback program as an integral part of operations management in a sustainable manner to achieve 'a near zero risk target' (NZRT) to protect life, property and the environment from undesired consequences.* Here, the culture represents organizational policies, practices, communication mechanisms and general human attributes generally shared among the staff members that include grassroots employment through middle management to top policy makers and authorities responsible for operational and strategic decisions.

We can discuss the above technical framework in four steps or levels, plant operation ecosystem, risk and performance evaluation, identifying major plant functions, major activities for risk-conscious functions, and finally the feedback systems for corrective action and management programs, as follows:

4.5.4.2 Plant Operations Technical and Management Ecosystems

The first level, that is, the plant ecosystem, represents the fundamental policies, policies in general and quality, and derisking policy in particular, plants (the major elements are SSCs that include nuclear, electrical and electronics as part of the process and industrial systems and their coordination and integration that facilitate ensuring safety as well as deliverables), human and organizational practices, management hierarchy, procedures/documents, review and regulatory provisions. Apart from this, manual and automatic monitoring and protection systems along with the management of resources that include raw and finished products, trained and qualified staff, and waste also form part of plant operations.

There are two major aspects, one that refers to safety documents primarily the technical specifications that present the plant safety policy in the form of safety limits, limiting safety system settings, limiting condition for operation, organization chart stipulation for treating deviations from normal plant operations that are considered as potential or real threat, surveillance and maintenance policy such that plant operations can be ensured such that it with the stipulated condition or enveloping conditions in which plant operations have been demonstrated to be safe in the analysis. The other safety document includes safety analysis reports (probabilistic and deterministic) that demonstrate the safety of the plant for the design basis considering the postulation of various initiating and the development of scenarios based on the response of the available safety and protection system in the plant. Other documents include emergency operations and procedures.

The second aspect of the plant constitutes a collection of process systems. Operations of process systems in a reliable manner are required to ensure the plant deliverable, be it electricity production, health care products, facilitate scientific R&D experiment, and engineering testing and irradiation requirements as applicable.

In the context of apart from the general O&M operations risk-conscious culture, the periodic safety review forms a major element that facilitates the assessment of the existing culture and input or feedback for areas for derisking such that operations management decisions can be implemented to work toward the target of near zero risk policy. As is generally recognized, here, the major input, as evident from the root cause analysis, will be on human factor improvement with respect to the complete plant life cycle right from siting design, operations, maintenance, aging management and finally decommissioning and waste management to reach a green field condition.

4.5.4.3 Integrated Performance Review

This is the second level of technical element in risk-conscious culture. The safety performance review has four basic elements viz., a) the safety performance indicators, b) the review of operation of major safety and process systems performance, insights obtained from the root cause analysis risk-significant events as also the events that adversly affect the plant performance, c) the analysis and presentation of major safety indicators that includes deterministic as well as probabilistic indicators and d) the recommendation for de-risking and improvement of plant reliability/performance. The analysis of indicators is performed to evaluate its applicability. The integrated risk insights are formulated systematically toward categorization into six groups, as shown in Fig. 4.2, keeping in mind the organizational and functional structure. The six major areas include design, operations, maintenance, safety services or safety upgrades, and finally, regulatory compliance, as shown at the third level.

4.5.4.4 Implementation

This is the fourth and elaborate and resource-consuming step in modeling the analysis and implementation of each technical function of risk-consciousness culture. Figure 4.2 shows the technical functions for each category. For example, the design function in operations management addresses the design and analysis of modifications employing an approach that reorients the traditional approach to risk-based or risk-informed approaches to design. The integrated risk-based design approach (Varde and Pecht 2018) is deployed to capture not only to comply with deterministic but also probabilistic procedures and criteria.

For example, the integrated safety review along with the analysis of safety indicators suggested upgrades in the design of ECCS by improving redundancy in emergency pumping system configurations to achieve the flow and pressure requirement for a relatively larger mission time on the one hand and improve system availability through a probabilistic evaluation. The analysis identified an additional pump in both trains that inject emergency cooling water at the inlet and outlet of the reactor legs. Apart from this, the risk-conscious evaluation of emergency procedures also forms part of the risk-conscious culture. Additionally, it was also envisaged that the training checklist and related training module should also reflect this change or upgrade. A revised PRA analysis forms part of this upgrades. Before, the upgrade implementation starts, a regulatory approval with the above document that makes the safety case for regulatory review for approval.

Regulatory approval is in accordance with stipulations that might include requirements of quality, reporting criteria, system qualification and presentation of results of commissioning for regulatory review as part of consenting activities.

The probabilistic risk assessment of the plant is a fundamental requirement for the implementation of risk-conscious culture. The major insights from quantified estimates are used for the identification, prioritization, optimization, and sensitivity of unavailability at the system level, and core damage statements at plants provide an effective mechanism for change. The availability of a computerized model helps an effective approach for simulation and modeling to arrive at the final configuration for the cases being evaluated. Apart from the approach to the characterization of uncertainties, it is vital to evaluate the strength of the available data, which is critical to decision making and the evaluation of the impact of assumptions. The major activities that require considerations are depicted at the top block in Fig. 4.2, which include considerations of, apart from engineering modifications as discussed above, the scientific modeling for core considerations that is specific to an industry such as reactor physics and chemical provisions for corrosion-related degradation monitoring; rick-conscious supervision in the control room also functions as feedback generation for the changes, risk-conscious technical optimization where the focus is not only on risk information but also to relate the impact of human factors as root causes to arrive at optimized configurations for safety limits, safety system settings and limiting conditions for operations. The risk-conscious approach considers AI- or machine learning-based approaches for reducing human error, especially for emergency conditions, through intelligent operator aids that work by reducing the cognitive load through better awareness and alertness to operating staff. Extending the available risk-informed approaches to risk/reliability-based maintenance and prioritizing the cases for regulatory compliance and audits are effective options available to support operations management.

However, the most relevant mechanism to implement the risk-conscious culture is through reorienting the existing training and qualification program that prepares the staff for risk-conscious culture while meeting plant deliverables. The major areas of upgrades are modification of checklists to include the knowledge and insights from a PRA study systematically for various systems of the plant; simulator-based training for various initiating events by considering availability/unavailability of various safety systems; real-time experiments on simulator to have a better appreciation of operating envelop and emergency conditions; the periodic review of the events in the plant and associating and analysis of the risk factor, particularly human factors for better appreciation of decisions in a timely manner.

4.5.4.5 Six Major Pillars for Implementing Risk-Conscious Culture

The main objective of the RCOM approach is risk reduction and management through the application of structured, logical, measurable and resilient measures. There are four pillars for effectiveness implementation and success in terms of gainful utilization of this approach as follows.

Organizational Mandate

The organizational mandated driven through governmental and organizational policy is a must to implement risk-conscious culture. The policy or the stipulation should recognize the significance of risk reduction to support safety improvement, particularly through the improvement of human and organizational factors. Even though the nuclear industry has been recognizing the need for safety culture improvement and international organizations are developing and updating the safety culture framework by revising the codes and standards, these documents provide guidance on 'What to do' rather than 'How to do'. Here, the risk-conscious culture provides the framework to deal with plant-specific issues by employing a rational-based framework facilitated by a system to interpret numerical inputs into subjective feel factors such that the operational team can adopt these insights in support of decisions.

The organizational mandate should have provision of direct participation and responsibilities of top management toward ensuring adequate resources that include infrastructure, human resources, funding and provision for regulatory oversight.

Availability of Probabilistic Risk Assessment Model of the Plant

There is a need for a comprehensive and robust human model to reduce the uncertainty in understanding and modeling, monitoring human behavior in general and assessing human action, which includes physical and cognitive reliability during normal operations as well as emergency conditions. The CQB model of human reliability developed by Varde and Pecht has been updated and modified to cater to the operational environment. This model considers the role of consciousness, cognition, conscience and brain in an integrated manner to model human reliability. Keeping in mind the requirements of operational environment, the RCOM approach modifies the current structure of the CQB model such that consciousness has been considered the fundamental source/attribute to of human being and the cognition and consciences derive their functional energy from consciousness. Furthermore, the RCOM approach seeks to enhance the scope of this model to include precursor events/causes to human failure; furthermore, it also incorporates a mechanism to extend the attributes of the RCOM approach to performance-shaping factors in the CQB model.

A PRA model that qualifies for operational requirements should be (a) a model that incorporates intimate operating and as-built plant information that includes real-time plant configuration, plant-specific reliability data including CCF and human factor events derived from the incident reports, (b) quality PRA such that it adequately meets national and international quality attributes (e.g., IAEA, ASME), (c) apart from base level approval, the model should have specific approval for application to address real-time operations and maintenance aspects, such as plant configurational changes, test and maintenance schedule evaluation, risk-based in-service inspection program, and (d) the PRA model should have a regulatory approach for real-time applications.

Apart from risk insights from plant-specific PRA and incidences, the RCOM approach requires that there should be administrative and management provision to ensure that the training and qualification activities update and include this new insight from events and accidents from historical records and new events such that learning lessons can part of the plant-specific culture.

One of the major objectives of risk-conscious culture is to develop a resilient culture that ensures operations of systems such that systems and services capability, before, during and after disruption, are preserved to an acceptable level in terms of performance, specified mission reliability toward meeting risk objectives at the plant as well as at the safety functional level. To achieve this objective, the training and qualification scheme should extend the scope of training to reflect the recovery procedures, mitigation measures and accident management such that safety function maintenance and retaining their capability become a mind set for the operational screw.

Training and Qualification Program Targeted for Assuring Human Reliability

The reorientation of training programs, keeping in view the general observation that human factors are a major contributor to events and accidents, is a vital element in creating risk-conscious cultures. The overriding priority for derisking is central to the ethos of the plant, and the same should form the guiding thread across the training objectives and aims. As mentioned above, the availability of an approved probabilistic risk assessment study at a full scope level 1 model is a prerequisite to develop a risk-conscious culture. Apart from this, the technical insights and deterministic criteria that often form part of PRA in working out the success and failure criteria right from part and material constituting the component, component, systems and structures, and finally the plant, are key to defining the respective risk profiles that are required to create the risk model. Furthermore, an elaborate program focusing on training on human factors and common causes of failure is central to training programs.

Each training module should be subjected to revision and restructuring based on the knowledge and insights available in a considered and conscious manner as follows:

1. An organization policy in restructuring the training and qualification program with an orientation of risk-conscious culture.
2. Over encompassing role and significance of human factor in working toward near zero risk policy and achieving a target figure of CDF: $< 10^{-6}$/yr.

3. Apart from the individual redundant component, there should be considerations of the potential threat of common cause failure.
4. Availability of an approved risk monitor for facilitating the simulation of various risk profiles.
5. Availability of a full scope simulator as part of training and qualification program.
6. Availability of expertise in training and qualification procedures and evaluation programs.
7. Periodic and independent review/regulatory training programs ensure that the spirit of training, particularly human factor considerations, form part of qualification and authorization.
8. Provision of periodic physical and phonological health checkups and consideration of spiritual practices (Yoga, meditation, etc.) for ensuring enhanced cognitive, consciousness and conscience attributes.

The focus of risk-conscious culture is not limited to human reliability analysis but to further improving the human factors in perpetuality (i.e., not limited to the training period) to extend this human factor aspect for authorized staff.

Root Cause Analysis

Root cause analysis is an important element of an operational feedback system. This is an established tool in the nuclear industry. Often, it forms an integral part of precursor event analysis. The major procedural element of RCA starts with the formation of team, information collection, making a timeline in chronological order, interview with concerned staff, and analysis of the event employing one of the established methodologies/techniques, such as cause-consequence diagram, event sequence analysis, failure mode and effect analysis, fault tree and event analysis, and barrier analysis, depending upon the nature of the investigation to be performed. Finally, the objective is to arrive at the root causes responsible for the subject event.

The available documents also list risk factor categories that include design, maintenance, housekeeping, error re-enforcing conditions, procedural, communication, incomprehensible goals (tight timeline, complex task, new condition not covered in training), etc. (International Atomic Energy Agency 2015). However, for the success of the RCA program in such a manner that the RCA program can gainfully be utilized as part of risk-conscious culture, additional conditions need to be met as follows:

(a) Framework and Policy: The RCA policy framework and guidelines for the implementation of the RCA program are critical elements. This framework provides the metrics for the categorization of incidents events for their safety significance, the administrative setup for the initiation of the RCA program, the timeline, the responsibility and authority at the administrative, management, and RCA team levels, and guidelines on the level of depth or detail needed.
(b) Categorization Approach: The basis for the categorization of events in risk-conscious cultures is the potential likelihood and consequences of the event. This requires mapping of the events on the PRA model of the plant.

(c) The Levels of RCA: The level resources invested in an RCA should be commensurate with its risk-significance. Accordingly, the RCA can be performed at three levels: Level 1 RCA—preliminary analysis, Level 2 RCA -having medium risk-significance and Level 3 RCA—risk-significance is high. The experience suggests that a Level 1 RCA takes from 10 to 15 days, Level 2 takes up to two months and Level 3 takes from 2 to 6 months or more.
(d) Level of Details: While Level 1 may not require much detailed or in-depth analysis, Level 2 might require resources of a typical RCA approach in line with conventional RCA approaches. However, the Level 3 RCA might require starting from detailed PRA simulation and a reactor simulator, for example, to understand core neutronics or thermal behavior; finite element analysis to understand the stresses in various sections of a given component; thermal hydraulic analysis to understand system behavior for a given failure scenario; and detailed state of the art investigation for electronic system failure, for example, an electrolytic capacitor failure or a microchip failure might require investigation employing a state of the art lab facility that offers investigation right from an optical microscope to a photon emission microscope or laser or chemical decapsulator to reach the failure site.
(e) Infrastructure and facilities: The major facility, as mentioned in previous para, includes a reactor simulator, reliability lab for PRA modeling and analysis, a scaled down loop facility, prognostics/diagnostic labs, and an environmental testing lab to simulate and analyze the failure.
(f) Since it is not possible to have all the facilities and expertise in the operating organization, the RCA program requires networking with other labs in the organization or outside the organization. For example, a polymer O-ring failure investigation might require support from either the manufacturer and an outside consultant or a university support.

The point being made above is that some special incidences might require detailed investigation, and the typical steps of RCA might not be adequate and require understanding the physics of failure approach.

Quality Assurance

Quality, i.e., 'conformance to specification', forms the bottom line for any excellence, including the development of risk-conscious culture. Lack of quality assurance or organizations' inadequate attitude toward quality assurance, quality management and quality control has been found to be one of the root causes for the incidences or risk issues. All the activities/aspects that are considered for risk-conscious culture should be subjected to quality standards, be it technical specifications, documentation, operating procedures, maintenance schedules, regulatory compliance review, training and qualification program, the PRA model of the plant or a computerized maintenance management system and so on. To facilitate the development of quality assurance and management programs, the prerequisite is the development of specifications.

The specifications should be drawn by the experts and provide the right attributes for specifications. For example, quality specification for maintenance schedule should be based on the bench-marked duration for the activity, the shutdown schedule should provide the statement of risk of not meeting the set time and target keeping in view the uncertainty associated with the jobs. Similarly, the design modifications should show the tolerances and allowances for the parts and operational limits for achieving the set performance or declaring the part unusable. This approach of quality assurance is proactive and reduces failures and improves system reliability while reducing risk and consequences of failure.

As discussed in the previous section, a PRA considered for risk-conscious culture should be subjected to a quality assurance program with respect to all the major activities and will be discussed in the following section.

Documentation

Often, the quality consciousness of an organization is assessed by the standards and systems followed for documentation. The European ISO 9001 standard for documentation is one of the examples of quality assurance in documentation and certification. In the evolving scenario, advanced systems involving the cloud for storage and intelligent inference-based approaches are being used for the efficient uploading and retrieval of records. For a risk-conscious organization, it is required that daily reports, work permits, incidents reports, root cause analysis, reports, reliability analysis and probabilistic risk assessment, surveillance and testing reports, emergency drill reports and the reliability database as well updated system reliability indicators reports are stored in such a manner that makes the eliciting the information in support of risk assessment and updating the documents.

These six pillars are expected to facilitate an improved ecosystem for developing a risk-conscious culture and organization, as most of the factors that contribute to risk, as evident in a PRA model, can be effectively addressed.

Annex: Table Typical Generic Operational Activities

S. no.	Precursor description	Activity type	Human CQB attributes for precursors			Human-factors J_c	PE rate ε/yr	Time window t/T	Potential consequence of PE	Consequence Level (Table 4.1)	Risk factor for PE (R/yr)
		D/E/O /M/S	(Q_1 * w_1)	(Q_2 w_2)	(Q_3 w_3)		P_f	t/Tw	PE_{PC}		
1	Control room supervisory activities	O	4 * 0.3 = 1.2	4 * 0.6 = 2.4	2 * 0.1 = 0.2	3.8	1.5	0.67	Reactor trip/unavailability risk	L-1 1.45	0.12
2	Management of emergency scenario	O/M	8 * 0.6 = 4.8	1 * 0.3 = 0.3	1 * 0.1 = 0.1	5.2	1.0	0.91	Failure to recovery	L-5 10	0.75
3	Field supervision	O	2 * 0.6 = 1.2	6 * 0.2 = 1.2	2 * 0.2 = 0.4	2.8	0.1	0.5	Judgmental/error	L-1 1.2	$3.4 * 10^{-3}$
4	Procedure execution—Op	O	3 * 0.5 = 1.5	4 * 0.4 = 1.6	3 * 0.1 = 0.3	3.4	1.0	0.59	Individual attribution	L-2 2	$8 * 10^{-2}$
5	Process diagnosis	O	5 * 0.5 = 2.5	5 * 0.5 = 2.5	0 = 0	5.0	1.0	0.77	DG Diagnosis failure during Class IV PS Failure	L-1 1.7	0.13
6	Correlating the evolving scenario during a significant event	O	6 (0.8) = 4.8	4 (0.2) = 0.8	0 (0) = 0	5.6	1.0	0.84	Loss of Team capability to comprehend the scenario	L-3 3.5	0.32

(continued)

(continued)

S. no.	Precursor description	Activity type	Human CQB attributes for precursors			Human-factors J_c	PE rate ε/yr	Time window t/T	Potential consequence of PE	Consequence Level (Table 4.1)	Risk factor for PE (R/yr)
		D/E/O /M/S	(Q_1 * w_1)	(Q_2 w_2)	(Q_3 w_3)		P_f	t/Tw	PE_{PC}		
7	System state identification	O	7 * 0.7 = 4.9	3 * 0.3 = 0.9	0 (0) = 0	5.8	1.0	0.91	Identification for cause of the event	L-2 2.0	0.22
8	Field activities—status	O/M	3 * 0.3 = 0.9	2 * 0.5 = 1.0	5 * 0.2 = 1.0	2.9	2.0	0.67	Local or reactor availability risk	L-2 1.5	0.116
9	Training lectures	S	5 * 0.5 = 2.5	3 * 0.4 = 1.2	2 * 0.1 = 0.2	3.9	2.0	0.5	Organizational cultural gap	L-2 3	0.24
10	Training and qualification—checklist	S	5 * 0.4 = 2.0	1 * 0.5 = 0.5	4 * 0.1 = 0.4	2.9	1.2	0.56	Organizational cultural gap	L-2 3	0.12
11	Training and qualification simulator experiments for	S	5.0 * 0.6 = 3.0	3 * 0.2 = 0.6	2 * 0.2 = 0.4	4.0	1.5	0.84	Organizational cultural gap	L-2 3	0.30
12	Engineering modifications	S	7 * 0.6	2 * 0.3	1 * 0.1	4.9	1.0	0.56	Planning error	L-3 4.0	0.22
13	Analysis and development	O/S/M	8 * 0.8	2 * 0.2	0 (0)	6.8	1.0	0.56	Individual and training issues	L-3 4.0	0.3

(continued)

(continued)

S. no.	Precursor description	Activity type	Human CQB attributes for precursors			Human-factors J_c	PE rate ε/yr	Time window t/T	Potential consequence of PE	Consequence Level (Table 4.1)	Risk factor for PE (R/yr)
		D/E/O /M/S	(Q_1 * w_1)	(Q_2 w_2)	(Q_3 w_3)		P_f	t/Tw	PE_{PC}		
14	In-service inspection	O/M	3 * 0.5 = 1.5	3 * 0.2 = 0.6	4 * 0.3 = 1.2	3.3	2.0	0.77	Inadequate inspection has potential for loss of structural integrity	L-3 4.5	0.57
14	Maintenance of hardware Mechanical and electrical	M	2 * 0.2 = 0.4	2 * 0.4 = 0.8	6 * 0.4 = 2.4	3.6	1.8	0.77	CCF potential	L-3 5.0	0.5
15	Maintenance of digital systems and software	M	5 * 0.4 = 2.0	2 * 0.3 = 0.6	3 * 0.5 = 1.5	4.1	1.5	0.91	CCF potential	L-2 4.5	0.50
16	Instrument calibrations	M	3 * 0.1 = 0.3	4 * 0.3 = 1.2	3 * 0.6 = 1.8	3.3	2.0	0.91	CCF potential	L-3 4.5	0.54
17	Mechanical maint -pump	M	3 * 0.2 = 0.6	2 * 0.3 = 0.6	5 * 0.5 = 2.5	3.7	1.2	0.84	Potential for fire due to assembly error	L-2 2.5	0.185
16	Logging and reporting	O/M	1.0 * 0.5 = 0.5	1 (0.3) = 0.3	8 (0.2) = 1.6	2.4	4.0	0.77	Human and organizational error	L-1 1.5	0.22

(continued)

(continued)

S. no.	Precursor description	Activity type	Human CQB attributes for precursors			Human-factors J_c	PE rate ε/yr	Time window t/T	Potential consequence of PE	Consequence Level (Table 4.1)	Risk factor for PE (R/yr)
		D/E/O /M/S	(Q_1 * w_1)	(Q_2 w_2)	(Q_3 w_3)		P_f	t/Tw	PE_{PC}		
17	*Coordination with agencies—control room*	*O*	*5 * 0.6 = 3.0*	*4 * 0.3 = 1.2*	*1 * 0.1 = 0.1*	*4.3*	*2.0*	*0.9*	*Individual/organizational attribute*	*L-3 4.0*	*0.61*
18	Communications and coordination *during* control room in emergency	O	4 * 0.3 = 1.2	3 * 0.2 = 0.6	3 * 0.5 = 1.5	3.3	2.4	0.9	Organizational cultural gap: error in communication	L-5 7.4	0.57
19	Emergency management actions in control room	O	7 * 0.5 = 3.5	2 * 0.3 = 0.6	1 * 0.2 = 0.2	4.3	2.0	0.84	Organizational cultural gap	L-4 6.0	0.87
20	Human–machine Interface/interaction as part of recovery procedure during emergency	O	6 * 0.4 = 2.4	2 * 0.4 = 0.8	2 * 0.2 = 0.4	3.6	1.7	0.9	Individual/organizational	L-5 8.0	0.89
21	Updating documents and engineering drawing	E	3 * 0.8 = 2.4	1 * 0.7 = 0.7	6 * 0.2 = 1.2	4.3	1.8	0.77	Potential for missing as-built information leading to error in O&M	L-2 2.7	0.32
22	Shift change over—charge exchange in shifts	O	5 * 0.3 = 1.5	4 * 0.4 = 1.6	1 * 0.2 = 0.2	3.3	2.0	0.9	Organizational cultural gap/individual gap	L-2 2.5	0.18

(continued)

(continued)

S. no.	Precursor description	Activity type	Human CQB attributes for precursors			Human-factors J_c	PE rate ε/yr	Time window t/T	Potential consequence of PE	Consequence Level (Table 4.1)	Risk factor for PE (R/yr)
		D/E/O /M/S	(Q_1 * w_1)	(Q_2 w_2)	(Q_3 w_3)		P_f	t/Tw	PE_{PC}		
23	Interpretation of results of analysis—control room	O	8 * 0.5 = 4.0	6 * 0.3 = 1.8	0	5.8	2.0	0.9	Individual capability limitation	L-1 1.5	0.32
24	Plant areas upkeep	O	1 * 0.5 = 0.5	2 * 0.2 = 0.4	7 * 0.3 = 2.1	3.0	2.1	1.3	Organizational cultural gap	L-2 2.0	0.33
25	Plant status monitoring—control room	O	3 * 0.5 = 1.5	4 * 0.4 = 1.6	3 * 0.1 = 0.3	3.4	2.0	1.1	Events induced due to inadequate awareness and attention	L-2 2.5	0.37
26	Events and significant events analysis and corrective actions	O/S	8 * 0.5 = 4.0	2 * 0.5 = 1.0	0.0	5.0	2	0.84	Organizational cultural gap/Individual gap	L-3 3.5	0.58
27	Regulatory compliance	S	3 * 0.2 = 0.6	5 * 0.6 = 3	2 * 0.2 = 0.4	4.0	3.0	1.0	Organizational cultural gap	L-3 3.8	0.92
28	Safety documentation	S	4 * 0.4 = 1.6	4 * 0.5 = 2.0	2 * 0.2 = 0.4	4.0	3.0	0.9	Organizational cultural gap	L-2 2.5	0.54
29	Operations documentation	O	3 (0.3) = 0.9	4 (0.5) = 2.0	3 (0.2) = 0.6	3.5	1.0	0.67	Organizational cultural gap	L-1 1.5	0.07

(continued)

(continued)

S. no.	Precursor description	Activity type	Human CQB attributes for precursors			Human-factors J_c	PE rate ε/yr	Time window t/T	Potential consequence of PE	Consequence Level (Table 4.1)	Risk factor for PE (R/yr)
		D/E/O /M/S	(Q_1 * w_1)	(Q_2 w_2)	(Q_3 w_3)		P_f	t/T*w*	PE_{PC}		
30	TechSpec compliance	S	6 (0.7) = 4.2	4 (0.3) = 1.2	-	5.4	1.7	0.9	Individual/organizational cultural gap	L-2 2.5	0.41
31	Procedure execution—maintenance	M	5 * 0.6 = 3.0	4 * 0.3 = 1.2	1 * 0.1 = 0.1	4.3	1.5	0.77	Individual limitation/training adequacy	L-2 2.8	0.28
32	Maintenance of tools and support	M	3 * 0.4 = 1.2	2 * 0.4 = 0.8	5 * 0.2 = 1.0	3.0	1.0	0.9	Organizational cultural gap	L-2 2.5	0.14
33	Inadequate or improper Testing standby safety systems	O/M	5 * 0.4 = 2.0	3 * 0.4 = 1.2	2 * 0.2 = 0.4	3.6	1.0	1.0	Potential disabilities that affect equipment/system availability	L-4 7.0	0.51
34	Quality control and quality assurance	S	6 * 0.3 = 1.8	3 * 0.5 = 1.5	1 * 0.2 = 0.2	3.5	1.9	0.9	Compromised organizational culture	L-3 5.0	0.6
35	Fuel management	S/O	7 * 0.6 = 4.2	2 * 0.2 = 0.4	1 * 0.2 = 0.2	4.8	1.2	0.9	Threat to safety of fuel	L-3 5.0	0.52
36	Resilience activities related external events	S	8 * 0.5 = 4.0	2 * 0.5 = 1.0	0	5.0	1.4	0.9	Reduction in safety margins	L-5 8	1.0

(continued)

(continued)

S. no.	Precursor description	Activity type	Human CQB attributes for precursors			Human-factors J_c	PE rate ε/yr	Time window t/T	Potential consequence of PE	Consequence Level (Table 4.1)	Risk factor for PE (R/yr)
		D/E/O /M/S	$(Q_1 * w_1)$	$(Q_2\ w_2)$	$(Q_3\ w_3)$		P_f	t/Tw	PE_{PC}		
37	Staffing quality and positions	O	7 * 0.5 = 3.5	3 * 0.5 = 1.5	0	5.0	1.0	0.9	Organizational attitude	L-3 4	0.36
38	Spare inventory for safety functions	M	3 * 0.3 = 0.9	3 * 0.4 = 1.2	3 * 0.3 = 0.9	3	2.0	1.0	Equipment maintainability issues	L-3 4	0.48
39	Reactor physics and chemistry services	S	8 * 0.6 = 4.8	2 * 0.4 = 0.8	0	5.6	1.1	0.77	Analysis, trend monitoring and core safety activities compromised	7	0.66
40	Financial and administrative system/support that has potential to impacting plant safety	M	8 * 0.8 = 6.4	2 * 0.2 = 0.4	0	6.8	0.15	0.77	Threat to safety management	8	0.13

References

American Psychological Association (2019) Mindfulness meditation: a research-proven way to reduce stress. American Psychological Association

Bhattacharya S, Burman RR, Paul S (2019) The concept of measuring happiness and how India can go the Nordic Way (Commentary). Current Sci 116(1)

Buddhist-Meditation-Techniques-Practices (2022) [Online]. Available: https://mindworks.org/blog/buddhist-meditation-techniques-practices/. Accessed 25 June 2022

Corcoran WR (2004) Defining and analyzing precursors, accident precursor analysis and management: reducing technological risk through diligence. The National Academic Press. Open Book http://www.nap.edu/10766

Elmallah AM (2014) The facts behind three mile Island accident, nuclear energy engineering technology, Thomas Edison State College. Thomas Edison State College, USA

Fallon F (2022) Integrated information theory of consciousness. Internet Encyclopedia of Philosophy [Online]. Available: https://iep.utm.edu/integrated-information-theory-of-consciousness/#:~:text=Initially%20proposed%20by%20Giulio%20Tononi,according%20to%20the%20phi%20metric. Accessed 24 June 2022

Freud's model of the human mind. J Psyche [Online]. Available: http://www.journalpsyche.org/understanding-the-human-mind. Accessed 26 Jan 2021

Gil L (2021) IAEA bulletin: learning from Fukushima Daiichi: factors leading to the accident. IAEA (2021) [Online]. Available: https://www.iaea.org/bulletin/learning-from-fukushima-daiichi-factors-leading-to-the-accident

Goldman DJ, Schwartz GE (1976) Meditation as an intervention in stress reactivity. J Consult Clin Psychol 44(3):456–466

Hart W (2020) Vipassana meditation—the art of living, as taught by S.N. Goenka. Vipassana Research Institute, Igatpuri, India (2020)

IAEA (2022a) Culture for safety. IAEA [Online]. Available: https://www.iaea.org/sites/default/files/culture_for_safety_leaflet.pdf. Accessed 24 June 2022

IAEA (2022b) Safety and security culture—human and organizational factors. IAEA [Online]. Available: https://www.iaea.org/topics/safety-and-security-culture. Accessed 23 June 2022

IAEA (2022c) Safety Culture Continuous Improvement Process (SCCIP), IAEA [Online]. Available: https://www.iaea.org/services/review-missions/safety-culture-continuous-improvement-process-sccip. Accessed 2024 June 2022

Investigation Committee on the accident at the Fukushima nuclear power stations of Tokyo Electric Power Company [Online]. Available: https://en.wikipedia.org/wiki/Investigation_Committee_on_the_Accident_at_the_Fukushima_Nuclear_Power_Stations_of_Tokyo_Electric_Power_Company. Accessed 24 June 2022

International Atomic Energy Agency (1986) The chernobyl accident: updating of INSAG-1. IAEA, Vienna

International Atomic Energy Agency (1991) Safety culture, safety series 75-INSAG-4. IAEA, Vienna

International Atomic Energy Agency (1992) INSAG-7—the chernobyl accident: updating of INSAG-1. IAEA, Vienna

International Atomic Energy Agency (1996) Convention on nuclear safety. IAEA, Vienna

International Atomic Energy Agency (1999) Basic safety principles for nuclear power plants 75-INSAG-3 Rev. 1. IAEA, Vienna

International Atomic Energy Agency (2002) Safety culture in nuclear installations, IAEA-TECDOC-1329. IAEA, Vienna

International Atomic Energy Agency (2004) Precursor analysis—the use of deterministic and PSA based methods in the event investigation process at nuclear power plants, IAEA-TECDOC-1417. IAEA, Vienna

International Atomic Energy Agency (2005) Risk informed regulation of nuclear facilities—overview of the current status, IAEA-TECDOC-1436. IAEA, Vienna

International Atomic Energy Agency (2006) Code of conduct on the safety of research reactors, IAEA, Vienna, 2006. IAEA, Vienna

International Atomic Energy Agency (2015) Root cause analysis following an event at a nuclear installation: reference manual, IAEA-TECDOC-1756. IAEA, Vienna

International Atomic Energy Agency (2019a) Human and organizational aspects of assuring nuclear safety—exploring 30 years of safety culture. In: Proceedings of international conference, 22–26 Feb 2016. IAEA, Vienna

International Atomic Energy Agency (2019b) IAEA safety glossary, 2018th edn. IAEA, Vienna

International Atomic Energy Agency (2020a) International Atomic Energy Agency, safety culture practices for the regulatory body, IAEA-TECDOC-1895. IAEA, Vienna

International Atomic Energy Agency (2020b)Performing safety culture self-assessment IAEA—safety report series no. 83. IAEA, Vienna

International Atomic Energy Agency (2020c) IAEA releases 2019b data on nuclear power plants operating experience, IAEA, 25 June 2020. [Online]. Available: https://www.iaea.org/newscenter/news/iaea-releases-2019b-data-on-nuclear-power-plants-operating-experience. Accessed 24 June 2022

International Atomic Energy Agency (2022) Safet culture. Vienna. https://iaea.org

IRSS code of conduct

Jang S, Park S, Jae M (2017) Development of an accident sequence precursor methodology and its application to significant accident precursors. Nucl Eng Technol 49:313–328

Kastenberg WE (2011) Ethics, risk and safety culture: reflection on Fukushima and beyond, paper based on lecture presented on August 2, 2011. University of California, Berkeley

McLeod S (2015) Freud and the unconscious mind. Simplypsychology.org [Online]. Available: https://www.simplypsychology.org/unconscious-mind.html#:~:text=According%20to%20Freud%20(1915)%2C,and%20stored%20in%20the%20unconscious. Accessed 24 June 2022

Musikanski L, Cloutier S, Bejarano E, Briggs D (2017) Happiness index methodology. J Soc Change 9(1):3–4

National Academic Press (2014) Lessons learned from the Fukushima nuclear accident for improving safety of U.S. nuclear plants. National Academic Press (USA), 29 Oct 2014 [Online]. Available: https://www.ncbi.nlm.nih.gov/books/NBK253947/. Accessed 24 June 2022

Prabhu HR, Bhat PS (2013) Mind and consciousness in yoga—Vedanta: a comparative analysis with western psychological concepts. Indian J Psychiat 55(2):182–186

Ramsay K (2019) The three levels of human consciousness. Achology (2019) [Online]. Available: https://medium.com/achology/the-three-levels-of-human-consciousness-6d9a59fed577#:~:text=We%20can%20view%20consciousness%20as,preconscious)%2C%20and%20the%20unconscious. Accessed 24 June 2022

Rempe JL, Knudson DL (2013) TMI-2 a case study for PWR instrumentation during a severe accident. Idaho National Lab, USNRC, Idaho (2013)

Safety and security culture

Sarvapriyananda S (2013) Vedanta definition of consciousness. A discussion with Deepak Chopra

Ura K et al (2021) An extensive analysis of HNH Index. Centre for Bhutan Studies [Online]. Available: http://www.grossnational.com/wp. Accessed 26 Jan 2021

Varde PV, Pecht MG (2018) Risk-based engineering. Springer, Singapore

Wikipedia, Intelligence [Online]. Available: https://en.wikipedia.org/wiki/Intelligence

Wikipedia (2022) Culture, Wikipedia, 2 June 2022 [Online]. Available: https://en.wikipedia.org/wiki/Culture. Accessed 24 June 2022

Chapter 5
Risk-Based Engineering

5.1 Introduction

Risk-based engineering (RBE) provides the basic risk framework for the development of risk-conscious operations management (RCOM). The subject, i.e., the RBE concept, model and methods have been developed in much detailed by Varde and Pecht (2018). This chapter will introduce the essential features that are of specific interest to risk-conscious operations management along with suitable modification to some chapters to align with the requirements of risk-conscious culture. As we know, the traditional approach to safety has been the deterministic approach, and even today, this approach is considered the primary approach to the design, operation and regulation of nuclear plants. In recent years, probabilistic risk assessment (PRA) has increasingly been used as part of risk-informed approaches, particularly for application in regulatory applications. In the risk-informed approach, the PRA is used essentially as a complementary or supplementary role to the deterministic approach.

The risk-based engineering model proposed by Varde and Pecht considers an integrated framework to model along with other features, which will be discussed in this chapter. The aim is to make the overall approach relatively more flexible and rationally based than the deterministic approach; however, at the same time, it is more conservative to the existing risk-informed approach. Since the primary responsibility of safety rests with the operating organization, the organization risk-conscious culture should maintain higher risk considerations and the risk-informed approach that is developed to serve primarily for regulatory reviews.

P. V. Varde, *Risk-Conscious Operations Management*, Risk, Reliability and Safety Engineering, https://doi.org/10.1007/978-981-19-9334-3_5

5.2 Why Risk-Based to Risk-Conscious Culture

Why risk-conscious approach? When the foundation for risk-based engineering has already existed. The insights, goals and criteria derived from PRA and the deterministic approach form the foundation of risk-based engineering. In regard to merging with the operational environment, it becomes a challenge to explain the complex probabilistic and statistical knowledge base for communication in operations and maintenance environments, be it management, shift, or grass root level staff. For example, what to infer from the statement of core damage frequency and its estimate as 1.0×10^{-5}/reactor-year, or even the unavailability value of 1.0×10^{-3}/demand. These estimates without the subjective explanation would not generate a conscious perception or intelligent, emotional response that gives sense of risk, low, medium, or high or prohibitive risk levels and how to react in operational ecosystems, be it action, planning, modification, or even regulatory review. Of course, to begin with, we moved from the subjective notion of risk to an objective estimate of risk through the PRA framework, and now we are again explaining the subjectivity associated with the numerical estimate derived from PRA to effectively use this information to support risk reduction and management measures. For example, for a reference plant the CDF statement 2.0×10^{-6} per reactor-year, here even though objective or quantitative assessment provides part information, if one has to make it more meaningful we can explain the subjectivity associated with this statement like

(a) This is an estimate of limited scope Level 1 PRA performed for only internal events, and reactor core as the source of hazard/radiation and for full power operation.
(b) The confidence in the estimate is further explained by the width of the uncertainty band; for example, there is a major difference in the two statements of uncertainty expressed by the lower and upper bounds, even though the median value is the same as follows:

    Case *A* CDF: $9.0 \times 10^{-7} < 2.0 \times 10^{-6} < 6.0 \times 10^{-6}$. Error Factor ~ between 2 and 3
    Case *B* CDF: $1.9 \times 10^{-7} < 2.0 \times 10^{-6} < 2.1.0 \times 10^{-5}$. Error Factor ~ 10

    The median (central value) is the same in cases *A* and *B*, i.e., 2.0×10^{-6}/reactor-year. However, the error factor, which is defined as the ratio of the upper bound to the median value or the ratio of the median value to the lower bound, is in the range of ~ 2–3 in Case *B* and 10 in Case *B*. That is, the width of the CDF distribution in Case *A* is much lower, i.e., uncertainty is lower than in Case *B*, which is 10. The uncertainty associated with Case *B* is lower and acceptable, while for Case *B*, the uncertainty is unacceptable; therefore, the CDF expression should not be treated as credible, as the uncertainty is 10, which is normally not acceptable.
(c) The CDF estimates of nuclear power plants compiled from the available open literature fall in the range of 1.0×10^{-5}–1.0×10^{-7}/reactor-year, and the CDF for the reference plant is 2.0×10^{-6}, which is within the range and creates a

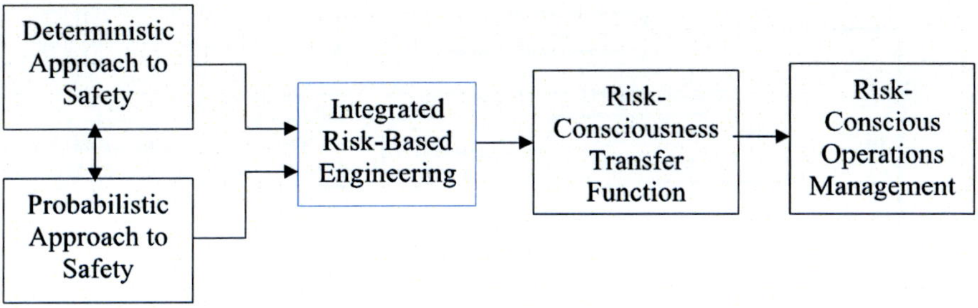

Fig. 5.1 A major element of risk-conscious operations management

sense of where we are. This provides a basis to say that our plant has low risk potential, which is a linguistic expression and provides a sense of feel factor, of course with the assumption that the data, model and overall approach are standard.

(d) What sense it makes when CDF changes while updating the PRA model based on last 1 year of data from 2.0×10^{-6} to 2.8×10^{-6} CDF; this is 40% increase compared to the reference value which is very high. If the operation and maintenance (O&M) team is considering a relaxation in reducing the frequency of, say emergency core cooling system from quarterly, i.e., once in a quarter of a year to half-yearly, and if the increase in CDF is 40%, then this relaxation cannot be considered, as the international guidelines provide an increase in CDF to 1% with applicable conditions. An intuitive explanation provides an effective feel to the O&M team.

Furthermore, the risk-conscious culture has a human factor at its core and assumes that most of the failures can be connected to human fundamental roots, be it either in Design, Installation, Commissioning, Operation and maintenance. This means that in risk-conscious culture, the root cause analysis of hardware failure is performed to further down in detail to reach human roots through well-known human factors such as error of omission, error of commission, human consciousness, cognition and consciousness to identify the roots responsible for either direct human error or hardware failure. The experience of many incidents, hardware and human failure provides motivation for developing a framework for investigating the role of individual and organizational consciousness in incidents/accidents.

The pictorial representation showing the interrelationship of the deterministic IRBE and risk-conscious framework is shown in Fig. 5.1.

5.3 The RBE Framework

The integrated risk-based framework, as shown in Fig. 5.2, shows, apart from the interrelation between deterministic and probabilistic approaches, elements such as

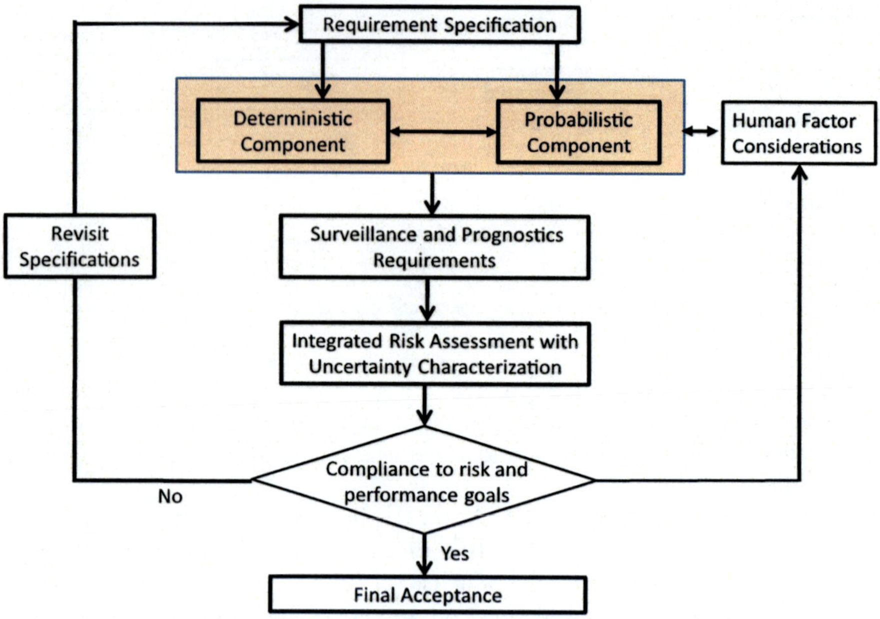

Fig. 5.2 Integrated risk-based engineering framework (Varde and Pecht 2018)

human factor considerations, the role of uncertainty and monitoring, diagnostics and prognostics for evaluating any requirements against set goals and criteria.

The first step is to define the problem and specifications in terms of purpose objective requirements of failure criteria, acceptable level of uncertainty and impact assessment by sensitivity analysis, availability of data, deterministic and probabilistic goals and criteria, and so on. In IRBE, the deterministic and probabilistic approaches complement and supplement each other. However, one of the characteristics of the risk-conscious culture is that it is more conservative than the risk-informed approach as it is the evaluation being done by operational agencies and must be more conservative the risk-informed approach to underline risk-conscious culture as operating organization has primary responsibility for safety. If the evaluation is being performed to reflect change in operational policy such as revision of technical specifications or incorporation of modification or safety upgrades, then a Level 1 full-scope PRA meeting specific quality attributes developed at the organizational level or from international sources such as IAEA (International Atomic Energy Agency 2016) or ANS/ASME (Trusted Information Sharing Network 2008) for the specific applications. The PRA and Accident Analysis document being considered for the application should be a duly approved regulator. In case some quality attributes are not meeting, then this should be identified and recorded such that a sensitivity analysis can be performed, or additional compensating provision can be made in the application.

The integration of human factors should be part of the evaluation along with an acceptable level of uncertainty. This means evaluation of not only human actions but also the applicable emergency operating procedure for adequacy of human action considerations considering plant layout, provisions considered in the evaluation, for example, fixing a leak, availability of connecting point and supporting logistics, etc. There is a separate section when human actions and organizational aspects in IRBE are translated into the risk-conscious culture.

Whenever a change, be it operational policy, retrofit, modifications or procedure, is incorporated, the real-time feedback on the performance of the system and related aspects need to be monitored and recorded for trend monitoring such that corrective action can be initiated to avoid or mitigate the consequences associated with the failure of the system. Online data are often used for condition monitoring, in-service inspection, diagnostics and prognostics and health management. This feedback monitoring ensures, despite all the design and operational considerations, that if the system has some new mode of failure, then the monitoring and feedback of key parameters attract the attention of plant staff for necessary action.

The compliance of the expected performance target be it design, operational or risk as specified in the requirements specification should be demonstrated within the given uncertainty band. For example, the increase in CDF should not be more than 1%, and the uncertainty bounds should be well in the acceptable level with an error factor < 3.0. This requires simulation of the Level 1 PRA model with the new configuration. The change in CDF and uncertainty compared to the reference Level 1 PRA should comply with the above condition.

If all the goals and criteria are met, then the change can meet the specifications required for final acceptance of the proposal/application. If some targets are not meeting, then the whole procedure must be revisited for reconsideration and re-evaluation.

The major areas covered in integrated risk-based engineering need more elaborations for a better appreciation of this framework.

5.4 Deterministic Safety Assessment

Before we go to deterministic safety assessment (DSA), we will discuss the essential features of defense in depth at the core of the deterministic approach to the safety of nuclear plants. The scope of the section deals specifically with aspects of the deterministic approach that are directly relevant to risk-conscious operations management of nuclear plants.

5.4.1 Defense-In-Depth Philosophy

Traditionally, the defense in depth forms the primary approach to safety for nuclear power plants. This approach originated in Military Establishments as a strategy to stop, delay or mitigate the damage potential of enemy forces (Trusted Information Sharing Network 2008).

If we look at the very high level of safety standards implemented and maintained in the nuclear industry on the one hand and the societal contribution of the 443 nuclear power plant operating world over and contributing to over 10% of the world's net consumption of electricity on the other, it is time we pay a tribute to the safety philosophy followed in the nuclear industry, where the defense in depth primarily provided the robust foundation for safety. Adoption of defense in depth in recent applications, such as cyber security, financial risk management and critical infrastructure in general, is a testimony to the success of this approach.

There are many definitions of defense in depth; however, for the purpose of nuclear plants, we use the expression for defense in depth as 'All safety activities, whether organizational, behavioral or SSCs related are subjected to layers of overlapping provisions, so that, if a failure should occur it would be compensated for or corrected without causing harm to individuals or the public at large' (International Atomic Energy Agency 1997).

The defense in depth is implemented in nuclear plants by adopting a twofold strategy that involves design based on the multiple barriers and levels of protection and maintenance of these barriers as part of the operations management strategy. Here, the conservative approach and the deployment of redundancy, diversity, fail-safety and considerations of human factors formed the major features of defense in depth. The multiple levels of barriers to radiation hazard are (a) fuel metrics, (b) cladding, (c) coolant system pressure boundary, (d) containment and associated system and (e) accident management to protect the public from radiation hazard. The multiple levels of protection are (a) Level 1 prevention of abnormal operation by employing the principle of conservatism and ensuring quality assurance in design, construction, installation, commissioning and operation, (b) Level 2 control of abnormal operation and detection of failures by control, limiting, protection and surveillance features, (c) Level 3 control of accidents within design limits through engineered safety features and emergency operating procedures, (d) Level 4: control of severe plant conditions including prevention of accident progression and mitigation of consequences by employing the complementary measures and accident measures and (e) mitigation of radiological consequences or significant release of radiological materials by invoking the off-site emergency response measures. Figure 5.3 shows the layers of multiple barriers and layers of protection implemented in a nuclear plant.

Defense in depth has evolved from the traditional approach where the three levels of prevention, detection and mitigation to additional levels, as discussed above; deployment of redundancy and diversity, overly conservative; to use of best-estimate code; and use of surrogate criteria for core parameters; application and treatment of common cause failure and paradigm shift to include use of probabilistic risk assessment insights as part of risk-informed approach.

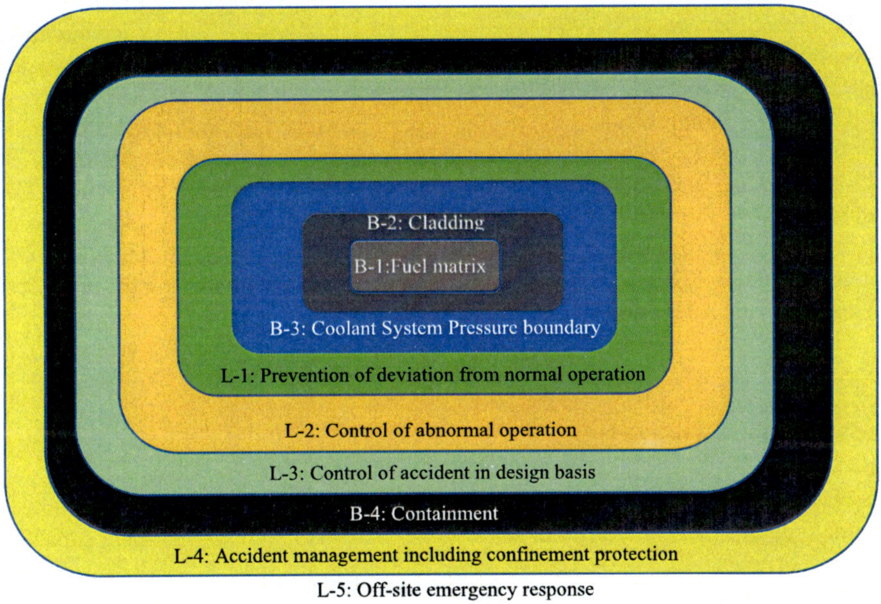

Fig. 5.3 Multiple barriers and levels of protection were implemented in a nuclear plant

5.4.2 Deterministic Safety Assessment

Traditionally, deterministic safety assessment (DSA) forms the primary approach for safety assessment for the siting, design, operation and regulation of nuclear plants. The major feature of this approach is (a) the identification of applicable initiating events, i.e., failure, particularly process systems that have potential safety implications, (b) formulation of safety acceptance criteria for systems structures and components SSCs and (c) analysis of the response/performance of the safety system assess its adequacy for demonstrating that the safety of the plant is ensured. The design basis of the plant translates into the design philosophy of the plant by identifying the initiating events demonstrated by analysis, testing and modeling that the plant can withstand these events such that the radiation protection and technical criteria are not exceeded. The accident sequence that involves the initiating event and response of the safety systems is designed conservatively to characterize the worst-case scenario and referred to as the design basis accident, for example, demonstration of safety of the plant considering the maximum credible initiating event major LOCA and response of the safety systems, viz. reactor shutdown and decay heat removal system and engineered safety features, emergency core cooling system and in case the containment system. The safety demonstration requires that the fuel, clad, coolant and primary envelope acceptance criteria not be violated.

However, deterministic safety assessment has also evolved due to accumulated operating experience in nuclear plant operations, the development of advanced materials and alloys and the availability of advanced knowledge of material behavior under varying conditions and the availability of advanced technology and efficient computer codes. For example, a better understanding of material behavior, particularly fuel, clad and core structural material, in terms of reduced uncertainty has helped address the issue of overconservatism in design and safety. Furthermore, learning from major nuclear accidents, particularly TMI (1978), Chernobyl (1986) and Fukushima (2011), has provided invaluable lessons for improving the approach to deterministic safety analysis (Ohba et al. 2021).

The following section will provide a brief overview of the major features of DSA.

5.4.2.1 Approaches to DSA

As discussed, DSA has moved from a primarily conservative approach to the best estimate due to the availability of knowledge, insights and computational capacity and now of course intelligence inferencing from a large database through data mining and analytics. Given that the organization has to make a choice constrained by (a) availability of codes, (b) assumptions to be made for system or safety system availability and (c) considerations of initial and boundary conditions for the analysis. There are four options available per the IAEA safety standard (I. A. E. Agency, International Atomic Energy Agency 2019) on DSA, viz. Option-1 Conservative: DSA is conservative with respect to computer code, assumption for safety system unavailability, initial and boundary condition; Option-2: Combination of Conservative and Best-Estimate, i.e., computer code is a best estimate code, system unavailability and consideration of initial boundary condition is on conservative basis; Option-3: In this option uncertainty in estimate forms part of the analysis along with having a best-estimate computer code, assumption of availability of safety system remains conservative while initial and boundary conditions are based on best estimate and Option-3: DSA is realistic where the entire analysis is primarily best estimate, be it computer code, safety system availability and initial and boundary conditions.

5.4.2.2 Plant Status

The DSA of the plant starts at the conceptual stage to support the siting activities and moves to preliminary design and final design by updating and expanding the new information, requirements and data. The higher-level objective of DSA is to demonstrate the safety of the plant against the set of initiating events considered in the analysis.

The scope of the safety analysis covers the entire range from normal operation to accident conditions, which includes anticipated operational occurrences, design basis events and design extension conditions, as shown in Fig. 5.4.

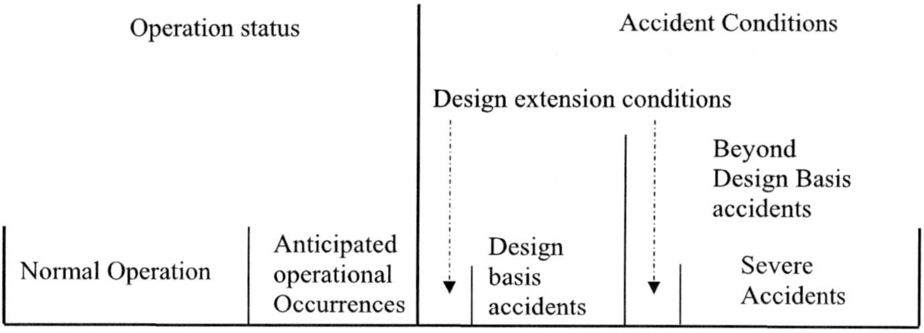

Fig. 5.4 Range of plant status right from normal operations to an unlikely scenario of severe accident

5.4.2.3 Selection of Initiating Event

The objective of selecting the initiating event task is to create a list of initiating events that are as complete as possible. The list should include internal generated due to process system failure or human error inside the plant. It might also include hazardous events such as internal fires and internal flood events as also external events like Seismic, flood and impact of external object like airplane. To ensure that the list is as complete as possible, the procedure includes (a) reference to a standard list of initiating events at the national and international levels, (b) reference to safety documents from similar plants, (c) if the safety analysis is being performed for an operating plant, then the reference to the plant record on incidents, near-mis-events and plant component failure and human error databases, (d) a logic tree approach where the information is drawn from the failure events leading to initiating events and (e) analysis of safety system failure that might lead to an initiating event.

Once a draft list is made, the list is subjected to screening qualitative criteria for expected frequency and potential consequences. The events having risk quotients in terms of likelihood and consequences are selected for the safety analysis. Here, the initiating events are divided into two categories, viz. anticipated operational occurrence that includes anticipated transients without scram, anticipated transients and design basis/accident conditions. If any initiating event has a frequency > 1.0/year, then the event is classified as an anticipated operational occurrence. For further categorization, guidance available in safety standards/guidelines should be used.

5.4.2.4 Acceptance Criteria

Before work on detailed safety analysis is performed, acceptance criteria are developed to assess the acceptability of the safety analysis for safety demonstration. The results of the analysis, for example, fuel integrity analysis for normal operation and accident conditions, are compared with the acceptance of criteria and the available margin between the results of the analysis and acceptance criteria. The acceptance criteria could be qualitative or quantitative. There are three categories of acceptance criteria per IAEA Standard SSG-2 (Rev. 1): (a) *safety criteria*, which deal with radiological consequences expressed in activity levels or doses during normal operations and accident conditions, *referred to as high-level radiation criteria*, or could be related to the structural integrity of successive barriers, referred to as *technical criteria*, against the release of radioactivity (I. A. E. Agency, International Atomic Energy Agency 2019). To ensure the integrity of the cladding, the limit on maximum temperature, ratio of departure from the onset of nucleate boiling, minimum oxidation temperature, etc., are some of the example of safety criteria for ensuring the integrity of the clad. (b) *Design Criteria:* performance conditions or limits are set for systems, structures and components as part of the design basis. The limit on main coolant system pressure and temperature and enthalpy, the minimum flow and pressure requirements for normal operation and emergency condition. Furthermore, this leads to flow and pressure requirements for emergency core cooling systems for a mission time, which are examples of acceptance criteria for emergency core cooling systems as part of the design basis accident LOCA. (c) *Operational Criteria*: Additionally, referred to as Limiting Condition for Operations (LCO), which lays down governing rules and criteria for operations during normal operations, anticipated operational occurrences and emergency conditions.

The safety criteria on the barriers, viz. fuel and clad coolant systems, form input for defining the definition of core damage in Level 1 PSA, while the criteria related to the potential for radioactivity release from the reactor pressure boundary to the containment and containment source term provide the basis for 'release' potential and 'risk' to the member of the public as part of Level 2 and Level 3, when along with the consequences, the frequency of release is also characterized.

Similarly, the failure criteria of SSCs, such as the safety system shutdown cooling system and decay heat removal system, and the emergency safety features, such as the emergency core cooling system and containment system, provide the basis to assess the integrated assessment of failure probability in PRA.

5.4.2.5 Assessment of Plant Response and Use of Codes

Generally, SSC design is performed in three stages, viz. (a) Conceptual design, (b) preliminary design and (c) final design. There are three major elements to ensure a sound design: (a) the application of well-established national and international standards, (b) wherever possible use of the best-estimate computer codes, (c) a design procedure review and management system are in place and (d) a quality assurance program that includes documentation.

Safety margin assessment and application to account for uncertainty related to (a) material property, (b) lifetime accumulated stresses and degradation, (c) lifetime shocks, such as seismic, flood and external object impact, (d) robustness of model, (e) testing insights and (f) human factor affecting any of the above and postulation of human factor requirements during operational phase. The safety margin assessment complies with the code/standard guidelines.

At every stage of design, it is subjected to safety review and should involve operations experts. Furthermore, the final design should be subjected to peer review and independent verification.

Plant response in terms of performance of safety system or human factors against the identified set of initiating events and design basis against the design criteria that in turns are related to barrier integrity forms part of safety demonstration. Considerations of the initial and boundary conditions form an integral part of the safety analysis. The category of safety analysis might involve (a) reactor physics and neutronic analysis, (b) source term analysis for various accident conditions and evaluation of doses in the plant and in the public domain, (c) thermal as well as hydraulic analysis and (d) structural analysis.

The current practice is to perform reliability analysis for systems during the initial stages of design of the system, and later as the design is matured, a PRA analysis is performed. PRA activities initiated right since conceptual stages provide valuable insights in terms of system configuration optimization, identifying strength and weakness in design and postulated activities in operations.

5.4.2.6 DSA and Operational Interface

As the plant prepares for commissioning the design, staff extends the support for preparing the operations. An approved safety analysis and related document provide the basis for the development of operational documents, such as (a) plant technical specifications, (b) commissioning reports, (c) identification of emergency operating procedures, (d) training and qualification documents, such as design reports, design and layout drawings, P&ID diagrams and operating manuals and (e) monitoring and surveillance programs.

DSA is reviewed periodically to account for changes that include operational policy, modification and retrofits and to comply with the new safety regulation.

5.5 Probabilistic Risk Assessment

Probabilistic risk assessment (PRA) is an analytical activity to assess the risk from a plant by postulating a potential accident scenario. PRA is performed at three levels, viz. Level 1 PRA, which deals with reactor system modeling and analysis for predicting the core damage frequency; Level 2 PRA, which deals with containing scenario modeling by postulating a breach in successive safety barriers, including the primary pressure boundary for estimating radiation source terms from the containment; and Level 3 PRA, which considers the various source terms identified and whether conditions to evaluate likelihood and consequences for the source terms and estimate the potential risk to members of the public from the plant.

The general trend worldwide is that Level 1 PRA is being performed for most of the nuclear plants in general and nuclear power plants (NPPs) in particular. There is growing interest in Level 2 and Level 3 PRA. The availability of PRA standards, quality guides and documents (International Atomic Energy Agency 2016; American Society for Mechanical Engineers 2009) has further provided the needed boost to PRA application to nuclear plants. A trend is also seen to extend the scope of Level 1 PRA moving from limited scope—for full power operation, with internal events, reactor core as source of radiation to full-scope Level 1 PRA that includes low power and shutdown conditions, external events, such as Seismic, Flood and Aircraft impact. The methodology and techniques have been considered for real-time applications (United States Nuclear Regulatory Commission 1995). The major boot for PRA activities world over has come as PRA techniques are being actively considered for risk-informed applications. In IRBE, PRA along with the deterministic approach is used such that both complement each other.

Major functional steps involved in PRA are as follows:

5.5.1 General Considerations

The organizational mandate and a policy decision to initiate a PRA study are necessary for the success of a PRA project(s). Having a PRA program in an organization is a resource-consuming option that requires the availability of human resources with adequate qualifications and expertise for PRA tasks, lab facilities for setting up reliability data and information management, root cause analysis, accelerated life testing, etc. As part of good practice, it is required that a PRA project report be submitted to the organizational leadership for approval. This report should explain the justification for initiating a PRA project, such as implementing a risk-informed approach in support of operations and regulation, assessing design weakness, identifying critical human factors, evaluating the contribution of common cause failures and supporting regulatory approvals. The project report should present the objective, scope, resources needed, a preliminary list of initiating events, reliability analysis program, mechanism to ensure quality assurance of PRA, requirements of training in PRA, a brief overview of content of final PRA report and expected benefits in terms of risk reduction in operations management.

For example, the objective of the PRA project could be to perform Level 1 PRA to evaluate core damage frequency, identify safety-significant accident sequences, identify and prioritize risk-significant SSCs, CCFs, Human Actions, etc. The scope can be defined in terms of limited or full-scope or to include Level 2 and Level 3 PRA along with Level 1 PRA; consideration of reactor core and spent fuel pool as the source of radiation, and other factors discussed in previous para (Yang 2012). The PRA quality largely depends on the availability of quality and adequacy of plant-specific data. It must be noted that in a PRA project, major efforts are required to collect and compile the plant-specific SSC performance and human error data. The evaluation of uncertainty requires expertise and tools and methods, and these aspects should also form part of the PRA project report (Winkler 1996).

5.5.2 Overview of Major Steps Involved in PRA

The PRA project starts with submission of a project report by a PRA team member comprising reliability/PRA experts, data analysis, design or operation team members led by a project leader who has good experience with PRA modeling-associated requirements. The performance of PRA is a resource-intensive activity; therefore, the support, involvement and approval of the leadership in an organization is a key element for the success of the project. The PRA team, along with PRA modeling, should also extend the scope to include the development of risk-informed/risk-based applications. This serves two purposes: (a) management will always look forward to some applications, such as risk-informed In-service Inspection, that might bring down inspection efforts, and (b) through the application, the model tends to be validated, reflecting the as-built features of the plant.

The major steps involved in the performance of Level 1 PRA are as follows:

5.5.2.1 Plant Familiarization

This step is applicable to only operating plants. Plant familiarization is the first procedural step for initiating a Level 1 PRA study. Intimate knowledge of plant layout, design, operations and maintenance practices helps the study to be more specific to plants. There are three major steps to support the familiarization of the plant: (a) study of the plant deterministic safety analysis report, technical specifications, design manuals, P&ID, operating manuals and emergency operating procedures, (b) lecture program by the design and operation staff, reactor physics and neutronics experts and health physics experts and (c) plant walkdown (International Atomic Energy Agency 1992a).

The study safety analysis provides a list of initiating events, accident analysis considering the plant safety function response and provides information on qualitative and quantitative physical criteria: information on the provision of barriers and level of protection and their capacity, often in terms of measurable parameters.

The technical specification provides safety limits, safety system settings, limiting conditions for operations and surveillance schedules for safety-significant SSCs. The design manual along with the P&ID provides information on the design basis for each system, system configurations, layout details and so on.

The operating manuals provide plant operational information for system maneuvering, interpersonal communication and authorization protocols related to system specificity, and if two or more systems are interconnected, then the requirements of communication and authorization from the control room. The study of emergency operating procedures provides information on emergency options and man–machine interactions and verifies that the provisions envisaged in emergency operating procedures match site observations and are suitable for postulated emergency conditions.

The lecture program on respective subject experts followed the study of documents to complement and consolidate the process of learning. The lecture program also provides a platform for interactive sessions with the experts and interteam member discussions and clarifications. The best part is it opens a channel of communication and interaction between plant design experts, operations staff and the PRA team in a familiar ecosystem.

Plant walkdowns are organized and organized gainfully, either as a first step or in between or after the individual lecture schedule depending on the requirements and choice of the experts, and the PRA team provides knowledge of the as-built and operated status of the plant.

The drafting of the project report and plant familiarization are the two tasks that can go together simultaneously.

5.5.2.2 Formulation of List of Applicable Initiating Events

The objective of formulating a list of initiating events is to make the list as complete as possible by postulating all the internal as well as external transients and conditions. A preliminary hazard analysis of the plant provides, as a first step to kick-start the work on the development of a list of initiating events (International Atomic Energy Agency 1993). In the absence of the a hazard analysis of the plant, the analyst can work on three major approaches to formulate the list of initiating events. First, use the generic list of initiating events for similar plants from generic sources or other plant specific sources; second use the plant operating experience as input for plant specific initiating events and third, use an analytical approach like master logic diagram for updating the applicable list of initiating events. Of course, if the PRA is being performed for a first of a kind design, then the job becomes slightly more challenging; however, the logic model of the plant and other insights along with the information available from generic sources will help in drafting a list of initiating events. The initial list is subjected to screening analysis to ensure which event should be considered for detailed analysis using the screening criteria and what are those events they do not need detailed modeling consideration and can be closed by deterministic or probabilistic considerations. A final list of initiating events and the information on the safety function of the plant that includes safety systems and engineering safety features and human actions along with success criteria or failure criteria enable the PRA team to start the work on accident sequence modeling.

5.5.2.3 Accident Sequence Modeling

Event tree modeling is one of the most elegant and well-accepted graphical techniques to propagate plant responses to an initiating event by inducing the success/failure of plant safety functions (International Atomic Energy Agency 2010). The plant safety function mainly includes the response of the safety system and human action arranged as top header events in chronological order. The success and failure on the

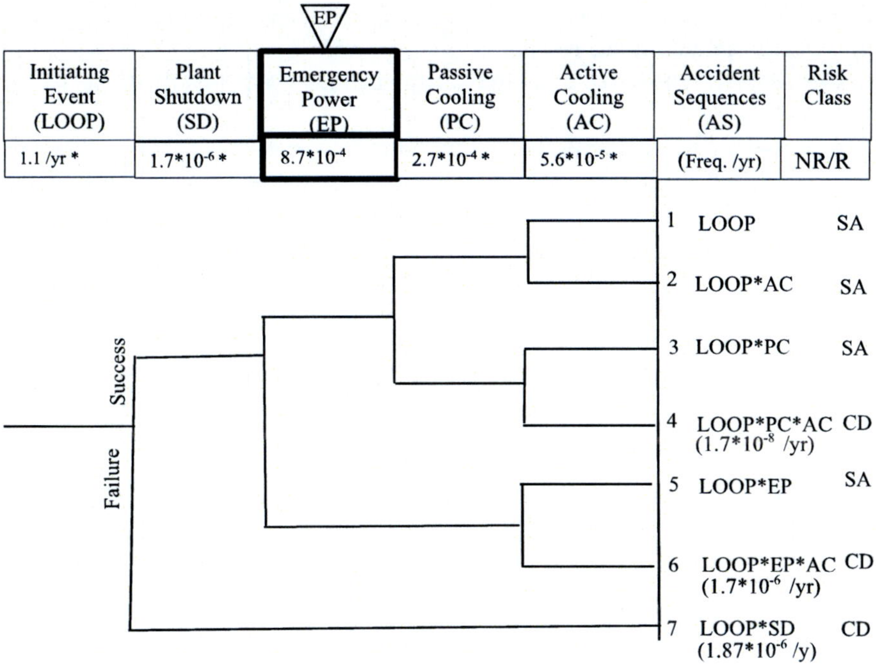

Fig. 5.5 Accident sequence modeling employing the event tree approach

event tree are represented by the branching point; a branch moving or following the same line indicates 'success', while downward branching represents the 'Failure' of the function. Event tree event trees are developed for all the initiating events considered for detailed analysis, as shown in Fig. 5.5. The output of the accident sequence modeling is a list of accident sequences.

The event tree is an inductive graphical technique to propagate the accident scenario based on the plant response to an initiating event. The safety functions are arranged in a chronological sequence, and the success/failure of these safety functions determines the stages of the plant state. There are two shutdown cooling systems for the plant, the passive system, which is automatically activated after the plant trip and caters to the requirement of decay heat. The second active shutdown system is valved in after the temperature of coolant drops to a level such that extended cooling can continue. This event tree has been developed to demonstrate the accident sequence development for an initiating event—loss of off-site power (LOOP) to the plant. The top three accident sequences (LOOP, LOOP*AC, LOOP*PC) indicate that there is no risk (NR) to the plant as shutdown and cooling functions are met successfully. The last three accident sequences (LOOP*SDC*EC, LOOP*EP, LOOP*SD) have risk implications, as in two accident sequences, cooling of the plant is adversely affected due to unavailability of electric power, and in the last accident sequence, the shutdown function failed and poses a risk to the plant. The instantaneous plant response is to shut down the reactor.

For example, let us see an event tree for an initiating event, loss of off-site power (LOOP), as follows:

As seen in Table 5.1, except for the Emergency Power Supply System, other data have been assumed for the purpose of demonstrating the process of evaluating the accident frequency for the plant. There are six accident sequences (AS). AS 1 has all the safety system success, while in the case of AS 2, 3 out of the two AC or PC systems, one system caters to the cooling function. Furthermore, in the case of AS 5, PC will be available (and AC will not be required immediately; here, it has been assumed that power will resume before it is time to AC to be switched on). Therefore, the conclusion is that AS, 1, 2, 3 and 5 terminate into the SA (safe) state. AS 4, 6 and 7 result in core damage (CD). AS 4 and 6 due to loss of cooling in shutdown mode, both AC and PC are not available. AS 7 takes the plant-to-core damage state as the automatic shutdown function failed to shut down the plant. Since AS 4, 6 and 7 contribute to core damage, the equation of core damage frequency (CDF) is as follows:

$$\mathrm{CD} = \mathrm{CDF} = \sum_{i=1}^{n} \mathrm{AS}_i$$

In this case, $n = 3$.

Therefore, the equation for

$\mathrm{CDF} = (\mathrm{LOOP} * \mathrm{PC} * \mathrm{AC} \;\; U \;\; \mathrm{LOOP} * \mathrm{EP} * \mathrm{AC} \;\; U \;\; \mathrm{LOOP} * \mathrm{SD})$; U is the symbol of union.

As per Boolean logic (idempotent law).

$A + A = A$, which implies that events 4 and 6 are the same, as both involve the same IE LOOP and safety systems (PC AND AC). Therefore, the CDF equation reduces to:

$$\mathrm{CDF} = (\mathrm{LOOP} * \mathrm{PC} * \mathrm{AC} \;\; U \;\; \mathrm{LOOP} * \mathrm{SD})$$

Table 5.1 Input data for event tree analysis

S. No.	Safety system (header event in ET)	System nomenclature	Probability of failure/demand	Remark
1	Plant shutdown	SD	1.7×10^{-6}	Assumed for illustration
2	Emergency power system	EP	8.7×10^{-4}	Estimated
3	Passive cooling system	PC		Assumed for illustration
4	Active cooling system	AC		Assumed for illustration

$$CDF = 1.7 \times 10^{-8}/\text{yr.} + 1.87 \times 10^{-6}/\text{yr}$$

The CDF contribution to core damage due to LOOP events has been estimated to be:

$$CDF_{LOOP} = 1.887 \times 10^{-6}/\text{yr}$$

$$\sim 1.9 \times 10^{-6}/\text{yr}$$

As per the Level 1 procedure, CDF is a statement of risk where the consequence is severe damage to the reactor core. Now let us consider that the regulatory target for CDF is 1.0×10^{-5}/yr. Additionally, given that the LOOP contribution to CDF ranges from ~ 10%, one has a better understanding that the CDF target requirements might be met as a sufficient margin is available for other events.

5.5.2.4 System Analysis

System analysis addresses the unavailability modeling of engineering systems, particularly safety systems, engineering safety features and safety support systems. The role of these systems is to remain in standby mode and actuated by normally automatic command-generated control and protection system-based sensing of failure or inadequate operation of the process system. As seen, in the event tree shown in Fig. 5.5, the safety systems are arranged, as seen in the header boxes on top in an order they are required to meet the safety function. The plant shutdown system (SD) and shutdown cooling system (SDC) are safety systems and are required whenever even in the normal operational phase of the plant they are required to shut down and maintain in the shutdown state, while emergency power (EP) for the purpose of the case being discussed, comprising emergency diesel generator sets, is a safety support system. The emergency cooling system (EC) is an engineering safety feature in the plant and is required only during emergency conditions when the normal process cooling system and shutdown cooling system fail to function, which is a low probability scenario. The details of the Emergency Power System Fault Tree in Fig. 5.6 have been taken from Fault Tree 3.16 in Chap. 3.

As part of the system modeling program, the unavailability analysis is performed, normally employing fault tree analysis. Therefore, as part of the PRA program,

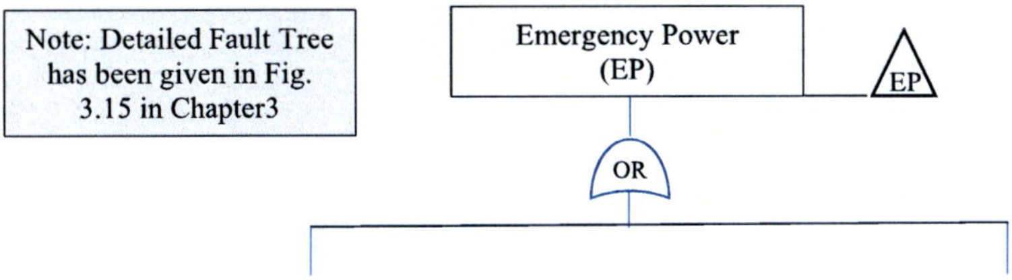

Fig. 5.6 Fault tree for the emergency power (EP) system in the PRA

system analysis is performed for the safety systems shown as the header row of the event tree of system modeling. In this section, a simplified fault tree is shown for the emergency power system of the plant.

The emergency power (EP) system comprises three Diesel Generator (DG) sets in parallel configuration. The individual DG set has capacity such that even one DG set operation is adequate to cater to the shutdown loads of the plant. The definition of failure criteria is required for system modeling. The system failure criterion is 3-out-of-3 DG failure, or conversely, the system success criterion is at least 1-out-of-3 operation such that all shutdown load requirements are met. The second requirement is defining the TOP event in the fault tree. The fault tree for the EP system is shown in Fig. 5.6.

The fault tree is a top-down approach; therefore, we start from the top event 'Emergency Power' (EP). In this modeling, the emergency power failure can occur due to (a) DG-Independent failure as shown in the left intermediate event OR (OR-gate) (b) due to CCF of all the DGs simultaneously.

Common Cause Failure CCF Modeling PSA. There are two types of CCF modeling: (a) explicit CCF and (b) implicit CCF. The explicit CCF events are those that are modeled as dependent events, e.g., power supply to redundant pumps; these are called systematic events and are epistemic events. These events occur basically due to lack of knowledge. The other, i.e., implicit CCF fall under the category, is due to random or unforeseen phenomenon and in the analysis; it is accounted for separately. Here, the adequacy of the defense provision for CCF is checked.

In fault tree analysis, we stop going down when we arrive at the basic event shown by the circle.

We write the equation of the Top event as follows:

$$
\begin{aligned}
T(\mathrm{EP}) = [&\mathrm{DG}-1\times\mathrm{DG}-2\times\mathrm{DG}-3+DG-1\times\mathrm{Breaker}-2)\\
&\left.\times\mathrm{DG}-3+\mathrm{Breaker}-1\times\mathrm{DG}-2\times\mathrm{DG}-3 \atop +\mathrm{Breaker}-1\times\mathrm{Breaker}-2\times\mathrm{DG}-3 \right]\\
&+\left[\begin{array}{l} DG-1\times\mathrm{DG}-2\times\mathrm{Breaker}-2\\ +\mathrm{DG}-1\times\mathrm{Breaker}-2\times\mathrm{Breaker}-3\end{array}\right.\\
&+\mathrm{Breaker}-1\times\mathrm{DG}-2\times\mathrm{Breaker}-3\\
&+\mathrm{Breaker}-1\times\mathrm{Breaker}-2\times\mathrm{Breaker}-3]\\
&+\mathrm{BUS}-\mathrm{CCF}+\mathrm{INSTALLATION}-\mathrm{CCF}\\
&+\mathrm{MAINTENANCE}-\mathrm{CCF}
\end{aligned}
$$

5.5.2.5 Data Collection and Analysis

A system for data collection and analysis should be developed at the organizational level involving senior operating staff who have a direct sense of plant system performance, data analysis, uncertainty and sensitivity analysis expertise. The system of data collection should be working in perpetual mode not only during the PRA project stage to have a risk monitor or living PRA model but also during the operational stage. The major sources of plant-specific data are plant daily summary records, work permit records, monthly reports, annual reports and incident reports. The data collection team scanned these records and provided input for the data analysis team. There should be realization or consciousness in operating staff about the understanding about the operational performance and risk data and expectation of the analysis team as to what data are required and in what form (Kumar et al. 2015).

5.5.2.6 Quantification of Plant Model

The data analysis team processes these data into individual component failure, data accumulated operating time/demand, and updates the point value of failure data in the plant reliability data and information management system. The data categorization involved classifying these data into safety-related (unsafe failure) or operational failure (safe failures) and independent or common cause failure, human error or hardware or software failure. This categorization makes the use of these data effective for use as input into the plant PRA model.

The uncertainty analysis module in the reliability data and information management system enables the uncertainty and sensitivity analysis team to update the uncertainty estimate of the SSCs, including CCF groups. In system analysis and quantification of event sequences, apart from reflecting the as-built and operated features, the use of the data should be as accurate as possible. Effort should the data be used from generic sources, and then a conservative approach that provides more pessimistic results should be used. There should be sound rationale and basis for estimating human reliability and should be demonstrated through simulation, mock sessions and work studies on plant configuration and layout in creating near real-time scenarios. The common cause failure groups play a critical role and dominate the final unavailability estimates. Again, a conservative approach should be used for quantifying common cause failure estimates.

Let us return to the system quantification task. The data collected for the basic events are shown in the fault tree by circles, such as DG-1, Breaker-1, DG-2, Breaker-2, BUS-CCF and so on. For the present case, let us assume that the analysts arrived at failure probability from some identified generic source. Additionally, we assume that DG-1, DG-2 and DG-3 failure probability is same $= 5 \times 10^{-3}$/demand; as also Breaker-1, Breaker-2 and Breaker failure probability is also same $= 2 \times 10^{-3}$/demand.

CCF for BUS-CCF, INSTALLATION-CCF and MAINTENANCE-CCF arrived from a generic data source as 2×10^{-4}, 7×10^{-5} and 6×10^{-4};

The unavailability of the EP system using these cut-sets has been estimated to be 8.7×10^{-4}.

The CCF factor is a dominant contributor to the top event. The failure probability of the top event is plugged into the event tree top header event, EP. Along similar lines, the system failure probability for other events, SD, SDC and EC is also plugged into the event tree. Furthermore, it was found from the plant record that the failure frequency of off-site power is 1.0/yr. This value is used for the LOOP in the event tree.

5.5.2.7 Common Cause Failure

The current state of the art and empirical models have been discussed in the chapter on dependability analysis. However, the Common Cause Failure is one of the important dependable failures that challenges the experts, ability of the available tools and method, to arrive at procedure where we might have better consensus. There are some reasons for this uncertainty: (a) the sparsity of common cause failure data, particularly at the plant level; (b) the empirical models that are available are based on limited experience and do not have scientific validity and a basis for the numerical factors; and (c) there is no general consensus that provides satisfactory credibility and scrutiny of the review and regulating body.

It is clear that it is a challenge to have CCF estimates for the CCF group of components from plant-specific sources using plant-specific data. Therefore, traditionally, CCF estimates are derived from independent failure data by considering CCF as the fraction of independent failure, e.g., the β-factor model for CCF analysis. The β-factor model belongs to a family of Greek-factor models that further include the α-factor, Multiple Greek letter model []. Given that in many cases CCF dominates, the final results and insights from PRA, it is challenging to improve the CCF science on analysis of CCF.

The silver-lining is there, in a sense at the national and international levels, toward improving the overall science of CCF analysis (United State Nuclear Regulatory Commission 2007; International Atomic Energy Agency 1992b; OECD-Nuclear Energy Agency 2022). The Working Group Risk of Nuclear Energy Agency (NEA/WGRISK) spearheading a project on the international common cause failure database, and the major features of this project are the development of taxonomy for the CCF data and the categorization and classification of CCF data resourced from the participating organizations/countries (CSNI-Nuclear Energy Agency 2020). The integration of knowledge based available in the public domain and the available literature has the potential to reduce uncertainty in CCF prediction. The INIS database has been compiling the publications on common cause failure that helps as input on state of the art on CCF [worldwide]. Considering the higher impact factors of protection logic employing digital technology, extensive efforts are being made to arrive at a consensus approach to CCF modeling for digital systems where the software component poses specific challenges. The risk-based engineering approach proposed by Varde and Pecht provides a state of the art on digital system modeling challenges to address CCF for digital systems (Varde and Pecht 2018).

It is known that CCF taxonomy has two major elements: root cause of failure and coupling mechanism. Toward providing a scientific and rational-based approach to model CCF, the physics-of-failure approach is considered a promising approach to CCF modeling. This approach is related the coupling mechanism and the root causes of failure through physics of failure modeling (Zhou et al. 2018).

5.5.2.8 Uncertainty Analysis

In Level 1 PRA, a major part of uncertainties is characterized by the confidence bounds. Here, the system and plant model level uncertainty are characterized employing the Monte Carlo approach (Durgarao et al. 2009). Furthermore, if uncertainty is found to be more than the acceptable level, an uncertainty analysis is performed to evaluate the impact of uncertainty individual cases that include SSCs, assumptions, human actions and common cause groups by performing the sensitivity analysis on the overall results of the analysis (Iman 1990).

From the operational perspective, uncertainty characterization and management require, apart from the epistemic component, the subjective component of uncertainty. In fact, the operational ecosystem poses a special challenge in making decisions under uncertainty. Major part of the uncertainty creeps due to part of critical information that is subjective or often imprecise and requires deeper understanding correlation and analysis. The situation becomes more challenging during emergency scenarios. A look at the accident scenario management, during the major accidents, had posed challenges often due to abstract setting, be it any action to be taken in time, public communication, coordination between plant and government bodies and so on (Rachel et al. 1998).

Advances in computational technology and AI, for example, have proven promising to support the development of tools and methods, such as artificial intelligence-based operator advisory systems (Varde et al. 1998), robot-based operations in inaccessible areas (Moore 1985) and smart sensors with diagnostic features (McIntyre and Hauser 2011). However, the approaches for public communication still pose a challenge, as seen in the TMI, Chernobyl and Fukushima accidents (Yang et al. 2015).

Aspects of emergency preparedness and public communication have been discussed in Chap. 8 on risk-conscious operations management.

5.5.2.9 Human Factor Considerations

Human factor considerations form part of deterministic as well as probabilistic evaluation (http://www.courses-online.org/mod/book/view.php?id=181). Generally, the deterministic approach focuses on point estimates and the uncertainty if captured by the factor of safety, which tends to be conservative, i.e., highly pessimistic, limiting the gain in terms of margins for manual actions, particularly for the postulation of accident scenarios. While for a given design, deterministic considerations might provide information on the type of human–machine interface, time required for a given action, plant procedures, such as technical specifications and emergency procedures, the probabilistic approach provides an effective framework for integrating human performance in the risk model of the plant, as it considers the quantified estimates of uncertainty bounds, which is vital for propagation of uncertainty from human action to the system and finally at the plant level.

Keeping in mind the requirement of risk-conscious operations management culture, where there is an emphasis on what caused human-to-committee error, the analysis of fundamental causes of human error already committed, on the one hand, and the question is 'how to analyze human error data' and estimate the human error probability. PRA uses human error data and integrates these data into the risk model. PRA model of the plant to predict the risk level, i.e., estimate of the core damage frequency/year for a plant. The procedure for the analysis of human error employing the CQB approach developed by Varde and Pecht (2018) and the adoption of this approach for the enhanced-CQB (eCQB) model for RCOM is discussed in Chap. 7. This chapter provides a brief overview of human error rate prediction while discussing in-depth detailing (a level below the conventional root causes such as error of commission, error of omission, lapse and mistake) to reach fundamental generic root causes for human error, however, the root causes of human error.

5.5.2.10 Interpretation of Results

The statement Core Damage Frequency/yr along with the uncertainty bounds is one of the major results of Level 1 PRA. Other insights are (a) unavailability of individual safety systems, (b) common cause contribution to the net core damage frequency, (c) impact of human error on the overall CDF, (d) risk-significance of individual SSCs that enables identification and prioritization of SSC-based risk-importance measures, (e) insights into the strength and weakness of the design and operations and (g) recommendations for further risk reduction in the design and operations of the plant.

However, these results carry subjectivity and require supporting or background information, particularly to appreciate the quantified results of PRA in support of decisions or risk-conscious decisions. Often, the statement of CDF needs to be compared with the identified goals and criteria. Additionally, for better appreciation this statement should be compared with the real-life other, nuclear plants, sources of energy systems or community services. For example, comparison with CDF statements of other plants in the country and other nations would provide a better perspective on the CDF of the plant in question. The comparison will be more effective if, along with the point estimate of CDF and associated uncertainties, forms part of the comparison, as higher uncertainties make the results less meaningful.

For example, the consequence of core damage frequency in terms of immediate fatalities (e.g., 1 death) is 1.0×10^{-6}/year. This statement should be compared with accepted societal services such as risk involved in Air Travel, use of Elevators, Railways travel and other energy systems, such as thermal power plants, hydel plants and gas-based plants, for improved insights or interpretation.

5.6 Role of the Deterministic and Probabilistic Approach in IRBE

The deterministic analysis provides failure criteria as an input for probabilistic analysis. For example, for a loss of coolant accident scenario, the coolability criteria and accordingly the injection flow requirements are derived from deterministic analysis. In IRBE, the specific requirement is to provide apart from the point value of the parameter, such as emergency flow and the uncertainty bounds of the parameter (such as minimum and maximum flow). In fact, the probabilistic model is developed based on the given plant configurations that satisfy the objective and performance function of the plant. The probabilistic goal could deal with complying with system level unavailability (1×10^{-3}/demand) and change in core damage frequency (CDF) goal (change in CDF $< 1\%$ of the reference value). The acceptable radiation level criteria may require considerations of ALARA (as low as reasonably achievable) principles.

5.7 Requirements of Monitoring and Surveillance, Diagnostics, Prognostics and Health Management

Furthermore, the IRBE procedure requires input on plant surveillance provisions, which include testing frequency, online monitoring, condition monitoring and diagnostic and prognostic capability and requirements at the plant, system and component levels. Any proposal under consideration would require these aspects to be evaluated for closely monitoring the change such that degradations and failures could be detected in time. The identification of any failure by monitoring the identified precursor signature in advance provides an opportunity for corrective action such that the consequences of failure can be avoided. The surveillance and monitoring capability also plays a critical role after the change has been affected to obtain feedback on subject system performance. Therefore, the plant proposal for a change/modification should include either the information about the existing provision's capability to track the performance and health monitoring of the systems where the change has been affected and demonstrate that these provisions have sufficient capability for required surveillance or should propose additional monitoring and surveillance provision such that there is confidence that any deviation in performance or incipient failure would be detected in advance and the plant is capable of implementing a health management program.

In fact, the availability of technology with respect to Risk-informed ISI, Surveillance test interval and allowable outage time optimization, risk/reliability centered maintenance, review of emergency operating procedure for its applicability by risk-informing the strength and weakness is an advantage for extending these techniques to risk-conscious framework such that deeper aspects of human errors can be addressed. These insights can further be used to develop strategies for reducing human error, including the deployment of automation, the implementation of inherent and passive designs, the deployment of intelligent operator aids, the elimination of the root causes responsible for errors, etc.

5.8 Integrated Risk Assessment

Integrated risk assessment is an iterative process where all the assumptions and uncertainty bound at the component, system and plant levels are evaluated to ensure that the change complies with the risk and performance goals so that the subject change can be accepted. The results obtained from the integrated model are subjected to validation with applicable deterministic and probabilistic performance criteria, achievable goals, or regulatory stipulations. If the evaluation does not meet the set goals and criteria, the complete process is revisited until an acceptable solution is obtained. Even though the steps in IRBE have been listed in chronological order, the nature of the complete implementation involves many parallel activities with recursive tasks. The problem in hand should be formulated in a crisp and clear manner so that the objective and scope function can be defined in a clear and unambiguous manner.

Many steps in IRBE are governed by the nature of the issue in hand, e.g., the problem in hand deals with new design, in support of existing plant operations (change in configurations or operating policy) or in support of regulatory review. Hence, the nature of the evaluation and resources required will change depending on the scope and objective of the analysis.

The analysis should reflect the assumptions, constraints, limitations of the analysis tools and methods, data, etc., and validation of the same by performing the sensitivity analysis. As mentioned, uncertainty modeling, be it a deterministic or probabilistic aspect, is an integral part of IRBE. Similarly, the system boundary should be clearly marked, and in case there is some interfacing connection with other systems, then this needs consideration as part of the analysis.

The quality of a deterministic and probabilistic analysis is critical and has a direct bearing on the results of the analysis. For example, only validated best estimate codes should be used in the analysis. The analysis should take note of the code and standards followed for designing these systems. If any part of the system does not meet the standards, then the provision of supporting additional safety criteria should be demonstrated in the analysis. In case a change is affecting plant technical specification, then compliance should be checked. It should be ensured that human error events that form part of the analysis are covered in plant emergency operating procedures. If certain human actions are not considered in available plant emergency operating procedures, then the recommendation of the analysis should reflect this requirement.

Similarly, quality assurance for PRA is also critical, particularly when a real-time application is being developed. Subjecting various procedural and logical steps to establish risk assessment quality attributes is vital to obtain the required confidence in decision making. The available framework for implementation of a quality checklist for each of the procedural elements, including initiating event selection, system modeling, human reliability analysis and common cause failure analysis, includes international standards/documents such as the IAEA-TECDOC-1101 on Quality assurance in PRA (International Atomic Energy Agency 1999), IAEA-TECDOC-1804 on quality attributes for Level 1 PRA (International Atomic Energy Agency 2016) and ASME/ANS standards (ASME 2008). Apart from this, national guide and standards also enable characterization of the quality of the applications.

The plant system description, like any other analysis, forms an important element of IRBE. This includes systems descriptions, associated drawings, system modes of operations, description plant/system logic and expected behavior of the system in various modes, such as normal operation and transient and emergency conditions.

The deterministic principles and criteria provide the basic framework for ensuring safety. The role of the probabilistic approach should be to consolidate but not dilute these principles. For example, if the deterministic principles provide for a redundant system to cater to certain safety functions, the probabilistic approach should provide the probabilistic evaluation should provide quantitative statement reliability or availability of these redundant provisions.

Other important deterministic or design/operation and maintenance aspects that need to be considered for evaluation include (a) safety code or plant technical specification applicability, (b) assessment of consequences, e.g., in terms of leak size, containment scenario and radioactivity release and hence adequacy of emergency provision with respect to on-site and off-site emergencies, (c) considerations of in-service inspection insights with respect to structural integrity assessment, (d) credit to be given for condition monitoring or prognostics provisions and status indications, (e) assurance against human performance and training-related aspects, (f) requirements related to the test override provision, (g) demonstration of built-in system capacity and plans and procedures to assess plant coping capability for a certain scenario and (h) requirements of plans and provisions for regulatory review and oversight.

The objective of data collection and analysis is to generate reliability estimates for hardware components and common cause failures. The data should facilitate simulation with a range of test intervals to aid optimization. For example, steady-state unavailability data or data on demand failure probability do not allow dynamic modeling. Additionally, test interval optimization requires dynamic equations with respect to incremental changes in time intervals; hence, the use of steady-state values is not appropriate. Most of the components involved in the safety systems are modeled as standby-tested components. Accordingly, the standby-tested model is used to determine the component failure probability (International Atomic Energy Agency 1992a).

The insights obtained from deterministic analysis in the form of available system redundancy, diversity, cooling criteria, or system failure criteria form the inputs to the probabilistic methods for risk assessment. Keeping in mind the complexity of the problem and the problem definition, the Level 1 PRA model of the plant should be used. There are two major aspects of PRA model simulation—event tree analysis and fault tree analysis. Event tree modeling is performed to generate accident sequences associated with a given initiating event and plant response in terms of success or failure of actuation of safety systems or human action. Fault tree analysis is performed to determine the safety system failure probability.

Judgment or decision making should not be based on point estimates. It is advisable that uncertainty analysis should be performed for real-time applications. Uncertainty analysis accounts for randomness in the variables, lack of knowledge, or data or inadequacy of the model employed for probabilistic modeling. Uncertainty analysis provides the upper and lower bounds of the estimates along with the median and mean values of the parameter. However, for this illustration, uncertainty and sensitivity analyses have not been performed.

Sensitivity analysis is performed to evaluate the impact of various assumptions on the overall results of PRA. Apart from this, sensitivity analysis is also performed considering the range of uncertainty bounds for selected components/human actions, for which the uncertainty is relatively high.

Technical review is performed at two levels, viz. peer review and regulatory review. The data and assumptions that form input for the analysis are an important aspect of this review. The peer review is performed by experts dealing in the respective areas but not the individual or group involved in the analysis. The role of the probabilistic expert is to assess the model and data accuracy and the overall representation of the system. The design and operation and maintenance aspects are checked and verified, keeping in mind the overall objective of the change. It is also verified that the fundamental safety aspects, such as defense in depth, are not diluted, and therefore, no compromise in redundancy or diversity level takes place. It is also ensured that the provision of fail-safe criteria is not compromised.

5.9 Compliance to Risk and Performance Goals

Finally, this review includes aspects such as assessment and implication of system unavailability and CDF estimates and its comparison with the target goals. A checklist procedure is used to ensure that all the safety issues have been addressed. For example, the major metrics that form the inputs for decision making could include, say for a given LOCA scenario, are as follows:

1. To ensure that the provision of defense in depth is not diluted, the following are typical checkpoints:

    (a) The provision of redundancy meets the design intent.
    (b) The provision of diversity meets the design intent.

(c) Fail-safe criteria are maintained and meet the design intent.
(d) System success criteria for applicable postulated scenarios are met.
(e) Provision exists to detect latent critical failures, or the likelihood of this event is low.
(f) Alternative provisions still provide the same backup as before the change.
(g) The results of in-service inspection and condition monitoring did not show any symptoms of significant degradation.
(h) There are no new issues with respect to human actions and procedures.
(i) The provision of surveillance and prognostics are commensurate with the change.

2. Probabilistic criteria

(a) The data used for the analysis are from plant-specific sources.
(b) The increase in system unavailability is (e.g., < 5%).
(c) The increase in CDF is (e.g., < 1.0%.
(d) Sensitivity analysis did not show any significant impact of any single assumption.
(e) Uncertainty analysis bound for the system unavailability and CDF is well within the acceptable range.
(f) PRA attribute analysis provides adequate confidence in the PRA model, data, assumptions and uncertainty in the final results.

After obtaining the regulatory clearance, it is management's responsibility to ensure that the recommended changes are implemented. This means modifying the technical specifications, changing system procedures and schedules, and training the staff wherever needed and required. Management should ensure regulatory compliance in words and 'spirit'. Generating plans and procedures for follow-up and trend monitoring is an important part of a change implementation program. The follow-up program should ensure that adequate provision exists for prognostics and health management or a condition-monitoring program for critical functions in the system. Finally, the activities related to documentation and records should comply with various communication protocols.

References

ASME/ANS RA-S Standard for Level 1/Large Early Release Frequency PRA for NPP Applications. ASME (2008)

American Society for Mechanical Engineers 1 (2009) American Society of Mechanical Engineers. ASME

Course-online.org (2022) Deterministic versus probabilistic models in HHRA (human health risk assessment). Courses-online.org. [Online]. Available: http://www.courses-online.org/mod/book/view.php?id=181. Accessed 27 June 2022

CSNI-Nuclear Energy Agency (2020) Collection and analysis of common cause failure analysis due to modifications in nuclear power plants, NEA/CSNI/R(2019) 4. CSNI, Paris

Durgarao K, Kushwaha HS, Verma AK, Srividya A (2009) Uncertainty analysis based on probability bounds (P-Box) approach in probabilistic safety assessment. Risk Anal 29(5):662–675

I. A. E. Agency, International Atomic Energy Agency (2019) Deterministic safety analysis for nuclear power plants, IAEA Safety Standards Series No. SSG-2 (Rev. 1). IAEA, Vienna

Iman RL (1990) Methods used in probabilistic risk assessment for uncertainty and sensitivity analysis. In: Cox LA, Ricci PF (eds) New risks: issues and management. Advances in risk analysis, vol 6. Springer, Boston, MA. https://doi.org/10.1007/978-1-4899-0759-2_47

International Atomic Energy Agency (1992a) Procedure for conducting level 1 PSA for NPPs, safety series No.50-P-4. IAEA, Vienna

International Atomic Energy Agency (1992b) Procedures for conducting common cause failure analysis for probabilistic safety assessment, IAEA-TECDOC-648. IAEA, Vienna

International Atomic Energy Agency (1993) Defining initiating events for the purpose of probabilistic safety assessment, IAEA-TECDOC-719. IAEA, Vienna

International Atomic Energy Agency (1997) Implementation of defense-in-depth for next generation light water reactors, IAEA-TECDOC-986. IAEA, Vienna

International Atomic Energy Agency (1999) A framework for quality assurance program for PSA, IAEA-TECDOC-1101. IAEA, Vienna

International Atomic Energy Agency (2010) Development and application of level 1 probabilistic safety assessment for nuclear power plants, SSG-3. IAEA, Vienna

International Atmic Energy Agency (2016) Attributes of full scope level 1 probabilistic safety assessment (PSA) for applications in nuclear power plants, IAEA-TECDOC-1804. IAEA, Vienna

Kumar S, Joshi NS, Mishra V, Varde PV (2015) Experiences and insights in development of probabilistic safety assessment of research reactors in BARC. In: Current trends in reliability, availability, maintainability and safety. Springer, India, pp 497–507

Mcintyre C, Hauser E (2011) Using smart instrumentation. Control Eng. [Online]. Available: https://www.controleng.com/articles/using-smart-instrumentation/. Accessed 27 June 2022

Moore T (1985) IAEA Bulletin Robots for Nuclear Power Plants. IAEA. [Online]. Available: https://www.iaea.org/sites/default/files/publications/magazines/bulletin/bull27-3/27304393138.pdf. Accessed 27 June 2022

OECD-Nuclear Energy Agency (2022) International Common-cause Failure Data Exchange (ICDE) Project. [Online]. Available: https://www.oecd-nea.org/jcms/pl_25090/international-common-cause-failure-data-exchange-icde-project. Accessed 27 June 2022

Ohba T, Tanigawa K, Liutsko L (2021) Evacuation after a nuclear accident: critical reviews of past nuclear accidents and proposal for future planning. Environ Int 148

Rachel B, Barkan Z, Erev I (1998) Accidents and decision making under uncertainty: a comparison of four models. Organ Behav Hum Decis Process 74(2):118–144

Trusted Information Sharing Network (2008) Defense-in depth. [Online]. Available: www.tisn.gov.au/. Accessed 25 June 2022

United States Nuclear Regulatory Commission (1995) Use of probabilistic risk assessment methods in nuclear regulatory activities; final policy statement, 60 FR42622. USNRC, Washington DC

United State Nuclear Regulatory Commission (2007) Common-cause failure database and analysis system: event data collection, classification, and coding (NUREG/CR-6268, INEL/EXT-07-12969, Revision 1). USNRC, Paris

Varde PV, Pecht MG (2018) Risk-based engineering—an integrated approach to complex systems—specific reference to nuclear plants. Springer, Singapore

Varde PV, Sankar S, Verma AK (1998) An operator support system for research reactor operations and fault diagnosis through a connectionist framework and PSA based knowledge based systems. Reliab Eng Syst Saf 60(1):53–69

Winkler RL (1996) Uncertainty in probabilistic risk assessment. Reliab Eng Syst Saf 54(2–3):127–132

Yang J (2012) Development of an integrated risk assessment framework for internal/external events and all power modes. Nucl Eng Technol 44(5)

Yang Y, Jin L, Li J, Fang C (2015) Crisis communication about nuclear accidents with psychological approaches. J Risk Anal Crisis Response 5(3):169

Zhou T, Droguett EL, Modarress M (2018) A physics-of-failure approach for common cause failures, age related degradation. In: Proceedings of the 14th probabilistic safety assessment and management, (PSAM-14). CA, USA

[Online] Available: https://worldwidescience.org/topicpages/c/common-cause+failure+analysis.html

Chapter 6
Risk Simulation

6.1 Introduction

Risk simulation for the purpose of risk-conscious operations management addresses modeling and assessing the response of the complex engineering systems to the postulated scenario through a computerized recursive process involving hardware and human failures toward predicting the risk. Risk simulation can be static or dynamic. In static risk simulation, the risk levels are assessed against the identified set of conditions, which remains effective for a simulation session, for example, the duration of the postulated event, the status of components, failure criteria, assumptions and other system conditions as selected by the analysts. The dynamic simulation, on the other hand, addresses changing conditions in temporal and spatial dimensions using real-time data or simulated real-time data such that risk levels of the plant are obtained as a trend of risk over a period depicting the response of the plant to changing responses of systems and safety functions.

The risk simulation procedure requires software systems to simulate the plant condition and the computations of risk levels based on plant conditions and on system, subsystem and human interactions against the set failure and success criteria. For example, in a nuclear plant, the loss of coolant accident (LOCA) condition is simulated on a plant simulator. The trend of safety parameters is monitored after the breach in the coolant system boundary due to the main coolant pipe line rupture followed by uncontrolled leakage from the system, automatic initiation of the injection system, adequacy of cooling conditions and trending for the important parameters, such as adequacy of core cooling in terms of system solidity, core temperature and pressure, radioactivity levels in various areas of containment and outside and so on, such that the operating crew has information on the safety status of the plant.

The risk assessment simulation procedure involves modeling the input from deterministic conditions as above and computing the risk by employing a probabilistic risk assessment procedure. The output of risk simulation is risk levels and associated uncertainty in the estimate for the identified vector or parameter conditions for a

P. V. Varde, *Risk-Conscious Operations Management*, Risk, Reliability and Safety Engineering, https://doi.org/10.1007/978-981-19-9334-3_6

static simulation and trend of risk and associated uncertainties for a dynamic simulation for the plant. Since the subject topic is risk-conscious operations management, this chapter will treat the risk simulation in an integrated manner by considering (a) accident scenario development in a plant simulator control room, (b) an operator support system through which the operator interacts with the plant and (c) a risk-consciousness framework for scenario development.

Accordingly, the approach to risk simulation involves processing the information and data in three major facilities or models, a plant simulator for simulating a scenario, a risk monitor for the evaluation of risk and an operator support system that captures operator response. The risk monitor is based on an approved and qualified probabilistic risk assessment model of the plant. Since the accident scenario is developed on the plant simulator, there can be questions regarding the correlation between the scenario developed in the simulator environment that may or may not represent the actual scenario that might be witnessed when the event happens in real time. Plant events can broadly be divided into the anticipated operational occurrence, e.g., loss of off-site power (LOOP), which is expected to occur more than once in the life of the plant, and second, the accident scenario, such as loss of coolant accident (LOCA), which is not expected to occur, but the plant design basis is such that it has adequate defense for managing and mitigating the postulated accident scenario. Even through plants have provision to manage and mitigate these scenarios, as seen in real time, there is more experience and understanding of the crew with respect to the anticipated transients compared to the accident condition; therefore, simulation can be an effective strategy for training to identify the strength and weakness in the plant hardware and management systems. Off-course there are hard questions like 'How strong the correlations between the plant and simulator environment', particularly in terms of stress levels as simulator is a model of the plant and 'not the plant itself' therefore there are inherent limitations in terms of uncertainty associated with the insights and results obtained in simulator environment. Here, let us understand that even though we can be conscious of this limitation, there can be efforts to reduce this gap and ensure that the results capture the uncertainty commensurate with the gap areas. This is where this chapter proposes an approach where the three modules work in an integrated manner, as well as identifying the areas of potential gaps, such as stress levels and limitations of simulators in terms of plant coverages. There is also an assumption that apart from certain high cognitive stress moments, the rest of the operations, such as planned actions, may not involve the stresses. Furthermore, the use of a plant simulator is expected to bring down, especially in terms of carrying out procedural and diagnostic components, as it has a preconceived knowledge base in terms of transient identification (a highly stressful period), diagnostics of plant configurations and equipment failures, which might help reduce the cognitive load on the operator. Apart from this, the other observation in terms of real-time plant operation is that there are moments during the management of anticipated transients or system leaks or a unique failure that challenges the operator to arrive at an appropriate strategy to bring back the plant to a normal state or otherwise not allow the plant to lead to accidental conditions, e.g., recovering the plant from loss of off-site power failure (LOOP) to normal operation and avoiding the station blackout condition. The argument is that there are some opportunities to collect the data and insights from normal transients and challenges that certainly provide a strategy to use these insights for a simulator environment. The above description culminates in an Integrated.

6.2 Integrated Risk Simulation Framework

There are essentially four major modules to support integrated risk simulation framework (IRSF) in RCOM, viz., a) a module for storing cases on plant incidences & experiments/mockup tests, data on performance of safety systems that is updated periodically, b) a risk monitor (or a computerized PRA model of the plant), c) a full scope plant simulator and d) an intelligent operator support system or atleast a plant data and information system, as shown in Fig. 6.1. The availability of a probabilistic risk assessment model is a prerequisite for the development of a risk simulation environment. This model is adopted to develop a risk monitor of the plant that essentially requires two major inputs, the quantified accident sequences and minimal cut-set equations, along with their uncertainty estimates, a database of the SSCs, human failure probability and common cause failure that forms part of PRA modeling and their respective importance measures. Apart from this, the IRSF approach requires a plant simulator to facilitate the performance of sessions for the initiating events and desired safety system failures to facilitate input for the risk monitor in terms of data and insights, particularly on human performance and common cause failure. In IRSF, the operator support system that generates advice for the operator is interfaced with the simulator such that the adequacy of the operator response in terms of time, and finally, stress levels can be obtained. Risk simulation for complex engineering systems requires a model, and insights are a process required to begin with modeling and assessment of the subject plant and its SSC response considering its process operation, the potential for various deviations, the plant control provisions to bring the plant back to normal, the performance of the plant safety and protection system in case the plant deviates from the normal operation envelope, the application of mitigation measures and finally the accident control measures such that the risk to the plant staff and members of the public is as low as possible. The integration of human factors with plants is an essential component of any simulation, as human actions are critical to the risk management features of the plant. The plant control room and the support features that include man–machine interaction, availability of computerized information system and operator support systems are critical to arrive at improved assessment prediction through the simulation experiments.

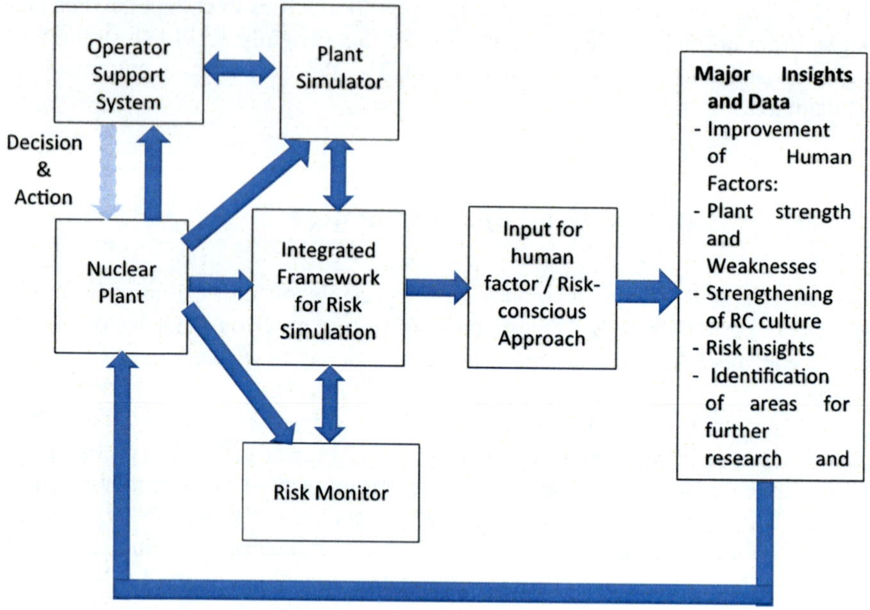

Fig. 6.1 Integrated risk simulation framework in RCOM

6.2.1 Plant Experience and Experiments

In integrated frameworks apart from the identified modules, the data and information gathered directly from plant documents and records often form a perpetual requirement that might include changes, specific layouts, site feature communication and line of authority, and inventory management-specific test and maintenance schedules that may not form part of routine training. There is an opportunity to obtain this knowledge during hands-on round-the-clock training. The general feature of a plant, i.e., plant documents and records such as design and operations manual P&ID, design and commissioning documents, technical specifications, emergency operating procedures and incident reports provide adequate information and training for the operating crews along with hands-on training. Apart from this, the documents on periodic surveillance tests, mockups and emergency drills prepare/refresh the staff about emergency safety management for the postulated scenario. The initial step is to perform a theoretical study to assess the situation and obtain information on available safety margins in terms of plant capability and the stress levels imposed by the new situation or additional considerations. If the results of the theoretical study provide adequate confidence, the study becomes part of the plant safety analysis, and the matter is closed. However, there could be a situation in which additional experiments are required to obtain the required confidence. If the experiment can be done to assess the safety of the plant for the postulated condition without compromising the safety of the plant, then this could be the best option, as the available new insight will provide more confidence in the results of the experiment, as the experiments were done under real-time plant conditions meeting all the specifications of the safety case. However, there are situations when the use of plant systems to perform the experiment is prohibitory due to safety implications or for any other reason. In such cases, an experiment can be set up to independently simulate the plant condition, and this insight can be used after a review.

The experiment with the actual plant often becomes prohibitive due to (a) safety implications and (b) it introduces an interruption/disruption in the normal process operation of the plant. In such situations, the availability of a simulator might provide an option to understand plant behavior. In fact, the simulator might even offer some features and insight that might not be available with the plant. However, for experiments that involve risk assessment and assessment of risk margins, design basis, beyond design basis; the simulator experiments are not an option but mandatory, as these experiments cannot be done on the actual plant.

For example, the plant data on studies related to off-site power failure provide a better option, as this is an anticipated transient, and plant records can provide good insight into the plant's capability in terms of coping time and so on. However, if some LOCA aspects are to be analyzed, then the simulator provides a good option, provided that the simulator has the capability to adequately model the scenario. The availability of a simulator can also help in understanding the human factor aspects, which otherwise are difficult to postulate. Of course, at the end, the simulator is also a model of the plant and not the plant, and this aspect should be adequately factored in while looking at the results of the experiment.

6.2.2 Risk Monitor

Risk simulation is facilitated by a risk monitor. Risk monitors are a computerized model of the plant. A Level 1 PRA model of the plant in a risk monitor environment facilitates the evaluation of accident frequency and associated consequences for a core damage scenarios. The IAEA definition of risk monitor is '*A plant-specific real-time analysis tool used to determine the instantaneous risk based on the actual status of the systems, (structures) and components*' (International Atomic Energy Agency 2019a). For risk simulation experiments here, the risk-based approach has been employed. In risk-based approaches, one of the major elements is the integration of deterministic and probabilistic approaches. Accordingly, the deterministic considerations, e.g., the component and system failure criteria and definitions, support the PRA modeling and are used in an integrated manner to ensure that the approach is holistic. Furthermore, the safety margins as considered in the deterministic approach are implemented in terms of uncertainty bounds in PRA. Furthermore, for holistic risk simulation, deterministic insights and criteria, including structural, thermal hydraulic, reactor core physics, nuclear radiation protection and any structural aspects, are employed as part of risk evaluation in an integrated risk-based engineering framework. The deterministic aspects and insights are generally obtained from safety documents of the plant, additional plant-specific experiments or studies, and simulator test results or lessons learned from a similar plant (Varde and Pecht 2018).

In this section, the aspects directly related to risk simulation, such as probability distribution, Monte Carlo simulation, Boolean algebra and fault tree, event tree and Markov modeling and the requirement of reliability data and uncertainty analysis and sensitivity analysis, and finally, the risk monitor that enables the simulation process through a user-friendly interface are discussed.

6.2.3 Plant Simulator

Even though there are varying definitions of simulation, the one that serves the purpose of this chapter and at the same time maintains its generality is *A simulation imitates the operation of real-world processes or systems with the use of models. The model represents the key behaviors and characteristics of the selected process or system, while the simulation represents how the model evolves under different conditions over time* (TWI 2022). However, for complex systems such as nuclear plants, aviation systems, space systems, process systems, transportation, electricity generation and transmission systems, electronics concept demonstration, etc., where the simulator plays a role in the demonstration of the technology even before the real-time systems came into existence such that the corrections/modifications can be implemented when the products are launched. Furthermore, the simulator versions are also refined/modified to reflect, as for as much as possible, the design and operational features of the real-time systems. Accordingly, the design of the simulator is the first part of the simulation technology, and conducting the experiments is the second phase when the simulator is inducted into the operational phase of the plant. Hence, we will have to split the definition of simulation into two parts: first, what is the simulator? and second 'what is simulation?'.

A research reactor nonreplica, full-scope simulator operational at Bhabha Atomic Research Centre developed for training purposes, was also designed such that it facilitates sessions for human factor studies during postulated emergency sessions along with plant mock trials and experiments (Joshi et al. 2019). Details about this simulation environment will be discussed in the following sections on the subject. The simulator is a model designed and developed to reflect, to the extent possible, the operational behavior of real-time systems and plants. This is achieved by systematically incorporating and integrating the process and control, including the operational features of the man–machine interfaces. Domain expertise is at the core of simulator development, and one must be very careful, as a simulator is a model of a real-time system and not a real-time system. The developer should have a conscious mind that junk-in-junk-out. Accordingly, the involvement of operational experts, design teams, safety experts, members from control and protection, thermal hydraulic experts, radiation protection experts and of course training and human factor teams is critical to the quality of the work. Even though the primary purpose of the simulator is to support training and qualification programs, increasingly, simulators are being designed and developed, keeping in mind the requirements of risk analysis in general and human factor studies.

Depending upon the scope and objective of development, the simulator can be full scope or limited scope. The full-scope simulator is a complex and comprehensive machine that allows not only the simulation of process and control parameters for normal operation as well as deviations from normal operations for design basis conditions but also facilitates the study of beyond design basis conditions.

The simulation of risk studies requires a simulator capability that facilitates the simulation of postulated initiating events that form part of the plant safety study and tracking of critical parameters, including additional parameters that may not be available in the real-time plant. All the postulated scenarios and their management should be validated and verified against the available operational procedures and emergency operating procedures of the plant.

The traditional approach to the design and operation of the plant is basically conservative in nature, and at times, the conservative assumptions appear to weaken the safety case. The simulator experiments can provide vital insight to understand the available margin, and following the current standards and reviews can demonstrate that conservative assumptions can be addressed to improve plant safety. For example, the plant has two shutdown systems: one gravity-inserted set of cadmium absorber fast-acting shut-off rods and the second moderator dumping system that removes the moderator from the core to take the system to subcritical conditions. The conservative approach allows credit for the shut-off rod system, as it is fast acting and makes the reactor subcritical faster compared to the dumping moderator dumping system, where the reactivity removal rate is slower than the shut-off rod system.

Detailed experiments can be performed in a simulator by shutting down the reactor on a shut-off rod system alone and moderator dumping each separately to compare the adequacy of moderator dumping alone to shut down the reactor. These experiments were performed against two postulated initiating events, viz., loss of off-site power and loss of coolant accident. The core physics and thermal hydraulics analysis showed that these two cases are the bounding cases. A detailed analysis included a simulator experiment to understand the core reactivity, coolant, and fuel temperature trends for different scenarios.

It could be demonstrated that for LOOP and minor LOCA conditions, even moderator dumping alone can meet the core safety criteria. For major LOCA, two factors were assessed: (a) the probability of major LOCA is insignificant, and (b) the reactor will become subcritical well in time. Furthermore, the analysis for coolant voiding, which again is a low probability scenario, needs to be examined. The risk modeling part will be discussed in the next section.

6.2.4 *Intelligent Operator Support Systems*

There is an active interest in the deployment of machine learning or artificial intelligence (AI) approaches in support of nuclear plant operations and management. Some of these include the following:

6.2.4.1 Operator Support System

Higher cognitive stresses induced during emergencies might lead to error due to a paucity of time for action, lack of awareness or knowledge, personal attributes or attitudes. These situations require an operator advisory provision in the plant control room to reduce the chances of error through computerized automated input and advice by online abstraction, diagnostics and ergonomically designed man–machine interfaces. These provisions for improving the alertness quotient with respect to plant evolving status or aspects of communication, in support of timely actions, particularly during emergency conditions, help operators in decision making. To address these situations, the traditional computational approach might have applicability in a limited sense and requires a system that has the capability to model and execute nonlinear algorithms often using imprecise or fuzzy data and information. Furthermore, this system should have the capability to learn based on available training patterns or knowledge using real-time data from the plant. In fact, the system should mimic a trained operator attribute (Varde et al. 1998).

One of the approaches to reduce the chances of human error or aid decision making during emergency scenarios is the deployment of an intelligent operator support system that has been designed employing intelligent algorithms, such as artificial neural networks and knowledge-based systems, such that it acts as an operator aid that assists and presents solutions as the situation in the plant evolves over a period of time and helps the operator either bring back the plant to normal conditions or assist the operator in implementing mitigative measures such that the consequences of the accident can be minimized. This subject has been covered in detail in Chap. 10 of this book.

These intelligent operator advisory systems can interface with the reactor simulator of the plant for assessing the human operator response as part of an integrated approach for the simulation of risk scenarios. For example, simulating a session with a loss of off-site power (LOOP) scenario coupled with emergency power system failure partly or fully provides risk insights for a postulated accident sequence obtained from the PRA model of the plant, viz., LOOP * EPS * HE (LOOP and emergency power system failure (EPS) and human error in bringing the plant to a normal state or performing mitigative action). This approach, apart from training the staff in handling on anticipated transient handling, provides valuable lessons on human capability and behavioral aspects in the control room situations of the plant.

6.2.4.2 Fuel Handling Activity Planning and Management

Fuel management activities in a plant can broadly be divided into two major areas. First, fuel handling operations to replace the irradiated fuel with fresh fuel and discharge the irradiated fuel into a spent fuel bay ensure that the coolability condition/specification is not violated. Second, theoretical simulations are used to generate in-core fuel management for optimizing fuel utilization while also ensuring reactor safety. Both approaches are elaborate and complex and require the application of risk-conscious framework strategies to reduce errors or undesired consequences. For in-core fuel management, the traditional linear programming strategy may not produce the desired results. Here, the artificial intelligence (AI)-based approach is required for in-core fuel management as part of evaluating fueling requirements for operational plants to meet the risk as well performance management objectives (Nissan 2019). Reactor fueling is a human–machine-intensive procedure that requires highly trained fueling staff, as the reactor core boundary is extended to fueling machine containment barrels. Any inadvertent action along with human error in aligning the alternate cooling system can pose challenges to reactor safety. Therefore, reliable automation based on the well-optimized configurations and actions of the fueling operations is at the core of ensuring the performance and risk management goals. Despite this, it has been found that human intervention volume is significant in fueling operations.

Here, intelligent approaches have been explored with the objective of reducing human error while executing fueling procedures as well as during the management of emergencies, if needed. These systems are generally implemented employing rule-based algorithms. The objectives are to ensure fuel coolability and performance functions such that the fueling schedule can be completed in time while ensuring that human error or inadvertent action does not pose threats to safety and availability targets (International Atomic Energy Agency 2021).

6.2.4.3 Robotics for Emergency Actions

There is a growing interest in the development and deployment of robots for emergency operations inside and outside areas of the plant to facilitate emergency action more effectively. Often, the higher radiation field, as well as for higher coverages, the robotics applications based on intelligent algorithms have been shown to be an effective strategy. In fact, even during routine or periodic surveillance/operations, such as accessing complex spaces and geometries such as servicing and maintenance of steam generators, maneuvering operations in glove boxes and shielded cells where the radiation field makes the operating cell area inaccessible are some of the examples of the application of intelligent approaches in nuclear plants.

This aspect is discussed here for the sake of giving the state of the art in the application of intelligent approaches to normal or emergency management.

6.3 Probabilistic Modeling and Data Analytics

The probability distribution forms the backbone of data collection, analysis and inferencing. There are two categories of numerical data that are encountered in real-time situations: continuous and discrete data. In fact, for operations risk and reliability management, there is a third category of data called qualitative or imprecise data. Modeling this category of data requires the application of fuzzy logic, which will be discussed in the following sections. Accordingly, in this section, there are two types of distributions: contentious and discrete. We will discuss the distributions that are directly relevant and useful for risk and reliability modeling in operations management systems. The major categories of continuous distributions are exponential, normal, log-normal, Weibull, etc., while the discrete distributions include Poisson, binomial and F-distributions.

6.3.1 Monte Carlo Simulation

The Monte Carlo approach is central to risk simulation. Estimation of the point estimate of the objective function considering the point estimates of the input variables is considered as a result assessed through the classical approach. There are the following major steps in Monte Carlo risk simulation:

(a) Defining the mathematical formulation of the problem along with the objective function
(b) Identification of the input parameter used in the formulation along with their appropriate probability distributions
(c) Randomly choose a probability value for each of the input parameters and obtain the estimate of the objective function
(d) Step c will be repeated N times (the value of N is chosen by the analyst and could range from 1000 to a higher number
(e) Since it is prohibitive to perform such a large iteration, Monte Carlo simulations are performed employing computers. Development of a software platform that facilitates the simulation is one of the key requirements for implementing Monte Carlo simulation
(f) Presentation of results that include the estimate of the mean for the objective function along with the uncertainty bounds.

The fault tree and event tree analysis, as discussed in Chap. 1, enable the development of the mathematical equation of the employed cut-set analysis. Here, we will discuss Monte Carlo simulation of an emergency power supply or Class III power system in a nuclear plant. Emergency power supply here assumed to comprise on-site diesel generator sets.

In this section, the aspects that are directly related to simulation are probability distributions, Boolean logic to generate the minimal cut-sets in the fault tree and finally operating features of Monte Carlo simulation and interpretation of results.

6.3.2 Probability Distributions

Selection of a probability distribution for a given input parameter is a complex task given a situation that you are dealing with the field data collection process and using these data to arrive at the probability of failure and characterization of the uncertainty in the data. There are approximate methods, viz. graphically analyzing the applicability of data and other regression approaches to characterize uncertainty in prediction. However, one is not sure what best is the best. Often in risk analysis assumptions about considerations of distributions are used as also some heuristics applied but the situation remains, and this could be one source of data/model uncertainty. This subject will be discussed in the data analysis section. In this section, we will discuss the frequently used probability distributions in risk simulation.

I. **Continuous Distributions**

6.3.2.1 Exponential Distribution

This is the simplest and most elegant single parameter distribution. The major part of risk analysis for a complex system is performed assuming that the failure data follow exponential distributions. There is an over-encompassing consideration that in the majority of cases, the plant is considered in the 'useful life period', i.e., the failure rate is constant. The rationale for this consideration is that the monitoring and surveillance, including the condition monitoring program along with preventive maintenance approach to management, might not allow the failure rates to increase without any check. However, that does not mean that for all the data this rational might apply. Hence, analysts must take a conscious approach to see the failure data trend for a specific component to enable considerations of a constant failure rate. For example, there are situations where the root cause analysis is initiated to understand frequent failure or failure posing potential for higher consequences. For the subject issue discussed here, analysts should be careful about choosing the period of data collection such that frequent failure data do not mask the final estimate, as the low failure period does not mask the frequent failure—a major drawback of averaging methods.

The graphical approach might provide confidence to the risk modeling team about the applicability of the exponential distribution. The procedure is simple and elegant. Plotting the failure event data on an exponential probability paper and the straight line passing through the data point will further provide the required confidence for the applicability of the exponential distribution. Annexure A.2 illustrates the procedure for evaluating the applicability of a model to a given dataset.

The probability density function (pdf) *for the exponential distribution in time t is given as*:

$$f(t) = \lambda * \exp(-\lambda t) \tag{6.1}$$

where λ is the failure rate (e.g., failure/hour) or failure/year)

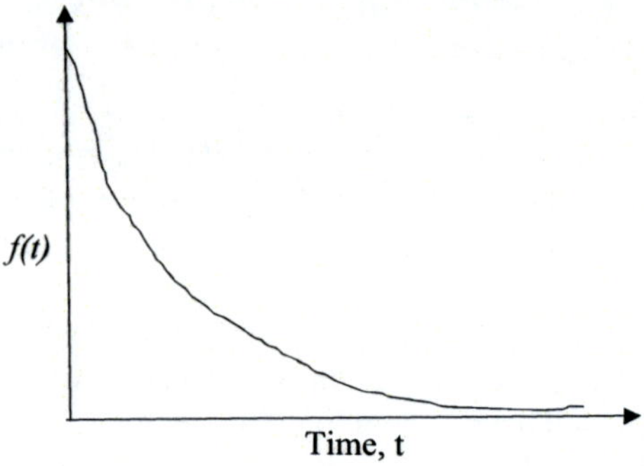

Fig. 6.2 Probability distribution function (pdf) for exponential distribution in time domain

t is the mission time for operation of the system (Fig. 6.2).

The cumulative distribution function (CDF) can be obtained by integrating Eq. (6.1) for limits 0 to 1 as follows:

$$F(t) = 1 - \exp(-\lambda t) \tag{6.2}$$

The data collected from the plant will provide the point estimate of the failure rate; the mission time is the time required to bring the system to a safe state that can be obtained from the plant safety record. For example, an injection cooling pump must operate for at least 24 h to ensure the safety of the reactor.

The uncertainty bound for the exponential distribution assumes that the data follow the χ^2-distribution and is estimated as follows:

$$\frac{\chi^2_{\frac{\alpha}{2};2\gamma}}{2T} \leq \lambda \geq \frac{\chi^2_{1-\frac{\alpha}{2};2\gamma}}{2T} \tag{6.3}$$

The left-hand side estimates the lower limit of the distribution, while the right-hand side estimates the upper limit, often referred to as the lower and upper bounds of λ, more frequently as 90% bounds, as shown in Fig. 6.3, 5% on the left side and 95% on the right side. The significance of this is that these probability bounds show the uncertainty in the estimate. The greater the spread of distribution is, the greater the uncertainty. In fact, this insight is very important, as higher uncertainty often makes the data meaning less, and a decision must be made whether to use these data. For example, if the distribution is derived with insufficient failure data points, the uncertainty will be higher, which will reflect a higher spread in the distribution.

6.3.2.2 Normal Distribution

The normal distribution is a widely used distribution, be it production process quality control, consideration in design variables, health care systems, scientific studies

Fig. 6.3 Normal distribution

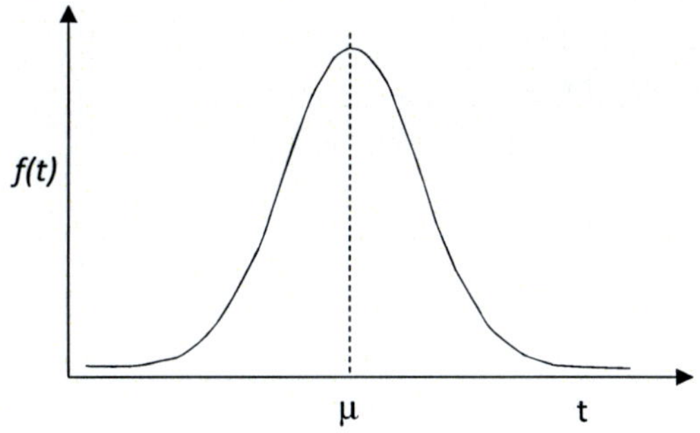

and so on. The normal distribution is characterized by a symmetric bell-shaped distribution. The pdf of the normal distribution is shown in Fig. 6.3.

A normal distribution is a two-parameter distribution characterized by μ and σ and written as $N\ (\mu, \sigma)$. The pdf of the normal distribution is given as follows:

$$f(t) = \frac{1}{\sigma\sqrt{2\pi}} \exp\left[-\frac{1}{2}\left(\frac{t-\mu}{\sigma}\right)^2\right] \tag{6.4}$$

where μ: mean of the distribution

σ: standard distribution.

Estimation of the cumulative distribution function for the above form employing the conventional integration approach is complex. Therefore, a modified form of the normal distribution called the standard normal distribution is used in practice. The procedure to transform to a standard normal distribution involves using a standard normal parameter 'z' as follows.

We substitute z for $z = \frac{t-\mu}{\sigma}$ in the equation and differentiate w.r.t. t, we obtain

$$\mathrm{d}z = \frac{1}{\sigma}\mathrm{d}t$$

$$\sigma \mathrm{d}z = \mathrm{d}t \tag{6.5}$$

The cumulative distribution function can be written as follows:

$$F(t) = \int_{-\infty}^{+\infty} \frac{1}{\sigma\sqrt{2\pi}} \exp\left[-\frac{1}{2}\left(\frac{t-\mu}{\sigma}\right)^2\right]\mathrm{d}t \tag{6.6}$$

$$F(z) = \int_{-\infty}^{+\infty} \frac{1}{\sqrt{2\pi}} \exp\left[-\frac{1}{2}(z)^2\right]\mathrm{d}z$$

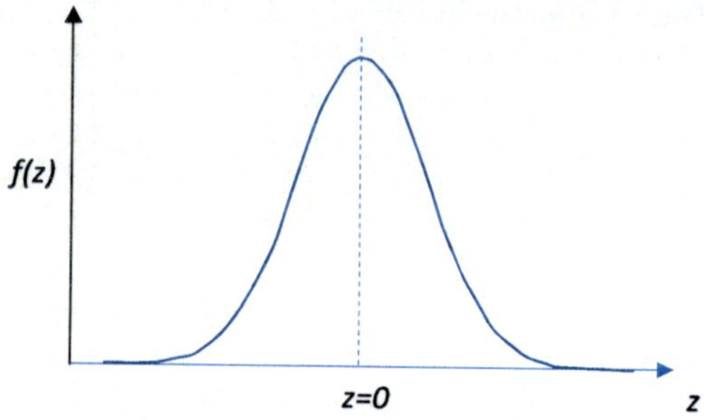

Fig. 6.4 Standard normal distribution

$$F(z) = \frac{1}{\sqrt{2\pi}}[\exp(-z)]_{+\infty}^{-\infty} \tag{6.7}$$

We obtain the standard normal curve with $z = 0$ as follows (Fig. 6.4).

The area under the curve can be obtained for various values of z in Annexure A.1.

6.3.2.3 Log-Normal Distribution

When the logarithm of the data follows a normal distribution, the distribution is called a log-normal distribution. In risk assessment, exponential and log-normal distributions are used more than normal distributions. Often, when the data are collected from more than one independent source, the experience has been that log-normal distribution works better. Log-normal distribution, unlike normal distribution, is skewed to the left side (Fig. 6.5).

The skew shape of the log-normal curve shifts from left to white as the value of standard deviation and mean, e.g., the black (1) curve has σ ~ 0.25; red (2) curve has σ ~ 0.5; and the blue (3) curve has σ ~ 1.0; for a μ value ~ 0.0;

The pdf of the log-normal distribution is given as

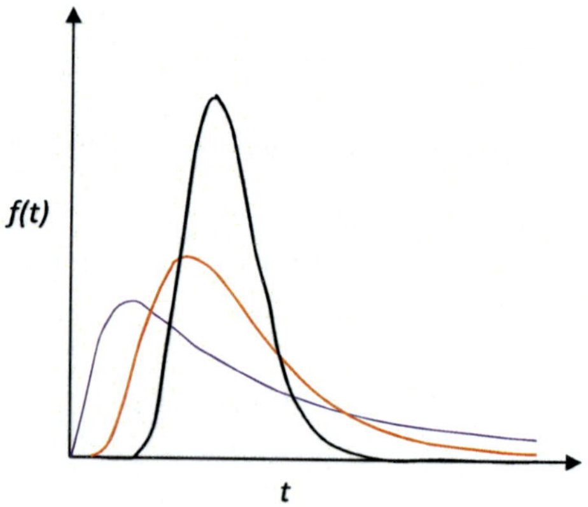

Fig. 6.5 Log-normal distribution

$$f(t) = \frac{1}{t\sigma\sqrt{2\pi}} \exp\left[-\frac{1}{2}\left(\frac{\ln t - \mu}{\sigma}\right)^2\right] \tag{6.8}$$

The mean, mode and median of the log-normal distribution are given as follows:

$$\text{Mean} = \exp\left(\mu + \frac{\sigma^2}{2}\right) \tag{6.9}$$

$$\text{Median} = \exp(\mu) \tag{6.10}$$

$$\text{Mode} = (\mu - \sigma^2) \tag{6.11}$$

Error Factor: Often, the log-normal distribution in risk assessment studies is modeled considering the error factor and median value, and the error factor is defined as the ratio of the 95% confidence bound and the median value or the median value and 5% confidence bound, i.e.,

$$\text{EF} = \frac{95\%\ \text{bound}}{\text{median value}} \tag{6.12}$$

or

$$\text{EF} = \frac{\text{median value}}{5\%\ \text{bound}} \tag{6.13}$$

6.3.2.4 Weibull Distribution

Here, we will discuss two-parameter Weibull distributions characterized by the parameter characteristic life, α, and slope parameter, β. Weibull distribution has found wider application, particularly in the area of life testing, aging studies and study time to failure modeling. This distribution also provides an elegant way to model life characteristics, even if the data follow normal, exponential and log-normal distributions. For example, if $\beta = 1$ for a given data trend, then the data follow an exponential distribution; if $\beta =$ between 3 and 4, then the data follow a log-normal distribution.

The pdf of the Weibull distribution is written as follows:

$$f(t) = \frac{\beta t^{\beta-1}}{\alpha^{\beta}} \exp\left[-\frac{t}{\alpha}\right]^{\beta} \tag{6.14}$$

where β: Slop/shape parameter

α: Scale parameter/characteristic life.

The cumulative distribution function of the Weibull distribution is obtained by integrating pdf and is given as

$$F(t) = \exp\left[-\frac{t}{\alpha}\right]^{\beta} \tag{6.15}$$

We have reviewed some selected continuous probability distributions that we deal with more frequently. There are some discrete probability distributions that we encounter when dealing with discrete data, such as the total number of demands on a diesel generator system and the number of failures, which are countable numbers unlike the continuum of values that we consider in continuous distributions. Keeping in mind the scope of this section, we will discuss the binomial and Poisson distribution.

Discrete or Time-Independent Probability Distribution

Failure events that can be counted and produce outcomes as failure, e.g., failure of diesel generator to start on a demand. There are causes to these failures; however, the degradation trend of the part that caused the failure exhibits a catastrophic or random phenomenon. It may not be possible to relate to the time that was required to cause this failure, as the normally diesel generator is an emergency system and remains in standby mode until either surveillance test or a power failure demand required the DG to start. Here, it is also not possible to determine when the culprit component transitioned to a failed state before the demand was put on the DG. Hence, these categories of components are modeled employing the time-independent failure. We will discuss two distributions, binomial and Poisson distributions, as follows.

6.3.2.5 Binomial Distribution

There are situations in risk assessment when out of total number of component n it is required to assess the probability of k component such that system success or failure probability can be estimated. Let us consider a control and protection system comprised of three parallel redundant configurations such that if one channel out of three goes out of service, the system is considered available/healthy; however, if two channels fail, then the system is considered to fail. Let us consider, second example, the primary shutdown system of a nuclear plant comprised of 14 shutdown devices. On every trip signal, all the rods fall and shut down the reactor successfully. However, the system safety philosophy should have a margin to allow two shutdown devices to fail such that with even 12 devices, the reactor can be brought to a shutdown state. This means that for system success, at least 12 shutdown devices should function to meet their success criteria. This means that if there are 3 shutdown devices that fail out of 14, then the system is declared to have failed. Hence, the failure criteria for the control channel are 2-out-of-2 channel and shutdown system 3-out-of-14 failure. These types of situations are encountered in risk assessment, and the binomial distribution enables modeling of these logics.

The pdf of the binomial distribution system is given as follows:

$$p(p; k/n) = \binom{n}{k} p^k (1-p)^{n-k} \tag{6.16}$$

p: probability of component failure
n: total number of components
k: number of component failures to satisfy the failure criteria of the system.

For example, the failure probability of one of the shutdown devices is 1e−3; as discussed, $n = 14$; and $k = 3$ (as up to two devices failure system operation is success).

The probability of system failure can be modeled as follows:

$$p(3; 14) = \binom{14}{3} (1 * 10^{-3})^3 (1 - 1 * 10^{-3})^{14-3}$$

6.3.2.6 Poisson Distribution

The events are regarded as Poisson distributed when the average rate of occurrence, failure, is constant and countable, for example, the number of failures in a given time, the number of defects per square foot on a steel sheet or the number of defects per meter length of wire. If we relate to operational problems, the average rate of component/part, say gaskets withdrawal from store or the number of fuels, being received in the reactor per month. This information is critical to spare part or inventory management.

$$p(x; \lambda) = \frac{\lambda^x \exp^{\lambda}}{x!} \tag{6.17}$$

where λ: average rate of event, i.e., failure

x: outcome of interest.

The electromagnetic relays are the store item, and the record shows that the probability of failure of relay is $3 * 10^{-4}$. A total of 2000 relays operating in the field. What is the probability of not having more than 2 relays to support the inventory management program?

The failure rate of relay λ is $2000 * 2 * 10^{-4} = 0.4$

The event of interest $x = 2$

$$p(2; 2000) = \frac{(0.4)^2 \exp^{0.4}}{2!}$$

$$p = 0.12$$

This figure suggests that the probability is high, and we need a higher figure to obtain the required confidence that adequate inventory will be maintained. Let us consider if we consider 10 relays; similar computation can be done, and we obtain $4 * 10^{-11}$. That means that given the consumption, an inventory of 10 relays will be adequate for a given type of relay.

6.4 Probabilistic Risk Assessment

There are three major aspects to system or event sequence modeling at the qualitative level. The development of reliability or unavailability graphic models employing, for example, the even tree, fault tree be it static or dynamic model to model plant configuration. Once the graphic models are developed, the next step is to generate the event sequence as part of the event tree and cut the set as part of fault tree modeling. Boolean logic is used to generate the minimal combination of events and components or human actions to analyze the failure or unavailability model at the event tree and fault tree, respectively. The next step is quantification of models using the reliability data, such as failure rates and demand failure probabilities, where the reliability data collection, analysis and plugging these data into the event tree and fault tree followed by computerized analysis is a vital step for frequency estimation at the accident sequence level unavailability evaluation at the system level. The following subsections discuss these steps as follows:

6.4.1 Boolean Logic

In unavailability analysis and probabilistic risk assessment (PRA), the generation of minimal cut-sets requires the processing of models through Boolean logic to obtain the list of minimal cut-sets of the component(s). The minimal cut-set (MCS) is the single or combination of the minimum number of components that can cause the failure event to occur. The minimal cut-set could be of first order: having a single component; second order—comprising 2 components; third order having three components and further higher-order corresponding order of the components. In fact, Boolean reduction can be done manually when the fault tree is simple and requires processing of 10 s of cut-sets; however, as the number increases, manual Boolean reduction is not possible, and computerized algorithms employing Boolean laws are employed. The out-cum of the Boolean reduction is a list of minimal cut-sets, as mentioned, which may comprise first, second, third and higher orders depending on the cutoff cut-set settings set in the algorithm. The summation of the minimal cut-sets, say generated during fault tree analysis, when equated with the top event provides the equation for Monte Carlo analysis, for estimating the system unavailability or

Table 6.1 Boolean laws for cut-set processing in fault tree and event tree

S. No.	Description	Logic	
1	Commutative law	i	$A.B = B.A$
		ii	$A + B = B + A$
2	Associative law	i	$A + (B + C) = (A + B) + C$
		ii	$A.(B.C) = (A.B).C$
3	Idempotent law	i	$A + A = A$
		ii	$A.A = A$
4	Absorption law	i	$A.(A + B) = A$
		ii	$A + A.B = A$
5	Distributive law	i	$A.(B + C) = A.B + A.C$
		ii	$(A + B).(A + C) = A + B.C$
6	Complementation	i	$A.\overline{A} = \phi$ (null event)
		ii	$A + \overline{A} = \Omega$ (Universal event)
7	De Morgan's law		$\overline{(AB)} = \overline{A} + \overline{A}$
			$\overline{A + B} = \overline{AB}$

initiating event frequency. The same process when applied for event tree analysis provides the list of accident sequences. All the laws have been discussed using three events A, B and C (consider these events as the basic events in the fault tree for better appreciation). Table 6.1 provides the important Boolean laws.

6.4.2 Even Tree Modeling

Probabilistic risk assessment modeling essentially employs the fault tree technique for system level modeling, and the event tree technique for accident sequence analysis has been introduced in Sect. 3.4.3.8 in Chap. 3 on dependability. This section will discuss PRA model simulation such that this section backs up the risk-monitoring operation in an operating plant environment.

Reliability data analysis forms a critical component of the assessment and simulation of risk. For example, analysts often come to a situation when the data on component performance or certain attributes need to be evaluated to assess human factors as part of risk modeling or simulation. On the other hand, domain knowledge in the form of qualitative input is available. This input can be used to generate quantitative reliability data employing fuzzy logic, expert elicitation techniques or both together to address the issue of lack of data. This section will discuss the applicable features of fuzzy logic and expert elicitation for generating reliability estimates. Another situation is that performance data on certain hardware, software or human actions are available but not adequate. This section will also discuss the Bayesian updating of the priori (already available data). Furthermore, the role of life testing and prediction will also be presented because new components are existing components for which data are not adequate.

6.4.3 *System Unavailability Modeling*

In this section, we discuss the static fault tree approach complementing the fault tree analysis discussed in Chap. 3.4.3.7 and some portion reproducing, e.g., cut-set generation, which is vital for simulation, for the purpose of simulation for unavailability modeling. The phrase 'fault tree' or FT refers to a static fault tree, while the 'dynamic fault tree' is referred explicitly with the word 'dynamic' as a prefix to 'fault tree', i.e., a dynamic fault tree (DFT), which will be discussed in the following section. As mentioned earlier, fault tree analysis is employed for system unavailability modeling. Let us revisit the fault tree model for the emergency power supply system discussed in Chap. 3, as shown in Fig. 3.16. The failure criterion for the emergency power supply systems is a 3-out-of-3 DG system failure. This was possible by common cause failures (Bus-CCF, DG-CCF, Breaker-CCF, Maintenance-CCF, Installation-CCF).

The list of cut-sets is reproduced here to avoid and streamline the simulation demonstration. As part of preparing the input for the simulation, Table 6.2 shows the individual component/basic event description, component code, type of component, applicable probability distribution, mean and uncertainty bounds. As mentioned, a Monte Carlo simulation technique generally forms part of fault tree analysis.

In practice, fault tree analysis is performed by a PRA application software computerized tool that automatically generates a list of cut-sets, and once quantification is

Table 6.2 Input table for simulation

S. No.	Basic event description	Component code	Component type	Applicable probability distribution	90% uncertainty bounds
1	Diesel generator-1	DG-1	Standby	Exponential	5 and 95%
2	Diesel generator-2	DG-2	,	Exponential	,
3	Diesel generator-3	DG-3	,	Exponential	,
4	Breaker-1	BREAKER-1	,	Exponential	,
5	Breaker-2	BREAKER-2	,	Exponential	,
6	Breaker-3	BREAKER-3	,	Exponential	,
7	Electrical bus-CCF	BUS-CCF	,	Log-normal	,
8	Installation or site-specific-CCF	INSTALLATION-CCF	,	Log-normal	,
9	Maintenance induced CCF	MAINTENANCE-CCF	,	Log-normal	,

performed, then the output of the fault tree is the top event probability with the uncertainty bound. For 90% confidence, the 5 and 95% bounds form part of the statement of unavailability of the top event.

However, for the purpose of explaining cut-set generation, Boolean logic is employed for generating the minimal cut-set. The cut-set equation for system unavailability is derived as follows:

$$
\begin{aligned}
\mathrm{T(EP)} &= \mathrm{DG} - \mathrm{I} + \mathrm{DG} - \mathrm{CCF} \\
&= [(\mathrm{DG} - 1 - \mathrm{System}) \times (\mathrm{DG} - 2 - \mathrm{System}) \times (\mathrm{DG} - 3 - \mathrm{System})] \\
&\quad + [\mathrm{BUS} - \mathrm{CCF} + \mathrm{INSTALLTION} - \mathrm{CCF} + \mathrm{MAINTENANCE} - \mathrm{CCF}] \\
&= [(\mathrm{DG} - 1 + \mathrm{Breaker} - 1) \times (\mathrm{DG} - 2 + \mathrm{Breaker} - 2) \\
&\times (\mathrm{DG} - 3 + \mathrm{Breaker} - 3)] \\
&\quad + [\mathrm{BUS} - \mathrm{CCF} + \mathrm{INSTALLTION} - \mathrm{CCF} + \mathrm{MAINTENANCE} - \mathrm{CCF}); \\
&= (\mathrm{DG} - 1 \times \mathrm{DG} - 2 + \mathrm{DG} - 1 \times \mathrm{Breaker} - 2 + \mathrm{Breaker} \\
&\quad - 1 \times \mathrm{DG} - 2 + \mathrm{Breaker} - 1 \times \mathrm{Breaker} - 2). \\
&\quad (\mathrm{DG} - 3 + \mathrm{Breaker} - 3) + [\mathrm{BUS} - \mathrm{CCF} + \mathrm{INSTALLTION} - \mathrm{CCF} \\
&\quad + \mathrm{MAINTENANCE} - \mathrm{CCF}] \\
&= [\mathrm{DG} - 1 \times \mathrm{DG} - 2 \times \mathrm{DG} - 3 + \mathrm{DG} - 1 \\
&\times \mathrm{Breaker} - 2 \times \mathrm{DG} - 3 + \mathrm{Breaker} - 1 \times \mathrm{DG} - 2 \times \mathrm{DG} - 3 \\
&\quad + \mathrm{Breaker} - 1 \times \mathrm{Breaker} - 2 \times \mathrm{DG} - 3] \\
&\quad + [\mathrm{DG} - 1 \times \mathrm{DG} - 2 \times \mathrm{Breaker} - 2 \\
&\quad + \mathrm{DG} - 1 \times \mathrm{Breaker} - 2 \times \mathrm{Breaker} - 3 + \mathrm{Breaker} - 1 \times \mathrm{DG} \\
&\quad - 2 \times \mathrm{Breaker} - 3 + \mathrm{Breaker} - 1 \times \mathrm{Breaker} \\
&\quad - 2 \times \mathrm{Breaker} - 3] + \mathrm{BUS} - \mathrm{CCF} + \mathrm{INSTALLTION} - \mathrm{CCF} \\
&\quad + \mathrm{MAINTENANCE} - \mathrm{CCF}.
\end{aligned}
$$

The representation for the above event requires identification of the parameter and associated distribution for the purpose of simulation. For example, the exponential distribution is a single parameter distribution with λ as the parameter of the distribution. Accordingly, input data for a given component, let DG-1, will require a failure rate value, while the log-normal distribution is two-parameter distributions represented in the fault tree analysis y median value and standard distribution of log-normal distribution (μ_1, σ_l). For simulation purposes, the log-normal distribution is represented by the median value and error factor. The error factor is the ratio of the 95% or upper bound to the median value or the median value to the 5% or lower bound. A normal distribution is also a two-parameter distribution and is expressed as N (μ, σ).

Once these values are assigned to basic events in the fault tree, the fault tree is ready for Monte Carlo simulation. Generally, the number of iterations is set to at least 1200 to arrive at a smooth distribution for the top event.

6.4.4 Dynamic Fault Tree

To demonstrate the applicability of the dynamic fault tree for PRA application, we will consider the essential features of a case of DFT discussed by Varde and Pecht (2018). The dynamic fault tree is an extension of the static fault tree methodology with time requirements. It is recognized that dynamic fault tree evaluation contributes to improved nuclear plant safety by focusing on improvement tests and maintenance activities. Considering that, as part of dynamic PRA, it was required that a safety system unavailability (i.e., the ECCS) is to be evaluated for certain component status changes, human intervention, or actuation of a recovery mechanism. After the initial epoch, i.e., at time $t = 0$ (t_0), to any time t_n ($n = 1, 2, 3 \ldots$), these changed configurations over a period of time need to be evaluated. The case in point is shown in Fig. 6.4. For example, consider an example scenario in which after LOCA initiation at time $t = 0$, the ECCS comes on demand as per the initial configuration and injects coolant from accumulator-1; however, the accumulator-1 inventory is exhausted at the epoch of time t_1, and accumulator-2 is pressed into operation at t_1. The change in system unavailability is reflected at t_1 and remains constant up to t_2. As per the system design, accumulator-2 lasts up to time t_3; when the suppression pool water injection is initiated (either manually or automatically), the change in unavailability is reflected in the figure. This example illustrates the characteristics of dynamic modeling in the temporal domain.

The dynamic fault tree gates, priority AND gate (PAND), house event (HOUSE), spare (SPAR), sequence (SEQ) and functional dependency (FDEP) are shown in Table 6.3 along with their respective graphical representation, description and specific remarks (Fig. 6.6).

The available approaches for solving dynamic faults can broadly be divided into analytical and simulation approaches. The major analytical approaches are state-space (normally employing Markov analysis), numerical integration and Bayesian networks. Each of these approaches has its pros and cons and is employed according to the modeling and resource requirements. Hence, researchers have used one of the above approaches considering the specific characteristics of the problem to be solved (Fig. 6.7).

The modeling of a dynamic fault tree addresses (a) time-dependent component unavailability $Q(t)$ and (b) application of house events to the model. Case A addresses outages of equipment using the OR gate, and case B addresses house events under the AND gate to model more operational modes, as shown in Fig. 4.16.

As seen, the time-dependent unavailability $Q(t)$ and the house event enable modeling of equipment outage and analysis of various operational modes of equipment, as shown in cases A and B, respectively. The available literature shows that the Boolean operations performed in a dynamic fault tree for complex systems, particularly as part of a PRA study, often become computationally prohibitive. Hence, the simulation approach provides an alternative approach to dynamic fault tree analysis.

However, the accident scenario evaluation for real-time situations can be modeled more realistically with treatment of the scenario in the temporal domain and provision of more than two branching points. The branching in the dynamic event tree follows the demands of the specific scenario and happens arbitrarily but at a discrete point in the temporal (time) domain.

Probabilistic dynamics enable us to fully account for the temporal evolution of the interaction of dynamics and stochastic in the evaluation of the accident consequences and their conditional (the condition is the initiating event) probabilities. Due to the nature of dynamics, the number of event sequences can be complex and

Table 6.3 Dynamic gates

Dynamic entity/gate	Graphical representation of entity	Description of gate operation on input for generating an output and *failure criteria*	Significance/remarks
PRIORITY AND gate PAND gate	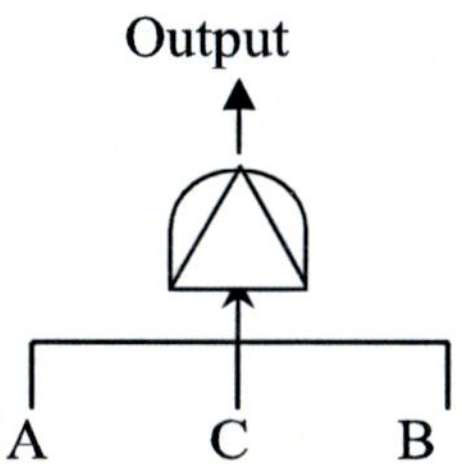	The output occurs if the input occurs in sequence as set by the analyst. For example, output A.B.C occurs, if the events A, B, and C follow the chronological sequence A then B and then C. Failure occurs at the gate output when the component at the input fails in this sequence, A, B, and the C	This gate has already been discussed in Sect. 4.5 on static fault tree analysis. This gate is normally used in dynamic fault trees
HOUSE or External event (H event)	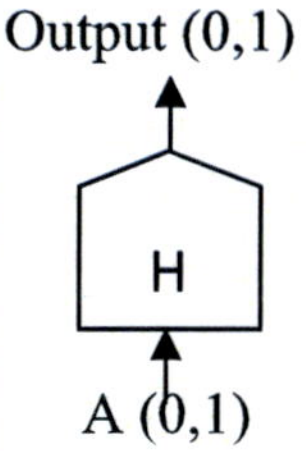	House events are extensively used in dynamic fault tree modeling for configuration management as a switch to make an input set to 0 or 1. These events are set to 0 or 1manually in the static fault tree and dynamically in DFT	House events in static fault tree are used to avoid duplication of the fault tree when selected assumptions are to be reflected in the analysis as per the requirements

(continued)

Table 6.3 (continued)

Dynamic entity/gate	Graphical representation of entity	Description of gate operation on input for generating an output and *failure criteria*	Significance/remarks
Spare gate (SPAR)	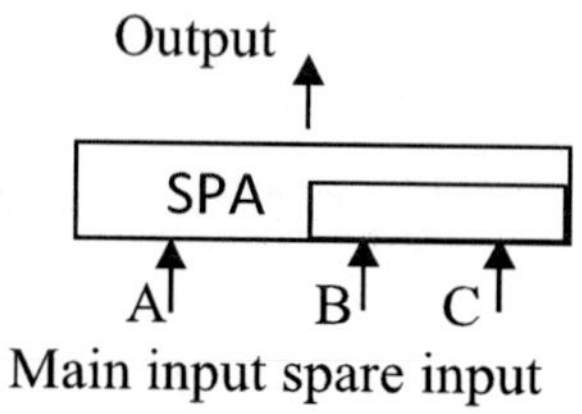	This gate can have one or more main input(s) and one or more spare input(s). The gate has provision for defining the minimum number of components required to deliver an output as gate failure. When the main component fails, then the spare is switched on and the system fails when the failure criteria are met	This gate represents a plant scenario where redundancy in the safety function is common and represents a dynamic condition. Failure of the main component and later the spare component led to the failure criteria
Sequence enforcing gate (SEQ)	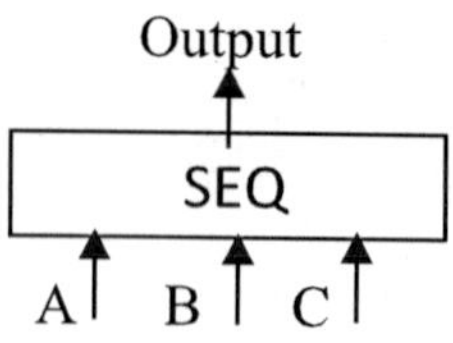	This gate is used when modeling requires failure to occur in a given predetermined sequence, to register a failure at the output of the gate	It is different from the PAND gate in that a sequence is predetermined unlike PAND gate where it follows a sequence from left to right
Functional dependency gate—(FDEP)	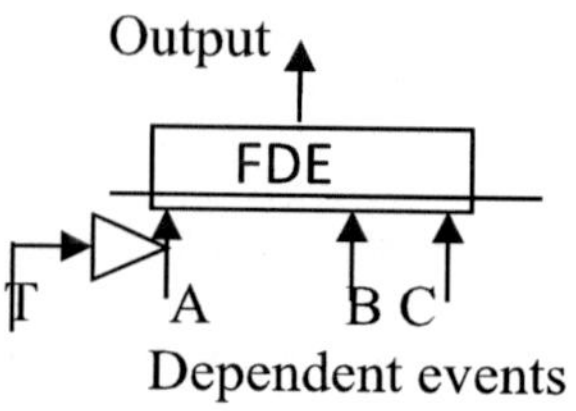	T is a trigger event. If there is a trigger, then the dependent event occurs and causes failure registration of the FDEP gate output	The trigger event could be a basic or output from a gate that causes other dependent events (here, A, B or C) to occur. Output from a Markov model can be used to cause the dependent event

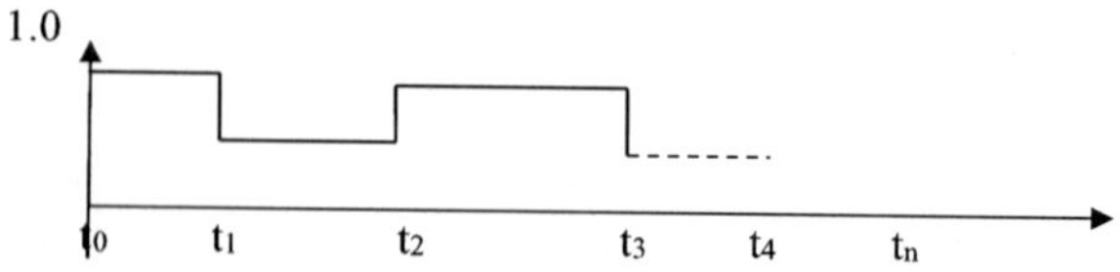

Fig. 6.6 Illustration of the output of a dynamic fault tree

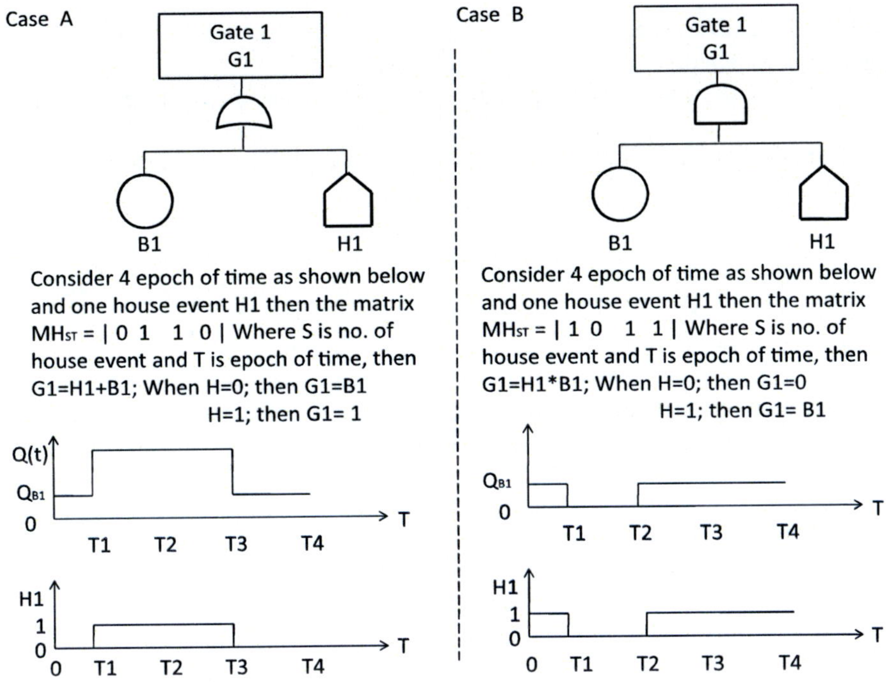

Fig. 6.7 Illustration of house event operations in a dynamic fault tree (Aldemir 2013)

very large to the extent that they might pose computational constraints. Probabilistic dynamics operates on actual time/state space; however, discretization is required for the evaluation to be numerically manageable (Hofer et al. 2002). The computational resources could pose a constraint for this dynamic evaluation. The dynamic event tree analysis, as part of an accident sequence evaluation, requires data inputs such as process variables, information to assess changes in component status, and human actions and their effects on the system toward modeling the transient/change in plant status. Therefore, quantitative criteria/heuristics and plant variables are required for the gainful utilization of resources.

We will consider the simplified methodology of dynamic event tree modeling formulated by Costa and Sui (Acosta and Siu 1993; Acosta and Sui 1991) to illustrate the modeling and analysis of dynamic event trees in this section. The dynamic event tree analysis method has been discussed to present the essential features and characteristics of dynamic event trees. The salient features of the dynamic event tree method are illustrated using an example of two systems *A* and *B*.

As seen in Fig. 6.8, only two systems, *A* and *B*, are considered for simplicity. However, in real time, there are more than two systems. At this point, it is important to understand that modeling and analysis of dynamic event trees requires computational techniques, which in turn require formalization of computational functions and the associated set of variables or parameters and decision heuristics/rules controlling the course of computations.

As we have seen, applications of the dynamic fault tree and the event tree approach are increasing as independent studies as well as part of risk assessment. We have also discussed the challenges these approaches are facing in regard to dynamic modeling, such as the requirements of computational resources, modeling of large fault trees, heuristics for fault tree reduction and its analysis, including identification of root and terminal events, selection of required sets of deterministic and probabilistic parameters in dynamic event tree analysis, and selection of time steps for modeling. However, extensive works have been reported in improving the dynamic approach either through the development of new structures (Guillaume et al. 2011) or through simplification or use of assumptions in the present approach (Bucci et al. 2008).

Dynamic fault tree and event tree analysis can be carried out either employing analytical approaches or using simulation methods. The existing four major methods are (a) state-space-based methods such as Markov modeling, (b) stochastic methods, (c) combinatorial methods (Shannon 1938) and (d) simulation methods generally based on the Monte Carlo approach. For example, (Bucci et al. 2008) used the Markov approach to capture dependency as part of dynamic modeling. (Rao et al. 2009a) applied the simulation-based approach to address the complexity involved in dynamic modeling in PSA. For new approaches to the analysis of dynamic modeling, readers are directed to the related literature. Table 6.4 presents characteristic set of

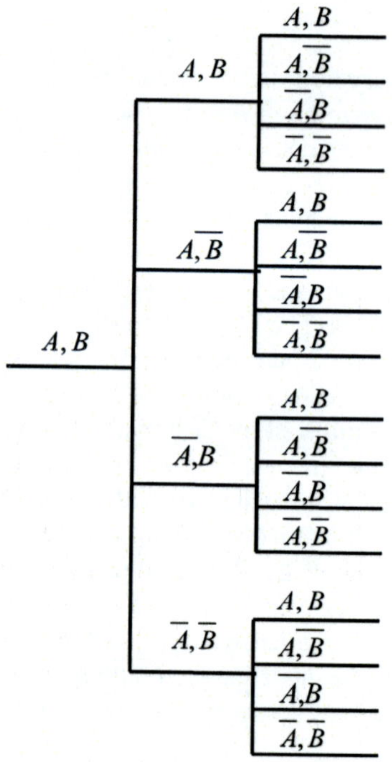

Fig. 6.8 Dynamic event tree illustration with two systems

variables defined by the DETAM application (Acosta and Siu 1993; Acosta and Sui 1991).

6.4.5 Reliability Data

Since the theme of this book is operations management, the focus of reliability data analysis will be to view the data collection, data analysis and use of these data in PRA or unavailability or maintainability modeling and simulation. Data collection and analysis is a complex task and often takes extensive, ~ >50% of the total project time if the objective is to use, as for as much as possible, plant-specific data. Of course, we have an optimistic situation in which the plant operating experience is adequate to generate plant-specific reliability data, at least for critical events such as safety system components and functions and some initiating events such as Class IV power failure, regulating system, transients. There could be another situation in which the plant is in the initial phase of operation and the available PRA model that was developed for regulatory approvals has been developed using data either from generic sources or from similar plant operating experience. There could be another situation in which the plant has been operating and entered a useful life period, but when we look at the availability of performance data, the situation is that the data are clearly inadequate. Say if the plant has operated for say 5–10 years than it is possible that the available data is not adequate. However, there is a clarity that the data in the generic sources are available. Therefore, a conscious decision needs to be taken into account all the aspects, including availability of resources, expertise, availability of computerized tools and regulatory requirements and so on to have a strategy that meet the set objective and provide the required quality in assessment and simulation. For example, uncertainty characterization and sensitivity analysis should be part of the project, as these requirements are resource consuming and often needed. Similarly, consideration of human error and the level of detailing is something that is again a resource-consuming task.

Obtaining data from generic sources is, to some extent, a straightforward task and saves considerable resources in a PRA project. When the operating experience is an either not there or at the best it is ~ 5 years, then it might be that this experience is not adequate. Here, when we investigate the insights and results of these assessments, one should know that these models might not represent the expected performance of the SSCs, which might lead to higher uncertainties in the results. For such studies, normally generic data are used (Table 6.4).

The second situation is when data are considered adequate. Here, the term 'adequate' is a subjective notion; every organization will have to make a conscious decision, and the basis of these decisions must be recorded, including the related assumptions. For example, for plants that are part of a fleet operating elsewhere with minor variations, an experience of ten years might be considered adequate, as the operating experience from other plants can be used for validation or verification, forgetting the required confidence. Here, mathematical techniques such as Bayesian

Table 6.4 Characteristic set of variables defined by the DETAM application (Acosta and Siu 1993; Acosta and Sui 1991)

Variables/rules	Description	Implementation in DETAM and further interpretation
Branching set	Plant parameters/conditions that determine the status of a plant and safety function, operators' identification of system status and the proposed action in response to the plant diagnosis	{XA, XB} is the branching set. Since there are two binary states, XA can have value of 0 or 1 for successes and failures, respectively. The same argument applies to XB When the number of safety functions increases, say to a number n, the branching set will be {XA, XB, XC … XN}
Plant status	Defined by branching variables and plant process variables	Each branch in the dynamic event tree has an end state, and the frequency value of XA, and XB determines the probability of the respective end state, i.e., the plant state determined by that branch under consideration. For example, for the two systems there are 16 end branches that lead to 4 plant states The plant status grows geometrically with the number of safety functions
Branching rules	Branching occurs at fixed points in time	In DETAM, the branching rule is a fixed interval (t_1, t_2) for XA and XB However, keeping in view the requirements of specific problems, the rules for branching could be, for example, change in status of the safety function or an event enforced externally, such as an operator action. Generalization of DET requires algorithms to interpret this information from plant variables. The advantage of this generalized model is that it avoids computation resources because, for example, the use of a fixed interval will require a small Δt to capture the instant of change in the plant/safety function status
Expansion rules	Cutoff criteria are used to terminate the accident sequence, below the predefined lower value of a frequency. This is to conserve computational resources	The cutoff criterion is one of the most used approach to arrest the uncontrolled growth of accident sequences. Other criteria could be termination of plant into a safe state as defined by either termination of the postulated initiating event at the root or ensuring availability/implementation of safety function or based on a crew decision for a given plant state

(continued)

Table 6.4 (continued)

Variables/rules	Description	Implementation in DETAM and further interpretation
Tools	Problem-specific models for plant behavior, stress, conditional frequency of operator state changes	There are two major categories of tools used in dynamic scenario evaluation—deterministic and probabilistic. The physical phenomenon for computation of a plant state for given plant transient or end state is performed by deterministic analysis, such as the value of some process variables, etc. The probabilistic tools (e.g., PRA software) provide the estimate of frequency for a spectrum of plant states

updating can be employed that enable updating the available plant-specific data with generic data.

It is recognized in PRA or system modeling that domain knowledge in the area of plant operation and maintenance, which includes design aspects, particularly when it is utilized in data collection and analysis, is critical, as the quality of data largely governs the quality of results. The insights into the strength and weakness of the plant provide an optimized approach to risk simulation and conscious utilization of these results in risk management.

6.4.5.1 Sources of Data

The major sources of data in an operating plant include daily logs, work permits, monthly reports, incident reports, periodic testing and surveillance program reports, maintenance records on equipment, annual reports and safety review reports. These sources, although vital and valuable, provide input that is voluminous, at times inadequate, vague and many times incomplete to convert them directly into reliability parameters, particularly in an environment where there is no systematic system of reliability data collection and analysis. The challenges are genuine, as the field staff often do not know what format the information is needed. Many times, the field staff are supplied with standard formats for data collection; however, there is always a communication gap when it is officially ok to submit the information. We simply put the requirement that the form should be completed and duly signed to have a robust system. It might not work as often all the information is not available while recording the event. If one puts the condition that whatever information is available that should be filled up and submitted as 'first information report'. It might work as for the communication part from field to analysts is concerned but in long run the experience is that this information and the system of collecting information is not adequate.

The whole discussion boil-down to having a system of data collection and analysis. The involvement of decision makers and middle-level management staff in grassroots-level staff with a communication protocol is central to data collection and analysis. Training and education of staff included provision for review of plant operational aspects, including incident reports, human factor aspects, learning from simulator training and experiments. The following section presents the major function of the reliability data and information management systems.

6.4.5.2 Reliability Data and Information Management Systems

Any organization with risk assessment and simulation programs must have an integrated data collection and analysis system. All the activities, including communication and automatic recording and logging of information using a computerized system, are key to successfully implementing the system. The following section discusses the essential features of the system:

Component Coding: The fundamental requirement for this system is to have a component coding system in place. Component code should be such that looking at code, it should be possible to comprehend the information about the subject component. A component coding system can have as many places as possible to provide the related information, such as the component from which the reactor system, the generic type of component, the specific category, the subsystem to which the component belongs, the failure mode applicable for the component and so on. The attempt should be to have a system that is self-explanatory and works with the plant staff, management as well as at national and international meetings and discussions. There are international systems for developing component coding, such as the IAEA reliability data database and coding system (International Atomic Energy Agency 1988, 1977a). These systems provide standard coding for component type and category, failure mode and so on. Figure 6.9 shows an example of an eleven-digit component code. Here, the first two digits are reserved for plant identification, two digits for the system name, two digits for the component type, two digits for the relay identification number on the P&ID diagram, two digits for component specification and the last digit for the component failure mode. Let us take the example of a 48 V protection relay in the digital plant protection system of the Dhruva reactor. The code for the 48 V protection relay in the digital plant protection system of the Dhruva reactor will be as follows.

The following table shows the list of IAEA standards for failure modes (International Atomic Energy Agency 1988):

S. No.	Failure mode	Code
1	All modes	A

(continued)

(continued)

S. No.	Failure mode	Code
2	Degraded	B
3	Fail to change position	C
4	Fail to remain in position	D
5	Fail to close	E
6	Fail to function	F
7	Short to ground	G
8	Short to draw high current	H
9	Open circuit	I
10	Plug/rupture	J
11	Spurious function	K
12	Fail to open	O
13	Plug	Q
14	Fail to run	R
15	Fail to start	S
16	Rupture	T
17	Other critical faults	X
18	Leak/external leak	Y
19	Dropped/uncoupled/over traveled rod	1
20	Fail to insert	2
21	Improper movement	3
22	Control rod failure	4
23	Overheated	5
24	Leakage (shell)	6
25	Leakage (tube)	7
26	Internal leak	8

In fact, the identification code provides a very transparent system to understand and analyze the minimal cut-set list and relate it to system functioning. The coding system also provides an effective and elegant mechanism for communication between the analysis team and the field staff by avoiding unnecessary confusion about the component's role in risk analysis, its type, rating, failure mode, etc. The management system should identify systems or subsystems depending on the operational requirements. For example, a digital plant protection system (PP) using an established coding system following the available national or international standards. Similarly, for the 'type', 'failure mode', the established database should also be used. These practices are required for standardization of the database.

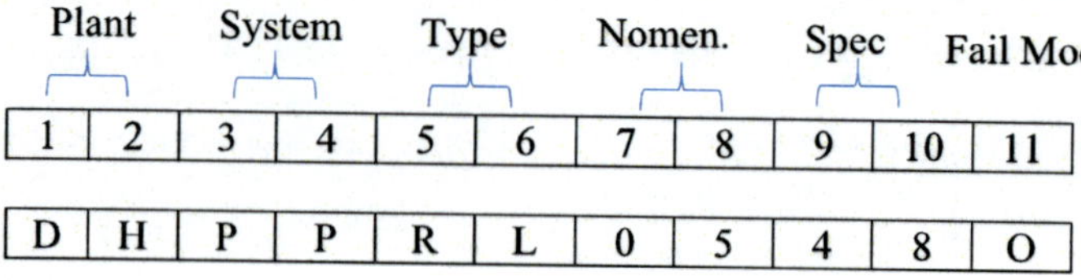

Fig. 6.9 Typical example of a coding system for SSCs in the reliability database

6.4.5.3 Requirement of Computerized Maintenance Management System (COMMS)

Reliability and risk activities should be reflected in the organizational structure. The leadership's support and periodic review are critical to the success of any risk and reliability venture. The allocation of resources, such as qualified staff, laboratory support, and a seamless protocol with the operations and maintenance group, plays an important role in delivering the quality of output. In fact, the reliability and risk program along with duties responsibilities and the communication, functional and authorization protocol should be part of the technical specifications.

Computerized maintenance and management systems (CMMSs) are available in many nuclear plants. This system can effectively be employed for reliability data collection and management programs. Essentially, the CMMS is maintained by the maintenance crew in the plant. For each SSC, there is a sheet that has fields that includes the SSC identification code, type of SSC, routine schedule, action performed, remark for any specific observation and so on. The CMMS is maintained and updated on a routine basis by the maintenance control center and ICCMS records, on a routine basis the input received from the field, on maintenance actions and checks overhaul, repair, modification, calibration, etc. Any deviation and/or failure/performance of SSCs during testing and real-time demand on the system is also recorded in the CMMS.

The format of the CMMS and the content should meet the requirements of reliability data and information management systems. For example, the timing details of failure, basis of declaring the failure, degradation level and other specific details required to evaluate the failure rate of SSC should be available. Similarly, there should be consistency in the CMMS and reliability database module such that the component boundary should be consistent. This will bring in clarity as to what has been included and what has been left out when we name an SSC. For example, the component boundary for a pump might include the pump, motor, power supply and main supply breaker, control system relays, pump foundation, pump suction and discharge valves. In fact, there should be a continuous window of discussion open for the exchange of technical details as part of the main framework.

Apart from the CMMS, the reliability data and management system also has a separate module for the incident database. This database facilitates recording and management of incidents, which includes the first information report, system and SSC involved, categorization of the incident based on technical specification of the plant into incident, event and significant event, SSC or human actions, and software that caused the incident. An incident report is assumed to be completed when the direct cause/root cause of the incident has been identified. Some safety characterization is also done in the incident/SSC that determines the potential safety implication, such as CCF category or potential for CCF to determine the precursor-CCF category, categorization of cause of human action, e.g., error of omission/commission, lapse, mistake, oversight toward determining the root cause of the incident. Often, the reliability data and management system facilitate aging, software and quality-related failures with the objective that the plant management receive feedback toward improving the program.

The system should have two modes of operation, one for the system administrator and the other for the users of the database. While the administrator should have access to the programming module for modification, changes to ensure that the system is updated periodically, ensuring the system database is updated. The user protocol enables only use of the database and estimation of new parameters, if needed, while updating of the database should be possible by the system administrator.

6.4.5.4 Major Reliability Modeling Features

The major modes and types of components are continuous or operational, standby, and periodically tested. The reliability data management system should facilitate the estimation of failure rate λ, demand failure probability q and associated uncertainties. Based on the new data received from the CMMS, the surveillance and test exercise, the incident database, etc., the system should update the existing data with new inputs, including uncertainties.

The database should facilitate routine operations such as combining more than one type of SSC in the same category to improve the population and data points such that the effect of this exercise can be seen by analysts. For example, there are two types of relays, 48 and 24 V, in the database. However, due to the paucity of data for 48 V relays, the analyst is interested in increasing the population of relays by combining the two types of relays. The database should allow these manipulations, which otherwise become very cumbersome manually. Often, it is required to update plant-specific data with generic data. Therefore, the Bayesian updating program should be implemented in the database, where apart from the mean/median value, the uncertainty bound can also be evaluated.

The database should be capable of facilitating the categorization of common cause failures and associated information, such as coupling mechanisms and direct and root causes; human error and associated causes, such as errors of omission, commission, lapse, and mistakes; institutional failures that might include procedural error, communication error, training inadequacy, risk-conscious culture-related issues, quality, aging, administrative and management issues related to compliance, monition, implementation and attitude toward complying with safety committees and plant recommendations.

Finally, the database should have a report generator module that facilitates apart from generation and compilation of reliability database, report of CCF of a given type, category of CCSs, human and institutional failures, etc.

6.4.5.5 Bayesian Updating and Uncertainty Analysis

The objective of employing Bayesian updating in risk assessment and simulation is to address the inadequacy of plant-specific failure data, particularly in situations where on the one hand, the data from plant-specific sources do not provide the required confidence in the data and, on the other hand, the data are available from generic sources, such as from data handbooks or from other plants. In this context, mathematical techniques such as the Bayesian approach provide an effective mechanism to use generic data, referred to as 'priori' knowledge, and use plant-specific evidence to update the a priori information to provide an estimate of the updated statement of the failure data.

The basis of the Bayesian theorem is the conditional probability of two exclusive events, say A and B. The intersection of the two events A and B, i.e., $A \cap B$, can be written as:

$$\begin{aligned} A \cap B &= P(A).P(B|A) \quad \text{i)} \\ &= P(B).P(A|B) \quad \text{ii)} \end{aligned}$$

Before we proceed, let us formulate the problem for PRA. There is a situation in which we are modeling an emergency power supply system comprising diesel generators. However, given the number of tests and actual demands, we have from record that there were a total of 200 demands and 3 failures. Clearly, these data, given the demands and failure, may be adequate to give the analysts the required confidence to use these data for estimating the emergency power supply system failure probability. However, in the IAEA reliability database, the emergency diesel generator failure probability for similar data specifications, in terms of power, failure mode, application, etc., the failure probability median value of 5×10^{-3}/demand with confidence bound of 5% as 2×10^{-3}/demand and for 95% as 7×10^{-3}/demand, the database shows that these data have been collected for over 30 years.

Again, we return to the problem formulation as follows:

The a priori value is $A = 5 \times 10^{-3}$/demand.

The evidence is $B = 3$ failures in 200 demands.

The objective function is to estimate the posterior, i.e., $P(A|B)$. From equations i and ii, we see that we can equate these two equations as follows:

$$P(A).P(B|A) = P(B).P(A|B) \tag{6.19}$$

The posterior probability that Probability of An in light of evidence B can be expressed as

$$P(A|B) = \frac{P(A).P(B|A)}{P(B)} \tag{6.20}$$

or

$$P(A|B) = P(A)\left[\frac{P(B|A)}{P(B)}\right] \tag{6.21}$$

This equation gives the posterior probability, i.e., $P(A|B)$. This theorem is called the fundamental Bayesian theorem and, in reliability engineering, is employed extensively for updating the available data and associated uncertainties.

We have the axiom of probability, the a probability of a set a, i.e., $P(A)$ and probability of its complement, i.e., $P(\overline{A})$, when added, will be equal to 1; accordingly, we write

$$P(A) + P(\overline{A}) = 1 \tag{6.22}$$

We have events A and B, which are mutually exclusive; therefore, for posterior probability, we can write as follows:

$$\sum_{n=1}^{N} P(A|B) = 1 \tag{6.23}$$

Multiply both sides of the equation with $P(B)$

$$P(B) = \sum_{n=1}^{N} P(A).P(B|A)$$

According to the product axiom, we have

$$P(B) = \sum_{i=1}^{N} P(A.B),$$

Again, using the product axiom, we can write $P(B)$ as follows:

$$P(B) = \sum_{n=1}^{N} P(B).P(A|B)$$

Therefore, the equation for posterior probability can be written as

$$P(A|B) = \frac{P(A).P(B|A)}{\sum_{n=1}^{N} P(A).P(B|A)} \tag{6.24}$$

This is the Bayesian updating equation. In risk assessment and risk simulation, we often deal with continuous variables, such as failure rate, frequency of event. The equation for frequency events, λ, will be as follows:

$$P(\lambda|B) = \frac{P(\lambda).P(B|\lambda)}{\int_0^1 (\lambda_i).P(B|(\lambda_i)\mathrm{d}\lambda_i} \tag{6.25}$$

There are three major sources of uncertainty: data uncertainty due to uncertainty causes due to inherent randomness or inadequate data or knowledge and model uncertainty due to inadequate model representation. Uncertainty due to a lack of or inadequate data or knowledge is called epistemic uncertainty, while uncertainty caused by randomness in data is called aleatory uncertainty. The aleatory uncertainty cannot be reduced beyond a point, and it is an inherent characteristic of data, while the epistemic uncertainty can be reduced by obtaining more information, data about the subject failure phenomenon and in this way, epistemic uncertainty can be reduced to an acceptable level.

One of the conventional approaches in risk assessment to assess randomness-induced uncertainty in data is by probability bounds or confidence bounds. We mainly encounter two types of data, i.e., the failure rate for a continuously operating system that is expressed in either failure per unit time, e.g., for a failure/hr. The second category of components is the standby type of component, which starts only after a test or actual demand. Accordingly, we will present here the mathematical representation of the confidence bound for continuously operating and standby components.

The consideration of the constant failure rate and approximation of the χ^2 distribution for an exponential distribution is the two major assumptions for dealing with time-dependent operating systems. The confidence interval for failure to run/hr. data is given as

$$\frac{\chi^2_{\left(\frac{\alpha}{2,},2n\right)}}{2T} \leq \lambda \geq \frac{\chi^2_{\left(1-\frac{\alpha}{2},2n+2\right)}}{2T} \tag{6.26}$$

where λ: failure rate (per unit time)

α: level of significance
n: degree of freedom
T: time or accumulated test duration.

For $\alpha = 0.1$, the confidence interval is 90%, and the upper bound and lower bound are 95 and 5%, respectively. If $\alpha = 0.05$, then the confidence interval is 95%, and the upper and lower bounds are 97.5 and 2.5%, respectively. Annexure A.3 provides the χ^2 (Chi-square) factors as a function of α or levels of significance and n, i.e., degree of freedom.

The standby system failure rate is expressed as failure on demand. As discussed earlier, the time-independent distribution or discrete distributions are used to estimate demand failure probability. For example, the binomial distribution is employed to estimate the probability of exact r-failure-out-of-N total demands. The valid demands are the demand on test or actual demand when the component was qualified for the service.

For demand failure data, the confidence limits are estimated using the F-distribution as follows:

$$\frac{1}{1+\left[\frac{(N-r+1)}{r}\right].F_l\left[\frac{\alpha}{2},2.(N-r+1),2r\right]} \leq p \geq \frac{1}{1+\frac{N-r}{r+1}\cdot\frac{1}{F_u\left[\frac{\alpha}{2},2(r+1),2(n-r)\right]}} \tag{6.27}$$

The left side of p gives the lower confidence bound, while the right-side term gives the upper bound of probability.

where N: total number of demands

r: number of failures

$^*F_1$: F-distribution—lower bound, for the parameters, $\alpha/2$, 2. $((N-r+1), 2r$, degrees of freedom; obtained from the F-distribution table

α: Level of significance

$^\#F_u$: F-distribution—upper bound, for the parameters, $\frac{\alpha}{2}, 2(r+1), 2(n-r)$, degree of freedom; obtained from F-distribution table in Annexure A.2.

These bounds for failure to run and demand failure probability can be used in the fault tree analysis along with their point estimates for the basic nodes to estimate the top event—system failure probability and associated uncertainties.

6.4.5.6 Fuzzy Approach to Reliability

Often, analysts encounter situations when the data collection and analysis program is adversely hampered due to unavailability or inadequacy of data where even limited data are not available, particularly for critical SSCs, human action or a new aspect or initiating event. The situation is further aggravated when the requirement of uncertainty is a must for assessment or simulation. In such situations, qualitative information based on operating experience, if available, can be utilized for estimating the failure probability or possibility of failure and associated uncertainty by employing the fuzzy logic approach.

In fact, the traditional approach to risk modeling suffers from one major binary limitation while dealing with failures, as we either consider SSC as either failed or successfully operating. There are many situations during modeling and simulation in which scenario modeling requires various levels of degradation, such as degradation levels 1 and 2, which lie between success and failure. The binary or crisp logic does not permit modeling of more than two states, as out of 0 and 1, the system can be either in 0 or 1, i.e., fail or success state.

Fuzzy logic, unlike crisp logic, allows the characterization of system performance at any level between 0 and 1. For example, it is possible to grade SSC performance between 0 and 1, such as fail, degraded, acceptable, ... success facilitated through membership function μ_x for event x expressed as a continuum of fractions between 0 and 1 as μ_x (0, 0.1, 0.3, 0.4, ... 0.1.0). For example, expert elicitation techniques are often used in risk modeling. The objective here is to convert linguistic variables such as good, excellent and not adequate to quantitative estimates of reliability. The fuzzy approach is used to elicit qualitative opinion / judgement of operations experts and use this input to obtain the possibilistic or probabilistic estimates on the subject matter. Furthermore, fuzzy logic intuitively allows the assessment of uncertainty employing, for example, α-cut methodology.

6.4.5.7 Sensitivity Analysis in Support of Risk Management

Sensitivity analysis in risk assessment is extensively used for evaluating the impact of assumptions on the results at the system and plant levels. It is also necessary to assess the impact of uncertainty in the data, model, etc., on the results of the assessment and simulation results. The sensitivity analysis is performed, particularly where the SSCs or human actions appear to have uncertainties, and there is a need to check the acceptability of these uncertainties by analyzing the impact of these uncertainties on the main results of the analysis.

6.5 Risk Monitoring

The traditional approach to engineering asset management, which primarily includes management and operation and maintenance activities, in a nuclear plant is based on engineering judgment. Here, the defense-in-depth and conservative criteria form the basis of decisions. Ideally, it is expected that the decisions should be based on the performance of systems, structures and components (SSC), which should also take the plant safety margin into consideration. A risk-based operations and maintenance management system (RBOMMS) has been developed for a research reactor by extending the scope of risk monitor to cater to estimation of safety indicators, such as unavailability, Change in CDF, in support of optimization of test intervals, allowable outage time and related parameters. The results of a level 1 PSA study in terms of minimal cut-set equations along with other engineering plant information forms a major input for this system. The system is envisaged to be operational in plant control room conditions in support of routine decisions.

This paper presents the development details of RBOMMS. The future regulatory requirements for this system along with verification and validation procedures have also been discussed. It is expected that the use of this system in real-time scenarios will provide an improved understanding of safety on the one hand while enabling enhancement of plant economics on the other. The asset management approach for any complex system, such as a nuclear plant, is based on qualitative criteria and guidelines. For this paper, the term asset management exclusively refers to the operational and maintenance activities of the plant. The operational actions are governed by the plant technical specifications and other approved documents, which are prescriptive in nature. The test and maintenance schedules are formulated based on experience in the design, operation and maintenance of the systems and components. In a few cases, the manufacturer's recommendations also form the basis for the planning and scheduling of maintenance.

Even though safety forms the prime driver for operations and maintenance management in complex systems such as nuclear plants, emphasis is also based on plant economics, which is essential to plant availability and the cost of operation. Safety limits are stipulated by regulatory bodies, while economic factors are given importance due to market competition. Since maintenance is directly associated with the safety and cost of operation of the plant, it needs to be optimized in such a way that even though safety takes priority, the cost factor should be optimized to address complex issues. The number of inspections carried out on the systems directly influences the failure probability or unavailability of the system; hence, the overall risk of the system needs to be optimized with the number of tests or maintenance.

The state of the art in 'PSA-an analytical approach' enables the representation of performance parameters of the component, safety, and human action in the form of reliability and safety (International Atomic Energy Agency 2019a). The PSA insight couples with design and operation activities a framework called the risk-inform approach. This approach provides a probabilistic, systematic, and logical framework for the optimization of parameters in decision making and gives an integrated plant performance view for safety performance. The outcome of this approach is used for several applications in decision making.

6.5.1 Scope and Objective for Risk Management

The development objective of RBOMMS is to estimate, apart from risk-based plant configuration management, (a) technical specification evaluation and optimization, which means assessment of surveillance test interval and allowable outage time and comparing the same with the existing deterministic rules, (b) identification and prioritization of technical issues, c) decisions based on comparisons between operational risk and shutdown risk, d) uncertainty characterization and sensitivity analysis, (e) shutdown scheduling and (f) implementation of a mechanism to compare with international criteria in support of decision making (TWI 2022; Joshi et al. 2019; Varde et al. 1998; Nissan 2019; International Atomic Energy Agency 2021).

The major features of the approach are (a) development of level 1 PSA for the research reactor for full power as well as shutdown state of the plant, (b) preparation of core software specifications that takes into account GUI (graphic user interface)

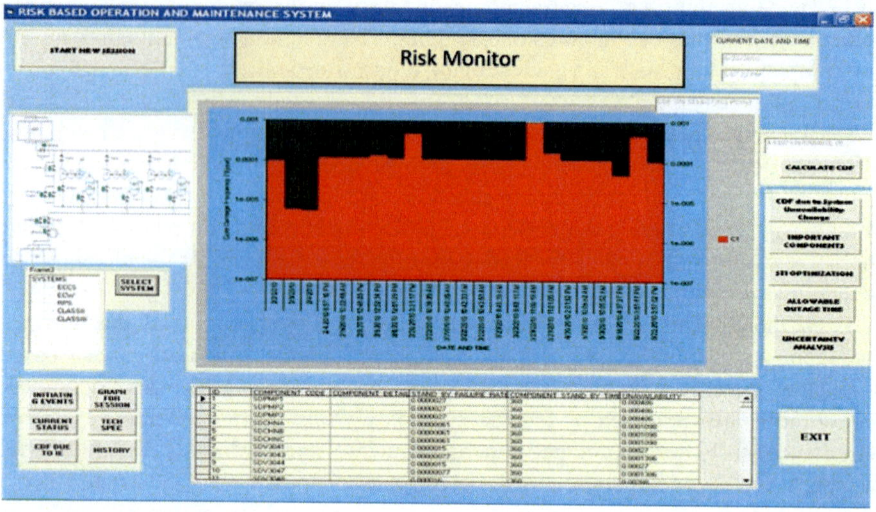

Fig. 6.10 Main graphical user interface (GUI) of the risk monitor (Electronic Corporation of Indian Limited 2018a)

requirements and plant operational requirements, (c) development of software with sample cut-set equations, (d) updating and modifying the output of PSA studies considering online RBOMMS applications, (e) finalization and incorporation of unavailability equation at the system level and accident sequence list at the plant level in the core module of the software, (f) development of the RBOMMS GUI, (g) incorporation of plant reliability data for quantification of parameters and (h) test run as part of validation and verification of the system (Fig. 6.10).

6.5.2 *Acceptance Guidelines*

Basic quantified criteria and guidelines form the backbone of risk-informed decisions. However, national criteria stipulated by the regulatory authority are requisite. However, in the absence of national stipulations, international criteria can be adopted as guidelines with adequate interpretation and margins (Varde and Pecht 2018).

The guidelines for CDF are as follows:

- When the calculated increase in CDF is very small (less than 1E-6 per reactor-year), the change will be considered regardless of whether there is a calculation of the total CDF (Region III). While there is no requirement to calculate the total CDF, should there be an indication that the CDF may be considerably higher than 1E-4 per reactor-year, the focus should be on finding ways to decrease rather than increase it. Such an indication would result, for example, if (1) the contribution to CDF calculated from a limited scope analysis, such as the IPE, and, if appropriate,

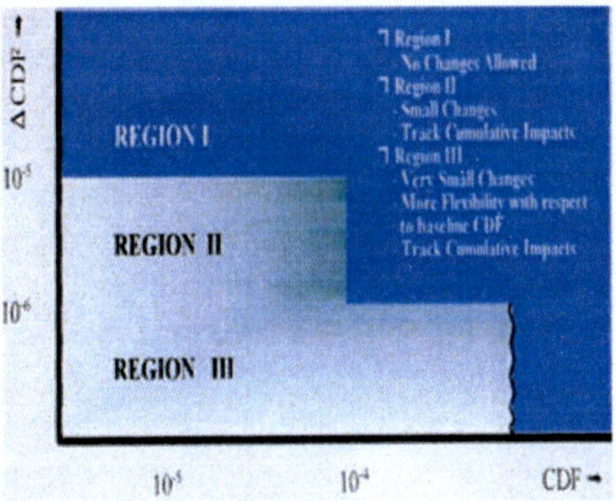

Fig. 6.11 Acceptance guidelines for CDF (Varde and Pecht 2018)

the IPEEE, significantly exceeds 1E-4; (2) there has been an identification of a potential vulnerability from a margins type analysis; or (3) historical experience at the plant in question has indicated a potential safety concern.

- When the calculated increase in CDF is in the range of 1E-6 per reactor-year to 1E-5 per reactor-year, applications will be considered only if it can be reasonably shown that the total CDF is less than 1E-4 per reactor-year (Region II).
- Applications that result in increases to CDF above 1E-5 per reactor-year (Region I) would not normally be considered (Fig. 6.11).

6.5.3 Implementation Procedure

1. First, a level 1 PRA model of the nuclear plant needs to be developed. This provides a holistic and logical view of the plant. For the same component, a database needs to be developed.
2. Using the PRA model's fault tree results, i.e., the minimal cut-set equation, calculate system unavailability. Calculate the core damage frequency (CDF) using system unavailability and initiating events.
3. Uncertainty analysis and sensitivity analysis are performed, and uncertainty analysis is important for any plant PRA. The main objective of this is to obtain a measure of uncertainty in PRA results. The sources of uncertainty in the PSA results are uncertainties in the modeling parameter, data input, lack of knowledge, etc. Sensitivity analysis is performed to determine the sensitivity of a component at the system level and the effect of system unavailability and initiating events at the plant level.
4. After uncertainty and sensitivity analysis, the obtained PSA results were compared with NUREC acceptance criteria guidelines for CDF and various system criteria.
5. Prioritization of the component using the importance measure. The importance of the component at the system level is determined using the Fussel-Vessely method. This helps to prioritize the component according to the percentage

contribution to the system unavailability or total risk. Using this prioritization, one must select the component, so the change in risk cannot be more than acceptance criteria.

6. After the importance measures calculation, select the component(s) for changing the plant configuration during operation/maintenance/shutdown. Then, the risk is compared with the acceptance criteria. If the risk is under the acceptance limit for a specified period, then the configuration can be changed. With the same configuration, the allowable outage time is also calculated.
7. Now calculate AOT for operating as well as for standby component. This tells the basis for which component can take out of the system without affecting plants operation. After a change in AOT, the risk is calculated and compared with acceptance criteria.
8. In sequence, configuration control and AOT calculation optimization of the surveillance test intervals for the standby systems of the plant can help to increase both the safety and availability of the system. The frequency and strategy for equipment testing should be controlled to confirm equipment functional operability, to minimize the risk from unavailability due to undetected failures and to minimize the risk from unavailability due to tests and test-induced failures. The effect of change in STI is on system unavailability, men-rem, cost of operation, and personnel training. The parameters affecting system unavailability are the test interval, testing time, human errors during maintenance/testing, and degradation/aging of the system/components that occurred during maintenance/testing.
9. After obtaining the optimized STI, the system unavailability and CDF are calculated using the new STI, and the results are compared with acceptance criteria. If the results are acceptable for both the system and plant levels, then the economic risk is optimized with STI.
10. Now, after STI optimization, the planning and scheduling activities can be finalized based on resource allocation. The planning is based on optimized STI results, and the STI is calculated using the unavailability of components/systems or using risk.
11. Based on the importance ranking, the surveillance test interval and allowable outage time allocate the plant and manpower resources for the job during shutdown planning and scheduling.

6.6 Simulator in Risk Simulation

6.6.1 Introduction

A full-scope nonreplica simulator facilitates, apart from training, the conduct or simulation of neutronic and thermal hydraulic experiments to support R&D deterministic and human factor studies. There are three major constituent components of the simulator: the graphic and control configured on the visual display units, the control console for plant operations, and the session development module, which configures the sessions to be run or fired to the simulator. The trip and alarm display and recoding facility has features for assessing the chronological order of events. Major experiments are related to the evaluation of accident sequences, especially those that require further studies on human factors or assessments of existing procedures and systems. The simulator facility has provision for assessment of cognitive stress monitoring and evaluation. The simulator facility enables assessment of the response of the operator to anticipated transients, such as loss of off-site power and design of experiments for postulated accident conditions, e.g., loss of coolant (LOCA) or station blackout conditions, in a limited sense, as the simulator after all is a model of the plant and not the plant itself. The major process and safety and safety support reactor systems selected for the modeling simulator were the main, secondary and tertiary systems, off-site electrical supply, regulation, process and control, control room mimics, control console that facilitates power maneuvering operations, trip/alarm and automatic logic and safety systems that include automatic reactor shutdown and cooling systems, emergency core cooling systems, Class 2 (AC) and Class I (DC) for required motive and control logic.

6.6.2 Simulator Architecture and Major Features—An Overview

The simulator hardware consists of set PCs and a data server PC interconnected on a fast ethernet network. Figure 6.12 shows the simulator architecture. The simulator process and control models generated in pSimSui a simulator kernel (Electronic Corporation of Indian Limited 2018a). The plant database related to configuration, control and protection logic; parameters are also maintained in this system. The major features of console include instrumentation and display that facilitate reactor manual start-up, power maneuvering and shutdown. The equipment includes recorders, switches, push buttons, analog meters, digital meters, demand power switches, etc. The input/output (IO) system consists of IO software used to drive the equipment of Console. IO Software sends inputs (equipment commands such as switch positions, push button, set power) to the simulation server and receives model outputs (such as reactor power, moderator level, pump and valve status) from the simulation server periodically. The HMI system runs plant mimics with dynamic data that can be displayed in various formats, such as graphical, tabular, and matrix formats. Data logging can be performed periodically. HMI Clients: HMI client software will be running on these systems. The development sessions based on the design of the experimental objectives and subsequent firing and monitoring of the response of the system and operator reactions are performed in the instructor station. The system can simulate record and evaluate the responses for R&D or training requirements.

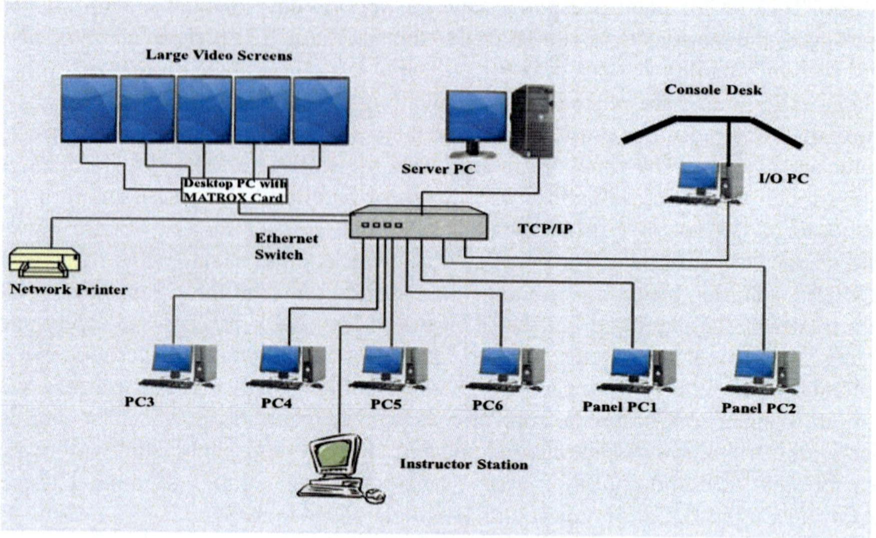

Fig. 6.12 Schematic representation of the simulator architecture

The 'pSimSui' kernel provides an integrated configuration and simulation environment, which allows engineers functions such as process flow, logic and control, electrical networks, database management, real-time execution clock, instruction station and graphic development modules. It is a PC-based simulation environment running on a Windows operating system that facilitates building simulation loads using different modules without writing any code. The is a special feature in pSimSui that facilitates interface with applications developed by user for a specific requirement. For example, in risk simulation, this facility was used to interface an operator support system and risk monitor to perform integrated operations. An efficient and elaborate GUI interface is provided that allows users to operate with ease various control operations while keeping an eye on the process dynamics (Fig. 6.13).

Few factors determined the final configuration and type of the simulator. To conduct effective training, the facility should resemble the actual control room as much as possible. Additionally, the space available to accommodate the simulator systems was limited, and the entire project had to be completed within the sanctioned budget. Keeping these factors in mind, the main control console was designed and fabricated similar to one in a control room with identical meters and switches to have a feel of control room. All the equipment on the control console is hard wired and retained as it is. To accommodate the control room graphic panels, a large amount of space was needed, which was the major constraint, so the graphic panels were replaced with LV touch screens. However, the mimic details on the graphic panels were retained and produced on the soft screens. The screens were installed in such a manner that the operator sitting at the console could view the entire mimic. In the case of the actual control room, the entire panel is not visible from the console. As

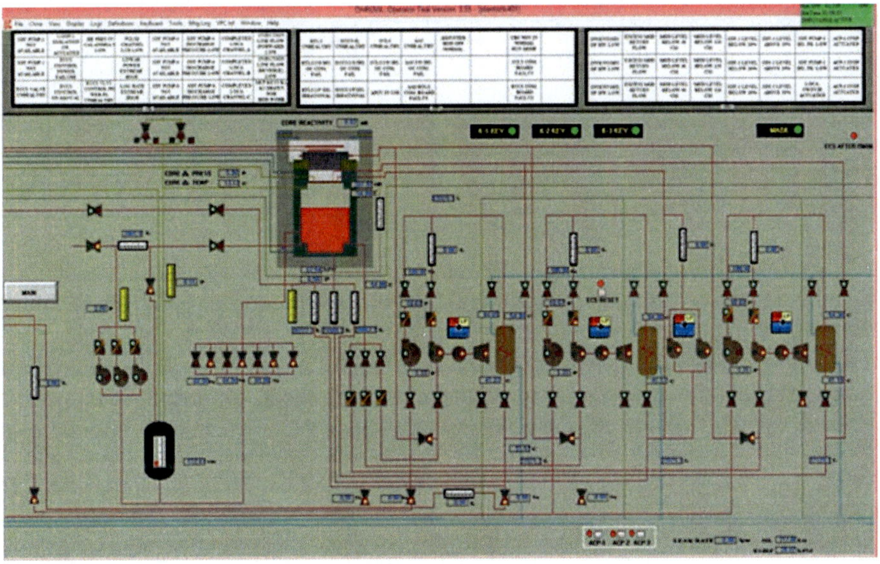

Fig. 6.13 Mimicking of the main coolant and moderator system

the control console was large in size, it was not possible to shift the console in the room into a single piece or two pieces. Therefore, a wire frame model was made to check the transportation from the lab corridors and doors, and based on the results, the console was fabricated in four parts and later joined at the site. The simulator control console is shown in Fig. 6.14.

Fig. 6.14 Plant simulator for design and execution of risk simulation

6.6.3 Core Neutronics Point Kinetics Model

The development of process system models was performed using the PF Tool. The models of core neutronics, reactor regulating system (RRS), reactor trip logic system, start-up logic system, alarm annunciation system and coolant activity monitoring system were developed in house and interfaced with the main module. Here, a brief description of the development of core neutronics and the RRS model is provided.

The core reactivity transient simulation required modeling for a strong neutronically coupled core. Therefore, a point kinetics model was chosen to simulate various reactivity-initiated transient conditions. Point reactor kinetics code is based on point kinetics model where a finite sized reactor is assumed to be a point reactor, i.e., change in neutron density under reactivity insertion reflects uniformly at all points within a reactor. The rate of change in neutron density and precursor concentration are described by a set of coupled linear ordinary differential equations given below.

$$\frac{\mathrm{d}n(t)}{\mathrm{d}t} = \frac{\rho(t) - \beta}{\Lambda} n(t) + \sum_{i=1}^{g} \lambda_i C_i(t) + S(t) \tag{6.28}$$

$$\frac{\mathrm{d}C_i(t)}{\mathrm{d}t} = \frac{\beta_i}{\Lambda} n(t) - \lambda_i C_i(t) \tag{6.29}$$

where n: core average thermal neutron density

C_i: core average delayed neutron precursor density
S external neutron source
Λ prompt neutron generation time
ρ reactivity
β_i fraction of i-th delayed neutron
g number of precursor groups

The fourth-order Runge–Kutta method is used to solve these equations. Reactivity calculations include reactivity control and safety devices, Xenon, voiding in channels and power level changes. The parameters displayed are neutron power (% full power), neutron power rate (%/second), Xenon load, and core reactivity.

The calandria is the reactor core, which is modeled mathematically as the core neutronics module, and from the Campbell Channel onward, it is the RRS module. The instantaneous power signal is fed to the RRS module through the Campbell Unit. It consists of a linear amplifier, log amplifier and differentiator circuits for generating the voltages corresponding to linear power, log power, linear rate and log rate. These four signals are then fed to comparator units to generate representative average signals and to reject the faulty signal. The average values of each of the four signals are then fed to the function generator, which computes the error signal between the actual reactor power and the Set power. Based on this, it generates the corrective signal for the control valve so that actual reactor power is controlled at the desired value, i.e., set power. Mathematical modeling of the RRS involved computation of the transfer function for each block (using the actual circuits deployed in Dhruva) and then conversion into time-domain differential equations. The set of equations obtained were converted into a FORTRAN code and coupled with the core neutronics module.

Subsequently, extensive real-time integrated testing of the system was carried out. During this phase, the observed steady-state error between the set power and actual reactor power and fluctuations in the control valve signal leading to hunting of the position of the control valves were the main problems encountered, which were overcome by systematic fine tuning of the RRS module.

6.7 Intelligent Operator Support System

An intelligent operator support system module communicates with the plant simulator. There are two major modules of the intelligent operator support system (OPAD): the transient identification module and the diagnostic module (Varde et al. 1998). The typical characteristics of reactor transients are (a) in a relatively short time, the plant systems and parameters change their state or value from the steady state to a new value, (b) each transient has a set pattern associated with it in terms of equipment states and parameter values, (c) in most cases, each plant transient can be identified by the initial couple of parameter statuses, while the following parameters in terms of time that register can be related to the root (initial registrations), (d) some parameters take little more time and may or may not form part of the transient and (e) the stabilized new state of the plant characterizes the end of the transient. At the onset of a transient, in a short time, i.e., an order of seconds but not more than a minute in most cases, the control is flooded with the trips and alarms along with the change in parameter values, challenging the operator's cognition to arrive at a conclusion with respect to the identification of transients. This situation requires a parallel processing of the transient pattern such that indexing can be performed to the nearest matching pattern. This requires an intelligent approach such that the algorithm facilitates training of the set of transient patterns. Here, transient identification requires matching the real-time pattern with the stored or learned patterns. Artificial neural network techniques have been found to provide an effective strategy for transient identification patterns. The ANN algorithm can be trained to identify the new state efficiently in terms of time and accuracy. Therefore, the transient identification function is performed by an artificial neural network (ANN) system, while the diagnostics are performed by a fuzzy rule-based system. Furthermore, any deviation from the new normal that new state of the plant requires a diagnostic capability. For example, during a loss of off-site power (LOOP) scenario, the reactor trips along with its associated main coolant circulation systems and plant transit to the shutdown state, which requires a shutdown cooling system to cater to reactor core cooling and to start and establish automatic operation of emergency power supply systems comprising two diesel generators to cater to plant emergency loads. If one of the two diesel generators fails to start, then the diagnostic module should perform diagnostics to identify the causes of diesel failure and generate advice for the operator for further action to resume the operation of the failed diesel generator or ensure the implementation of management functions such that the risk of station blackout is avoided.

The OPAD module forms an integral part of the risk simulation in two ways: first, the availability of an operator support aid reduces the chances of human error, particularly for handling emergency scenarios, and second, the human action conditional probabilities form part of the accident sequences in the probabilistic risk assessment (PRA) model of the risk monitor. The response of the operator is used as input for risk evaluation and thereby predicting the risk levels for the subject scenario.

For further details, Chapter No 9 is dedicated to this aspect; therefore, readers are requested to obtain more details on intelligent approaches to operator support systems.

6.8 Case Study: Reassessment of Shutdown Safety Margin

This case study addresses the application of a risk-based approach for the re-evaluation of design risk margins available for the built-in shutdown capability of the reference nuclear plant, including the role of human actions as part of associated accident sequence sequences (Varde et al. 2018). Furthermore, this case study will postulate the condition and causes that can lead to identified human error probabilities.

The reference reactor under consideration is a research reactor with a vertical core loaded with natural U as fuel cladded with aluminum and heavy water as coolant, moderator and reflector. The reactor power regulation is achieved by varying the moderator level in the reactor vessel with constant inflow and variable outflow (through feedback telemetry by adjusting the opening/closing of control valves in the moderator outlet or dump line). There are two shutdown system shutdown systems to ensure higher shutdown reliability: a primary shutdown system comprised of a set of shutdown assemblies called a shut-off rod, backed up by a moderator dumping system. In the case of a reactor trip/shutdown signal, a shut-off rod along with moderating dumping is actuated and ensures reactor shutdown. In fact, the safety analysis has demonstrated that the gravity insertion of the shut-off rod alone is adequate for the safe shutdown of the reactor, i.e., to make the reactor subcritical. However, the moderator dumping is ensured as a backup action through the control and protection logic.

Probabilistic risk assessment (PRA) studies reveal that the core damage frequency is significantly lower than the primary damage frequency, and the backup shutdown system is credited for reactor shutdown and complies with the general target criteria of core damage frequency. $1.0 * 10^{-5}$ per reactor-year, as applicable to old reactor systems; i.e., the systems belong to the first- or second-generation system, built during the late or early 1980s. The deterministic argument that moderator dumping is not as fast as the primary shutdown system achieved by essentially gravity insertion of shutdown devices ensures reactor subcriticality in less than 3–5 s. On the other hand,

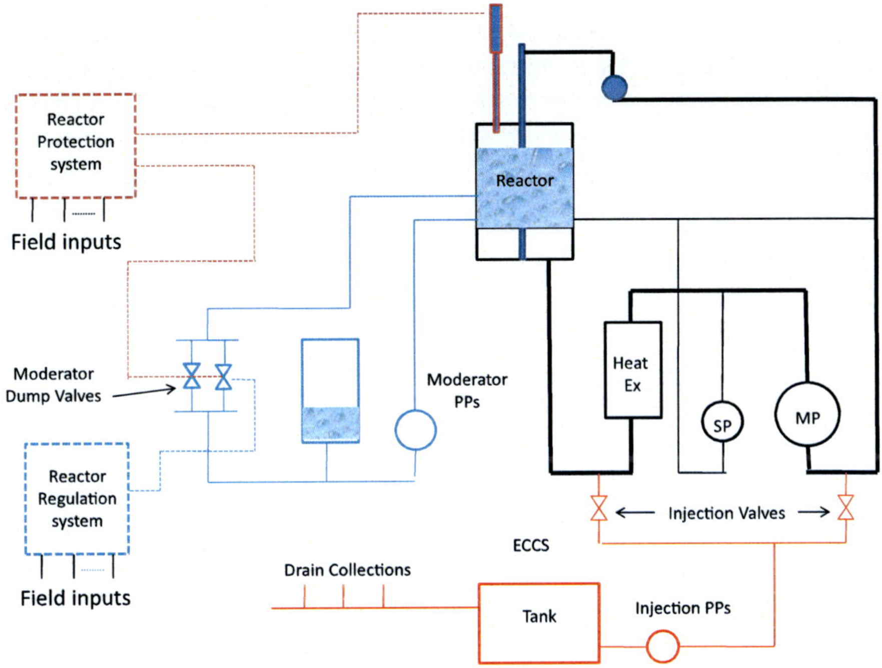

Fig. 6.15 Simplified schematic of the reference reactor showing the reactor vessel, main and shutdown cooling (black), moderator dumping system (blue); ECCS injection system (red); reactor shutdown devices and protection gray) regulation modules (blue)

the moderator dumping, even though it initiates a subcriticality process before even the shut-off rods enter the reactor core, remains effectively slow compared to the shut-off rod as the dumping rate is a function of the available head (Fig. 6.15).

The objective here has been to use the available safety margin built into the reactor to demonstrate the safety of the plant in general and complying with the probabilistic criteria, i.e., core damage frequency as mentioned above. This case study, even though it will cover a brief account of the simulations that involve experiments and modeling, will argue its case with the considerations of three safety significant postulated initiating events, viz., loss of off-site power (LOOP), loss of regulation accident (LORA) and loss of coolant accident (LOCA).

6.8.1 Postulation of Loss of Off-Site Power Scenario

LOOP is an anticipated transient and expected to occur more than once in the lifetime of a reactor. Figure 6.16 shows Simplified event tree for the loss of off-site power scenario (with credit for the backup shutdown system contributing to shutdown action along with the primary shutdown system. The deterministic flow coast down analysis performed supports the argument that moderator dumping has the capability to shut down the reactor safely, as evident from the clad and coolant temperature as well as from reactivity considerations can be seen the postulated failure of Shutdown Cooling System (ECS) and Primary Shutdown (PSD) form the major components of net results. The objective of this study was to assess the importance of moderator dumping by simulating the LOOP event with and without the consideration of moderator dumping. Here, the event tree shows that credit has been given to to primary shutdown as well as moderator dumping system.

6.8.1.1 Reference Level 1 PRA

The major output of a Level 1 PRA is the statement of state in the form of core damage frequency of the plant. The following graph shows the mean/median value of CDF: 1.1 * 10^{-5}/reactor-year. This estimation takes credit for moderator dumping. The reference core damage frequency distribution for the research reactor under consideration is given in Fig. 6.17.

This distribution represents the uncertainty associated with the median value of 1.1 * 10^{-2}/r-y, with 5% (or lower bound): 9.0 * 10^{-6} and 95% (upper bound): 5.4 * 10^{-5}/r-y. The major reason for the lower uncertainty in the main result is the consideration of a plant-specific database built using the experience on component performance ranging from 25 years (on coolant circulation piping) and ten years for general active and passive components.

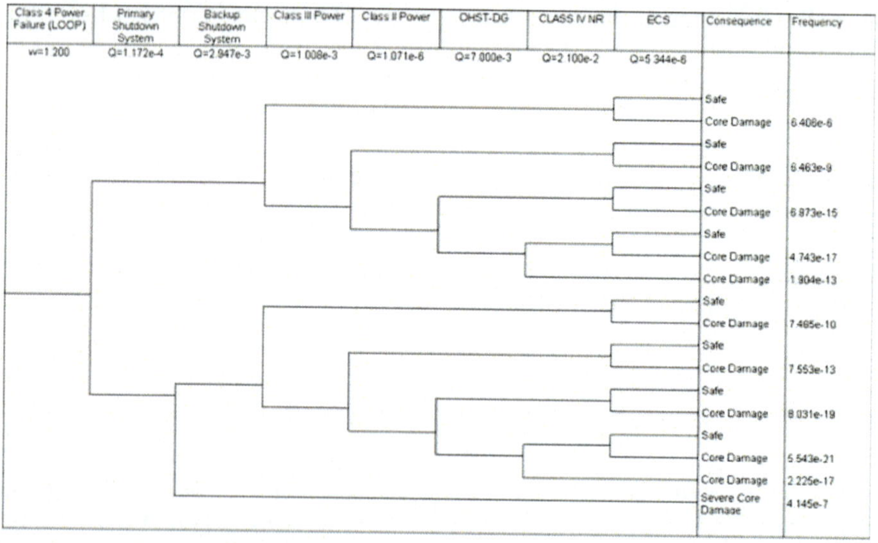

Fig. 6.16 Simplified event tree for loss of off-site power with credit for moderator dumping (backup shutdown system) along with primary shutdown system

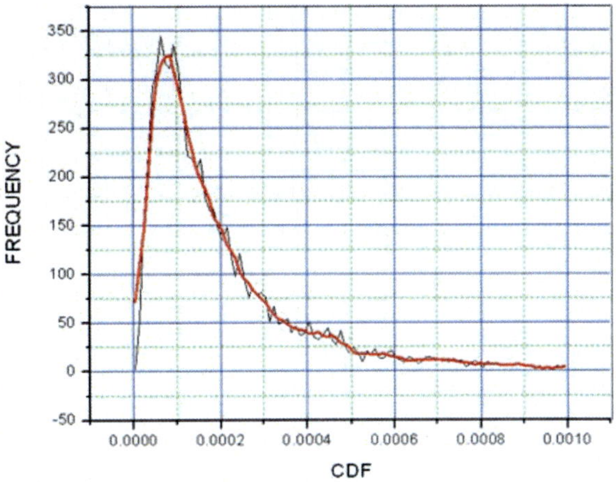

Fig. 6.17 Represents uncertainty in the estimation. The CDF distribution takes into account the contribution from other postulated accident sequences toward predicting the result

The PRA results with and without moderator dumping show that there reduction in CDF value is significant by almost two orders, i.e., decreasing by a factor of ~ 100, which is a significant contribution of moderator dumping, and it could play a role as a secondary shutdown system and not exactly as a backup shutdown system.

6.8.1.2 Deterministic Simulation of Reactor Core Parameters

Off-site power is required to cater to Class IV power loads (AC power to cater to plants 3.3 kV and 415 loads major loads are primary, secondary and tertiary coolant loads) for operating the plant. In the eventuality of interruption or nonavailability of off-site or Class IV power supply, the reactor control and protection system ensure that the reactor is tripped by the automatic action of primary and backup shutdown system and the reactor reaches subcritical state. Once the reactor is tripped, the self-sustaining chain reaction stops, and the heat production in the core is also reduced to a level where the decay heat component remains that requires automatic switching of the shutdown cooling system such that there is no interruption in cooling flow even in the shutdown state of the reactor. In the intermediate period when the coolant pump trips and the shutdown cooling flow is established, the stored energy of the fly wheel mounted on the main coolant pump ensures a flow coast down such that core cooling requirements are met.

To study the adequacy of moderator dumping alone, i.e., no credit was taken for the primary shutdown system, experiments were designed in a reference reactor simulator. The experiment involved blocking the primary shutdown system to study the adequacy of moderator dumping alone for safe shutdown of the reactor, i.e., no reactivity, temperature-related consequences. Simulation sessions were run for conditions when a reactor trip occurred due to moderator dumping along (no credit for the primary shutdown system), and the clad and coolant temperatures were recorded with time. Since this study was carried out by simulating flow coast down that is

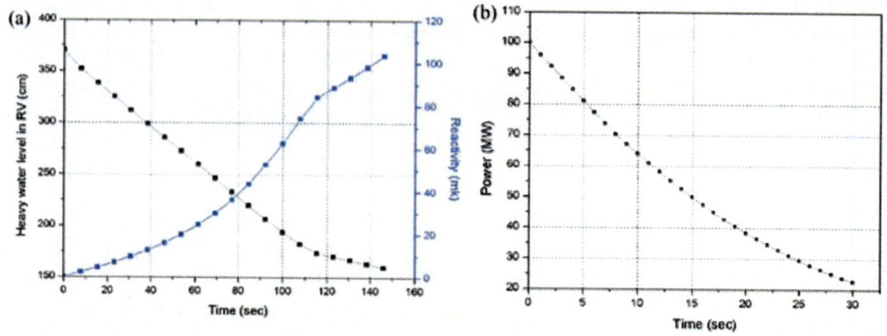

Fig. 6.18 Reactor tripping figure **a** shows the moderator dumping, i.e., drop in moderator level with time after the command goes to control and dump valve to open and dump the moderator heavy water from reactor vessel to dump tank (shown in black dots) and increase in core core-ve reactivity with time. Graph **b** shows the drop in reactor power plotted for time from $t = 0$ when the reactor trips until the 30th second. Within this time, the shutdown cooling pump starts and delivers flow, and finally, decay heat removal is established such that core cooling is ensured during the LOOP transient

initiated by the loss of off-site power, the reactor vessel level along with the insertion of negative reactivity and power levels with time. Figures 6.7 and 6.8 show the results of these simulations performed in the reactor simulator.

The postulation that moderator dumping alone can shut down the reactor without any threat to fuel/core safety for loss of off-site power condition when the reactor is tripping from full power of 100 MW. Therefore, it is concluded that for LOOP events, the moderator dumping system can be considered a secondary shutdown system (Figs. 6.18 and 6.19).

6.8.2 *Loss of Regulation Incident*

6.8.2.1 Background

The postulated initiating events that fall in the category of loss of power/reactivity regulation due to inadvertent actions or system failure basically due to random failure are grouped as incidents and not as accidents. This is due to the inherent safety features of the core, e.g., the negative coefficient of temperature, voids and reactivity per say, which makes the system safe due to self-limiting conditions. For example, the increase in temperature assists in limiting or avoiding the increase in reactivity; the same argument applies to reactivity and voids. Furthermore, human error-induced phenomena, particularly from reactor consoles performed during power maneuvering or start-up events, form part of the safety demonstration. However, the experience suggests that reactivity events have very low probability due to engineering provisions such as redundancy in regulation channels, diversity and fail-safe designs.

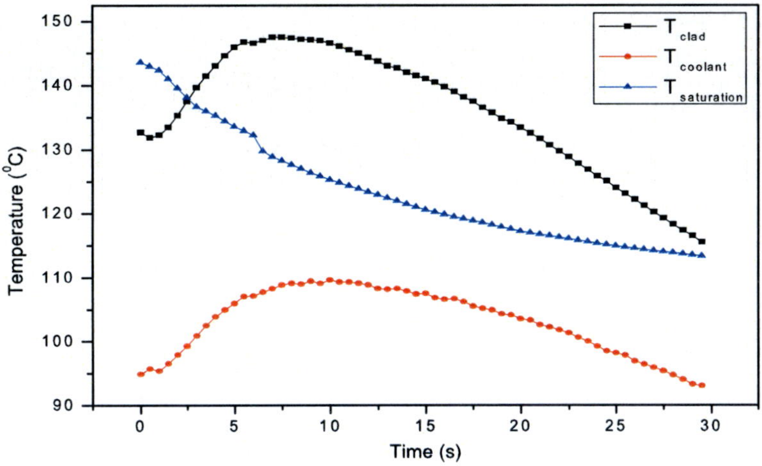

Fig. 6.19 The critical temperature trends, i.e., clad (black block) and coolant (red dots) temperatures during the transition from time $t = 0$ when the main circulation pump trips until the 30th seconds into the trip. The three values of clad, coolant and saturation temperature for a given pressure are shown in the graph. Even though the clad temperature exceeds the coolant temperature for approximately 28 s, the Jens-Lott equation is not violated, and no boiling takes place, which means that there are implications for fuel safety

In deterministic simulation, the assumption is that the transient is initiated from a neutronic power of 1.0 Kw range, as the possibility of reactor power going in the watt range low enough to initiate reactivity-related transients is negligible. Furthermore, in a long shutdown scenario, if the plant culture/safety specifications allow reactor start-up with a special procedure that involves regulator overview and review with special start-up regulating channels using fission counters such that reactor parameters can be tracked, monitored, and ensure safe automatic shutdown if the situation arises in the remotest possibility. Furthermore, the applicable reactor parameters and success and failure criteria for the shutdown system during simulation are modeled with normal operating range and conditions. The protection system initiates a reactor trip at 110% linear and 6% log-rate levels in all three channels of the regulation system. The transient is initiated with a moderator temperature and moderator level of 35 °C and 200 cm (dump level corresponding to technical specification on 50 mk subcriticality level), respectively. An effective delayed neutron fraction of 0.00804 considering the six groups of photo neutrons and a temperature coefficient of ~ 0.013 mk/°C was considered for the simulation. The reactor transient is terminated safely, which is evident from the maximum value of parameters recorded during the simulation (Fig. 6.20).

In this simulation, all the neutronic parameters, such as the reactor power, log-rate and variation of reactivity, during the simulation were found to have no impact on the physical technical specification limits. As seen in Fig. 6.11, the technical

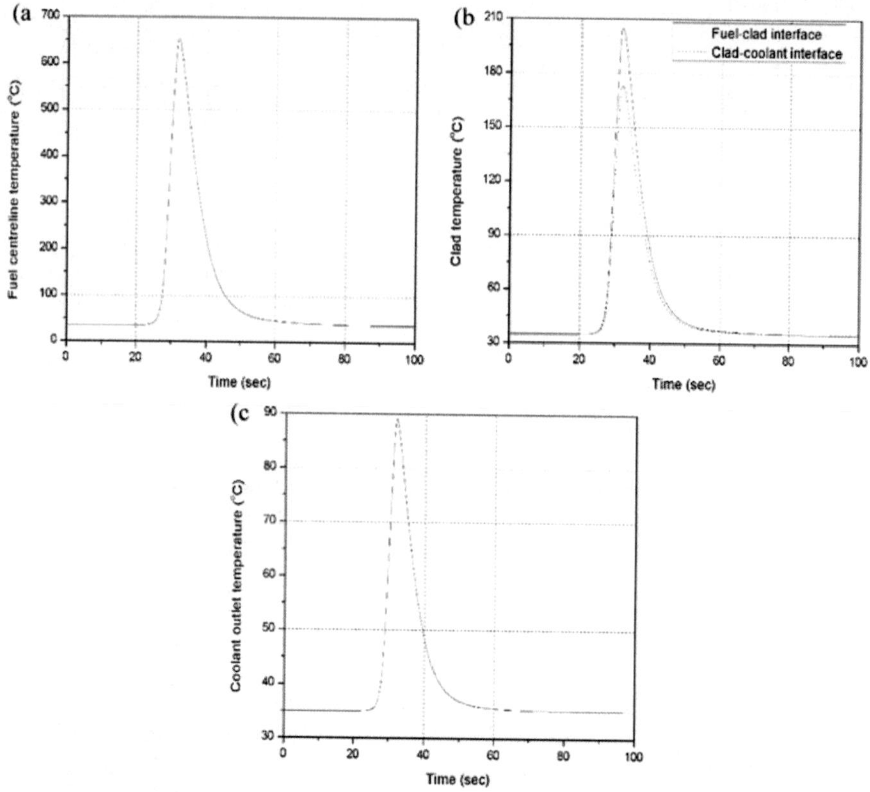

Fig. 6.20 Variation in max fuel centerline, clad surface and coolant temperature with time during LORI transient simulation

specification limits of 225 °C for the clad surface temperature and 90 °C are not violated.

6.8.2.2 Probabilistic Risk Model for LORA

The failure probability of triplicated regulating channels has been on the order of 10‾3/yr. and the primary shutdown and backup moderator dumping system initiate the reactor shutdown when the trip parameters reach the set value as mentioned above. Apart from this, to account for the possible temperature rise, the primary cooling and secondary shutdown or decay heat removal system also part of the header event in the event tree.

The simulations were performed for reactor shutdown with a primary shutdown system without and with crediting moderator dumping, and the corresponding CDF levels show that by crediting the moderator dumping, the shutdown capability of the reactor improves by a factor of over 200, which is a significant margin in terms of safety improvement.

The major conclusions of the above simulation study can be summarized as follows: (a) Given the current state of the art in deterministic and probabilistic safety, LORA is considered incident and not accidental, as LORA-induced accidents in reactor systems are highly unlikely; (b) the moderator dumping, even though slow compared to the gravity drop and initially spring-assisted primary shutdown system, can safely shut down the reactor. The critical parameters remain well within the technical specification stipulations.

6.8.3 Loss of Coolant Accident

6.8.3.1 System and Scenario Description—A Brief Overview as Applicable for This Case Study

Keeping in mind the plant and systems configurations, the LOCA event has been modeled for two scenarios, minor LOCA and major LOCA. When the postulated leak rate from the coolant circulation and related systems, such as the moderator system, is $\leq$ 1250 lpm, then it is termed minor LOCA, and when the leak rate is more than this value, i.e., > 1250 lpm, then the LOCA scenario falls in the category of major LOCA. The emergency core cooling system is automatically started by plant control and protection logic when the LOCA the valid LOCA signal is generated, and other conditions required for injection of leaked inventory are satisfied. In case the plant staff is satisfied that the system is experiencing loss of inventory and there is a threat to core cooling, the manual injection is affected by the control room. After an initial period, the operator assesses the situation operator can perform manual action, be it manual injection of light water from the alternate source, i.e., an overhead storage tank that has enough inventory to cool the core, if needed. Otherwise, the leaked-out inventory from the system can be circulated through the ECCS. To facilitate this operation, the plant has an emergency operating procedure applicable to different postulated scenarios.

6.8.3.2 Probabilistic Risk Assessment

There are two major components of LOCA modeling. System modeling where the failure probability of LOCA is evaluated using the fault tree approach. Second, the event tree modeling for LOCA involves two broad categories, viz., minor LOCA and major LOCA. In these two categories, there are also aspects such as location of leak and severity of failure that require event tree analysis, as the probability and consequences of a failure often provide insight into resources that might include gadgets, fixtures methods, expertise, etc., in support of the mitigation of consequences or emergency management such that plant and human safety can be ensured. The plant has a unique feature that contributes to the realization of higher safety levels. The moderator inventory in the system acts through the reactor vessel as the surge tank. That is, the reactor vessel overrides the coolant system and replenishes the leaked-out inventory of the coolant system. In fact, this function in NPPs is performed by an accumulator maintained in a poised state. This feature gives additional coping time depending on the leak size to the operating crew for emergency mitigation or management. Once this reactor vessel inventory drops below a level, the next step that injection of leaked-out inventory from ECCS is resorted. In the case of a highly improbable condition, if the ECCS fails to maintain core cooling, then the light water from the overhead tank is injected into the reactor core with manual action. These redundant aspects ensure the core safety that has been validated in the PRA model of the plant.

In LOCA scenario simulation the objective is to assess (a) the adequacy of moderator dumping alone to shut down the plant such that system subcritical by stipulated (technical specification requirements) and (b) the scenario do not pose challenges to core coolability and coolability criteria during operation, transient and shutdown is maintained.

The PRA simulation shows that the CDF contribution for the minor LOCA case without crediting a moderator dumping or backup shutdown system (BSS) and this accident sequence does not impact the net CDF. The accident sequence involving PSS failure for minor LOCA contributes $2.44 * 10^{-8}$/reactor-year. So probabilistically it becomes a hypothetical case as it involves all low likelihood events like LOCA, PSS and moderator dumping or BSS. Given that PSS and BSS are diverse and independent systems in terms of the probability of a very low probability of CCF between PSS and BSS.

Figure 6.21 shows the event tree for major LOCA (severe leak in nonisolatable region), which takes credit for PSS as well as BSS for reactor shutdown.

6.8.3.3 Deterministic Observation

For the case of minor LOCA, the PRA model demonstrates that there is no threat to the safety of the reactor either due to + positive reactivity addition and clad coolant temperature concerns. Safety reactor safety provision.

Plant designers have provided means to manage and mitigate the consequences of major LOCA. A rational and logical view keeping in view the plant safety and configurational provision considering that this reactor is a low-pressure and low-enthalpy system the probability of threat of serious consequences is negligible. For example, being a low-pressure system and the material of construction of piping is stainless steel with an operating experience of over 30 years, the probability of catastrophic or double-ended rupture is hypothetical. Second, even if there is a severance failure, the automatic provision of tripping the main coolant pumps will depressurize the system, and thereby, the leakage rate will drop significantly such that a major adequate coping time will be available for emergency as well as for planned action.

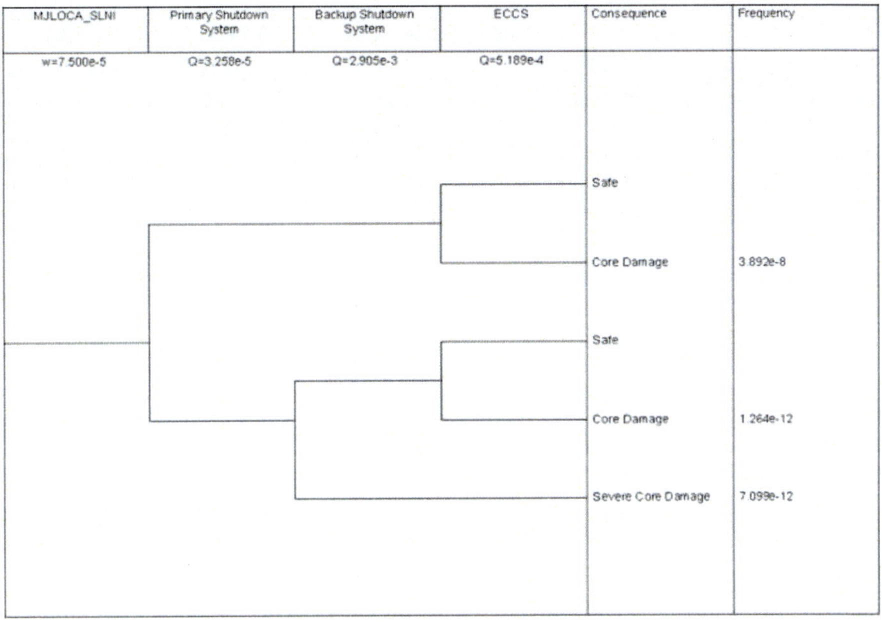

Fig. 6.21 Major LOCA (severe leakage, i.e., > 1250 lmp and location is nonisolatable region) where the credit for moderator dumping is shown as backup safe reactor shutdown. The analysis shows that if credit is not taken for moderator dumping, the results will not impact the net CDF probabilistically, as the contribution of only the primary system for shutting down the reactor leads to a CDF of ~ $4.0 * 10^{-8}$/reactor-year

To conclude, further, we summarize that there are two situations that have been investigated, viz., minor LOCA and major LOCA. In the case of a minor LOCA, the coolant system does not undergo a phase transition for at least 30 min, and deterministic analysis (crediting moderator dumping) shows that the fuel, clad and coolant temperatures remain well within the limit. Hence, consideration of moderator dumping for demonstrating the enhanced safety of the reactor is acceptable. However, the scenario of major LOCA, which involves catastrophic failure of the coolant inlet loop and common structure, i.e., inlet plenum, is different from that of minor LOCA. Although Dhruva, as a low-pressure and low-temperature system, is not expected to fail in a catastrophic manner (probability < 10^{-6}/yr), this needs to be evaluated probabilistically. The NUREG-800 (Varde et al. 2018) stipulates that the maximum leak size for low-pressure and low-temperature systems should be arrived at by considering the dt/4 model. Furthermore, plant safety features such as the automatic tripping of all the main coolant pumps on the LOCA signal and isolation of one loop are some features that reduce the chances of potential likely hood and consequences and with manual action isolation of other loops such that leaky locations in the areas more susceptible to depressurization of the system and reduce the leakage rate significantly, which in turn provides additional coping time such as the minor LOCA case. With these provisions and assumptions, the catastrophic failure of the coolant system will be beyond the design basis case. Furthermore, the probabilistic risk assessment of the plant where in the probability of major LOCA related risk, i.e., the chances of core damage frequency has been assessed to be much lower than the design basis criteria of 10^{-6}/reactor-year and the typical postulated accident sequences shows the core damage frequency to of the order < 10^{-8} which means this scenario can be considered as hypothetical case. Hence, the moderator dumping system can be considered adequate for LOCA situations.

6.9 Conclusions and Final Remark

The objective of this case study is to develop a procedure for using the available safety margins in the plant to demonstrate higher safety levels of the plant. The case in point is a reference research reactor that has two shutdown systems. The primary coolant system comprised a group of shut-off rods that alone have the capability to bring the reactor to the shutdown state by fast insertion of shut-off rods and bring the subcriticality level as per safety stipulations. The second shutdown is referred to as the backup shutdown or moderator dumping system. When there is a demand, the moderator is dumped from Reactor Vessel to a moderator dump tank. This action stops the chain reaction by removing the moderator and finally enables the system to reach a safe shut-down state. The reason why the moderator dumping system is considered a backup system and not a secondary shutdown system is that it is relatively slow compared to the primary shutdown system, which ensures subcriticality in a short time of ~ 1.0 s normal shutdown in 3–5 s. The major argument or basis of this analysis is that the moderator dumping system is also fast enough to shut down the reactor and requires a risk-conscious/risk-informed approach where the deterministic arguments are revisited to demonstrate the safety of the reactor.

The designers have ensured a higher level of safety by making these two shutdown systems exclusively diverse and independent to the extent possible. This means that there are no conceivable common cause failures that can adversely affect the two systems as plants in line with the defense-in-depth provisions the fail-safe and single failure criteria, further consolidating higher safety levels.

To generate an argument for qualifying the moderator dumping as a secondary shutdown system, this case study was performed. The three typical scenarios involving postulated initiating events are loss of off-site power (LOOP)—an anticipated occurrence, loss of regulation incident (LORI) considered as incident in the present safety parlance, and loss of coolant accident (LOCA) an integral part of the design basis accident. It is assumed here that these three cases are over-encompassing for the list of initiating events.

The aim here is to demonstrate the safety of the plant through the safe shutdown achieved by moderator dumping along. The above case studies show that the LOOP alone has no safety consequences demonstrated through the deterministic and probabilistic simulations. The LORA scenario was considered to be initiated at a power level of 1.0 kW, and the core clean condition was at 200 cm. It was found that the stored energy in the fuel has no consequences in terms of reactivity and fuel physical parameters such as temperature and coolant temperature with moderator dumping alone. For LOCA in line with plant safety philosophy and documents, two cases, viz., minor LOCA (when the leak rate is $\leq$ 1250 lpm) and major LOCA (when the leak rate is > 1250 lpm), have been considered. For the scenario simulation, the moderator dumping alone has been credited. For minor LOCA without any interpretation and argument and given the 30 min time available to the operator, moderator dumping alone can assure the safe shutdown of the reactor, and there is no threat to fuel or system integrity.

Even for major LOCA, the available provision and relevant assumptions that include (a) for low-enthalpy and low-pressure systems such as Dhruva, as per international standard/documents (United States Nuclear Regulatory Commission 1987) a break size of Dt/4 (D and t are diameter and thickness of maximum size of pipeline in the primary system) is adequate for demonstrating the safety, (b) experience of plant operation of over 35 years vindicates the considerations of safety, (c) plant safety provisions like automatic depressurization of the system by tripping of main coolant pumps on LOCA signal, (d) effective built in strategy to isolate the loops from control room, etc., effectively reduces the probability of large rupture and with this the risk of major LOCA can be considered to be hypothetical case. Given this background, the moderator dumping can be considered to be qualifying as a secondary shutdown system.

The above demonstrates the applicability of the integrated risk simulation framework (IRSF), which involves deterministic as well as probabilistic risk assessment modeling to evaluate the available safety and use the available safety or risk margin for higher safety demonstration.

Acknowledgements The author would like to thank Dr. Tej Singh, Dr. Tanay Mazumdar and Dr. J. Jain, who are also coauthors with me for the publication 'Risk-based approach toward design evaluation and reassessment of safety margin' Life Cycle Reliability and Safety Engineering, Vol 7, Issue 4, December 2018, Springer, which forms basis for this case study and for helping me the development of this case study. I also thank Shri N.S. Joshu is responsible for the conduct of experiments on the Dhruva Simulator. I also thank Shri Mayank Agarwal, who as my students helped me develop the Dhruva Risk Monitor, which provides support for risk simulation. Thanks to the team of ECIL, which mainly includes Mr. Subba Ram, Ms. Pratima, Mr. Pradeep, Ms. Prashanthi and Mr. Praveen, who were involved in the development, installation and commissioning activities of the simulator and still helped on the issue of integrating the operator support system with the simulator.

References

Acosta C, Sui N (1991) Dynamic event tree analysis method (DETAM) for accident sequence analysis, final report NRC-04-88-143. USNRC, Washington DC

Acosta C, Siu N (1993) Dynamic event trees in accident sequence analysis: application to steam generator tube ruptures. Reliab Eng Syst Saf 41:135–154

Agarwal M, Varde PV, Agarwal M, Varde (2009) Extending the scope of risk-monitor to implement risk-based operations and management system. Int J Reliab Qual Saf Eng
Aldemir T (2013) A survey of dynamic methodologies for probabilistic safety assessment of nuclear power plants. Ann Nucl Energy 52:113–124
Boudali H, Crouzen P, Steolinga M, Dynamic fault tree analysis using input/output interactive Markov chains
Bucci P, Kirschenbaum J, Mangan A, Aldemir T (2008) Construction of event tree/fault tree models from a markov approach to dynamic system modelling. Reliab Eng Syst Saf 93:1616–1627
Cepin M, Mavko B (2002) A dynamic fault tree. Reliab Eng Syst Saf 75:83–91
Corcuera PA, A full scope nuclear power plant training simulator: design and implementation experiences. University of Cantabria, Cantabria
Cruz-Cruz R (2010) Development of modern power plant simulators for operator training center. Jose Tavira-Mondragon and Vol IWCECS 2010, October 20–22, 2010, San Francisco, USA. In: World congress on engineering and computer science 2010, San Francisco, USA
Electronic Corporation of Indian Limited (2018a) Simulator manual. ECIL, Hyderabad
Electronic Corporation of Indian Limited (2018b) ECIL manual on simulation suite pSimSui. ECIL, Hyderabad
Electrotechnical Standard Association (2000) Dependability management. Part 3: application guide. Section 9: risk analysis of technological systems, IEC 60300-3-9. Finnish Electrotechnical Standard Association, Finland
Gareth WP (1996) The characterization of uncertainty in probabilistic risk assessment of complex systems. Reliab Eng Syst Saf 54:119–126
Guillaume M, Jean-Marc R, Jean-Jacques L (2011) Dynamic fault tree analysis, based on the structure function. In: Annual reliability and maintainability symposium 2011. RAMS 2011
Hofer E, Kloss M, Krzykacz-Hausmann B, Peschke J, Sonnenkalb M (2002) Dynamic event trees for probabilistic safety assessment [Online]. Available: http://www.eurosafe-forum-org/sites/default/files/euro-2-2-10-trees-probabilistic.pdf
Hu Y, Parhizkar T, Mosleh A (2022) Guided simulation for dynamic probabilistic risk assessment of complex systems: concept, method, and application. Reliab Eng Syst Saf 217
Huang W, Chen J, Guo J, Zhen W (2018) Development and application of probabilistic safety assessment in Daya Bay nuclear power station, Beijing, China. In: 18th international conference on structural mechanics in reactor technology (SMiRT 18), Beijing
International Atomic Energy Agency. Selection, specification, design and use of various nuclear power plant training simulators, IAEA-TECDOC-995. IAEA, Vienna
International Atomic Energy Agency (1988) Component reliability data for probabilistic safety assessment of nuclear power plants, IAEA-TECDOC-478. IAEA, Vienna
International Atomic Energy Agency (1989) Application of PSA to research reactors, IAEA-TECHDOC-517. IAEA, Vienna
International Atomic Energy Agency (1992) Manual for reliability data collection for research reactor PSA, IAEA-TECDOC-636. IAEA, Vienna
International Atomic Energy Agency (1997a) Reliability data collection and analysis for research reactors, IAEA-TECDOC-930. IAEA, Vienna
International Atomic Energy Agency (1997b) Generic component reliability data for research reactor PSA, IAEA-TECDOC-930. IAEA, Vienna
International Atomic Energy Agency (2001a) Training course series 12—reactor simulator development: workshop material, IAEA, Vienna, 2001. IAEA, Vienna
International Atomic Energy Agency (2001b) Applications of probabilistic safety assessment (PSA) for nuclear power plants IAEATECDOC-1200. IAEA, Vienna
International Atomic Energy Agency (2004) Use of control room simulators for training of nuclear power plant personnel, IAEA-TECDOC-1411. IAEA, Vienna
International Atomic Energy Agency (2019a) IAEA safety glossary—terminology used in nuclear safety and radiation protection, 2018th edn. IAEA, Vienna

International Atomic Energy Agency (2019b) Status and evaluation of severe accident simulation codes for water cooled reactors, IAEA-TECDOC-1872. IAEA, Vienna

International Atomic Energy Agency (2021) Conduct of operations at nuclear power plants, preprint IAEA safety standard series no. 76. IAEA, Vienna

Joshi NS, Sharma S, Kumar J, Varde PV (2019) Experience of development of Dhruva simulator. In: 4th international conference on reliability, safety and hazard (ICRESH-2019), IIT Madras, Chennai, 10–13 Jan 2019

Kadambi NP (2013) How deep is deep enough?—operational safety: defence in depth. Nucl Eng Int. www.neimagazine.com

Koivisto R, Douglas D (2015) Principles and approaches in ethics assessment: ethics and risk: ethical assessment of research and innovation. In: A comparative analysis of practices and institutions in the EU and selected other countries FP7/2007–2013. European Commission, Europe

Kumar J, Singh T, Joshi NS, Varde PV (2020) Dhruva simulator: core neutronics and thermal hydraulics modules. Life Cycle Reliab Saf Eng 9:45–61

Lee JC, McCormick NJ (2011) Risk and safety analysis of nuclear systems. Wiley

Majdar A, Nematollahi MR (2008) Development and application of a risk assessment tool. Reliab Eng Syst Saf 93:1130–1137

Marko D, Ilic S, Glisovic J, Catic D (2015) Dynamic fault tree analysis of lawnmower. In: 9th international quality conference, center for quality, Faculty of Engineering, University of Kragujevac

Nissan E (2019) An overview of AI methods for in-core fuel management: tools for the automatic design of nuclear reactor core configurations for fuel reload, (re)arranging new and partly spent fuel. Designs 3(3):37

Rao DK, Kushwaha HS, Verma AK, Srividya A (2007) Quantification of epistemic and aleatory uncertainties in level-1 probabilistic safety assessment studies. Reliab Eng Syst Saf 92:947–956

Rao KD, Gopika V, Rao SV, Kushwaha H, Verma A, Srividya A (2009a) Dynamic fault tree analysis using monte carlo simulation in probabilistic safety assessment. Reliab Eng Syst Saf 94:822–883

Rao KD, Gopika V, Sanyasi Rao V, Kushwaha HS (2009b) Dynamic fault tree analysis using Monte Carlo simulations in probabilistic safety assessment. Reliab Eng Syst Saf 94(4):872–883

Researchgate (2000) Nuclear reactor control room simulators: human factors research and development. Cognitive Technology and Work

Shannon C (1938) A symbolic analysis of relays and switching circuits. Trans AIEE 57(12):713–723

Soob W (1994) Uncertainty analysis in multiplicative models. Risk Anal 4(4):571–576

Sorman J (2021) Designing a risk monitor-challenges and solutions [Online]. Available: http://www.relcon.se. Accessed 25 June 2021

TWI (2022) Simulation [Online]. Available: https://www.twi-global.com/technical-knowledge/faqs/faq-what-is-simulation. Accessed 30 June 2022

United States Nuclear Regulatory Commission (1987) Standard review plan for the review of safety analysis reports for nuclear power plants. NUREG-0800. USNRC, Washngton D.C.

Varde PV (1999) Off-site power reliability analysis for Dhruva. BARC, Mumbai

Varde PV (2008) Development of a risk-informed approach in support of maintenance management for nuclear plants. In: International conference on advances in performance and safety of complex systems, India. Publication Macmillam India Ltd., India

Varde PV, Pecht MG (2018) Risk-based engineering. Springer, Singapore

Varde PV, Singh T, Mazumdar T, Jain A (2018) Risk-based approach towards design evaluation and re-assessment of shutdown safety margin. Life Cycle Reliab Saf Eng 7(4):215–234

Varde PV, Sankar S, Verma AK (1996) FUELEX—fueling expert system for research reactor. Indian J Indus Saf Mumbai

Varde PV, Verma AK, Sankar S (1998) Development of an operator support system for research reactor transient identification and fault diagnosis using PSA based knowledge based system. Reliab Eng Syst Saf 60(2)

Vaurio JK (1995) Optimization of test and maintenance intervals based on risk and cost. Reliab Eng Syst Saf 49:23–26

Weiland P, Lustosa LJ (2009) Modeling operational risks of the nuclear industry with Bayesian networks, 2009, September, 27–October 2, 2009. In: 2009 international nuclear Atlantic conference—INAC, Rio de Janeiro, Brazil

Chapter 7
Human Factors in Operation

Our normal state of mind is such that our thoughts and emotions are wild and unruly, and since we lack the mental discipline needed to tame them, we are powerless to control them. As a result, they control us. In addition, thoughts and emotions, in turn, tend to be controlled by negative impulses rather than positive impulses. We need to reverse this cycle to enjoy the lasting happiness that we are seeking.
The Dalai Lama's Book of Transformation,
– Harper Collins Publishers 2000

7.1 Introduction

Operations management, as mentioned earlier, is an intensive human interaction function that includes, apart from man–machine interactions, interactions among plant staff and other agencies, such as maintenance, fueling crew, engineering services, health safety, higher administration and management, during normal operations in general and emergency conditions. In fact, the emergency scenario brings in some additional elements of interaction that include interaction and communication apart from internal staff, the staff away from plant, security and at times fire agencies, members of public and city administration and might involve international communication. In all previous major accidents, aspects of public communication, the national culture and the inadequacy in implementation of emergency management programs were some of the factors identified to be some of the factors responsible for the disaster. The Fukushima incident was concluded to be the man-made disaster made in Japan (The National Diet of Japan 2012).

There is a consensus that human factors have been one of the major contributing factors for incidents and accidents; therefore, it is a well-recognized fact that human reliability needs to be enhanced (International Atomic Energy Agency 1989). The case in point are three major accidents, viz. Three Mile Island (1989) USA, Chernobyl (1986), Erstwhile Russian Federation and, more recently, Fukushima (2011), Japan (Bendix 2019). Why is it that even with the advances in computational technology,

P. V. Varde, *Risk-Conscious Operations Management*, Risk, Reliability and Safety Engineering, https://doi.org/10.1007/978-981-19-9334-3_7

improvement in training methods (e.g., induction of *simulator-based* training, model-based approaches, etc.), advances in human reliability evaluation methodology and technological advances such as inherently safe reactors employing passive features, automation, robotics, remote monitoring, sensing and so on,human factors remain a concern.

This is not to constitute that human is problematic area, not at all. It is the human capability that enabled the nuclear industry to demonstrate and maintain the highest level of safety compared to other energy systems. Humans, most of the time, go more than 99.999% right in their routine task, which makes the plant safe while ensuring reliable societal commitments. The fact of the matter is that safety critical systems, such as nuclear reactors, even though recognized for maintaining the highest level of safety standards, continuously work to improve human factors. In fact, the plants that have a CDF of 3.36×10^{-6}/r-yr can be argued to have reached six-sigma levels. Even though the six-sigma procedure for the assessment of error-free operations might differ, the PRA procedure that is required for evaluation is more rigorous than the six-sigma and complex procedure to achieve. Of course, to have a fair debate, it is important to know what six-sigma philosophy is. Here, it can be understood as an approach to the quality of the product, operations, services and management with an objective function fault or failure. To achieve the one-sigma target, the fault or failure should not be more than 3.36×10^{-6}/demand. Now, the six-sigma approach has many advanced versions and speaks about the success of this technique. Dabbawallas in Mumbai has been achieving this target year after year (Shutterstock 2022) (Fig. 7.1). Now, it is subject of pride as to how we compare high-end specialist applications such as nuclear plants and the services offered by dabbawalas. However, then there are other factors which are not in favor of dabbawallas. However, at least for normal operation scenarios when stress levels are normal, a figure of achieving 1×10^{-6} demand could be a target in our risk assessment culture. For example, a nuclear plant with a full-scope Level 1 PRA can claim to have achieved one sigma or better when the CDF is $\leq 3.36 \times 10^{-6}$/r-yr.

There is a consensus that human error probability needs to be reduced by its normal operation and accident condition. The reason is that there are cases when the accident initiating point was an error committed during normal operation, e.g., in TMI-2, the entry of resin into the feed water system led to blockage of the strainer element and the action by the plant crew to clear the strainer by using the process air that led to closure of the turbine outlet valve. This was typical of a normal operation scenario. However, this mistake led to the propagation of the scenario, and finally, the crew's inability to identify the stuck-open relief valve at the pressurizer, which created the LOCA accident, is something that reflects the plant culture. The lack of capacity for interpretation from the available information in the control room was at the root of the accident (NEI, 2022). The point is that human action in normal operation has a component for deviation from normal operation or introducing the transient that, if not corrected, might lead to incidents. Hence, a *target value of* 10^{-6} *by even considering advanced provisions such as operator aids or by ensuring redundancy, diversity* for any action at the individual or organizational level should be chased on a perpetual basis.

Fig. 7.1 Dabbawalas—An organized system of tiffin service provider. It is a dedicated service that involves collecting homemade food boxes from the client's residence to the office place enduring all harsh weather, crowded places often carrying loads on their head or using bicycles, and public services that mainly include local train and bus services in the city without mistakes or failures. They have achieved this glorious and prestigious six-sigma award as their failure rate meets the six-sigma target of less than $3.6 \times 10-^{6}$ failure. Courtesy www.shutterstock (The Six and Sigma Story: Mumbai Dabbawalas 2021)

The available literature on human reliability shows that most human error models identify human attributes or symptomatic parameters that lead to human error, e.g., error of omission, error of commission, mistake, lapse and so on, and often leave at not going to the root causes of these failures at the level of humans. Here, the question that arises is 'Why do these types of errors occur in the first place?'. These questions can only be answered if we have a human model developed from the first principles where the constituent components apart from material or physical, physiological, etc., the nonmaterial entity, viz. spirit/consciousness, that is fundamental to human existence or being are also considered part of human or a human being. The associated symptomatic aspects of humans, such as thoughts, emotions and attitudes, that also form a critical component of human behavior will also be discussed in this chapter. The *CQB* model proposed by Varde and Pecht is extended further, keeping in mind the requirement of operational ecosystem modeling (Varde and Pecht 2018).

Furthermore, the PRA framework uses human reliability estimates in risk calculations. However, a more fundamental question is 'What are the internal precursor or fundamental root causes that lead to human error'. Therefore, the human factor analysis should target two objective functions, one to understand the *fundamental root cause of the failure* and the second to evaluate the probability of contribution of these constituent human features responsible for the human error. The identification of fundamental root causes of failure enables assessment of the root factors and helps implementation of correction/modification/change in policy that can be adopted at the organizational level as well as at the individual level to improve risk-consciousness levels and thereby effectively reduce risk.

Keeping in mind the above, the objective of this chapter is to (a) extend the CQB to a human model to account for the precursor causes, (b) provide a line of rationales for use of CQB model for evaluation of human reliability by combining the internal and external (to human) factors and (c) explain the applicability of six-sigma philosophy for the target human reliability should be $1 \times 10^{-6}/demand$ for safety functions like developing a culture for derisking through operations precursor management during normal operations, such that preventive measure—as part of defense in depth can be strengthened when stress levels are normal or low.

Traditionally, human error, based on the nature of work, can be divided into error of omission and error of commission. The Cambridge dictionary definition has been derived from a system of accounting; it is required to translate the spirit of these definitions for complex engineering setup. Accordingly, error of omission is defined as A wrong action (mental or physical) that consists of not doing something you should have done or not including something such as an amount of fact that should have been included unintended. Error of commission has been defined as A wrong action (physical or mental) that consists of doing something unintended. Please note that the physical and mental components of human actions are part of both, i.e., error of omission and error of commission.

In the literature, a general observation has been made, i.e., errors of omission are likely to be more common than errors of commission. However, every ecosystem based on the nature of work might have different experiences. In fact, for safety critical systems, the error of omission or commission is governed by factors such as stress levels and situational awareness. Hence, more data and experience are required to generalize this statement.

Error omission is referred to as procedural error or generally where the cognitive activities are required to formulate/implement knowledge based on any procedural activities. As part of real-time actions, when it involves observing and validating the feedback for each step taken from the plant signals, it requires a knowledge-based approach; therefore, in the action part, error of omission due to requirements of knowledge or rule-based analysis error of omission is also applicable. Therefore, error omission might also lead to error of commission where the action itself involves interpretation and reasoning as part of knowledge-based action while handling a situation in real-time field activities. The error of commission is 'where physical actions are involved, e.g., error across human–machine interface.

If we assume that there are three major types of human error, viz. Lapse, Slip and Mistake. The slip is associated with the routine task that involves skill-based action, while the lapse is associated with skill-based, part rule-based actions while mistakes.

7.2 Human Factors in Design

Human factor considerations are an integral part of the design of not only man–machine interfaces but also process and safety systems. Although the chapter title is a human factor in operations, for the sake of continuity and completeness, a brief overview of the design consideration with respect to the human factor was an effective approach for discussing the operational aspects. The control room designs evolved from Generation I, which was based on an analog valve-based logic-based control static mimic graphic display panel starting with a reactor/process plant built from 1950 until 1970, to Generation II, which was essentially based on a solid-state logic static graphic and a very minimal or essential video-based system and automation as the main feature of the plant, and Generation III, which employs digital technology for control room design, operator aids and close circuit monitoring and surveillance and software-based systems in nonsafety critical applications and, of course, the use of artificial intelligence systems such as smart sensors and transmitters, diagnostic features and operator support systems and risk monitoring and improved security features. The advances in AI, fuzzy logic-based technologies, condition monitoring and diagnostic features available to operators in Generation-III systems. Generation-IV technology is evolving to use the potential use of digital data mining algorithms, faster processors than FPGAs and the deployment of online physics of failure (PoF) and prognostics and health management (PHM) for material life and failure prediction. Advanced reactor core modeling and analysis techniques can be expected to be some of the spinoffs from the advanced software and improved models in terms of accuracy and better coverage of complexity of problems.

There are well-established codes and standards available for complex human–machine aspects; for the entire range of plant conditions and states, the guide meets the requirements (International Atomic Energy Agency 1995; Kitamura et al. 2005). The accumulated operating experience and the learning from major accidents and incidents have provided a vital knowledge base to design the human–machine interface in general and the control room in particular. Postulation of accident conditions and the requirements of human action in stressful situations forms the basis of system configurations such that an adequate time window is available for taking action. Technological advances have made it possible to move from Generation-I control rooms to Generation-IV control rooms. This includes soft video dynamic displays, in place of hard-static (The India Forum 2021) mimic panels, application of smart sensors and instrumentations, wireless sensor network systems, employment of virtual systems, etc. Figure 7.2 shows the control room of a 1000 MW PWR in India employing digital technology for the human–machine interface.

Keeping in mind the scope of this chapter, the discussion in this chapter will focus on the design of man–machine interfaces, particularly the control room and higher level requirements of human factors, in plant configuration safety with an emphasis on safety system design, better and more efficient computational tools and methods that might include intelligent operator advisory systems that have the potential to

Fig. 7.2 Control room of an Indian 1000 MW nuclear power Gen III plant (The India Forum 2021)

reduce operator stress by reducing the cognitive load, particularly during emergency conditions (Varde and Pecht 2018).

One of the major efforts in control room design requires human factor considerations that address the postulation requirements of normal plant operation as well as during emergency situations. The fundamental conditions for control room design include (a) high reliability and availability of all control and safety provisions to know the reactor status and manipulate the plant equipment remotely from control to ensure plant safety and security, (b) availability of highly trained and qualified staff for supervisory activities, consultations and decision making, (c) ensuring reliability and controls and power supply, (c) maintenance of habitable conditions for control room staff and (d) operability of the communication network within the plant as well as with the locations/services required during emergency conditions.

Plant control room designs have built-in features that provide adequate coping time or an available time window for operator action/intervention during emergency conditions as part of a defense-in-depth strategy. Despite the progress of human factor considerations in nuclear plant design, human factors have been reported to be one of the major contributors to incidents and accidents. Even though at times the direct cause might appear to be human error for the accident, often the root cause analysis reveals that inadequate ergonomics, dormant or inadequate design faults and other features that include communication features, inadequate time-window for the actions, human–machine interface, environmental and operational issues, organizational and social aspects that eventually force human error and contribute to accident. Therefore, a holistic view is required to understand the other factors that contribute to so-called human error, particularly during emergency conditions. The available literature shows that there are many approaches available for human reliability analysis, some belonging to Generation I and others to Generation II, where the improvised models and approaches are employed to reduce uncertainty in the final estimates of human reliability. Of these approaches, many either do not have a human model or depend on symptoms or data derived based on limited data from plants, simulator experiments, etc.

RCOM was specifically developed by employing the CQB human reliability model and then integrating an improved framework for operational plant and simulator data collection analysis to evaluate human reliability. The following section addresses the adoption of the CQB human reliability model developed as part of the risk-based engineering framework. The CQB model has been extended to the RCOM approach.

The traditional approach to control room design was based on analog technology with a primary focus on safety such that the plant human–machine interface facilitates reactor operations in normal as well as in accident conditions (Sm 2021). The many standards at national and international organizations, such as IAEA and IEC, of course in India, the AERB and guides are used to design the control room or to make any changes in existing control room design. The IEC standard IEC60964 is one of the standards used in control room design along with other codes. However, the deployment of digital technology in the design of new and advanced plants posed a challenge, particularly from a safety assurance point of view. Even though digital systems bring along many advantages, advanced methods are required to characterize the (WHO 2020; Savarese and Lund 2017) potential of these systems to have adequate assurance from safety and reliability considerations. The deployment of new components, such as VDUs, smart sensors, transmitters and computer-based systems, requires considerations of evaluating dependency on, apart from hard hardware, software systems.

Often, the experience and insights developed on plant simulators are considered for developing specifications to support the design of control rooms for new plants or even upgrading the requirements for converting the existing control room design to digital control rooms. In fact, the simulator-based approach can provide a holistic insight that includes not only man–machine interfaces but can also develop human factors and environmental specifications for control room design to support decisions for normal operation as well as emergency control design. To further make these designs, the insights and data available from PRA are employed to meet the features of risk-conscious design.

7.3 Adoption of the CQB Human Model in the RCOM

The human model has been inspired by the Bhagavad Gita that provides the science and philosophy of human existence and a fundamental model that can be adopted in the material and spiritual world of existence (Ved 1996). Bhagavad Gita has classified the complete existence into two major parts: higher or transcendental existence and lower or material existence. There are three major modes of action: (a) goodness, (b) passion and (c) ignorance. There are four fundamental elements of human existence, Consciousness, Cognition, Conscience and Brian,hence, the C^3B or CQB model for human factor modeling is considered part of risk-conscious operations management.

In the CQB model, consciousness is postulated to be fundamental to existence or human beings as well as to the functioning of cognition, conscience and brain faculties, as consciousness is inherent and life force for human beings. The consideration of conscience captures the human qualities, viz. integrity, sincerity and dedication, value system or morality, which are particularly critical to security aspect modeling. Based on the assertions in the Bhagavad Gita, it can be argued that the heart is the center or source of (given their mode and level of goodness/passion/Ignorance) that controls the qualities associated with conscience. Furthermore, with the consciousness and conscience levels of sense, humans interact with the outside world and create their own perception and formations (in Sanskrit: sanskaras).

Furthermore, there are five factors responsible for the human action adopted from the Bhagavad Gita (Prabhpada Interpreter-Translater and Geeta xxxx), viz. (a) Adhisthana (resolution), (b) Karta (the performer) [body and mind], (c) Six Sense bases, (d) Prathak Chesta (different Endeavors or purposes) and (e) the Supersoul or the conscience.

7.4 Anatomy and Physiological Processes in Cognition

Cognitive stress modeling forms part of most of the existing human reliability methods. In CQB, cognitive process modeling involves considerations of the nervous system that includes the brain and the spinal cord as part of the central nervous system and peripheral nervous system that deals with the modeling of cognitive processes in the brain and communication of neural communication among the faculties in the brain as well as with the rest of the body. This section presents an overview of the relevant anatomy and physiology of the brain and the rest of the nervous system.

7.4.1 General

This section will provide a brief overview of the physiology of the brain and associated neural systems. What is control room for a complex engineering plant, the brain is for human body. The brain is the most complex system in this universe, as it has the two most complex side material aspects, i.e., physical and physiological, and the second most complex side deals with spirit, such as mind, consciousness and further higher components.

7.4.2 The Neuron

The human brain comprises 100 billion neurons and its connection, which rages in trillions called synapsis that communicates among other neurons and many other components of the brain (Hoffman 2021). The brain can be divided into two major parts, the area in the cortex region and the nervous system, which connects the brain to the rest of the body and is responsible for autonomous motor functions in the body. The nervous system facilitates communication between the physical mode and the brain. Figure 7.3 shows a neuron (along with a microtubule and its location in the nucleus), a basic building dynamic block that facilitates communication of controls and acts in the group as brain memory. There are three major components of a neuron nucleus, axon and dendrite. The cell body is similar to the terminal chamber for receiving the input signal from the dendrite and after required processing depending on the weight and type of processing passes on to the other connected neuron through the synapse. The voltage or chemical transmitters communicate through the synapse on the dendrite tip of the connecting neurons. The dendrite acts as a terminal node for communication through the synapse. The axion facilitates a communication duct between the two neurons. In the nervous system, a synapse is a structure that permits a neuron (or nerve cell) to pass an electrical or chemical signal to another neuron or to the target effector cell. Synapses are essential to the transmission of nervous impulses from one neuron to another (Synapse 2021).

7.4.3 Role of Consciousness in Cognition

The Eastern philosophy postulates/asserts that the source of consciousness in human beings is not within the body but from outside of the body, and what we call consciousness is nothing but a reflection of material existence in reflected consciousness (Ved 1996). There are extensive research and philosophical explorations in the Western world on the core of the subject, i.e., 'unity of consciousness' (Stanford Encyclopedia of Philosophy 2021). On the other hand, modern research on biological and computational science suggests that consciousness is produced through computational activities of neuronal interaction and that the source of consciousness is the microtubules in the axon. However, modern research that addresses objective orchestrated objective reduction (OOR) postulates that consciousness depends on biologically orchestrated coherent quantum processes in the collection of microtubules within brain neurons (Hameroff and Penrose 2014). The Orch-OR also introduces a new concept of beat frequencies of faster microtubule vibration as a possible source of the observed electroencephalographic (EEG) correlates of consciousness and concludes that consciousness plays an intrinsic role in the universe.

Further details about the source of consciousness and the sit of consciousness are discussed in Chap. 2.

7.4.4 Brain and Nervous System

The nervous system in humans can broadly be divided into two major parts: the central nervous system and the peripheral nervous system. The central nervous system has two major components, the central brain and the second spinal cord (Neurophysiology and Rehabilitions 2022). Figures 2.6 and 2.7 in Chap. 2 show side and plan views of the human brain. The major areas of the brain are the frontal, parietal, occipital, cerebellum, temporal lobes and medulla. These brain areas are involved in cognitive functioning depending on the level and conscience components of consciousness. The frontal lobe is responsible for memory, analytical, logical processing, thinking, decision making and other cognitive activities that support leadership functions. The parietal lob is responsible for sensations, touch, motor and movement, spatial awareness, etc. The occipital area is responsible for visual function, while auditory function is performed in the temporal lobe. The area interfacing with the frontal lobe and parietal lobe is referred to as the central sulcus. The adjoining areas on the frontal lobe and parietal lobe are referred to as the primary motor cortex and somesthetic association areas, respectively. The medulla performs some of the autonomous functions, such as regulating the blood pressure, heart rate and breathing while acting as part of the brain stem and passing neural messages from/to the brain and spinal cord. The coordination of muscular activities, body balance and motor function is performed by the cerebellum. The hindbrain contains the pons, medulla oblongata and cerebellum. The medulla oblongata transmits information from the spinal cord to the brain and regulates life support functions such as respiration, blood pressure and heart rate. The pons acts as a neural relay center, facilitating information flow on the left side and right side of the brain and vice versa. The pons also facilitates the balancing function of the body, sleep and arousal and in the processing of visual and auditory information.

The thalamus relays information to the cerebral cortex, the outer layer of the brain, while the hypothalamus controls the pituitary glands in the body. These formations and perceptions or thoughts and feelings form the basis of actions, physical and mental, depending on what the formation or knowledge and training has for him.

Therefore, the information processing in the Sangam procedure is performed at two levels, viz. at the stimuli level and the brain level, to form a perception of the job.

7.4.5 Human Reliability Considerations in RCOM

The objective here is to develop human attributes that are essential to characterize the risk culture at the individual and organizational levels. In the second step, these attributes will characterize the fundamental root cause of human error for application in PRA study.

Another interesting fact that enables the formulation of human modeling is musical art and science, which is considered a natural component of human consciousness that has the capacity to make a human in the most blissful state and connects to its natural frequency or rhythm. Here, it is assumed that blissful or eternal happiness is the fundamental requirement that has the potential to take humans toward the goal of perfection. In operations ecosystems, after safety documents such as technical specifications, safety reports and safety systems PIDs, procedural documents such as emergency operating procedures (EOPs) and other routine operations and maintenance procedures occupy a significant quantum of operational tasks that mainly include operations and maintenance procedures, including periodic surveillance testing of safety and support systems. Human reliability is critical to the successful completion of the procedure. Therefore, it is vital that the probability of human failure should be identified for each step and subtask in the procedure. The plant technical and regulatory framework requires that operational and maintenance jobs be performed using approved procedures. From these considerations, human reliability evaluation of the procedure might form one additional indicator for operational procedure validation and approval. The following section works with the first-principles approach for human factor modeling considering the adopted CQB model, as discussed in the previous section.

In RCOM, human reliability modeling is inspired by the Indian musical system. In this system, there are seven musical notes, *Sa, Re, Ga, Ma, Pa, Dh and Ni,* and again ends with *Sa,* that form the fundamental building blocks for composing classical and light music and songs, including devotional abhangs (spiritual songs). In the RCOM human reliability evaluation procedure, this system (model of Indian musical system) is adopted to suitably modify and apply the CQB human reliability model to develop a system of human reliability analysis for operations and maintenance, including emergency procedures, referred to as the Sargam human reliability methodology (SHRM) for procedural tasks in RCOM. In the Sargam framework, there are six external stimuli that correspond to five sense bases through which the human interacts with the outer world, as shown in Fig. 7.4. These six sense bases/or stimuli arranged into five orders can provide $6^6 = 46{,}656$ combinations of notes to express the task characteristics in a procedure.

The procedural structure comprised a main procedure (MP) with a title. For example, 'Class IV Power Supply Outage Procedure'. This is denoted by MP. The MP comprised more than one Subsection. Furthermore, each Subsection comprises more than one or more tasks. There is a gap between each task. Finally, each task comprised one or more Sargam notes. Each note represents a stimulus in the task. When the tasks are performed in sequence, they follow the series reliability model by reliability block diagram or fault tree model. The parallel or redundant tasks are represented by redundant or parallel modeling, as is the case with hardware reliability modeling.

The input on the stimuli is received by the individual sense base, and interval processing occurs at two levels to form a mechanical signature/pattern of the information, such as image, sound, taste and smell through the sensor and connected processing organs in the brain for making a sensory pattern and further processing

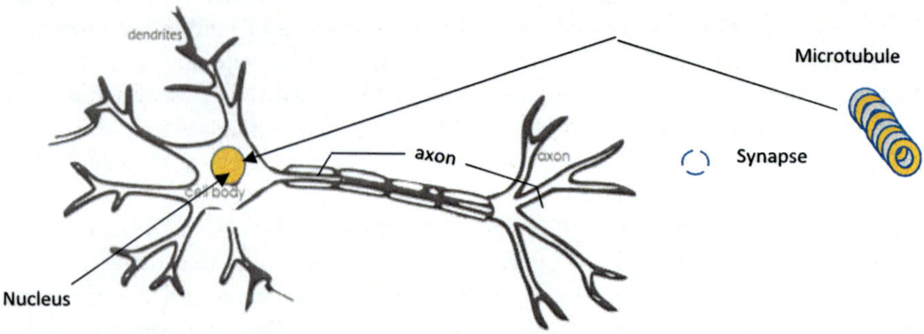

Fig. 7.3 A neuron structure depicting the nucleus in the cell body, microtubules, dendrites at the neuron tips and axons that act as communication ducts

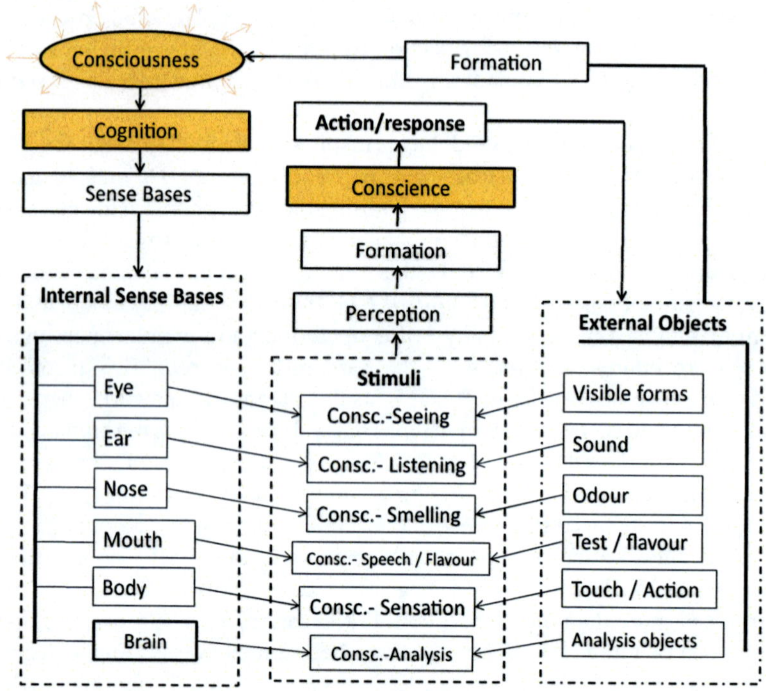

Fig. 7.4 The CQB Human Model (Varde and Pecht 2018)

for perception and emotions. The processing in the brain to generate perception and emotions makes the sensory information, like seeing an object to conscious seeing. The conscious component of seeing provides subjective as well as meaning to the object. This might include getting the geometrical attributes, color, a relative assessment of size, comparison with a pattern of similar object in the memory subjective attribute like perception about the image, emotions, motivation, etc. The objective and subjective patterns give meaning to the vision and purpose. The quality level of the conscious component in the brain to the physical object forms an input for further treatment about the adequacy of the human act associated with those stimuli.

In RCOM, the modeling of sense bases, which is a fundamental and integral component of human performance or human reliability, is inspired by Sargam. The *Sargam* framework has been adopted considering the six stimuli or the sense bases. The human interacts with the outside world employing through seven sense bases and experiences through six corresponding stimuli, i.e., Eye for Conscious-seeing, Ear for Conscious-listening, Nose for Conscious-Smelling, Mouth for Conscious-speech/test, Body for Conscious-neural sensation and Conscious-mind for expression for imagination, analysis, emotions, rationale and thoughts, intelligence and so on, as shown in Fig. 7.4. The discussion on the subject will continue in Sect. 7.8 entitled modeling of sense bases. The fundamental assumption is that music and recitation shlokas, the way they are pronounced, have a direct relation to human consciousness and memory. The message communicated with music directly fills the heart, memory and mind. Nevertheless, the use of shlokas from spiritual books, poems and pros are given in Annexure D. In RCOM, poetic or musical composition forms provide a special dimension for improving performance.

7.4.5.1 Procedural Knowledge Quotient

It is implicit and a nondeclarative component of memory functions in the human brain. Forms form the basis for procedural knowledge or personal experience. Maneuvering reactor control on the control console in the control room for reactor start-up, power raising and shutdown as per the predetermined sequence or logic while receiving feedback information from sensors showing the results of the action and initiating corrective measures if a new element or deviation is observed from the expected response from the reactor state. This function has been categorized as part of the cognition component, as it requires rule as well as knowledge-based processing for effecting smooth operation, intended response and, if needed, corrective action based on the system dynamic response.

Any error as part of implementing the procedural component is categorized as a 'mistake', even though the chances of lapse and slip also exist; however, the plant control logic has built-in automatic logic that calls for automatic correction action that takes the plant to a safe state.

Any mistake on consoles, of course, can be due to personal attributes; however, the direct cause for such mistakes rests with a lack of experience and calls for improvement revisit in training programs.

7.4.5.2 Logic and Reasoning Quotient

The left part of the brain is responsible for cognition with respect to logic, reasoning, rationales and analysis. This is a central understanding of the physical processes, phenomenon and dynamics of the reactor, subsystem and component/equipment state/condition that requires the application of logic and reasoning and inferencing and comprehension based on the information available in the control room. Certain events such as reactor transients flood control room with audio-visual alarms change in status of reactor and process and safety parameters and other communications in control room in a very short time followed by additional parameter registration over a period. This condition poses a cognitive challenge to come to the cause of the transient its source or location and further comprehension to enable planning of corrective action or even normalization. Logic processing and reasoning involve rule-based and knowledge-based analysis and therefore form part of cognitive functions.

The stability of cognitive faculties often poses a challenge as stress starts playing out and reduces cognitive capabilities. The dorsolateral prefrontal part of the frontal lobe in the brain is primarily responsible for processing and logic.

The capability of cognitive faculties for reasoning and logical conclusion of an individual can well be assessed in simulator or real-time plant conditions. The simulator sessions involve test runs with not only various accident conditions where the status of equipment can be changed to assess the operator's capability for correct decisions. Plant reports are a valuable source of evaluation of logic and reasoning capability; however, the training programme itself provides an opportunity to evaluate an individual's logic and reasoning quotient.

7.4.5.3 Perception and Formation

The perception and formation of plant and subsystem conditions and thereby arriving at the safety or risk implications or risk level based on the logic processing and reasoning during initial times sets the stages for comprehension at a higher level in support of decisions. The outcome of perception and formation or comprehension sets the stage for decision making.

7.4.5.4 Decision Making

Decision making is an executive function and involves broad-based situational awareness, the processing of information applying logic and rationality while trying to risk new evolving scenarios once decisions are affected in real time. During an emergency scenario, the risk reduction measure takes the center stage or primary goal of decision making. The cognitive or behavioral stability attribute of the operator, i.e., the efficiency and effectiveness of the processing of information in support of decision making, is critical to decision making.

In the context of the core decision matrices.

The frontal lobe in the brain addresses planning, while subfunctions such as executive and logical processing are performed in the dorsolateral region in the frontal lobe. The decision making process involves temporary memory that involves declarative as well as procedural memory capacity and processing capability.

Attributes such as the sensitivity of the stability of an operator to stress levels can be assessed in the Simulator environment by creating sessions that involve emergency scenarios with additional deviations that may require cognitive processing to come to decisions. Recording a control room scenario during anticipated transients, such as off-site power failure, or scenarios such as fire incidents might reveal useful information about the stability attribute of an individual.

7.5 Major Attributes Consciousness in RCOM

Although a separate Chap. 2 has been dedicated to this book, as consciousness has been considered fundamental and central to the title of this book, in this chapter, we will discuss the attributes of consciousness that form part of the human model, specifically sense/functional organs.

7.5.1 Conscious Formation

The physical phenomenon sense bases, viz., seeing, smelling, listening, bodily sensations become a 'conscious' phenomenon that forms perception after processing of the information or sensory input in the brain by different brain faculties.

For example, the dorsal stream and ventral stream perform parallel processing for visual stimuli. The processing of information or stimuli occurs at unconscious and/or conscious levels. The processing of the dorsal stream initially occurs at the parietal stream. This processing addresses the movement of the object, particularly relative movement. The processing of the dorsal stream is faster as it happens as an unconscious event and is meant to react, if needed. Any damage to dorsal stream faculty leads to deficiency or defunct in the processing of movement and relative speed processing issues. The ventral stream forms a conscious perception of the geometry, size, shape and color of an object. The brain faculties involved in the visual stream in sequence are the retina, thalamus, primary and secondary cortices. The primary cortices cognize the object orientation, boundary and edges. The secondary cortices identify the contrast between the object and background, also making contrast as the base identifies the contour of the object. Damage to any of the modules of the ventral stream can weaken the perception forming capability, such as not being able to recognize faces or landmarks.

The sound stimuli signal or information initially processes the eardrum through the auditory nerves and finally at temporal lobes for having a conscious perception of the sound. Similarly, the stimuli related to taste and smell are also processed by the taste buds on the tongue and nostril sensors, respectively, and the somatosensory lobes identify and compare it with the memory element patterns and make this information a conscious phenomenon and pass on to the other areas of the prefrontal area of the brain for reaction, such as arousal and other emotions. Further details are discussed in Sect. 7.2.

The conscious processing of information obtained from outside sources through sense bases has significance in real time in operational environments, as these patterns stored in memory form part of experience and characterize intimate connections with plant systems. Some examples can be discussed to have an improved understanding of conscious signals. Different types and sizes of pumps operate in a plant. The operator makes a visit to the plant to ensure that everything is normal. An experienced operator, when he visits the area, must make this feel of normal by comparing the current situation with the normal visual, sound and smell stimuli and an unconscious or conscious comparison with the normal patterns in his memory. For example, the sound emitted from a pump, feeling temperature on the bearing surface, any smell of oil or any abnormal smell are very subtle observations before any physical degradation and damage occur. Small leakage of compressed air or water carrying piping in the plant, making a hissing sound, might be an early indication of incipient failure.

The faint but noticeable burning smell in the control room or other control or power cable areas could be a strong signal of short circuits, loose terminals or degradation of the cable insulation, which might avert a potential series scenario.

This is where a conscious component for these stimuli is vital for reliability improvement and risk aversion. This aspect can often identify a better human component for connecting with the machine. Finally, a conscious connection with the plant can make a whole different part of the risk-conscious culture.

7.5.2 *Awareness Coefficient*

The awareness coefficient is the second important component that signifies the conscious connection with the plant. The sincerity, intelligence and diligence attributes contribute to enhanced awareness. The awareness in all the operating scenarios includes prevention of deviation, detection of off-normal conditions, prevention of accidents and mitigation and management of accidents. in terms of (a) the requirement and role of risk reduction measures on a sustained basis, (b) the time and space dimension for normal operation and emergency conditions, (c) plant policy during various plant conditions, (d) duties and responsibilities, e) purpose of the individual's role, (f) knowledge base and requirements of periodic updating, (g) plant condition and models of operations, (h) knowledge base or source of information to perform action part of the job, (i) reporting requirements, (j) administration and technical support to supporting staff, (k) role of communication with peers and higher level management and so on; this list is not exhaustive but indicative. Even though the major part of ensuring higher awareness requires conscious processing, there could be a cognitive component to the technical nature of jobs that requires reasoning, pattern recognition, etc.

The assessment of awareness quotient needs to be initiated right since the training period during checklist stages, and subsequently it can be graded during the written exam and interviews. Furthermore, the walkdown exam conducted, as part of the qualification process, should emphasize the assessment of alertness and awareness quotient and monitored when the staff forms part of regular operation.

It is a challenge to postulate the awareness quotient of an individual trainee/staff member during accident conditions wherein the stress level determines the stability of behavioral stability. However, some indirect methods that have the potential to indicate peep into the human personality during stressful scenarios include tracking emotional response during situations when the individual is suddenly exposed to a new situation, new simulated testing conditions, behavioral attitude when the time window for a task is limited and the task if not performed correctly, which might attract penalty in terms of loss of promotion, reputation in operational ecosystems, general emotional behavior during day-to-day operations when subjected to conditions involving incident management, or management of anticipated transients, particularly when the plant systems do not respond as desired and some actions are required to understand the situation and use alternate or reserve provisions to normalize the situations. Such situations might require cognitive processing; therefore, the effect of stress on human behavior will be discussed in the next section on cognitive modeling.

It can be noted that the objective of this section was to identify the human factor precursor such that incipient human failures can be reduced by initiating the correction action program and thereby consequences associated with these failures can either be eliminated or reduced.

The metrics for evaluating the awareness quotient require dealing with subjective information, and there cannot be a general consensus on this evaluation; however, the following section presents an indicative and not exhaustive approach to the development of the awareness quotient by considering identified attributes.

The objective of the identification of awareness of human precursors addresses the assessment of the awareness capability of individuals in operational tasks. Assessment of the awareness quotient of an individual enables the supervisor or the management to assess the strength and weakness of the individual and accordingly make a level of compensating and task monitoring provisions such that error can be avoided. There are 10 major awareness areas that need to be assessed for an individual trainee.

- Deterministic and Probabilistic Risk Assessment
- Plant Operational Policy TechSpec
- Plant emergency operating procedures
- Time window available for emergency tasks and actual
- Time it takes to complete the task under normal conditions
- History of plant incidents and learning
- Major modifications

- Plant layout and physical features
- Plant design and operational aspects need
- special attention and monitoring.

7.5.2.1 Situational Awareness

In the context of human factor engineering, situation awareness is defined as 'the perception of the elements in the environment within a volume of time a space, the comprehension and projection of their states in the near future' (Endsley and Bolstad 1994). Furthermore, the situational awareness model, in the context of the control room of the plant, is an integrated system of a human-system interface that has three major elements: (a) the human operator, (b) the hardware and software systems interfacing with the plant and (c) operator aids, such as communication systems, design and operational documents, computerized operator support systems for the identification, diagnosis and recommendation of corrective actions and procedures. The awareness quotient for any human-system interface is defined as a function of the capability of human awareness, a hardware system that audio and visual, a digital and printed script of alarm logging in chronological order, monitoring and measurement systems and support systems, as discussed above.

In the CQB approach, only the consciousness component of situational awareness captures only the observational component through 5 conscious-sense bases that deal with perception and formation, i.e., Creating a mental model of the plant state or system state identification is performed as a conscious part of the awareness. This part of situational awareness is sensitive to the conscious processing of information, where brain faculties to some extent, such as temporary memory processing, developing a sense or feeling or risk levels and emotions are involved. Any deficiency or degradation of sense bases or its processing in the frontal lobe of the brain adversely affects this situational awareness capability. Some medical conditions that create anxiety, such as ADH or stress, also adversely affect comprehension.

7.5.2.2 Declarative Awareness and Memory

Declarative and procedural aspects of training form a major component of training. In a way, the declarative component addresses the descriptive knowledge about the plant and systems, such as plant layout, system design description, routine operational aspects, limiting conditions for operation and safety limits, limiting safety system parameter settings, various plant configurations and states, nomenclature of various components and systems, their capability and normal parameter values, their make, identification number, individual equipment protection interlocks their trip and alarm settings, color coding used in the plant, system USI code, interconnectivity between two or more systems, elevation of equipment, information about the paging and announcement systems and other modes of communication, plant hierarchy and authority and individual responsibility, etc. The human memory capability and the observation quotient play vital roles in determining the declarative quotient. Declarative memory is also referred to as explicit and has two major components, episodic and semantic. Episodic information comprises important events in the plant, such as incidents and first criticality, while semantic events deal with facts or descriptive features and process system functions, as discussed above.

The evaluation of declarative awareness is an integral part of checklists, walk-downs and interviews at various levels; therefore, this declarative awareness quotient can be graded for an individual trainee as part of training against well-established metrics. The short-term memory storage and recall capability along with the processing of information provide the foundation for cognitive components such as procedural information, reasoning, diagnosis and decision making, as discussed in the following steps.

7.5.3 Alertness Quotient: Concentration and Focus

Lack of concentration and focus has been found to be one of the major factors that lead to error of commission and error of omission. Psychological and physiological conditions have been found to adversely impact concentration and focus.

We must understand, a priori, as part of discussion on human error precursor, 'Why human error happens'? Here, we confront two contrary but established facts. First, 99.9% of the time humans know what their next move will be, right or wrong, only 0.1% of the time, particularly during emergencies when we are subjected to cognitive overload due to stress, this assumption slides to the pessimistic side. Second, by nature or through the process of evolution, the 'human mind's normal condition is disorder', as our brains are designed to notice, recognize and attend to a variety of random sensations, perceptions and, more importantly, thoughts, all of which are competing for attention. Our brains do this to keep us in touch with our internal as well as external needs and ultimately to keep us alive. Again, either through evolution or after the birth the brains develop, we become better able to manage this chaos and turn into a reasonable order by exerting attention control (Bailey 2020). From here, we arrive at one major conclusion: The 'concentration' and 'focus' attributes of 'attention' should be considered major attributes in precursor studies at the individual and organizational levels. Furthermore, we need to understand that if humans complete a task 99.9% of the time during normal conditions, it is not adequate, as the stress and other situational awareness factors during emergency scenarios might make humans more prone to errors.

One should not be surprised that the human error rate during normal conditions in the control room setup for tasks such as acknowledging an alarm and resetting after the parameter might be less than 1 in 10,000 opportunities. human error probability for this task is $\leq 10^{-4}$. Therefore, we conclude that for skill-based tasks, the human error probability can be in the range of 10^{-4} to 10^{-5}. This means that by further working on the precursor parameters, as shown in Fig. 7.5, achieving a figure of a six-sigma target figure of 3.6×10^{-6} normal operating conditions should not be a

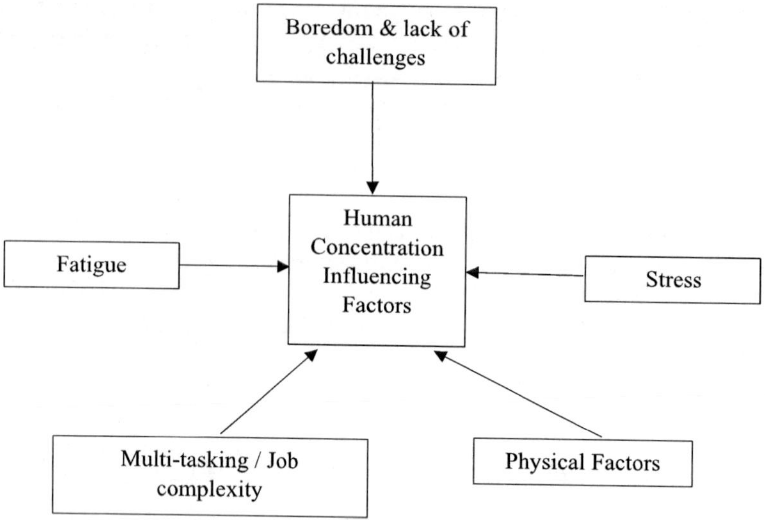

Fig. 7.5 Human concentration and focus affecting factors

major challenge. This experiment can be designed and conducted on a simulator to demonstrate the considerations/assumptions discussed in this para.

Unlike hardware systems, where the degradation levels measured online or offline or any new signature audio or video provided information about a precursor to an incipient failure, in humans at the individual or organizational level, there is no direct monitoring that provides information on precursor human failure for a specific task. Of course, the medical and psychological assessment initially as part of training and letter at a certain predetermined periodicity or a visit to a health care facility due to illness might reveal physiological or psychological conditions that might reveal certain disorders or conditions that form possible precursors to human error during operational ecosystems. However, the human attributes that form part of risk-conscious culture may provide effective input about the predominant precursor factors that could lead to human error.

7.5.4 Emotional Quotient

The right side of the human brain is responsible for the emotional component of human existence. The emotional component has been found to be a major factor that determines the level of organizational consciousness and is responsible for smooth management functions in the plant. The most important attributes of emotional stability are (a) the emotional stability of the individual in normal and stressful conditions, (b) the relationship of the individual with colleagues, superiors and management, (c) the capability to manage the staff working under the individual,

(d) the establishment or striking capability to connect with other individuals and (e) the capability to communicate. Often, it has been seen that in a management environment where teamwork forms the basis of task success, the emotional quotient becomes a more important attribute than the intelligent quotient.

Emotional quotient is a personality trait that can be assessed during day-to-day interaction, particularly in a control room type setup where coordination and communication, task management is a perpetual scenario. The immediate superior can easily assess this attribute. A lack of emotional quotient can hinder the successful and smooth completion of the task and is prone to human error and forms and important precursors. For team leaders, emotional quotient is an important attribute, as the team leaders' job is to complete the task where interaction, communication, management with other agencies and coordination are the prerequisites. The processing of basic emotions is performed at the amygdala and lateral orbitofrontal faculty of the limbic system in the brain. These faculties characterize the emotional stability of an individual vital for team management or team activities for the management and coordination of plant activities, particularly during deviation or accident management.

7.6 Conscience

7.6.1 Background

Conscience is a subtle quality that enables an individual to distinguish between right and wrong. The sense of ethics is central to the conscience. This attribute is often used to characterize the national conscience, organizational conscience or collective conscience, or individual conscience. In fact, the aggregate of individual conscience makes for collective organizational conscience. After ethics, the next important parameter is the attitude of the staff toward the duty they perform, other colleagues and society at large. Some factors provide vital insights into the collective conscience level of the staff. For example, how many 'employees work for salary and benefit' and how many for the 'love for what they are doing'. In the context of the plant environment, the conscience attribute is related to deliberate acts that compromise plant and operating organizations' mission of security, safety and performance objectives. These intended errors can be categorized as noncompliance and violation. The role of ethical behavior and moral values systems of the individual and collectively at the organization level is critical to assess the conscience of the individual and the organization.

There are four major subhuman attributes that need to be checked and verified during the recruitment stage of the candidate by carrying out background verification through intelligence information gathering. This is to eliminate the possibility of involvement of insider elements in the postulated threat scenario. The following section will discuss the four major human attributes considered relevant that could form potential precursors to safety as well as security.

7.6.2 Ethics

Ethics is defined as the moral principles that motivate the behavior of a person of conduct of the duty assigned to them. For modeling CQB, ethics has been considered a subset of consciousness. That is a conditioned form of consciousness that distinguishes good from bad in the context of human or societal wellbeing. In many cases, other attributes, such as integrity, honesty and morality, fall under the heading of ethics.

7.6.3 Integrity

Integrity is a self-imposed, demonstrated and consistent moral behavior of high standards that is recognized as an over-encompassing attribute of an individual. The major subattributes of integrity are commitment, honesty, dedication, conformance to organizational behavior and value systems and requirements. Integrity can be considered a vital component of dependability.

7.6.4 Honesty

Although the lack of honesty is a very uncommon behavior encountered in the plant, management should always remain vigilant to check this behavior. Honesty is a fundamental attribute that makes a staff dependable, as dependability occupies a significant dimension in safety and security. Honesty in performance, reporting, dealing and sharing information with management, colleagues, subordinates and outside agencies. Hiding or not providing full complete information/truth while reporting are signs of dishonesty at any level in the organization. This behavior/culture has the potential to cause more harm and might compromise safety and security, as correct information can only form the basis for corrective action such that the plant does not experience the same adverse situation again.

A general observation and, if needed, intelligence tracking during the recruitment stage followed by, if needed, during the service period might be of use for checking this tendency of an individual. In the RCOM approach, honesty makes for the ethics quotient of an individual.

7.6.5 *Morals—General Attitudes and Dedications*

The morality or moral values of an individual are the capacity to perceive what is right and what is wrong in each setup considering the organizational goals and objective. Morals define the overall standards of behavior in an organizational setup. The moral value system is basically governed by the intrinsic attributes or inner voice at the individual level that facilitate the process of judgment about right, appropriate and wrong with respect to a decision. For example, a consideration that 'whether my action will be in the larger interest of the plant or society', if yes, then the action is right. Conversely, the action that might not be in the interest of plant, but it will same me from embarrassment or penalty, is a wrong action. The individual should always put the plant's interest above the self-serving attitude to be moral. While some moral principles transcend time and space, others could be specific to a group or a society. For example, always speaking truth is a universal principle, but it can be argued that eating nonvegetarian food is right or wrong, as this behavior is governed by climatic, social and economical and religious considerations. For the purpose of the RCOM approach, the moral value system occupies a vital dimension, as keeping the conscience alive throughout one's personal life as well as part of a risk-conscious culture while perpetually maintaining a set of moral values is the key to consolidating a plant's safety, security and performance objective.

There are many other attributes of conscience, such as dedication for duty, organization and nation, happiness, sincerity, level of training and organizational consciousness level, that specifically are of interest from the point of security and will be discussed in other sections.

7.7 Reference Human Model in RCOM

The reference human is one that has been qualified for highest grades and authorized for identified jobs employing the following qualification and authorizing procedure:

(a) The candidate should have a Bachelor of Technology/Engineering Degree (with minimum honor, i.e., overall grading $\geq$75%) with an Excellent Conduct Certificate.

(b) The reference candidate who is inducted into a qualification program should be holder of A grade (>80% in his academics).

(c) An approved qualification procedure for plant manning that comprised (i) successful completion of all the checklists within prescribed time and obtained more than 90% grades, (ii) (c) cleared written test and obtained an overall grading of > 90 and not less than 70% in any individual subject and (iii) obtained an A + (> 90%) grade that referred to 'Excellent Grad' for his performance in an authorization interview conducted by a committee appointed by the regulatory agency.

(d) He has been subjected to an approved medical and psychological procedure that covers qualification of all his vital 6 sense organs. These tests should also include certification of all the identified health parameters that include physical, physiological, psychological and general wellbeing evaluation; and Intelligence quotient (IQ) tests. Each evaluation should meet the set parameters for the reference man.
(e) The final and integrated qualification requires evaluation in a simulator environment, particularly for responses to sessions on emergency procedures and anticipated transient execution and decision making capabilities. For the reference, human qualification grading requires a performance corresponding to the highest grading, i.e., A++ (available for perfect performance).
(f) An individual candidate is assigned a final grading to ensure how he compares to a reference man. Any specific areas need to be tracked, and the feedback is used for improvement.
(g) The authorization of the candidate marks the completion of the qualification procedure and enables a candidate to take on the responsibility he is authorized for.

In CQB, the reference fault tree enables visualization of the human in an integrated manner. This is in line with hardware systems where the engineering model is created by integrating the individual part of the system. The available human reliability approaches consider humans as entities that perform complex to routine and skilled tasks based on plant operational and safety requirements. The objective of creating the reference human model is to evaluate the human error rate data for routine conditions, i.e., for normal or lowest stress levels and excellent human–machine interfaces, all the stimuli are perfect and for a healthy (physically and mentally) operator. In short, the reference values are the estimates of Po terms in the human reliability. The conventional model generally works in two distinct cognitive and physical action capabilities for performing a task without many considerations on how, what, when and why and how individual organs and capabilities interact internally to enable a human to behave the way they do. Experimental or simulated responses form the basis for characterizing human performance or, more precisely, human reliability. The basic premise of CQB human reliability is creating a human model right from its individual internal organs and capabilities cognition, consciousness, memory and modeling of a complex brain working based on the available state of the art, with an objective to reduce the uncertainty in human performance prediction. The reason is, how without understanding and analyzing systematically, as we do for the hardware, software and other conditions in a plant, i.e., the plant is broken down into systems that make the plant and individual components and subsystems that make the system and subsystems for modeling the plant. Therefore, it was considered in CQB that by creating a human model by integrating the individual, organ, material, cognitive and psychological and conscious process, an improved human reliability approach can be created.

The organs, as discussed above, can broadly be divided into (a) internal organs, (b) external senses for interacting with the external world and (c) body organs that enable work to be performed physically and mentally as part of plant job requirements. We have already discussed six sense bases, Viz, eye, ear, nose, tongue, skin and mind, that enable interaction with the stimulus sight or vision, sound, smell, taste, sensation for touch/pressure and thinking, respectively. The work or functional organs are mouth for speech, leg for movement and hand for physical activities. Apart from these, there are two organs, viz. genital and anus, that are not directly relevant to our subject, but their relation to human behavior or wellbeing cannot be ruled out.

Apart from the above, there are, as discussed above, vital human internal organs that include the brain and connected parts, such as the spine, as part of the nervous system. Figure 7.6 shows an integrated fault tree representing the essential features relevant to creating the reference human reliability model for safety critical systems applications. Even though this model has been created to evaluate the reference failure rate/probability for a plant or specific application, the elements of this model, such as sense base, action organs, the nervous system or specific areas of the brain for a corresponding stimuli evaluation, can be used separately.

To further understand the features of this fault tree, the following discussion provides the essential features of this modeling. This modeling follows a series system modeling, as all the individual faculties have been assumed to qualify a human fit for work. The reference value either at the level of the transfer gate or at the basic component level, based on the availability and adequacy of the data, was

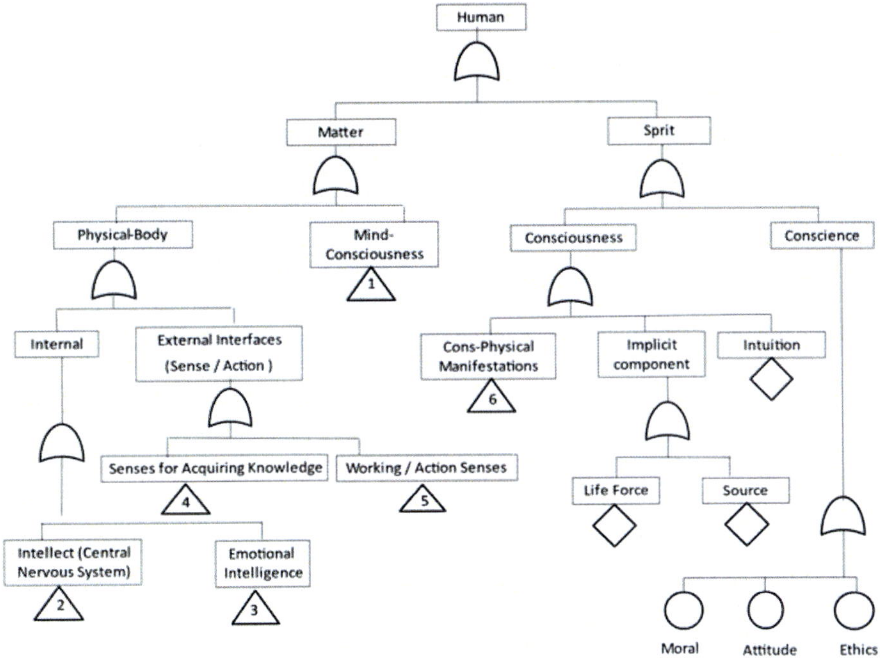

Fig. 7.6 Reference human fault tree

used. A diamond event in line with fault tree parlance is referred to as an undeveloped event. What it means, these either do not require further analysis, or due available state of the art no quantification and modeling can be done and this uncertainty has been managed by giving a qualitative treatment to justify termination, information is available but it is not adequate to have quantification of these end or basic events, the state of the art is not adequate but these events are relevant as there are evidence or accepted that they are part of human behavior for example, even though there is well-accepted fact that human consciousness if a reality but there is no general consensus or evidence on its source. However, for the purpose of human reliability modeling, it is accepted that breathing, heart bits/pulse rates and awareness and alertness are critical to accept symptoms of life force.

Certain assumptions have been made to keep in mind and take advantage of the current human resource management practices. The health checkup and periodic follow-up along with general self-observation as well as by others on an individual do not require detailed physical, psychological and physiological health of an individual. Here, generic data based on the real-time world can be adopted for quantification in a reference model of humans. Furthermore, a scope can be kept in line with other performance-shaping factors, such as stress, the human-pant interface and the context for individual situations.

It is a well-recognized fact that modeling humans is a complex task. It becomes particularly more challenging when the human model is created from its basic constituents, organs or faculties. Furthermore, sutler aspects, such as consciousness, spirit their source and its understanding, which affect human behavior, pose challenges. Although there is no claim in this work that the fault tree drawn is exhaustive and complete, the challenge is to develop a way such that on a practical level, we have a plausible ground. The objective here is to reduce uncertainty in the human factor assessment and bring the whole process in line with hardware system where the analysis involves integration of constituent components to represent the plant model. Here, the human reference model is developed by considering the available state of the art in understanding human behavior. In view of the above, there are many transfer events that carry the tag of undeveloped events, and further in some places, the nodes ended, which was assumed to be adequate for the purpose of its projected application, i.e., risk and reliability analysis.

7.7.1 Unmanifested States/Stages

The physical growth of the body with age and intelligence is realized in the form of cognitive capability, in fact, the physical symptoms of consciousness detected at the level of the brain and measured by advanced instrumentation, are referred to as manifested symptoms of certain phenomena occurring in our body. However, certain qualities, such as intuition or intuitive power, remain with the individual, and the behavior of any individual during an emergency remains dormant and cannot be detected. Furthermore, breathing and heart rate are symptoms of life, but we know very little about life and how they are related in terms of birth and death. Bhagavad Gita discusses the unmanifested stages/state of three modes of human material existence in life, viz. Goodness, Passion and Ignorant [Chap. 13 6-7th Mantra], that affect human behavior. For CQB, the human quality of Goodness has been considered. However, there are some qualities in the context of engineering operations that remain unmanifested and recorded formally as false egos that may not reveal the sensitivity of an operator's performance to higher stress levels. For example, different persons might exhibit different levels of deterioration in performance for the same stressful situations. These unmanifested qualities often contribute to human error. Second, moral qualities, such as ethics and honesty, are critical for security but may not manifest or become apparent until they adversely affect.

The following section discusses the major features and the treatment that was considered adequate.

7.7.2 Undeveloped Events in the Fault Tree

In the reference human fault tree, there are more undeveloped events. There are many reasons, such as (a) either no details were available, (b) the data and information at the given intermediate event can be accessed from a formal approach, such as an intelligent quotient procedure, (c) further details are not required and (d) even if the fault tree was developed in detail, the data on the basic component that results as part of going further down were not available. Some of these cases have been discussed as follows.

7.7.2.1 Physical and Physiological Health

The health checkups instituted as part of qualification and authorization and the periodic health checkups that form part of human resource monitoring and management are considered adequate to assign the reference reliability data in the fault tree. Generally, these nodes do not form the dominating or contributory factors to the human failure rate. The accumulated plant operating experience shows that physical and physiological health generally do not contribute to the routine random failure rate. The reference value for the contribution of health and physical and physiological contributions in the final calculation was less than 2×10^{-8}/hr.

7.7.2.2 Intelligence Quotient

Human resource management and monitoring during on-the-job training provide a robust mechanism toward ensuring cognitive capabilities for the analysis of the plant state and transients during routine operations as well as incidents. The qualification criteria and the, as discussed in the Section, i.e., graduation with 65% and above, the clearing of written and oral exams and further plant-level qualification procedures and follow-up through overseeing provisions and reauthorization after a preset period, are adequate to ensure above average intelligent levels for routine jobs as well as for incident management. The provision of monitoring performance during job experience and simulator experiments is an effective approach to assess the cognitive capability of operating staff.

Despite this, it has been found that human errors contribute significantly to accidents. Therefore, in the CQB model, the fault tree values are treated as reference values and other factors that challenge operators, such as stress level due to inadequate time windows, context or event complexity factors, overwhelming information, communication and coordination-related issues where an operator has to maintain equanimity of mind such that decision making capability is not adversely affected. These factors play out during emergencies; therefore, the appropriate performance influencing factors are determined by employing the CQB model. The reference data for the intelligent node have been $<10^{-9}$ h. Here, it is assumed that the IQ of an average operator is above 100.

7.7.2.3 Psychological Health

If an individual is suspected to have psychological conditions, such as irrelevant behavior, attitude issues and service-related irritants that are not corrected even after repeated advice, irresponsible behavior that has the potential for risk and reliability forms part of psychological uncertainty. As an administrative plant decision on the subject, individual operations engineers/staff are removed from regular plant operations and critical areas. This is done as a precautionary measure and unless certified other vise by expert or professional psychologists and further to the level of satisfaction of plant authority. Day-to-day professional behavior and periodic monitoring provisions, if needed, reduce the impact of this factor on plant risk. Based on the experience and expert provision and availability of many checks and exposure levels, a very low value of human error, i.e., 5×10^{-9}, has been considered.

7.7.2.4 Emotional Quotient

In the context of plant operations, there are two aspects associated with human emotions. First, from the point of view of the organizational aspect, the focus is leadership and the interrelationship among staff such that safety and reliability issues can be addressed effectively by employing various strategies, such as the development of team spirit, respect for each other and the realization that you part of a family and staff have a feeling of emotional security. Second, it addresses organizational provisions that ensure that the emotional level of employees is checked. For example, the promotions, salaries, perks and other incentives and social activities, such as getting together, meetings and seminars that serve both the technical and interpersonal relationships, where the staff receive recognition. Finally, there is a challenging aspect that needs to be checked: the emotional stability of staff. Often, it is observed that due to unbalanced emotions due to extreme conditions of pleasure, sorry or pressure, emotional aspects come into play that have the potential to jeopardize a plant's interest. Based on the observation and accumulated experience, no reference could be derived that reflects on the emotional aspects leading to incidents or events. Hence, this factor forms an integral part of the reference human model and has been assigned a failure rate of 1×10^{-9}. However, the uncertainty associated with this factor is relatively large, as there is direct reference available.

7.7.2.5 Organizational Factors

In this section, two factors, viz. human resource management, training quality, plant man–machine interface quality and plant overall risk consciousness, form the major operational performance influencing factors. Some of the features of human resource management, such as recruitment quality level, the organizational provisions for ensuring emotional security of staff, dependability of staff on salary, facility, perks and promotion policy, have been discussed. There should be an independent provision to ensure that the training adequacy level is overseed and to indicate the gap areas. There are some provisions based on which the performance of training and qualification provisions can be assessed. For example, adequacy of resources, such as periodic recruitment of new staff, the availability of a training section that offers plant documents, adequate and qualified staff that offers lectures and conduct checklist, schedule of plant visit periodically and hands-on training, organization of meeting and seminars for exposure on technical aspects as also incidents and events. The availability of a full-scope or simulator with a capability to create sessions, as for as much as possible on the emergency operating procedures, power maneuvering, operating systems and the recovering the plant due to failure of process or safety equipment. Training on maintenance management schemes and their role in plant safety can be offered in the plant as well as in the simulator environment.

7.7.2.6 Consciousness Levels

For the purpose of the reference human reliability model, the high level of consciousness that corresponds to a high level of alert and awareness that support the focusing and concentration of an individual has been assumed. Even though it can be argued, the low level of consciousness, induced by either, due to the repetitive nature of activity, physical and mental tired minds, environmental distraction due to harsh conditions, operational due to operational uncertainties, etc., is one of the major causes of human error. Based on the expert opinion, a value of $8 \times 10-^{9}$/hr. has been considered.

7.7.2.7 Conscience Qualities

For modeling the reference man, the insights available from general literature and the known areas, it was concluded that the quality attributes, such as moral, ethics, honesty integrity and conditioned attitudes, are rarest of rare based on the experience; hence, a perfect '1 was allocated for modeling a reference man. The second reason is that the factor associated with conscience has been given the highest priority, as it affects not only safety but also security issues, and this aspect has been apportioned in the final CQB model for human reliability. In fact, in many organizations, these attributes form part of performance assessment, which is treated as confidential. However, an internal risk, security and reliability assessment can effectively use these data for its end use limiting to the organization itself.

7.7.2.8 Cognitive Factors

Cognitive factors are central to human reliability analysis, as these factors dominate the results of the final estimate. As the procedure for human reliability, generally, the human nonperformance probability or human error model has two components, viz. Evaluation of Cognitive part and the Action Part. There are other approaches that consider the action component as integral to the cognitive component, and no separate evaluation is performed. The reference human model treats the cognitive component in relatively more detail by analyzing the individual faculties that affect cognitive performance. This includes the central nervous system, peripheral nervous system, associated building blocks or organs of these systems and cognitive processes involved in action parts, as seen in the fault tree. The major challenge in quantifying the individual basic or transfer events. The expert opinion and general insights on the periodicity of safety related along with general health care insight into these building blocks affecting human performance show that the probabilities of these faculties affecting human performance during normal operations are rare for an age group between 20 and 50 years with no comorbidities. Hence, based on their functions and expert opinion, the lowest failure rate is in the range of 10^{-4} and 10^{-9} /hr in the quantification of the reference human model.

7.7.3 *Other Undeveloped Events*

Human qualities such as fear, greed, physical and mental pain, susceptibility to distraction and hurt are apportioned under the general wellbeing psychological attributes. The incident root cause analysis often reveals these factors, which is why a quantification for some situations can be performed. There are certain aspects where even the neuroscience is working hard to understand the exact science and mechanism of phenomena such as sensory communication and various protocols, the way the chemicals or electrical signals generate, affect and fail, and the theory behind these aspects.

Why some have self-awareness, motivation and individual consciousness and in the same way why these attributes are different from organization to organization are very important but create reference human model, no direct reference on these attributes. Here, expert opinion and engineering insights have been used. The procedure involves identifying the individuals who are ranking very high with respect to these qualities and others who are very low. The same way identifies the organizations and ranks these organizations and uses a fuzzy approach to infer from the qualitative data.

7.7.4 Evaluation of Failure Probability Associated with Emergency Operating Procedures

The evaluation of the reference human error probability has two major applications. The first is the evaluation of procedural steps in support of the assessment of human reliability as part of the evaluation of plant emergency operating procedures (EOP), and the second is the evaluation of human error probability not scheduled in the plant emergency procedural steps. Here, the case of maintenance management or routine tasks that lead to unsafe or reduced reliability of plants, such as plant trips and failure of equipment, falls in the category of accident precursor events (APE).

7.7.5 Quantification of Stimuli for EOPs

In human error probability for procedural steps, the evaluation of Stimuli involved in each step provides the failure probability for the individual step. The aggregation of all the steps involved in that segment of a procedure provides an assessment of human error probability in the section or segment of the procedure. Finally, the final statement of human error for a given emergency procedure can be obtained by considering the eligible quantified figures in each step.

The procedure for failure of probability of an individual stimulus can be estimated by using the CQB model by considering the reference value of the stimuli, and if the knowledge base permits, its constituent components are modified by the operational performance influencing factors. For example, the value of the failure rate of the sense base Eye or Stimuli 'Sight' or 'Vision' can be estimated by plant records, simulator experiments or expert opinion. This reference figure can be modified by the applicable Operational Performance Influencing Factors (OPIF). If there are four notes in a step, then the failure probability for the step is the summation of all four stimuli failures toward estimating the failure probability in each step. Summation of all the steps in a task gives the probability of the corresponding task.

7.7.6 *Human Error Probability for a Precursor Event*

The precursor events are characterized by normal plant operation characterized mainly as routine tasks. Here, there are also two categories, viz. tasks that might have the scheduled timeline for completion of the task, but the available time window is either relatively large or does not have serious implications if the preset timeline is lost. Second, the category might include the procedures that, although routine due to systems safety/risk concerns, require completion of the jobs in a preset timeline; otherwise, it has safety implications or might call emergency actions such as fueling operations either on-power or off-power, where the maintenance and control of fuel/clad temperature is a safety concern even though the job is categorized as a routine task. In such cases, a precursor even analysis approach requires considerations of cognitive stress and level of consciousness that are critical to successful completion of the task. In the proceeding case that is a routine precursor task, the reference human error evaluation with random failures may work, but for routine events where there are safety constraints requires consideration of stress levels applicable for the various procedural steps involved in the completion of the task.

7.7.7 *Evaluation of Unavailability Stimuli for Reference Humans*

Each stimuli evaluation requires modeling for two major parts: the reference component and the second modifier. The reference component is required for unavailability evaluation for a reference human, while the modifying factors include qualifiers for ideal (as applicable to reference human as also deviation from ideal to lower grades).

We know that what is the control room for the plant, the brain is for the human being. All the consciousness and cognition components of the brain define the intelligence level and are central to all operational and maintenance activities. General awareness, alertness, analysis and logic processing often require parallel processing in support of general and technical management that includes leadership, coordination, communication and decision making. Therefore, before we start working on the individual stimuli evaluation procedure, we will discuss the brain, i.e., its structure, different faculties responsible for respective functions and finally coordination among and with the other part of the body through the nervous system network. Before we discuss intelligence and the causes responsible for supporting and affecting intelligence, we briefly discuss the structure of the brain and the faculties responsible for the tasks performed by the brain (Figs. 7.7, 7.8, 7.9, 7.10, 7.11, 7.12, 7.13, 7.14, 7.15, 7.16, 7.17, 7.18, 7.19, 7.20, 7.21, 7.22, 7.23, 7.24, 7.25, 7.26, 7.27 and 7.28).

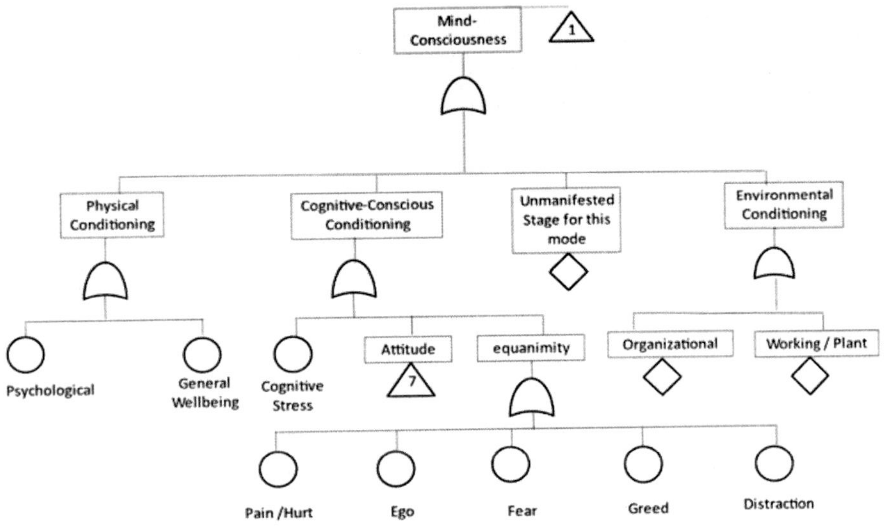

Fig. 7.7 Reference human fault tree—Continued (mind consciousness)

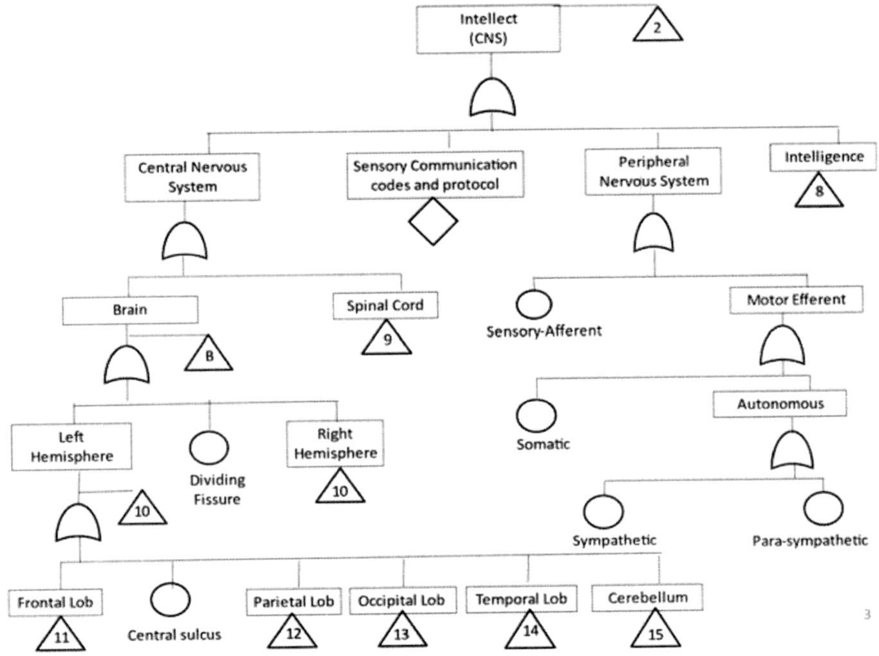

Fig. 7.8 Reference human fault tree—Continued (intellect)

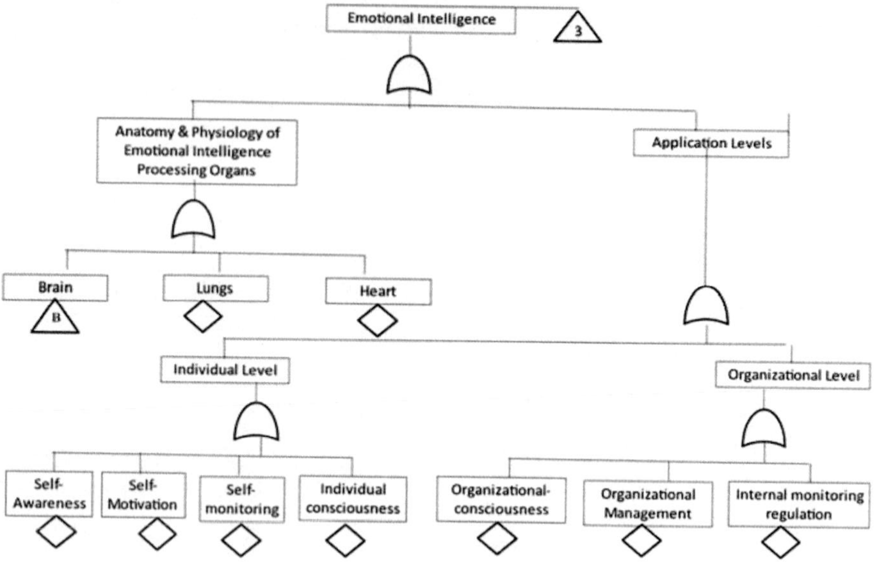

Fig. 7.9 Reference human fault tree—Continued (emotional intelligence)

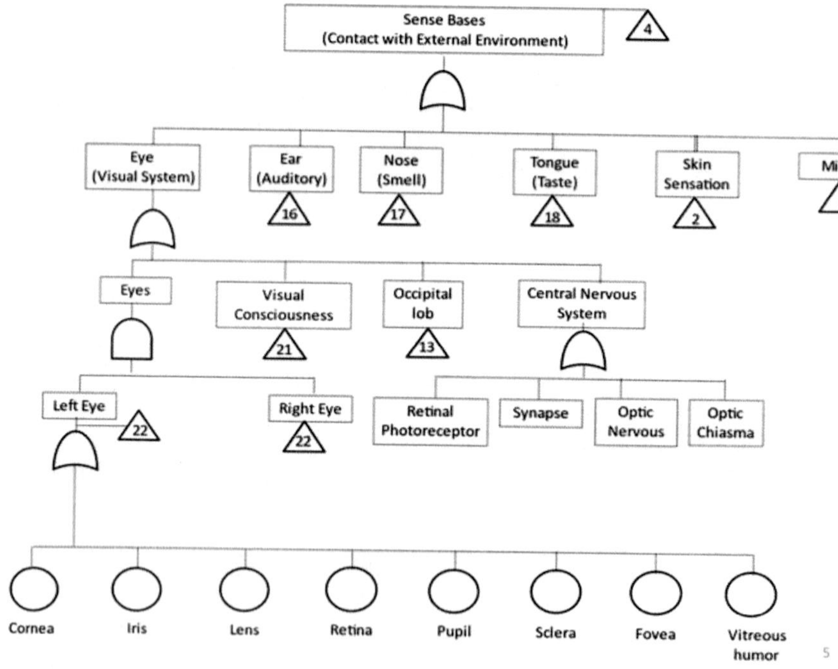

Fig. 7.10 Reference human fault tree—Continued (sense bases and eye)

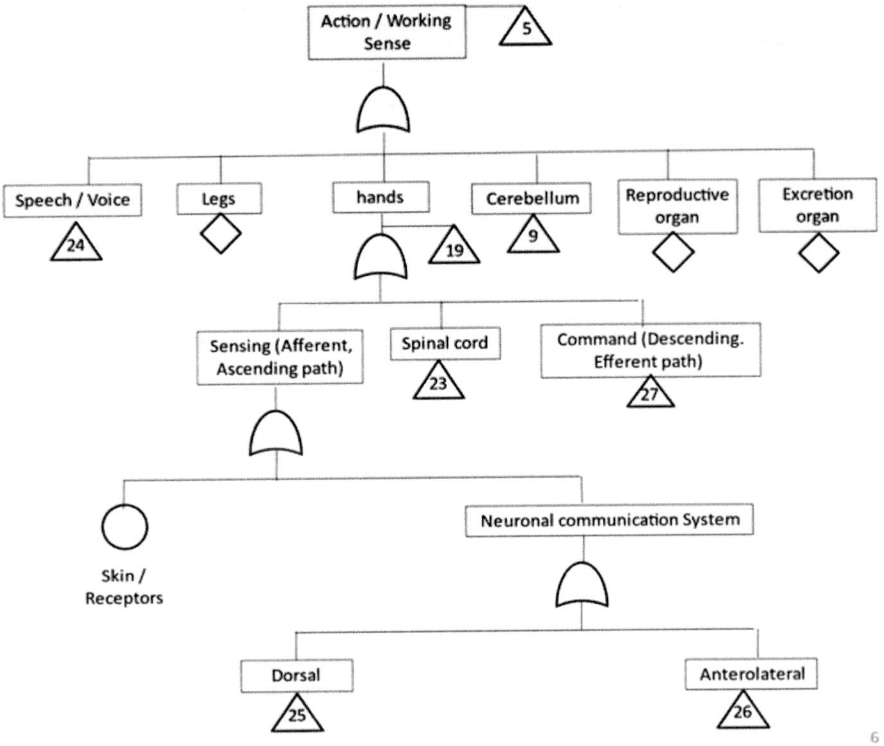

Fig. 7.11 Reference human fault tree—Continued (action/working sense bases)

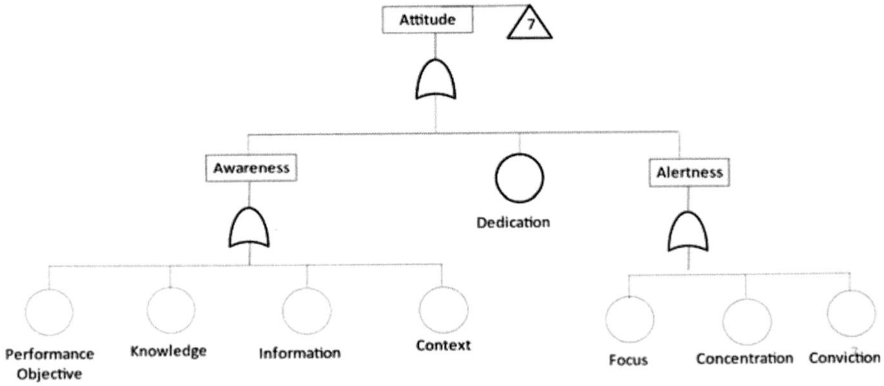

Fig. 7.12 Reference human fault tree—Continued (attitude)

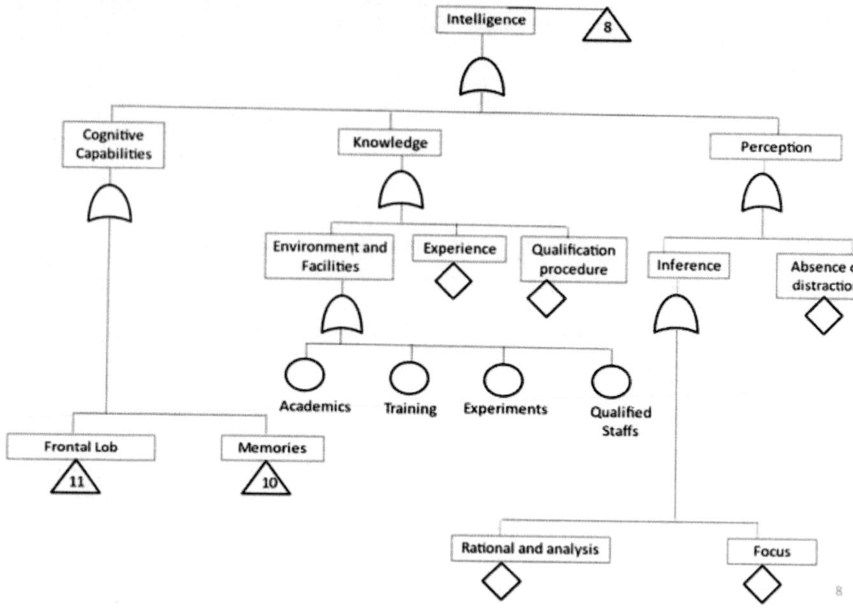

Fig. 7.13 Reference human fault tree—Continued (intelligence)

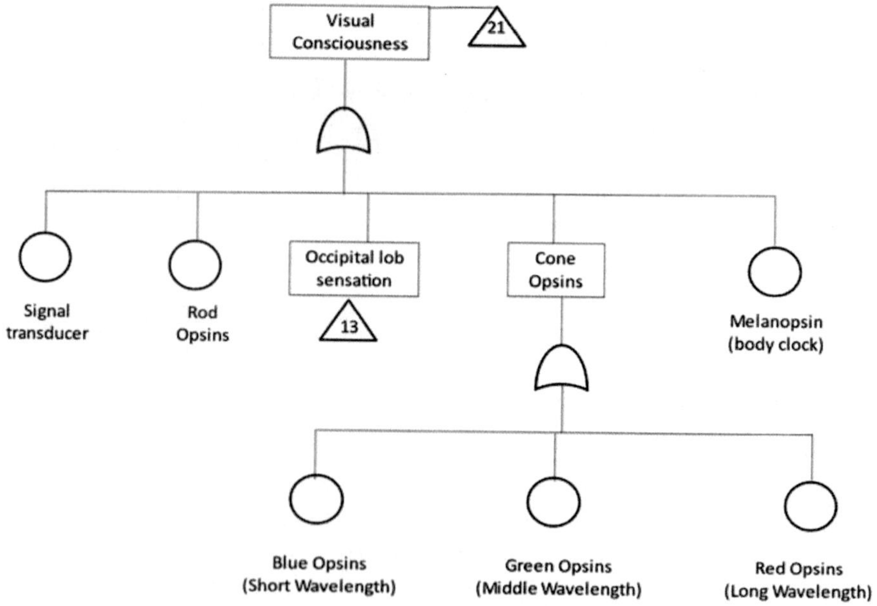

Fig. 7.14 Reference human fault tree—Continued (visual consciousness)

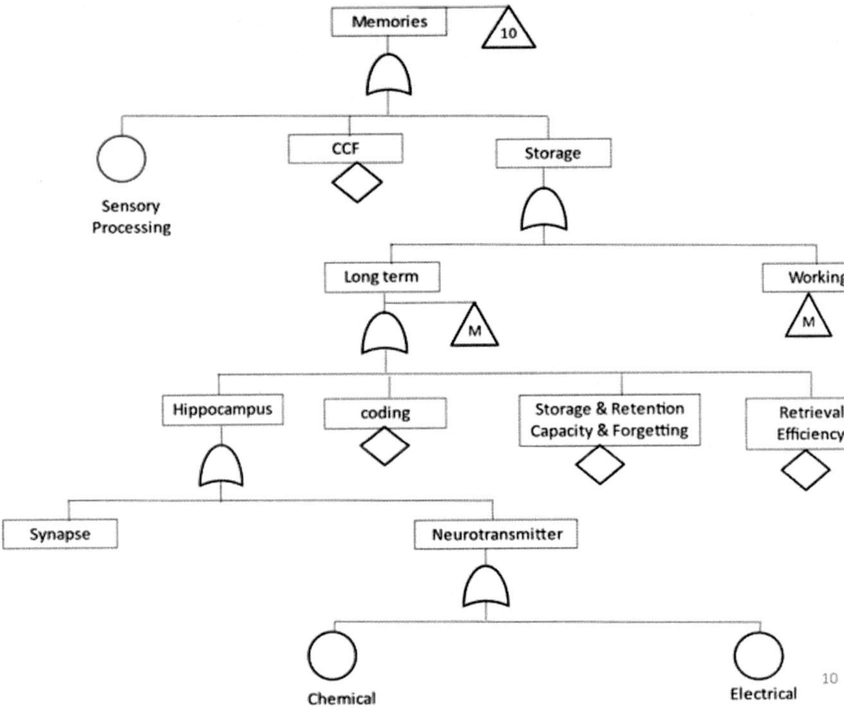

Fig. 7.15 Reference human fault tree—Continued (memories)

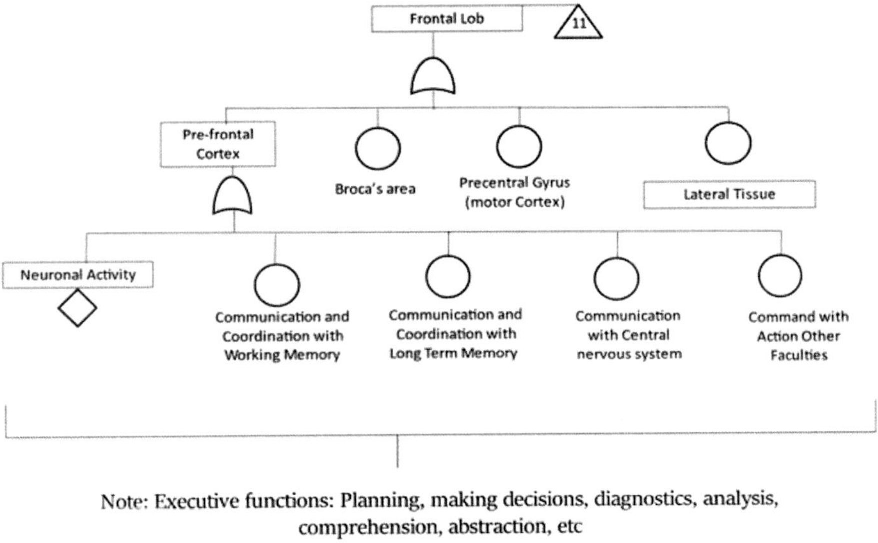

Fig. 7.16 Reference human fault tree—Continued (Frontal lob)

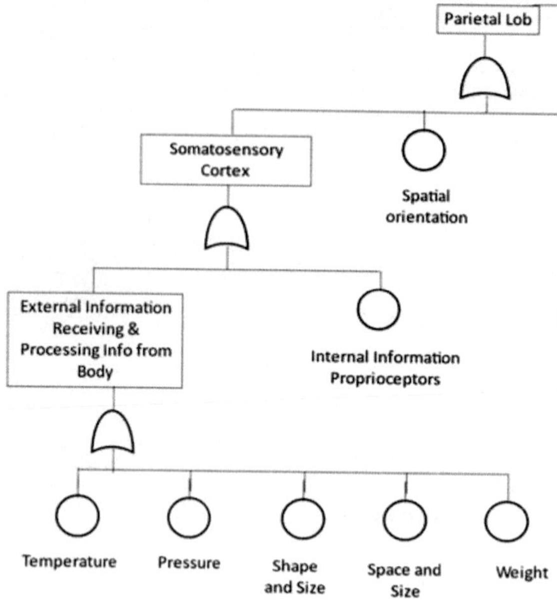

Fig. 7.17 Reference human fault tree—Continued (parietal lob)

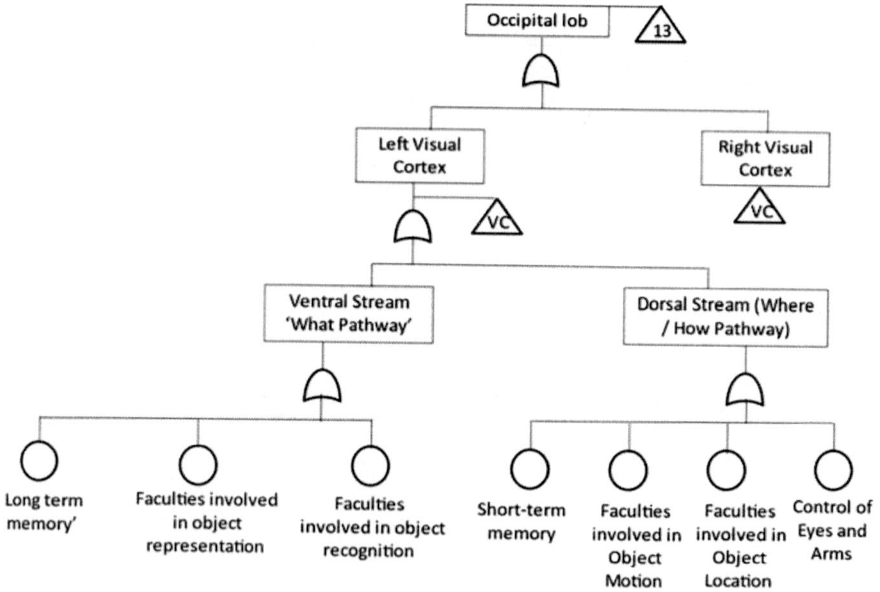

Fig. 7.18 Reference human fault tree—Continued (occipital lob)

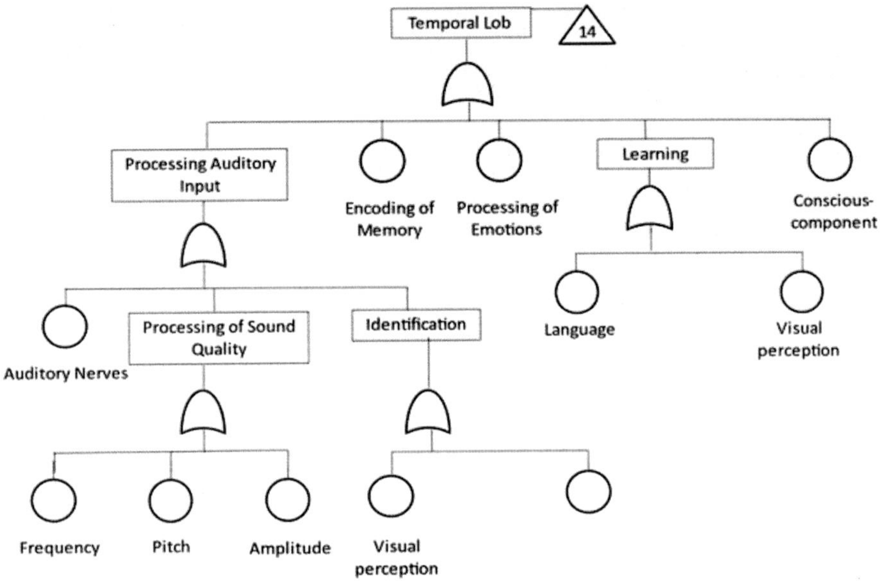

Fig. 7.19 Reference human fault tree—Continued (temporal lob)

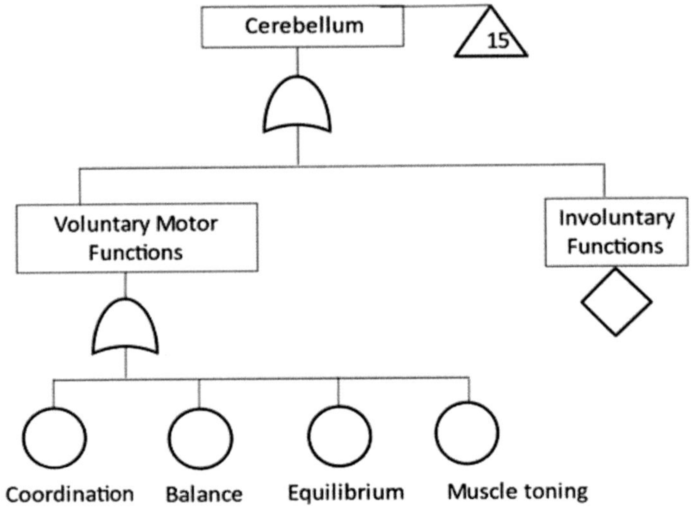

Fig. 7.20 Reference human fault tree—Continued (cerebellum)

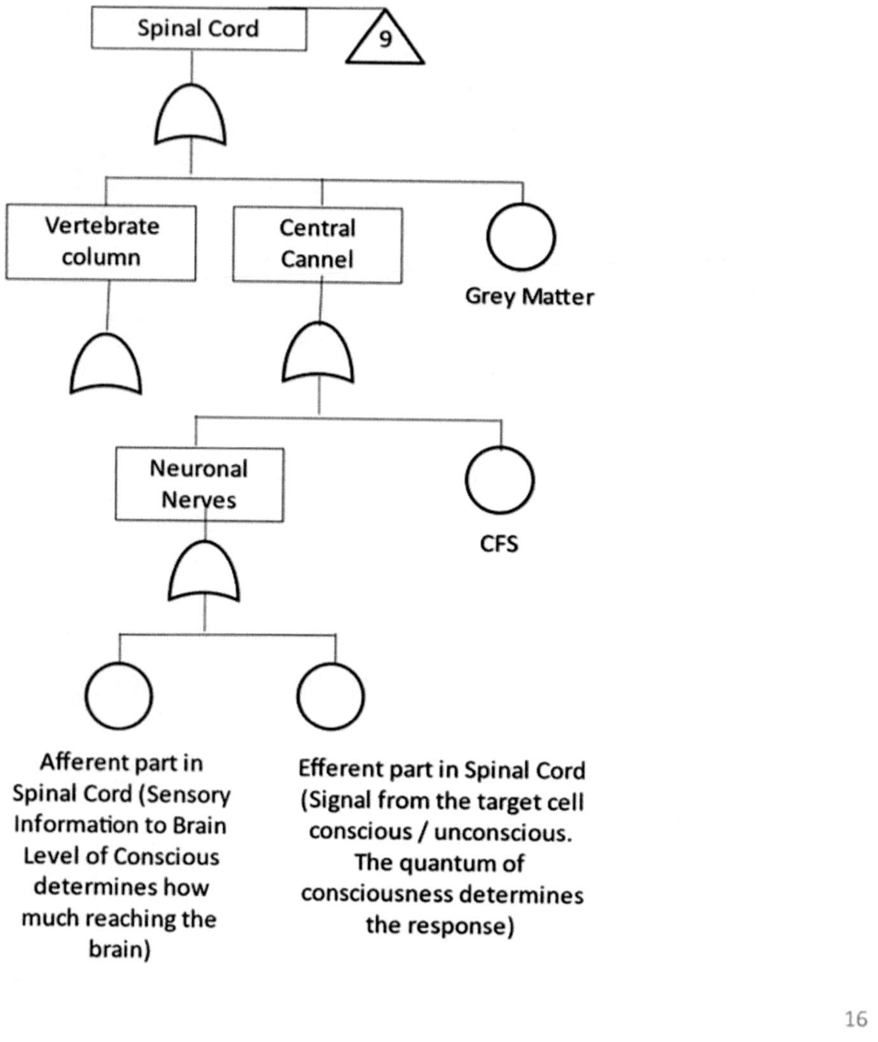

Fig. 7.21 Reference human fault tree—Continued (spinal cord)

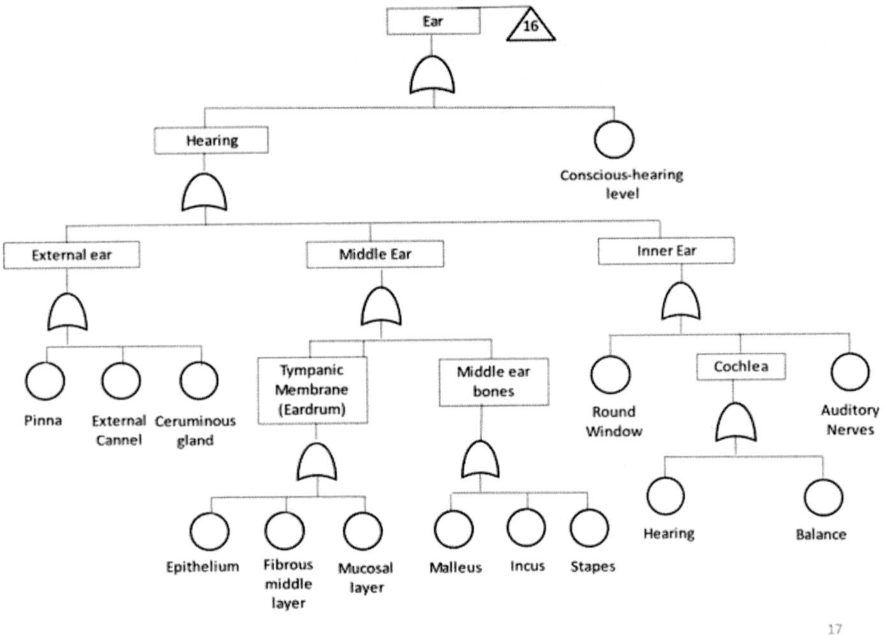

Fig. 7.22 Reference human fault tree—Continued (sense-organ—ear)

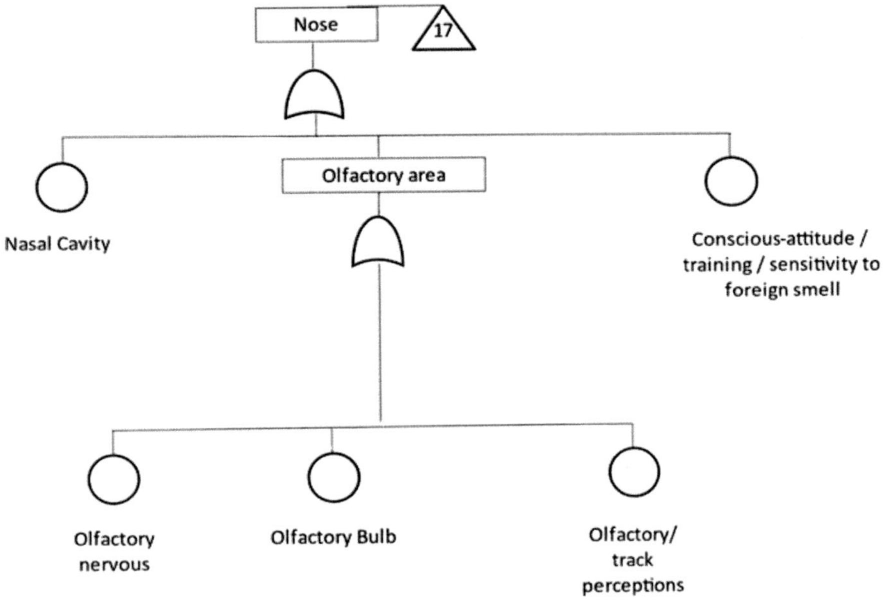

Fig. 7.23 Reference human fault tree—Continued (sense-organ—nose)

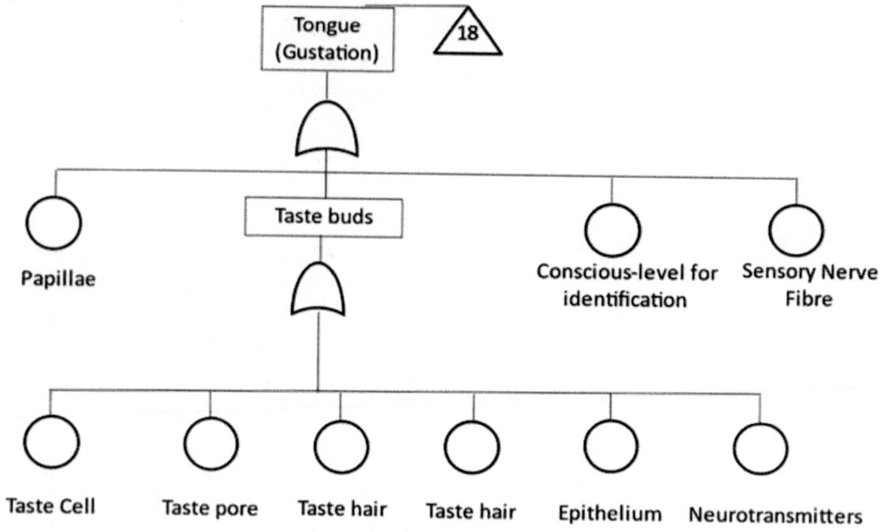

Fig. 7.24 Reference human fault tree—continued (sense-organ—tongue: Gustation)

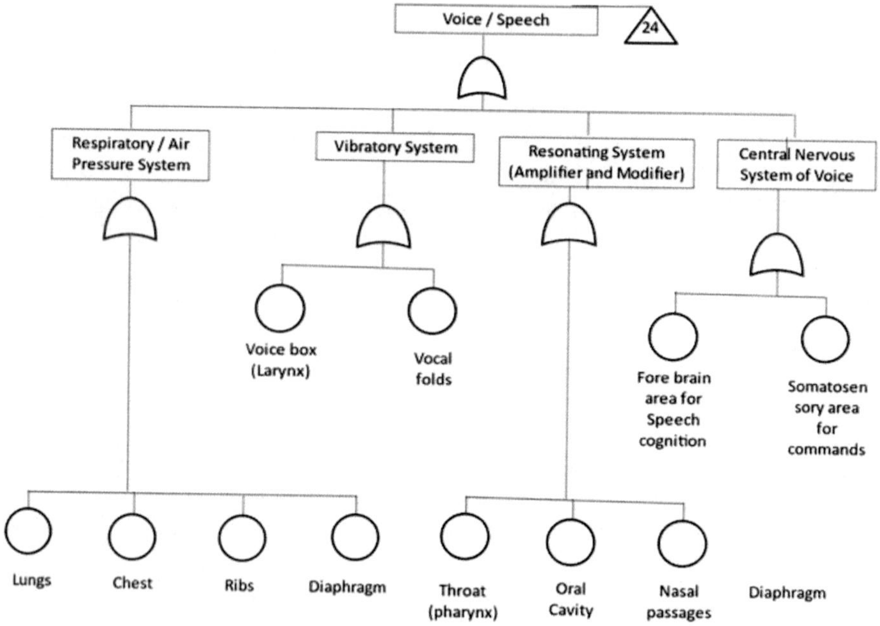

Fig. 7.25 Reference human fault tree—continued (sense-organ—tongue)

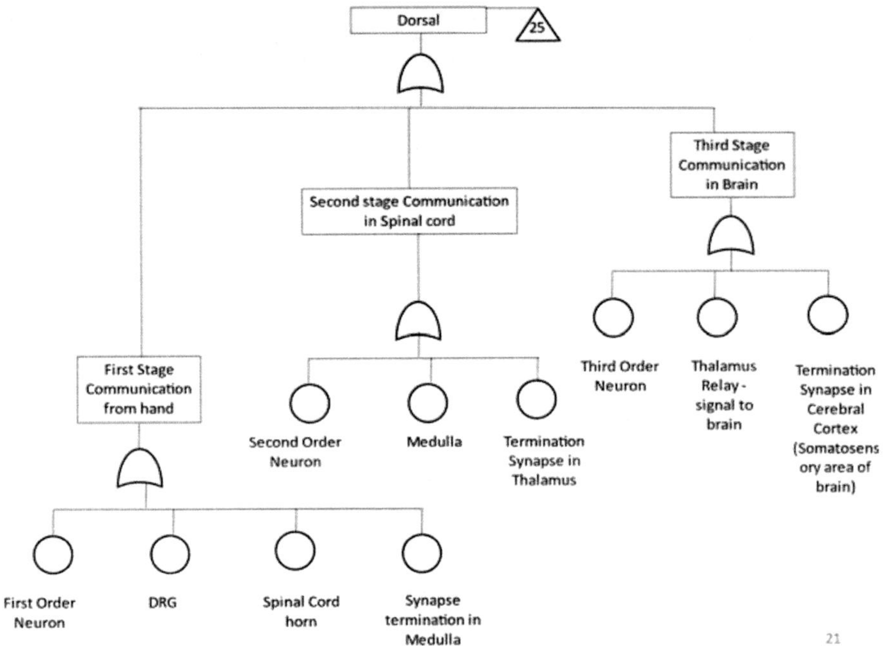

Fig. 7.26 Reference human fault tree—Continued (dorsal region neuronal communication)

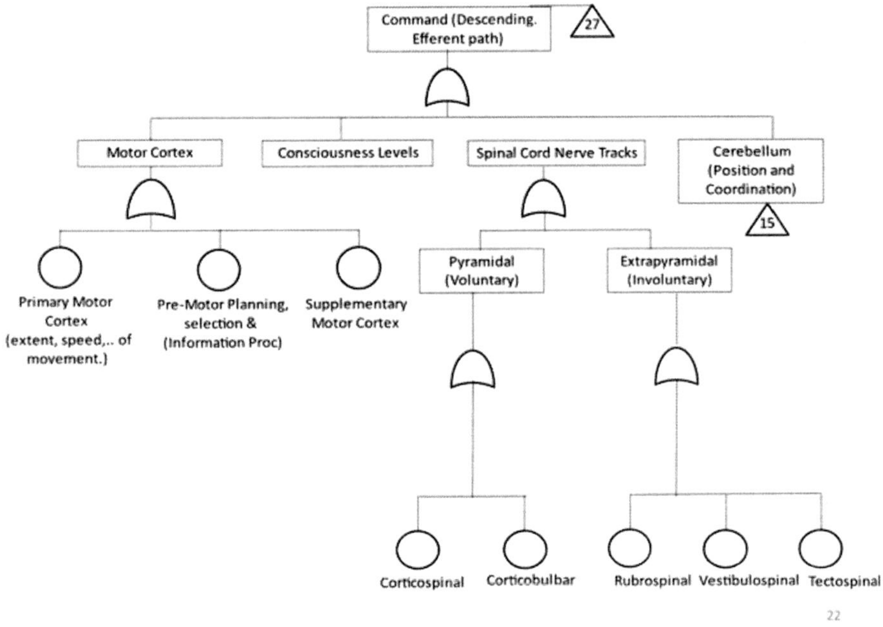

Fig. 7.27 Reference human fault tree—continued (descending path)

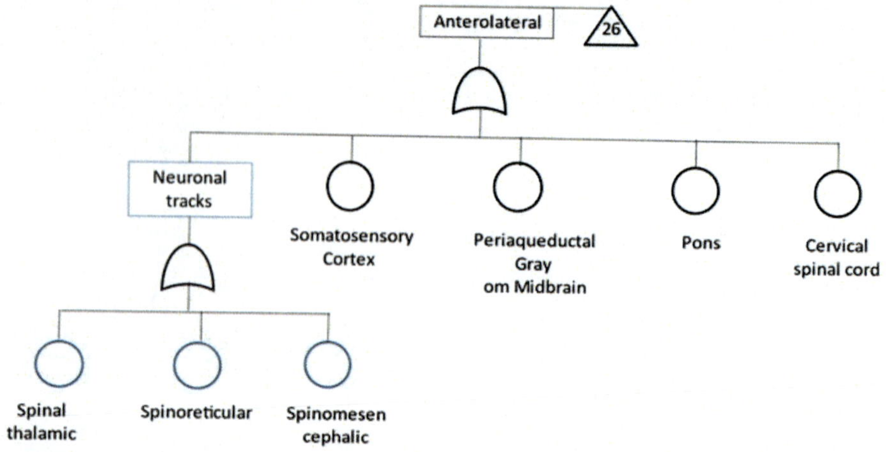

Fig. 7.28 Reference human fault tree—continued (anterolateral path)

7.8 Modeling of Sense Bases

7.8.1 General Features and Postulations

This section is an extension or updating of the CQB model proposed by Varde and Pecht. While performing a task, it is required that sense bases work in a cohesive and integrated manner with a high consciousness quotient such that the higher reliability of sense bases can be realized. In fact, qualities such as concentration, focus and even attention be related to consciousness realized through the sense bases, in CQB referred through Conscious-seeing and Conscious-hearing. It is recognized fact that the musical notes and their combination, when played/produced by an expert singer/musician, create a very harmonized and pleasing music that we all enjoy. In the CQB approach, when the sense bases, including the mind and action organs, work harmoniously, improved results are produced, be it supervision, execution of procedures or a maintenance task. A function of a group of sense bases is called an action, and a group of actions is referred to as a task and a group of tasks as a procedure. Here, any task completed successfully, particularly during challenging conditions such as emergencies, produces the same effect, a musical song or symphony.

The musical composition of rhythmic notes arranged in a certain way provides an effective mechanism for communication that is not only effective at the level of memory but also works at the consciousness level, where it inspires motivation and positive attitudes such as faith, compassion and other human qualities. Like a section of musical composition requires an appropriate combination of each note, in RCOM, a combination of one or more than one stimulus for the sense and action component constitutes a task. It is recognized that the rhythm, tone, ragas and musical component associated with the song and poetry works better for long-term memory and motivation in day to day as well as during challenging times compared to simple phrases and scripts. For example, Vedas or Shrutis (listening and reciting) were passed on from one generation orally. It was only later that all the stanzas were scripted. The adequacy of human action is a function of five elements, viz. (a) resolution: captures intensity, motivation, understanding to perform the task, (b) human the performer, (c) Six Sense bases, (d) objective and purposes and (e) the goodness/passion attribute to support decision making. There are three modes of human level, viz. goodness, passion and ignorance; goodness—dominated by higher self-awareness, attitude to work in the interest of organization and the nation and for benefit of mankind and an attitude for working on attaining higher spiritual qualities; second, passion—manifests as material qualities, such as administration and management, passion for work, ego and attitude that often involve attachment to power an urge to become a successful inherent weakness. Passion for work or any positive objective is good and benefits not only the self but also the group or society at large that come along like false ego, self-centeredness, lack of equanimity; and the last or ignorance—lower qualities, e.g., lack of knowledge, low alertness and awareness levels, lack of focus and concentration not fit for the job, and further, there might be challenges in terms of morals and ethics. For the subject area, i.e., the plant operational environment, we consider that most of the operations are performed in 'passion' mode, as this mode is appropriate and applicable, as the material component is the dominant component in this era of material, while the goodness mode requires perfection not only in the material sense but also in the spiritual sense. Therefore, a higher quotient of goodness and the lowest quotient of ignorance are at the center for judging the staff for the operational environment. One of these reasons might be the candidate causes for human error.

The objective here is to create action and a group of tasks (human action), which in turn will create grounds for human performance/reliability evaluation. Plant operational heuristics and knowledge can easily be captured by stanzas and poetic compositions for better learning and execution.

As discussed in Sect. 7.3 in the CQB, the 7 conscious stimuli are equated with 7 musical notes to create a system for the procedural part of reliability analysis in an operational ecosystem. The 7 conscious stimuli along with 2 action organs, viz. Hand and Feet, directly relevant to CQB are central to the human reliability model. These sense stimuli (sense bases) that connect a human to an external word are Sight (eye), Speech (Mouth), Hearing (ear), Smelling (nose), Taste (Tongue), Touch (Skin) and Thinking (Mind). We create a parity here with the 7 musical notes (Sa), (Re), (Ga), (Ma), (Pa), (Dh) and (Ni) in line with (Western musical system musical notes Do, Re, Me ....).

We consider the following devotional song in Marathi (an Indian language communication in Maharashtra) and its notes as follows:

The first stanza of the Abhang (devotional/Spiritual Song) goes like this (Fig. 7.29).

This devotional song is based on one of the many Indian Raga families called *'Bageshwari'*. Now, we know that the above devotional song comprised a different combination of 7 notes played with perfection in terms of a particular combination

We consider the following devotional song in Marathi (an Indian language communication in Maharashtra) and its notes as follows:

The first stanza of the Abhang (devotional/Spiritual Song) goes like this

---

"Majha Bhav Tujhe Charani,

Tughe Roop Maza Nayani"

The Musical Notes for the above stanza are:

Dha - (Ni)(Sa) - (Ni)(Dha)(Dha) - (Pa)(Dha) - (Ni)(Dha) -(Ma) - (Ga)(Re) - (Ma)(Ma)(Ma)

(Ma)(Ma) – (Ga) (Re)(Ni)(Re) – (Ga)(Ma) – (Ga)(Re)(Sa) – (Re) (Ma) – (Pa)(Dha)

The meaning of the devotional song is as follows:

'I dedicate myself to your lotus feet,
And bear ever your face in my eye.

Meaning 'O Lord my devotion to you is the only purpose of my life and existence'
Motivation: 'In spirituality, one sees the almighty as part of self' and dedicates all his attention
Inward to the self

Fig. 7.29 Musical notes in a devotional song composed by Sant Tukaram—a devotee to God

and sequence of notes arranged in a group with the required timing and gap, while some notes need more time, while the other notes are instantaneous or jumping notes on musical instruments. In short, when these notations are played properly by expert musicians, then it is considered a perfect job composition played perfectly on the instrument.

Similarly, a task having more than one subtask or procedure also comprised of generally more than one division or steps, each step involves one or more than one individual action involving a stimulus (notes) and the quality of each note in terms of timing, time gap determined by the perfection of knowledge of each step and the environment determines the quality and finally success of the job. As notations form, human expression based on their position in the row might comprise one to seven nodes. For example, if a human task requires only visual observation for success of the job.

The operating procedure, also at the conscious level, comprised individual tasks. In the Sargam model, the combination of notation also creates individual tasks, and the combination of tasks creates a procedure. The procedure is called success when all the individual tasks are performed according to the expectation of the task, and any deviation from the expected template can be considered a mistake.

Furthermore, processing of the information collected by the stimuli is processed in the brain through neural transmission. The brain faculties or cognitive faculties

Table 7.1 Adoption of the CQB approach for sense bases for input from the external world

S.N	Sense base	Stimuli	Consciousness-attribute	Stimuli notation
1.	Eye	Vision	Conscious-Seeing	CSe
2.	Ear	Sound	Conscious-Listening	CLi
3.	Nose	Smell	Conscious-Smelling	CSm
4.	Tongue/Mouth	Taste	Conscious-Taste	CTa
5.	Skin/Body	Touch	Conscious-Sensation	CSn
6.	Mind/Brain	Thought/logic	Conscious cognition	CCo

together, depending on the faculty requirements, work like a transfer function and covert the information into experience or formation or perception or understanding about the job that needs to be performed.

The major brain faculties are the frontal lobe, parietal lob, temporal lobe and occipital lob. The frontal lobe is involved in intelligent thoughts, planning, sense of consequences and rationalization. The occipital lob is responsible mainly for processing vision, while the temporal lobe processes auditory and memory and learning. The primary somatosensory cortex on the post central gyrus and Wernicke area is a general interpretive area. These lobs are responsible for imagination and give sense of time and space. The brain is composed of 100 billion neurons. The central nervous system comprises the brain and spinal cord, which provide a two-way sensory communication network between the brain and the rest of the human body.

Table 7.1 shows the sense sense-based, associated stimuli, conscious attributes and their notation or code as we discussed through the musical modes.

To the extent possible, the human model has been postulated to be similar to any other complex engineering systems; for example, a complex system such as a nuclear plant has a central control room that receives information from a complex set of networks that is processed at different electronic or electrical systems and presented to the operator. Some controls operate in closed feedback loops, such as autonomous systems in humans. Another mode of control is affected by the operator by affecting the suitable corrective action/activities. In humans, there are two distinct components, similar to engineering systems. One is sense bases and a network of information communication to and from the brain that interact with the external world as well as internal feel and autonomous activities. The second component involves physical or mental actions. These actions are enabled by five working/action senses to physically affect the change envisioned at the level of brain/body and provide feedback about the changes after action to the brain through external sense bases. The five action/working senses are (Table 7.2).

The autonomous system of the brain involves the somatosensory area and the cerebellum, which ensures balance and equilibrium of the action organs listed above through the central and peripheral nervous systems. The conscious and motivated involvement of action sense organs is required to perform a task by performing physical action/movement or the use of energy. Cognition and consciousness are

Table 7.2 Five external action organs that affect action or change in the external world

S.N	Sense bases for action/Work	Stimuli	Conscious-attribute	Stimuli notation
1	Tongue/Mouth	Speech	Conscious-Speech	CSp
2	Hands	Action	Conscious-Action	CAc
3	Legs	Movement/Standing	Conscious-movement and standing	CMo
4	Genital[a]	Reproduction	Conscious-Offspring	COe
5	Anus[a]	Body waste discharge	Conscious-Waste discharge	CWa

Note [a] The scope of the CQB procedure requires modeling items 1, 2 and 3, i.e., modeling the tongue for speech or communication, the hand for implementing the required physical action and the legs for movement. The last two items listed at 4 and 5, i.e., genital and anus, respectively, may not form part of the analysis. Furthermore, these items might be required for behavior modeling factors that affect job quality, such as distraction and wellbeing

at work while performing the physical action and ensuring that it was performed successfully.

7.8.2 *Task Modeling*

The broad architecture for implementation of an operations procedure is given in Fig. 7.30. The procedural hierarchy where task and further down individual steps forms the part of procedure. When these task descriptions as stipulated in the procedural sequential or parallel steps along with interaction with the plant come in contact with the trained operator/human, a 'human–machine interface' is manifested. Furthermore, human actions (mental/physical) are based on the consciousness and cognition cognitive-consciousness level. The response of the plant available through the human–machine interface or direct surveillance of the plant decides the outcome of the task in a procedure.

A procedure can be viewed as a collection of one or more tasks, where one or more steps comprise a task, as shown in Fig. 7.30. Of course, the procedural steps can also be there in parallel. The RCOM approach has been demonstrated by modeling at the task level and the objective unavailability assessment of a task.

Suppose in a long operations *procedure,* there is a *task* comprised of 7 *steps* as follows:

Task description: This is a generic task that requires physical action organs, e.g., opening/closing a valve supported by sense bases during normal operation. This procedure provides reference unavailability estimates for the task with performance attributes considering normal stress level, state of the art man–machine interface and experienced and trained staff.

The process comprised the following seven steps:

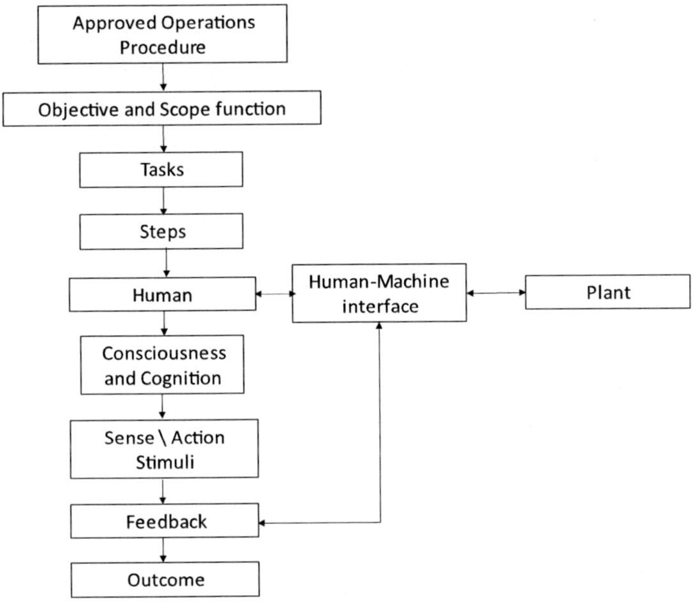

Fig. 7.30 Flow chart for implementation of the RCOM approach to task modeling

1. Taking note of evolving scenario/situation/context
2. Analyzing the situation
3. Define the objective and scope, and identify applicable plant procedure
4. Collecting additional information
5. Take physical corrective action
6. Obtain the feedback through audio, visual and further information collection
7. The result: Success/Safe or Failure (Plant trip or set back/Unsafe) and tag the task.

Step no 5 only deals predominantly with the physical action component along with reference levels of sense bases. The other steps require the involvement of a higher level of cognitive-consciousness levels for the sense bases. Nevertheless, all 7 steps are enabled by cognitive function supported by required consciousness levels. In a safety critical system, even though human actions are involved in most of the areas/locations in the plant, the control room is considered a place where most of the actions happen. Therefore, let us assume that the above steps are related to a scenario presented in the control room and that the operator takes note of audio/visual information available from the plant areas before he decides the course of action.

A typical control room scenario has been considered where trained authorized staff are posted to man the control room. The shift-charge engineer is responsible for decision making and activities required as part of management that includes diagnosis, analysis, coordination and communication. The Junior-Shift Engineer assists the Shift-Charge Engineer and performs coordination and communication function in the control room and with the other locations based on the instruction he gets from the In-charge. The final success of the job will be determined by the success desired response of the plant through control, logic or communication systems. In the following task modeling, it is assumed that (a) the man–machine interface, (b) the training of staff, (c) stress levels involved in the scenario and (d) the context meets the reference conditions or normal operations.

The above 7 steps can be characterized as shown in Table 7.3. It may be noted that this table is meant to illustrate this procedure only; hence, for specific tasks/procedures, the steps, level of details, objective function, maintenance and operational requirements and final feedback and long-term actions might differ. Furthermore, from the representation of the notation in each step, it is assumed that all the stimuli must function correctly for successful completion of the job in line with a musical system, where even if only note is played properly might adversely affect the quality of rendition. Hence, there are OR gate notation between each stimulus in a step.

Furthermore, each stimulus should exhibit the required quality of consciousness, e.g., concentration, focus or alertness, for adequate time such that intended results can be obtained. The wandering away of the mind of humans might affect focus, concentration and motivation. This requires, in simple words, consciousness and stability of the human minds involved in the job to be needed. The consciousness level and time metric should meet the required criteria or level for success of the job. Any deficiency in these two parameters might lead to failure of the task. The role of cognition is critical in the CQB procedure. The required level of cognition is determined by many factors, such as training, knowledge, reasoning and perception, and decision making. The importance of the notes is governed by the placement of the note in a group. For example, the note CCh assumes higher importance if it is first stimuli in a given step and further stimuli lower down the sequence. The modification of these stimuli will be governed again by the length of time and intensity of cognition required that will be affected by employing the modification factors in this group.

Let us consider Task 2, which addresses the analysis and exclusive cognitive activity, that is, how it has one stimulus, CCo (cognitive-conscious mind), with the highest weight, while other steps have more than one or even four stimuli with different weights. The tasks could include diagnostics, rationales, conscious planning, reasoning for understanding the situation, assessing the readiness of the task, awareness and alertness levels that call for conscious cognition at the level of mind.

Figure 7.31 shows the arrangement of the stimuli in a group that makes a step, and there are 7 steps that make the task in a given operations procedure.

The first line in Fig. 7.31 shows the arrangement of the complete procedure involving the requirements of stimuli in each step. The first group in the first line indicates the starting or initiation of the procedure, and the last group of notes in the second line shows the completion of the procedure. Furthermore, the combination of notes in the group follows the logic in the 'Success' domain. For example, the successful completion of the 1st step requires the success of individual notes as follows:

Table 7.3 Expression for the seven steps comprising the task for illustration purposes

Step	Step description	Human–Machine interaction	Characterization of human response in terms of the sense 6 and 5 action stimuli	Identification of notes/stimuli in Individual steps
1	Taking note of evolving scenario/situation	Shift-engineer takes note of the audio-visual, dynamic mimic and notes that there a reactor transient	The alertness and awareness level of the shift engineers determines the timely note of evolving scenario. This step is critical as early identification names the available time window for subsequent actions and normalization	(CLi)(CVi)(CCo)
2	Analyzing the situation	Performs diagnostics for the causes of the transient	Operators need to understand and identify the equipment/human action, or any external disturbance involved and accordingly plan the corrective action based on plant standard operating procedure. In conventional sense this task has knowledge base and rule-base component toward understanding the scenario to understand the direct causes of the scenario, diagnostics, logic processing, planning, etc	(CCo)
3	Define the objective and scope of the human action	To take the corrective action and identify the faulty component toward ensuring plant safety, reliability and availability through standard operating procedure that also involves action not only in control room but also in the area	Based on his assessment and analysis the operator performs technical management and supervisory function to cater to final objectives of ensuring reliability, availability but the safety remains with risk minimization remains the overriding factor. Identify the equipment, its failure mode and causes toward determining course of planning and further actions	(CCo)(CVi)

(continued)

Table 7.3 (continued)

Step	Step description	Human–Machine interaction	Characterization of human response in terms of the sense 6 and 5 action stimuli	Identification of notes/stimuli in Individual steps
4	Planning and collecting additional information	Collects additional information from control room and other areas of the plant in terms of available time and resources and actions needed. This also includes reference to standard operating procedure	The planning forms an integral part of any operational ecosystem. Keeping in view the objective function, it is required to retrieve the operational applicable authorization for the job, procedure and therein evaluation of each step such that postulated human actions and associated resources (this includes trained staff, tools, spares and gadgets) are in place also coordination and communication strategy is in place. Therefore, this step has knowledge-based plays predominant role. Even though other conscious stimuli might be involved, here, only mind and cognition are considered critical for successful completion of job	(CCo)(CSp)
5	Take physical corrective action	This might involve performing action in control room as well as in the plant site/areas. Coordination and Communication forms part of the actions	Handing over of the affected equipment following the applicable procedure to maintenance agencies. Identifying the right maintenance procedure its applicability to the current situation on hand, location, equipment, component, instrument, site evaluation, correct functioning of the tools and gadgets confirmation; determining correct action, required rule and skill for operation for local actions, that might involve some preliminary correction, adjustments, quick repair to long haul, replacement; installation of spare, alignment, checking from maintenance agencies, performance of operational procedures and bringing back the affected equipment to operational state and verifying the related parameters and reporting	(CSs)(CLi)(CVi)(CCo)

(continued)

Table 7.3 (continued)

Step	Step description	Human–Machine interaction	Characterization of human response in terms of the sense 6 and 5 action stimuli	Identification of notes/stimuli in Individual steps
6	Obtain the feedback through audio, visual and further information collection	Monitor the response of instrumentation and trips/alarms, process and digital instruments regarding variation in readings of the interest	Feedback on the performance of the affected equipment and system that includes status of equipment, parameters, correlation parameters with other interfacing systems and finally authorizing its operation to be OK or otherwise before initiating next procedure at higher level	(CCo)(CLi)(CVi)
7	The output is Cognitive Conscious Factor (CF) for the task	Ensures that plant returns to normal by ensuring the status of equipment and normalization of audio/visual indication and readings of the process parameters	Based on the operational checking and final qualification of the actions to be as intended, based on adequate checking in terms of time and for various modes, assigning the task as success and failure forms the final step. Recording of specific observation is a critical step for operational feedback forms a vital input for management or the operational team. It could be related to job analysis, revision of procedure, aging related issue, updating of the databases suitably, etc	(CCo)(CLi)(CVi)

Forward Steps

Step No	1st	2nd	3rd	4th
Notes:	(CLi)(CVi)(CCo) -	(CCo) -	(CCo)(CVi) -	(CCo)(CSp)

Resuming/Return/Steps

Step-No.	5th	6th	7th
Notes	(CLi)(CVi)(CSs) -	(CCo)(CLi)(CVi) -	(CCo)(CLi)(CVi)(CCo)

Fig. 7.31 The arrangement of notes in the group in the first and second lines shows the first part of the procedure in the forward direction, while the arrow in the second line indicates return toward normalization

$$S^1 = (\mathrm{CLi}) \cap (\mathrm{CVi}) \cap (\mathrm{CCo}) \tag{7.1}$$

Now, we assume the sensory consciousness factors denoted by 'SC'. Here, i denotes the individual ith sensory/action cognitive-consciousness intensity factor (IF). The generic representation at the group/step level in the success domain is as follows:

Let the sensory cognitive-consciousness factor be denoted by SC_i,

The success factor for the stimuli is given by

$$S^1 = \prod_i SC_i \tag{7.2}$$

However, the objective here is to evaluate the cognitive-consciousness factor (CF) factor for individual assembly toward arriving at CF for the step and eventually for the identified task in a given procedure as part of (IF) for the stimuli as part of evaluation. The effectiveness factor is a function of the importance ranking R_i and the weightage factor of the stimuli for the given step in a task.

$$\mathrm{CF} = f(R, W) \tag{7.3}$$

For example, the CF of $\underline{i}$th stimuli is given by

$$CF_i = (R_i * W_i) \tag{7.4}$$

For example, for the stimuli, the cognitive-consciousness factor for a given step comprising j numbers of stimuli is given as:

$$CF_{ij} = \sum_{i=1}^{j} CF_i \tag{7.5}$$

The CF for the task comprising k steps is given by

$$CF_T = \sum_{m=1}^{k} \sum_{i=1}^{j} CF_i \tag{7.6}$$

For example, for the case considered here, the task comprises 7 steps. Accordingly, the equation for the CF for this task can be written as follows:

$$CF_T = \sum_{m=1}^{7} \sum_{i=1}^{j} CFS_i \tag{7.7}$$

CFS is the Consciousness-Cognition Factor for the *i*th stimuli in an individual step. Additionally, if a step is represented by *S* and there are *k* steps, the equation for CF_T is given as follows:

$$CF_T = \sum_{m=1}^{7} S_k. \tag{7.8}$$

The equation for the CF for the given task is as follows:

$$\begin{aligned}&(\mathrm{CLi}) \cup (\mathrm{CVi}) \cup (\mathrm{CCo})] \cup [(\mathrm{CCo})] \cup [(\mathrm{CCo}) \cup [(\mathrm{CVi})] \cup [(\mathrm{CCo})(C\mathrm{Sp})]\}\cup \\ &\{[(\mathrm{CLi})(\mathrm{CVi})(\mathrm{CCo}) \cup (\mathrm{CCo})(\mathrm{CLi})(\mathrm{CVi}) \cup (\mathrm{CCo})(\mathrm{CLi})(\mathrm{CVi})(\mathrm{CCo})]\}\end{aligned} \tag{7.9}$$

There are a total of 11 sense/action bases, six sense bases and five action bases that facilitate human actions at the conscious-cognition level and affect physical actions. In RCOM, it has been assumed that any step in a step might directly require any of the four stimuli. It could be a combination of any four sense and action stimuli. This constraint is based on the experience in the field that given a procedural step, one stimulus is dominant, while other stimuli are supporting, and their role can be prioritized. This facilitates assigning the importance ranking of the assembly in a step. Furthermore, the weightage factor along with the importance ranking gives a CF estimate for a given step, as given in Eq. 7.4.

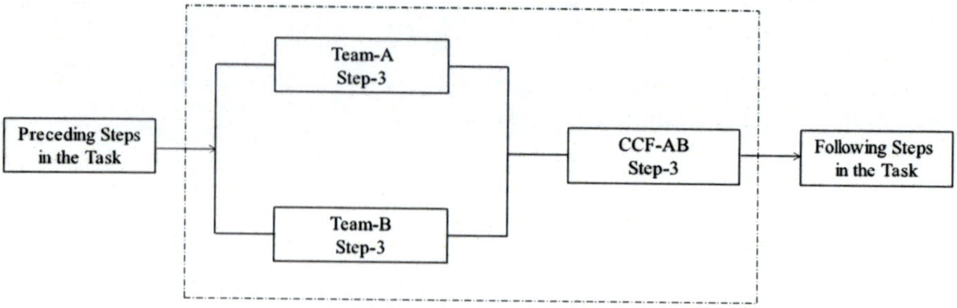

Fig. 7.32 Representation of redundant or backup provision for tasks in a procedural step

The procedural steps are assumed to follow a sequential or series model. However, in practice, many of the procedural steps that follow redundant/parallel provision can be referred to as recovery from mistake provisions during or after the completion of a task such that the failure probability can be reduced while meeting the targets of risk and reliability and the possibility of any adverse outcomes is reduced. Similar to hardware reliability, estimation procedural reliability can be evaluated for parallel actions or steps. Suppose there exists a redundant path for performing Step 3, as shown in Fig. 7.32, by having a provision of checking with the objective that mistakes in execution can be detected and corrected by an independent check performed by team B other than team A. Even though there are two independent teams in terms of human resources, tools, etc., there is a possibility of common cause failure (CCF) by both teams. Therefore, it will be more correct to factor in CCF in the modeling, as shown in Fig. 7.32. For demonstration, as mentioned above, Step-3 (CTh)(CVi) has been depicted considered.

7.8.3 Sense Base Characterization and Quantification

The modified CQB approach forms the basis for operational human reliability evaluation in RCOM. In the CQB approach, the human model is represented by Consciousness—a fundamental energy entity that defines alertness and awareness and emotional and happiness quotient, Cognition—that represents human intelligence that provides capability for rational, logical, memory and so on required for forming perception, analysis and decision making and finally the Conscience—an important attribute that captures ethics and moral behavior critical for safety and security. In CQB, it is assumed that all three Cs culminate and operate at the level of the brain and the human nervous system that uses sensory information from sense bases as well as the human body for input/output generation, which culminates in psychological as well as physiological and physical responses/actions.

There are two major components in human reliability quantification. Base or nominal human reliability for reference plant condition and second the modifying factors called human reliability influencing factors. There are two approaches for the evaluation of nominal human reliability, which is nothing but the human reliability for any action or response during the most ideal or optimistic conditions, e.g., when stress levels are low, the system-machine interface is the most advanced, the ideal organizational setup and environmental factors govern the condition of benign ground parameters. The second approach is the evaluation of human error probability for normal conditions employing ideal Cs, i.e., Consciousness level very high, cognition levels represented by intelligence quotient as very high and applicable for low stresses, and human and organizational conscience levels are withe very high or very low. All these summed up into brain and nervous system activities are normally inferred from the psychological and physiological health of the individual.

It has been assumed that this ideal CQB model tends to provide the reference or nominal human error probability in line or around the human error probability data collected during the reference reactor, i.e., normal reactor operation conditions, as mentioned above (Table 7.4).

These stimuli/notes are ranked according to their importance for the successful completion of the task. The most important note is called the primary note and has maximum weightage in the estimation of human performance. Subsequent notes are also assigned weightage considering their safety significance. The bracketed term is called the weightage factor.

The nodes are arranged according to the chronological order of their sequence in the task. It is required to designate the notes according to their importance in the task, viz. Primary, Secondary and so on. For example, in the control room, it is expected that the audio alarm draws the attention of the operator followed by a visual blinking signal. The sum of all weighting factors should not be more than 1.0.

For assigning the weights to any cognitive-consciousness factor, the condition is the addition of weight in step should follow the following rule:

Table 7.4 Task characterization using individual task Sense-base weightage-factor Conscious- and observed consciousness level (Fire Alarm)

Notes cognitive-conscious-importance ranking/Levels	Importance rank and associated weightage factor of stimuli in a task				
	(First Stimuli) $Wt =$ 0.5	(Second Stimuli) $W =$ 0.35	(Third) $Wt =$ 0.1	(Fourth) $Wt =$ 0.05	Ideal
VH (5)	√	√	–	√	1
H (4)	–	–	√	–	0
M (2)	–	–	–	–	0
L (0.5	–	–	–	–	0
VL ($\leq$0.1)	–	–	–	–	0

$$(W_S) = W_S \leq \sum_{i=1}^{n} (Wi) \leq 1.0$$

Equation 7.10 gives the cognitive-consciousness factor (CF) for a given step i comprising ≤4 *notes or stimuli.*

The RCOM procedure requires evaluation of the actual CF for a given step as well as the ideal CF as the ratio of the ideal CF_i and actual CF, CF_a, which is used for estimating the effective CF, CF_{eff}, for the given step as follows:

$$CF_{eff} = \frac{CF_i}{CF_a} \tag{7.10}$$

The value of ideal CF_i is obtained when all the stimuli have the best rating along with their applicable weightage factors, while the actual CF_a is obtained by using the real-time parameters, which is a subjective assessment. The ideal estimates of CF_i for a step comprising 4, 3, 2 and 1 are 5, 4.75, 3.21 and 2.5, respectively. The value of effective CF_{eff} is ≥ 1.0.

The above procedure is human-specific considerations. For example, if the reference value of the human failure rate for a given task is λr/hr;

Then, the updated failure rate λs/hr considering stimuli levels for the given step is given as follows:

$$\lambda_s = CF_{eff} * \lambda_r \tag{7.11}$$

The unique observation, as seen in the equation, is that the results reflect the modification of the reference failure rates employing the factor characteristics to human or human models. CF_{eff} accounts for the effectiveness of the sense or action organs. This step might be missing in existing human reliability modeling techniques. This equation implements the one of the governing considerations of RCOM that human should be treated in detail. Here, there is an opportunity to move little deeper to create a specific human model. To further augment the above discussion, the RCOM has the human root cause analysis (HRCA) concept, which allows one to go down to roots 6 stimuli and 3 major action organs that could become part of HRCA. The rest of the procedure involving modeling for stress, environment, etc., is discussed in the following section.

Furthermore, this model also allows failure of more than one stimulus in a step due to a common cause failure. For example, a lower consciousness level below a certain threshold might affect more than one stimulus. At the same time, one stimulus can form part of more than one task. Even though the nomenclature of the given stimuli is the same in all the tasks where it appears, the argument of Boolean reduction logic $A + A = A$ need not necessarily apply. One performance expectation of the same stimuli might be different. Therefore, if there is a requirement to use the Boolean reduction, then it should be supported or negated, as the case may be to have a realistic model of the procedure.

In an operations environment, it is not possible to have quantified data for the modeling of the real-time assessment cognitive-consciousness factor (CF). However, operations ecosystems have the advantage of having experts who can express their opinions on the effectiveness of real-time effectiveness of stimuli by linguistic expression. Additionally, the experience of HRCA can be used to understand the responsible stimuli for the failure and further down the associated roots that eventually led to the human error or failure.

7.9 Operational Performance Influencing Factors/Functions

In the CQB approach, the human reliability, apart from the core human model factor, i.e., Consciousness, cognition and consciousness play out at the level of the brain and through the nervous system in the physical body and are affected by human reliability influencing factors (HFIRs). These factors are referred to as performance-shaping factors or common performance conditions in other approaches. These factors have been represented into four categories: organizational, task complexity, system characteristics and environment.

Since numerical data are often not available, the use of operational expertise has been an effective approach to model these factors. In this context, the fuzzy logic approach provides an elegant approach to model the operating experience as most of the time it is required to use linguistic assessment for a given factor, for example, what is the adequacy of the system-machine interface and the opinion comes as excellent, very good and not good? The fuzzy approach is very effective in using this expert opinion.

The following section presents a few selected HFIR from the CQB model that are adequate to capture the operational aspects that affect human performance.

7.9.1 Organizational

Risk-consciousness levels in the organization, as per the RCOM approach, play a vital role in human reliability. There are four factors that have been considered: (a) training and qualification, (b) organizational responsibility, reporting, authority and communication hierarchy and (c) regulatory oversight and provisions.

The training and qualification programme affects the training and adequacy of the individual staff, and this aspect directly affects the performance of the staff not only immediately after training but throughout the length of the time by undergoing refresher training and requalification of the staff as long as he is part of the operating crew where his actions directly affect the plant risk levels. The clear definition of role and responsibility, line of authority, line of reporting and line of communication within the organization, particularly in a crew and at a higher level outside the organization, such as communication and reporting during emergency conditions and fuel management, are waste management tasks and are critical to smooth and safe or risk-conscious operation of the plant. The regulatory provisions mainly include reporting oversight, enforcement, compliance and authorization and thereby provide an effective mechanism toward ensuring adequacy of safety elements in the organization.

The quality levels and the factors that have been used for these factors are given in Table 7.1. Different quality levels have been determined employing stress and strength philosophy.

7.9.2 Task Characteristics

The task characteristics are primarily assessed in terms of parameters that enable modeling of postulated stress levels, complexity of job, operating state or condition of the plant or system that need attention and action. The availability of tools and fixtures, radiation-related tasks or normal area tasks, safety provisions in the plant, workplace or environment conditions, operational urgency and other interfacing parameters, such as quality or man–machine interfaces. Furthermore, the level of requirement of the knowledge base, rule base and skill approach and the training adequacy and coverage of the proposed job. Apart from this availability clearly written procedure or readily available procedure, the experience of the employee involved also characterizes/determines task characteristics. There could be many more factors, particularly during emergency conditions.

7.9.3 System

First, even when we discuss safety critical systems, different systems have different types of risk and potential consequences. The perception of risk and the risk management approach, for example, aviation, are different from those of nuclear systems. Even the chemical and process systems, if we investigate the totality as risk potential level systems are different, viz. the Bhopal Gas Tragedy involving the M/s Union Carbide plant, is a black given the number of deaths and long-term consequences to the member of public. Even though it is recognized that human error is generally a major contributor to accidents, human error in aviation has a special impact of human error, and most of the accidents are possibly attributed to accidents. If we look at the rail and road transport sector, we see that rail passenger and freight transport are more organized as a system compared to road transport. However, now in both sectors, the organizations have been moving and qualifying merely safety poster culture to design and operational changes, a superior level of training including deployment of simulators and deeper analysis of accidents and incidents to understand the root causes of failure, but in both sectors, furthermore, dedicated efforts are needed.

Even though even the nuclear industry, known for maintaining the highest standard of safety, has witnessed three major accidents, TMI, Chernobyl and Fukushima, many more rather small accidents have yet to learn and implement improvements and policies toward reducing accidents in general and major accidents. The unique aspect of nuclear plants is that any accident anywhere in the world impacts the industry. The initial reaction to the Fukushima accident occurred in Japan, but many other countries made the government sit back and review the policy.

This is where the implementation of risk-conscious culture not only in operations but at every step right since when the plant moves right from siting, to conceptual, design, commissioning and finally into operation, at every step improved understanding and recognizing the potential risk areas and consequent correction should be part of the risk-conscious culture. Of course, there are serious efforts going on world over to design inherently safe plants by ensuring passive safety features modern technology such as digital techniques and improved engineering and physics design. Here, there two important observations that need attention. It will take time to mature new concepts and technologies as well as the new failure modes and potential consequences of failure. This argument is applicable for third- and fourth-generation plants. By 2022, out of 436 operating nuclear plants and over 200 nuclear research reactors, most of the plants are based on Generation-I and Generation-II aspects. In fact, many plants as part of life extension from 40 to 60 years, incorporating new technologies, as has been evident in USA, i.e., of the ~100 plants, most of the plants have been authorized for 60 years operations. The unique observation is proposals are underway to further move from 60 to 80 years of life extension. However, the fact remains that we will have to live for many more years with older plants world over. Here, once we recognize that human factors are critical to the safety of the plant, the industry will have to take note of system shortcomings, particularly with the adoption of digital technology and its calibrated deployment. Reviewing operational policies such that the human factor is improved requires a paradigm shift. Some of the points, of course the list is not exhaustive but indicative one. The perception of training and qualification and its place in the organizational setup, incorporation of risk-informed/based decisions, concerns should not limit to reducing human error but go further down to understand the deeper aspects related to human, be it physical, physiological or psychological level at individual as well as at organizational level. One simple check will be the wellbeing happiness quotient of the organization, and individual members of the operations team might provide many insights that will point toward organizational strength and weaknesses. Furthermore, given that humans are fundamentally a spirit or consciousness, the aspects related to organizational consciousness could be an important factor in understanding the not that well-understood phenomenon of common cause failure (CCF). If we know that CCF is one big monster that eats away, all our defenses built are redundant and can be handled. This includes developing new sciences of understanding the CCF mechanism employing PoF models and not by rather superficial treatment that it gets employing the empirical models which are another source of uncertainty.

7.9.4 Environment

The environmental factor typically includes, with respect to situations such as the control room, generally maintaining benign ground conditions (temperature: 22 °C and relative humidity: 55–60%). Furthermore, plant area characterization includes these two parameters along with decibel level, approach and access to equipment and instrument panels. Floor conditions and general features, such as ground, basement, subbasements and substation conditions, also determine working conditions. Second, the context, such as normal operation, managing reactor operation recovery/restartup, or emergency management in the control room. If it is area activities, then maintenance activities, associated planning, availability of normal and emergency or protection equipment determine the environment.

If we consider the management environment, then the hierarchy, availability of documents, trend records, expertise for RCA, risk simulation, engineering services, provision for regulatory activities including management of compliance and communication with regulator, coordination with public, security and health physics including provision for communication and management with industrial safety and healthcare centers for speedy and effective recovery management for injuries and fatalities are integral to the environment (Table 7.5).

7.10 Quantification

7.10.1 General

There are two major approaches for quantification statistical approaches where the performance data on hardware, software and human aspects form inputs for the evaluation of the probability of an event. In many situations, the performance data are not available. In such cases, the inference regarding the likelihood of an event is based on qualitative information in the form of opinion or expert opinion. Here, the fuzzy logic approach is very effective for drawing inferences based on opinions. The quantitative models and methods to evaluate the probability of an event have been explained in Chap. 3 on dependability analysis. Furthermore, the approach to use quantified input to assess human reliability is well established; however, the salient features of the CQB approach based on the stress–strength model will be discussed here (Varde and Pecht 2018) for the same completeness of treatment.

Table 7.5 Human reliability influencing factors in terms of stress–strength characterization

HRIF	Strength parameters	Strength factors (Score)	Stress parameters	Stress factors (Score)
Organizational factors				
Training and qualification	• A system of ensuring highest efficiency in training and qualification • Hands-on training on design, operational and safety aspects particularly on off-normal and emergency scenario • Qualified and experienced training staff and faculties with a system of quality control of training activities • Simulator training on operational on power start-up/shutdown, power maneuvering, emergency operating procedures. Periodic upgrades of simulator • Periodically revised and updated, approved checklists, manuals, design drawings and availability of incident reports • An independent committee for qualification of staff • Adequacy of Highly experienced and committed training staff, Continuous improvements • Regular meetings on plant upgrades, incidents and its review and sharing of experiences on subject or similar/other plants • Any other aspect that can benefit plant safety and security aspects • Provision for mentoring and counseling of individual trainee by staff	Excellent (90–100%) Medium (70–90%) Low (<70%)	• For most of the qualified staff: – Checklist gradings are $\geq$ 90%; – Written exam, walkdown and interview gradings are $\geq$ 90% • Contribution of human error per year in incident/events: $\leq$1.0 • Regulatory review feedback $\geq$90% • Feedback on performance of staff after joining regular O&M $\geq$90%; (Stress Factor is <60% if all the attributes are complying)	low (<60%) Medium (60–80%) High (>80%)

(continued)

Table 7.5 (continued)

HRIF	Strength parameters	Strength factors (Score)	Stress parameters	Stress factors (Score)
Regulatory provisions	Regulatory provisions on: • Regulatory policy to ensure/check human reliability • Availability of PRA standard and HR guide • Periodic review of training and qualification programme and its effectiveness • Monitoring and trending of root causes of human error events • Specific guidelines/approach to check compliance and enforcement on human error and corrective actions Programme (All 5 elements strongly meeting then: excellent Only 4 elements are strong then then: Very Good Only 3 elements strong and tendency of the organization is to address other two elements then: Adequate If only less, then 3 elements are strong then: Deficient)	Excellent (80–100%) Medium (60–80%) Low (<60%)	• Regulatory grading on Safety culture/risk consciousness (high/average/low) • General Consciousness considering: Number of Safety-significant event: 0(High), 1(Medium) and >1(low) • Compliance levels: (≤2 item) • **Plant CDF Level:** • Reference Level (High) • CDF minor increase <1.0% (Medium) • CDF increase >1.0% (Low)	low (<60%) Grad. high Medium (60–80%) Grad Medium High (>80%) Grad: low (Grad: Grading)

(continued)

Table 7.5 (continued)

HRIF	Strength parameters	Strength factors (Score)	Stress parameters	Stress factors (Score)
Risk and reliability	Risk and Reliability Programme coverage includes • Availability of approved Living/updated Level 1 PRA • Availability of approved Level 2 PRA model • Availability of Level 3 PRA • Use of PRA insight in training and qualification, e.g., application of PRA in accident management, staffing, development of training and qualification Programme • Application of Risk-informed/Risk-based approach in plant risk reduction or management, like maintenance, testing, surveillance management, etc • RCA and HRCA for risk identification, assessment and management measures • A system for reliability data collection, analysis and management	Excellent (80–100%) Medium (65–80%) Low (<65%)	Risk and Reliability consciousness levels grades are as follows: The human factor contribution to the incidents and accidents Industrial Accident Rates (Minor injury]/Small [major injury]/Major [fatality] Nuclear Accident (Major/Medium/Minor)	low (<60%) Medium (60–80%) High (>80%)

(continued)

Table 7.5 (continued)

HRIF	Strength parameters	Strength factors (Score)	Stress parameters	Stress factors (Score)
Task characteristics				
Task complexity	• Approved plant technical specification • VDU-based multitasking human–machine interface • A reliable trip and alarm system • Availability of computerized operator support system with capability for learning and identification/prediction of incipient transient and performance of equipment diagnostics • Availability of safety parameter and display system • Availability of Risk Monitor • Highly trained, experienced and qualified staff manning control room • Distributed system for plant area coverages with audio/visual communication, each station has dedicated team of two staff members • Strong Supervisory culture for execution guidance coordination, communication with plant areas • Availability of approved engineering drawing preferably in digital format	Very High ≥80–100% Medium ≥70–80% Low ≤70%	• Predominantly Skill based but might involve Rule based component. Enough time window available for job • Time Criticality Level—1: Ratio of Time Window Available/Time Required ≥2 • Routine operations • Routine maintenance • Predominantly skill and Rule based but might involve knowledge-based component • Time Criticality Level—2: Ratio of Time Window Available/Time Required >1.2 • Shutdown Maintenance • Anticipated transients • Technical Specification violation • Controlling deviations • Predominantly knowledge and rule based but might involve skill-based challenges also • Time Criticality Level—3: Ratio of Time Window Available/Time Required ~1.0 • Logistic challenges	Low Medium High

(continued)

Table 7.5 (continued)

HRIF	Strength parameters	Strength factors (Score)	Stress parameters	Stress factors (Score)
Automation levels in the plant	• Systems have been fully and systematically optimized for automation with rationales and justifications by identifying and prioritizing the risk factors • Systems have adequate/required automation with rationales and justifications • System lacks or limited automation indicated by incidents and events	Very good Medium Low	• Automation is lacking and the same is reflected in incident rate where the RCAs are related to lack of availability of time, improper rule-based and knowledge-based errors • Automation exits but not optimized based on operational insights • Optimized automation and this fact are reflected minimal deviation or incident rates	High Medium Low
Environment				
Physical Conditions for human comfort (temperature, humidity and sound levels)	Ground benign Climatic conditions control the area environment Harsh Conditions	High Normal Low	Commensurate to operation of hardware and human Maintenance standards low and not in line with regulation/standards No provision for ensuring the ambience, not meeting regulatory and standard provisions	Low Medium High

(continued)

Table 7.5 (continued)

HRIF	Strength parameters	Strength factors (Score)	Stress parameters	Stress factors (Score)
Risk Consciousness	• Application of PRA and lessons and insights on Accident management strategy for training • A separate checklist on PRA forms part of training • Regular training and meeting on workshops on incidents and accidents in the plant and related plant elsewhere. (Yes/No) • Significant event rate in the plant—one in a year • Event rate <2 year • Availability of Severe Accident Management Program • Regulatory compliance performance—no risk related compliance pending	Very High $\geq$80–100% Medium $\geq$70–80% Low $\leq$70%	All points complied More than 70% Compliance Less than 70% compliance	Low Medium High

7.10.2 The CQB Mathematical Model

In the CQB approach, human reliability is conceived considering the strength of the plant provisions and factors that tend to reduce human stress during normal operations as well as emergency conditions. For example, the digital panel in the control room and on consoles is more effective for managing a dynamic scenario compared to the old system of creating static and hardware-based graphics and indications to show the status of the system/component and switches to remotely maneuver or change the status of the subject equipment. Furthermore, the availability of operator information and, more importantly, the diagnostic and transient identification system has significant potential to reduce operator stress. These plant provisions provide a strength component for human reliability modeling. Similarly, the stress level of the operator due to personal factors will come into play while managing plant operations during normal scenarios as well as emergency conditions.

Before we go ahead, let us discuss the administrative provisions available to aid the operator in supporting the operator in not only identifying the status of the plant and system but also taking action. The operations window forms or valve slips are issued to execute the job at site or in plant areas. To cater to emergency conditions, emergency operating procedures are available in the control room. The technical specification document provides the safety limits, limiting conditions and rules/conditions for compliance along with approved maintenance and surveillance programs for the plant. The organization chart along with roles and responsibilities for staff forms part of the technical specifications. The approved documents, such as emergency and operations procedure design manuals and drawing, are the operator's aid. Drawing. The operators manning the control room are trained, qualified and authorized staff with experience ranging from 2 to 20 years in operation, reauthorized periodically to ensure adequate safety in operations.

The above discussion provides fundamental requirements and background to create a human reliability model. In CQB, any challenge, be it available a limited time window for action, consciousness levels, cognitive loads and conscience levels, is treated with an ensemble of all the parameters into one parameter called the stress ratio (SR). Therefore, the human nonperformance probability as a function of stress ratio S_R and reference nonperformance probability Po can be written as follows:

$$P = f\left(S_{R,} P_0, A, B\right) \tag{7.12}$$

A and *B* are constant for job complexity and Training and experience quotient or quality, respectively. The design of the experimental approach was used to represent stress and strength parameters and their correlation with nonperformance factors, and finally, a weighted sum average approach was used to develop the stress and strength parameters and finally the stress–strength ratio, S_R (Varde and Pecht 2018). The nonperformance probability model developed employing an exponential distribution is given by Eq. (7.13) as follows:

$$P(S_R) = \left[1 - C_{sc}(1 - P_0)\exp\left(-\frac{S_R}{B}\right)\right] \quad (7.13)$$

The prerequisites for evaluating the nonperformance probability using the above model are as follows.

1. Identify parameters that represent the postulated stress level of the operator.
2. Group the above stress parameter and attaching the importance for each parameter in the group by allocating proper weight factor.
3. Repeat steps 1 and 2 for the strength parameter.
4. Determine the constant for training and postulated cognitive stress levels.
5. Evaluate the conscience levels and fuzzy number that represent the conscience coefficient ranging from 0 to 1. Since this parameter is more relevant for evaluating security risk, a binary number (0 or 1) can be used to remove the subjectivity in judgment.
6. S_R can also be used to represent the consciousness ratio as a cognitive stress ratio or any other parameter that enables the determination of the condition for stress and strength by employing the conditions.

The reference failure probability P_0 can be evaluated in this model based on data collected from plant records during normal operating conditions (when the stress levels are minimal and the situation does not force the limitation of a relatively short time window). The man–machine and trip and alarm conditions do not overwhelm the condition, and due to long exposure, the plant and its behavior are better understood by the staff.

7.10.3 Fuzzy Logic Approach for Human Reliability Analysis

Even though the plant performance review is performed by evaluating various quantified parameters to arrive at vital capacity factor, availability factor, number of plant trip and shutdown based on the quantified data from plant records. However, as we move toward evaluating the reliability and risk evaluation, situations often arise when either the data are not available or the available data are not adequate to arrive at the probability of events. Second, in decision making, the decision is often made employing qualitative and imprecise input, where the information that is available is often based on opinions such as 'Excellent', 'Very high', 'Very low', 'fair Low' or 'low' for a situation or parameter, which are imprecise statements but reflect the confidence of the experts. In operation ecosystems, it is often required to deal with imprecise information when we model expert opinions.

The probabilistic approach is not adequate, as this approach uses input as crisp binary numerical data. For example, if the valve is closed or open, the pump is on or off. However, it is required to make decisions using fuzzy or imprecise data. The following section will deal with the formulation of the solution when the input information is available in the form of expert opinion, which tends to be imprecise but valuable as expert opinion.

In Crisp logic, an element is either member of the set or not represented by 0 and 1. We assign 0 when the element is not a member and 1 when the element is a member of the set. For example, we have a set of intelligent people. Now, if we must assign whether Ram is intelligent or not, we have only two choices 'intelligent' or 'not intelligent'. However, in a real world, it will be inappropriate to use this crisp description and, of course, in decision making where we deal with imprecise and often vague information as input, the crisp logic falls short of treatment capability.

This is where the fuzzy logic approach to deal with imprecise information was proposed and pioneered by Zadeh (Verma et al. 2009) and is now being used extensively in varying fields ranging from electronics design to complex decision making. In a fuzzy system, membership is assigned to the elements in the set, and membership ranges from 0 to 1, i.e., an element can have any value, such as 0, 0.1, or 0.2. 0.3...0.9 and 1.0 depending upon how strongly or conversely weakly it is meeting the attribute for which the fuzzy set is being built. For example, if Ram is very intelligent, then a membership function 0.8, 0.9 or even 1.0 can be assigned depending on the judgment of the analyst. If Shyam is not as intelligent as Ram, the membership function of 0.6. can be assigned. If the analyst is very sure that a third person Deepak is not intelligent, then he can be assigned a membership function of 0.2, etc.

Fuzzy not only enables analysts to use imprecise data in support of analysis in decision making but also operates like probability distributions to represent data, evaluate interaction and union of two parameters or sets in support of inference. Fuzzy analysis generally comprises four major steps. These steps are (a) creating the sets with suitable membership function, (b) fuzzification of data, (c) formulation of fuzzy rule base, evaluation and inferencing and (d) defuzzification and presentation of decision (Verma et al. 2009). There are many geometrical functions used in fuzzy logic, however, considering the modeling requirements, we describe two triangular and trapezoidal functions as follows:

1. Suppose there are two parameters represented by Set A and B comprised of imprecise data. The fuzzy sets A and B are subsets of a universal set U. Based on the expert opinion, the two sets are represented using the membership function

Fuzzy set A is represented as

$$A = \{x, \mu_A(x)\} \quad x \varepsilon U$$

where μ_A represents the member function of fuzzy set A.

Accordingly, the fuzzy set A can be written as:

$$A = \sum_{i=0}^{n} \frac{\mu_A(x_i)}{x_i} \tag{7.14}$$

The Concept of α-*Cuts:* The relationship of an element x of a set X represented mathematically as $x \varepsilon X$ that is a member of a set A must assign a value that represents the membership strength in that Set, viz. the element of A_α of A is as follows:

$$A_\alpha = \{x \in X, \mu_A(x) \geq \alpha\}$$

There are many membership functions available in the literature; however, we will discuss two popular functions, the triangular and trapezoidal functions, as shown in Figs. 7.33 and 7.34.

Suppose the parameters for Low Medium and High are denoted as l, m and h, respectively. Furthermore, the above triangular fuzzy set is P and has been created for a parameter, say x. Then, the membership function for triangular representation can be given as:

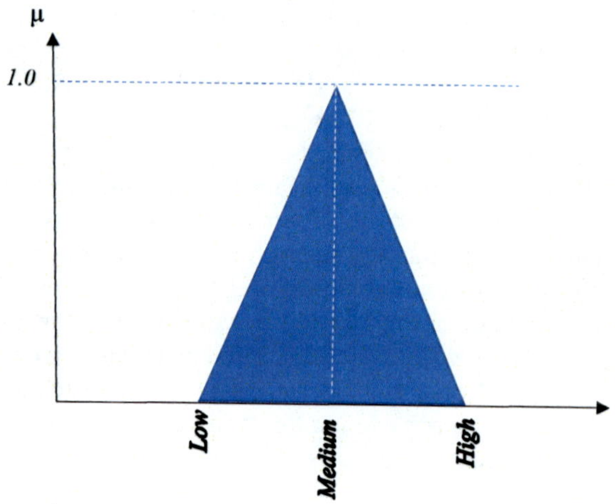

Fig. 7.33 Triangular fuzzy representation showing three levels of a parameter x

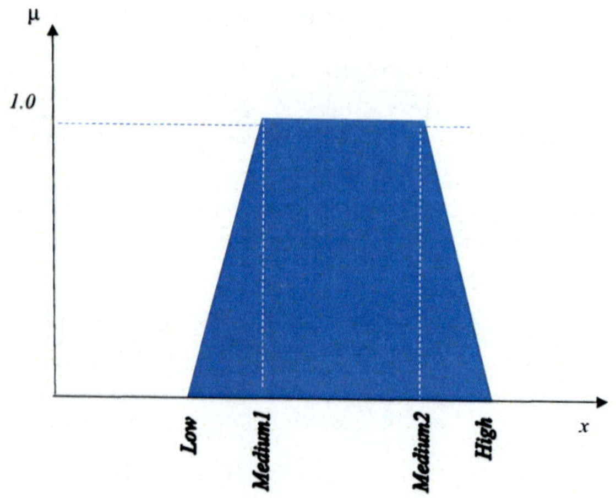

Fig. 7.34 Trapezoidal fuzzy representation with four levels

$$\left.\begin{array}{ll}\mu_P(x) = 0; & \text{when} x < l; \\ \mu_P(x) = {}^{(x-l)}\!/\!_{(m_1-l)} & \text{when} l \le x \le m_1; \\ \mu_P(x) = 1; & \text{when} m_1 \le x \le m_2 \\ \mu_P(x) = {}^{(h-x)}\!/\!_{(h-m_2)} & \text{when} m_2 \le x \le h\end{array}\right\} \tag{7.15}$$

The graphical representation of the trapezoidal function is shown in Fig. 7.34.

Let us represent the membership function Low as l, Medium1 as m_1, Medium as m_2 and High as h; consequently, the fuzzy set P can be expressed as.

$$P = (l, m_1, m_2, h)$$

The fuzzy expression for the trapezoidal function is written as follows:

$$\left.\begin{array}{ll}\mu_P(x) = 0; & \text{when x} < 1; \\ \mu_P(x) = (x-l)/\,(m_1-l) & \text{when l} \le \text{x} \le \text{m}_1 \\ \mu_P(x) = 1; & \text{when m}_1 \le \text{x} \le \text{m}_2 \\ \mu_P(x) = (h-x)/\,(h-m_2) & \text{when m}_2 \le \text{x} \le \text{h}\end{array}\right\} \tag{7.16}$$

7.10.3.1 Representation of Stress and Strength Parameters

In any safety critical plant, ensuring a high risk-consciousness level is a key and challenging attribute. One of the reasons why it is challenging is the nonavailability of a model and procedure to evaluate the levels or adequacy of the risk-consciousness level. However, there is a consensus in the organization that ensuring a high organizational risk-conscious level is a key and vital component to establish high standards of safety. The major challenge is that we need to model the organizational risk-consciousness level employing imprecise and qualitative parameters.

Fuzzification of the input variables A triangular membership function is employed to model the organizational attribute 'risk-conscious culture'. In CQB, the stress–strength modeling approach is used for evaluating the imprecise and linguistic parameters. Please refer to Table … for modeling of the stress and strength factors. The following figure shows the interrelation representation of stress and strength (Fig. 7.35).

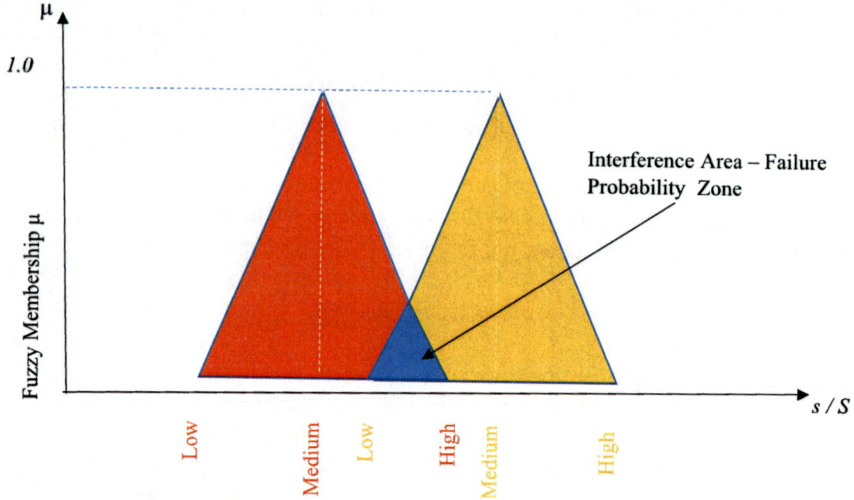

Red triangle represents *Stress = f (weakness in risk-conscious culture)*

Yellow triangle represents *strength due to compliance Strength = f (Compliance)*

Fig. 7.35 Stress and strength representation for organizational risk conscious

Suppose we have two fuzzy sets for risk-conscious stress and risk-conscious strength, represented by the symbols s and S, respectively. The function for stress and strength is represented by red and yellow, respectively, as shown in Fig. 7.36.

In each organization, the strengthening of risk-conscious culture is directly proportional to the organizational risk-consciousness strength and inversely proportional to risk-conscious stress factors. Considering the triangular function, the strength S is written $S = (L, M, H)$, also for stress $s = (l, m, h)$. Further assume that the fuzzy function has been created over a range from 0 to 10 units of stress levels. Furthermore, the stress and strength functions $s = (0, 3, 6)$ and $S = (4, 7, 10)$, respectively.

Fuzzy Inference

The second important step is fuzzy inferencing by developing the fuzzy rules. These fuzzy rules are derived from the fuzzy functions by using the information about the fuzzy elements in the set and their membership strength in the respective set. Typical fuzzy rules, such as crisp production rules in a rule-based system, have two parts, the antecedents and consequence, and their logical relationship is represented by.

If-and-then logic. For example.

IF

$$A = x$$

$$B = y$$

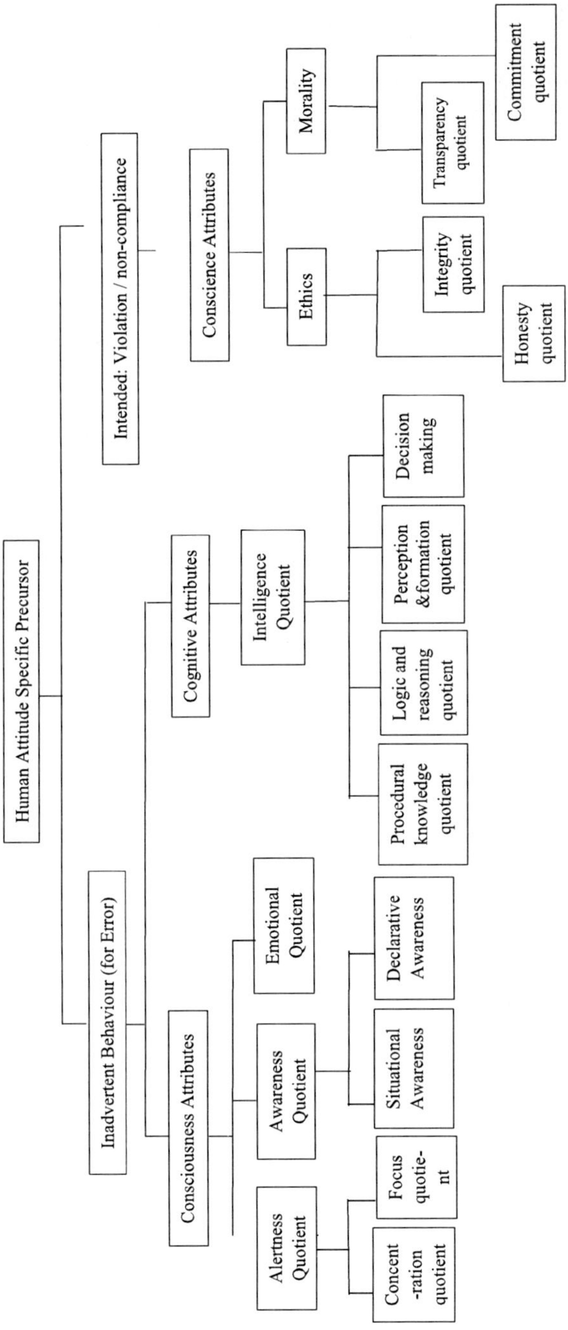

Fig. 7.36 Human attitude specific precursor

THEN

$$C = z$$

This relationship is read as 'If $A = x$ and $Y = b$, then $C = z$'. The expressions $A = x$ and $Y = b$ and $C = z$ are referred to as antecedent and consequence, respectively. Any fuzzy system comprises a rule base containing many complex rules and databases. The inferencing forms an integral part of decision making and generally requires more than one and often many fuzzy rules processing. For the case of the stress–strength model shown in Fig. 7.36, the interference area represents the failure probability zone in the form of a triangle. The area of the circle provides the estimate of failure probability for a stress–strength model; in this case, it is a risk-conscious culture. The centroid of the circle provides the center of the circle defined as 2/3 of the distance from the vortex to the midpoint of the opposite line. For the worst-case scenario, the ratio of strength to stress in this region will provide the stress ratio.

The overlapping area, in line with the interreference probability theory, represents the failure probability region. Accordingly, the fuzzy rule for the above representation pointing to failure is as follows:

$$\text{stress} s \geq L \text{ and } S \leq h$$

Then
$RC = Failure$

The parameter, interfacing area, i.e., area of the triangle represents the failure probability, centroid of the triangle further explains the failure characteristics of the risk-conscious culture. After normalizing the parameters with respect to the membership function of 0.1 as seen in the figure and assuming that the triangle is an equilateral for simplicity, we obtain the failure probability for the worst-case scenario for risk-consciousness culture to be ~6×10^{-3} and the centroid (at 2/3rd from apex to middle of the opposite) side as (0.05, 0.067) represented in terms of the normalized parameter as the above approach provides an effective and transparent mechanism that is based on the qualitative opinion of the experts on the subject.

7.11 Special Aspects

7.11.1 Human Root Cause Analysis

The traditional approach to root cause analysis either and often terminates at identification of human error and mode, viz. failure to turn the switch in time, delayed action, as commission error, missed a step in the procedure, action that leads to over travel or under travel of an up and down device as commission error, faulty identification of correct switch as mistake, etc. However, failed to go down to unearth as to What caused human or which faculty of human caused the failure as also the adequate factors that contributed to this failure, particularly the behavioral or the ones that challenged human capabilities during the testing times. In CQB, the reference human fault tree facilitates the performance of root cause analysis and thereby facilitates identification of the roots responsible for the human error. The reference human fault tree approach is a way to reach the root causes of failure and when these modifications employ the performance influence factors or make the job of the human to overcome the challenges. Apart from this, other factors that reflect organizational culture, such as training quality, are also used as modifiers for generating the human failure probability for a specific human action. In fact, the analysis of stress factors or other contributors, such as operational environment or organizational factors, is vital for identifying the susceptibility of an organ to affecting human faculties. Additionally, this information can be excellent feedback to review the reference human factor model.

7.11.2 Human Error Precursors

The precursor analysis addresses the analysis of operating experience, where the identified events are analyzed to assess their risk-significance or risk potential. The approach combines insights from deterministic mainly the root cause analysis of the identified failures and probabilistic risk assessment by mapping this event on the existing PRA model for assessing the potential of this event for initiating event(s) or safety system function failure. The identified factors in the event sequence are often referred to as precursors to accidents. Often, the human error(s) also form part of the accident sequence. For example, the accident sequence LOCA*ECCS*HE-Manual-Injection has been assessed to result in the core damage frequency. It can be read as, during 'Loss of coolant accident' if emergency cooling system (ECCS) fails to inject the coolant (say, due to injection pump failure) from underground tank (where all the leaked out inventory is collected), and operator fail to open a valve to inject coolant water from an Overhead tank manually (HE-Manual-Injection) than it is postulated in the analysis that this event might lead to core damage.

Now, in the traditional sense, the He-Manual-Injection is a precursor event to the accident sequences, and as per the operating experience feedback programme, a corrective action programme needs to be implemented to eliminate or reduce chances of error by modifying the existing system or changing the operational policy.

However, it is not adequate to stop at the Human error event as for implementing the corrective measures it is required as to 'what was the root cause of the human error' in injecting the light water manually. We might attribute slip, lapse or mistake as part of error of omission or error of commission, but it is vital to know what caused slip/lapse/mistake. In this chapter, as shown in Fig. 7.36, the precursor human factors were identified that affect human performance and finally form the fundamental root causes of failures. The same factor can also work as a precursor along with some more factors. These factors, as per the CQB framework, are (a) consciousness, (b) cognition and (c) conscience driven, and working at the level of the human brain physiologically and psychologically at the level of mind.

7.11.3 Human Factor(s) as Precursors to CCF

Human factors are more critical when they introduce the possibility of a precursor phenomenon or event that creates conditions for common cause failure in parallel or redundant trains sometime after introduction such that the error/fault remains latent and causes the redundant system to fail simultaneously.

Human error during operation and maintenance can introduce the CCF in a system comprising redundant components to ensure higher availability and reliability. For common cause failure to affect redundant components/systems, there are two prerequisite conditions, the root cause for the failure and the coupling mechanism. The latent common cause factors may surface in a safety system when a demand is placed by the control system to start/actuate the redundant components to perform the safety function or during the operational phase of the system. For example, the failure of ECCS injection pumps on demand or successful starting but eventually failure of pumps before the mission time.

Identifying precursors for CCF events is more challenging than identifying individual components, as one needs to have an understanding of the relation between the root cause failure and coupling mechanism and their joint capability to satisfy the failure criteria that lead to safety implications. For example, there is an increase in humidity and temperature in the instrumentation room where three cabinets for protection Channels A, B and C are located, leading to the unsafe failure of identical relays in more than one channel qualifying as a precursor CCF event. In principle, it may be possible and considered as good practice a priori to consider as common cause failure but requires design of experiment for evaluation of this scenario that might involve testing in temperature and humidity chamber to understand the synergy of the degradation mechanism between metal contacts, insulations or at cable junctions.

Another, relatively less complex situation involves adding inadvertently spurious lubrication oil (contained in a new lubricating oil drum) in four primary coolant pump bearings at approximately the same time that caused a bearing seizer of more than one pump. Similarly, calibration error for transmitters committed by the same crew for three instrumentation channels, inducing the same fault that has potential to CCFs.

To avoid the human induced CCF, as a safety practice, potential for a CCF assessment is performed. In this assessment, the human–machine interaction in operations as well as maintenance is performed to analyze the potential sources, including the precursor for CCF, and documented as a reference scheduling surveillance and monitoring program as well as identifying the factors or human actions that might induce CCF events.

To safeguard against these types of CCFs, the plant surveillance program requires a staggered approach for testing. In this scheme, each redundant component or channel is tested separately such that component testing is spaced wither on different days or different crews so that the possibility of inducing a common fault is reduced or eliminated.

Before, we go head let us get the perspective on the word, simultaneously'. Here, the term simultaneously is subjective and directly related to the successful completion of safety action timings. For example, in the reactor shutdown function, the failure criterion is that if 2 out of 14 fail to drop on demand, then the system as a whole will be considered to have failed. Here, considering the safety function requirements that make the reactor subcritical requires simultaneous insertion of rods, the word instantaneous means that even a delay of a few milliseconds may not be acceptable. Taking another example, there is a safety function that requires filling the overstorage tank for fire water injection using the redundant pumping system comprising 3 redundant pumps, and the failure criterion is failure of 2 out-of-3 pumps. Here, during an actual demand, only one pump started, and another pump failed due to intermittent fault; however, it started a moment later and operated successfully afterward. This caused a delay of a few milliseconds. Now considering the safety function that pumps water to an overhead tank with a minimum two-pump flow, if it does not pose any safety implications, then the definition of instantaneous action may spread more than a few milliseconds such that it does not affect the safety function capability, unlike the shut-off-rod system, where safety is adversely affected if the shut-off-rods do not drop literally instantaneously. Hence, in a pumping system, even if the second pump starts a moment later, it may not be called CCF.

7.11.4 Techniques for Improving Human Performance

We live in material world and our attitude, thinking, our approach to other living beings is to a great extent remains self-centered, motivated or driven by false ego, and attachment to self and everything we think we have in our control or possession and perennially generate craving for what we do not have. This life filled with contradictions, overwhelms our consciousness, or even digs into as subconscious-level. Consequently, they often get into mood-swing or the perpetual switching between normal and state of anger, conditioned into hate, worry, blame and other negative qualities that consume a huge amount of energy. In short, we are for away from a quality that determines our energy levels 'the state of equanimity' or stability of mind—vital requirement for transformation of individual self into a higher self. In CQB, the spirit or it can essentially mean consciousness forms an integral part of the characterization of human beings. Furthermore, this postulation asserts that if Spirit or consciousness does not form part of the human reliability model, the model cannot be claimed to be complete.

In spiritual parlance, higher consciousness is often related to self-realization. The process of self-realization itself transcends one into the path initiating one into the spiritual journey. What it means the realm of even realization makes a human being more and more stable as he is becoming free or detached from the negative virtues that were responsible for blocking the intelligence and spiritual qualities, which also include moral, ethics, honesty and integrity to name a few for the purpose of material domain requirements. What is the motivation or reward of self-realization in this material world? Of course, the growth of inner happiness is permanent as against happiness acquired by material gains. What does this happiness do? It directly improves performance and does not make one susceptible to a degraded or unstable state due to personal, environmental and societal challenges. This section will provide an overview of improving wellbeing and happiness in general.

Yoga, meditation and even physical exercise lift mood and generate a feeling of happiness. If anyone has any doubt about the effectiveness of these techniques, then let us try this. Sit in meditation or prayer posture for 30 min and see for yourself. You will feel that you feel better than what you were before sitting in meditation. Similarly, when you are stressed go for a good one hour walk with friends or even alone and you would know for yourself. In material world you require proof of anything you claim, I will tell my experience, in the plant simulator room we monitored brain waves using Electroencephalography technique. A clear calming of wave pattern during the meditative state could be observed when compared with normal working condition that validate the claim of effectiveness of meditation techniques. These patterns obtained through EEG monitoring sessions are vital to measure cognitive stress levels. It is a well-known fact that the monitoring of stress levels is carried out by tracking the blood pressure, pulse rate and CC TV Videos for body language. During reactor transient sessions, a general observation has been that the stress level increases, particularly if the session is laden with a complex scenario involving propagation of initiating events over simulated multiple failures.

7.11.5 Physiology of Happiness

There are four main chemicals that elevate the mood or make one happy when the faculty in the brain releases these chemicals through the neurotransmitters (brain chemicals) responsible for communication through the synaptic processes between neurons or between neurons and muscles in the peripheral nervous system and other organs. These four chemicals are as follows:

7.11.5.1 Serotonin

This is one of the key happy neurotransmitters in the human brain that has potential requirements for reaching the stage of equanimity. Equanimity is a factor normally associated with the mind to refer to and referred to as the 'equanimous mind', i.e., the mind that stops wandering (e.g., like a monkey) from one thought to another and often in search or to satisfy our quench for craving and aversion. Humans waste their majority of their thinking time and energy in jumping from pleasure and pain for themselves. If this time and energy is directed toward constructive jobs and activities, there can be phenomenal results for the individual, and most importantly, it may be reflected in their performance. The reason is that a simple stable or equanimous mind supported by stabilized brain faculties not only helps conserve energy but also uses the available energy for discharging one's duties and responsibilities more effectively. However, it may be noted here that happiness, referred to here, has two connotations. The short-term happiness achieved through material gains intuitively has a package of miseries delivered later. The second is the happiness achieved by performing the right mental or physical activities (Karma Yoga), e.g., performing your duty with full dedication, helping others your colleague, serving the family and the society at large bring long-term and permanent happiness. Similarly, employing the intellect or intelligence (Gyan Yoga) to find better methods for the challenges or for that matter trying to understand reality, facts figures and state of the art in the subject area can also be helpful for long-term happiness. Bhakti Yoga reflects the subtle quality related to love, dedication and renunciation that tends to support achieving perfectness in performance through attitude and behavior.

This perennial or stabilized happy condition of the brain achieved through the above yogic practices will stabilize your brain/mind. Since it is achieved through walking on the path of self-realization, it renders one this power for stability to the mind. The reason is what is important for an ignorant, i.e., untrained mind such as promotion, what will happen to my job, social rejection, fear of downgrading of family status and weakness that arises to overwhelming emotions, will disappear and the real self is in action without much fear and pressure, which causes cognitive or mental stresses that often have been found to be the causes of error.

Again, we come back to the subject, the importance of the serotonin on our body is more widespread in our entire body through nervous systems cells and that in turn through the peripheral nervous system connected to all the five sense and action organs mainly hand and feet. It affects eating habits (deficiencies in serotonin induce cravings for food that are often harmful or junk food that the tongue demands), digestion, sleeping patterns, bowl movement, etc. For example, if you are not sleeping properly, it will certainly hinder your performance while the plant is working. Similarly, other factors also distract one from doing their duty dedicatedly. In fact, serotonin is also helpful to cure injuries; however, excess serotonin causes chronic conditions such as blood clots and osteoporosis. Therefore, the balanced production of serotonin is critical to the human body. Here, we come to another important point, 'distraction'. Distraction due to mental or physical reasons has been found to be another major factor for human error.

The above discussion provides an opportunity for furthering our understanding of operator behavior during stressful conditions and training our minds for handling emergency operating procedures—a large bottleneck to address fallibility, sensibility to challenges or otherwise ensuring robustness of the brain during challenging conditions.

7.11.5.2 Dopamine

The brain and reward system activates the formation of dopamine, a neurotransmitter when one experiences contentment, mood lifting, feeling good or pleasure, due to certain events in the external or internal world. The brain faculty that deals with emotions called the hypothalamus, a small organ located below (Hypo) the thalamus, is responsible for the release of dopamine. In fact, dopamine also works and supports the mental processes involved in motivation, inspiration and in a sense in movement that support positive actions.

Deficiency of dopamine on a peroneal basis results in mood swings, depression and serious mental disorders. Therefore, dopamine is one of the constituents used to treat mental health conditions.

7.11.5.3 Oxytocin

The endocrine system in humans is responsible for the production and management of different chemicals or neurotransmitters. As in the case of serotonin, oxytocin is produced in the hypothalamus and released and regulated by the pituitary gland, a seed size organ located in the base of the brain. In fact, the pituitary gland releases many other hormones responsible for homeostasis in the human body. Oxytocin is associated with intense emotions that are governed by body organs such as cuddling, sexual behavior and produced or other events liked to emotional components in humans.

7.11.5.4 Endorphins

Additionally, known as happiness booster or pain reliever hormone. The nervous system is fully involved in the production and management of endorphin requirements of the body. Endorphin is also known as a stress management agent in the sense that it inhibits the feeling of pain by blocking the sensation to the brain/body, similar to the action external agent opioids administered in pain management. (Berry 2018).

The notable aspect in the context of the subject, i.e., RCOM, is that deficiencies of any one of the above or more than one adversely affects human performance. The good news is that the above happy hormones are an integral part of the body, and by performing healthy physical and mental exercises, they can be boosted. For example, simple and regular exercises, morning or evening walks, socializing, Yoga and meditations are very effective in the management of the level of these hormones.

So what the extended CQB offers is a mechanism for not only producing human error data but also (a) methodology for understanding human root causes for failure at the level of human anatomy or physiology, (b) how these root causes work at cognitive level particularly a how instable state accelerate potential for human error and (c) what can be done to make human less susceptible to underperformance and last how a validation of these techniques can be dome using modern scientific instruments.

Please see Annexure C on Yoga and Meditation Techniques for a further understanding of the role of these techniques for wellbeing, which includes the improvement of a happy positive attitude toward life and treatment/management health issues.

7.12 Remarks and Conclusion

Most of the human reliability methods are based on the data of either quantities or qualitative information/data available from either the simulator or expert opinions or the plant incident and event database. In short, the human performance data collected form the basis for human reliability estimates. The data are modified by various influencing factors to address cognitive stress level, organizational factors such as training, man–machine interface and environment, and quality. Even though these methods available in the literature and the compilation of handbooks offer a ready-to-use approach for the quantification of human reliability, the uncertainty associated with these estimates is often relatively higher, and at times, the estimates do not provide adequate confidence for their use in risk assessment or evaluating plant procedures or other operational applications.

A critical review of these available methods/techniques was performed with an objective to understand the requirements to develop an improved technique for human reliability. Based on this insight, most of the available reliability techniques, as accessed through the open literature, do not have a human model. Here, it may be recalled that for creating a risk model of a complex engineering, say a process/chemical plant or nuclear plant, the model is created by integrating the basic components to represent a system and further integrating the system to represent the plant. The risk assessment is carried out by evaluating the plant response for set-on undesired events called initiating events. However, the vital question was 'Why human model cannot, be created' by following the procedure that we use for hardware and systems; of course, human reliability final estimates also form part of this model. This was a major insight; accordingly, in the extended CQB, this model was created employing mostly a fault tree approach. Of course, there can be questions about the completeness and adequacy of representation, and yes, the most important one, certain aspects are not fully developed. However, this was expected as it is a well appreciated fact that human is a complex system and often there are claims that it cannot be comprehended. True. Here, the objective was to provide the best that can be done toward creating a human model. Human reliability without a human model is certainly not an effective approach. Furthermore, it was vital that the essential aspects capture how humans work, think and affect the required changes to ensure the safety and security of the plant. The second question was how a typical operator gets the information about the plant the way he does it and how he affects the changes. Here, the science material and spirit available in the Vedas and Bhagavad Gita along with the advanced scientific knowledge formed a vital source of insight. It was realized that cognition alone is not sufficient to create the human model. Something more vital is missing. There you are … and consciousness which is more fundamental to human existence should form part of human model. Accordingly, consciousness and conscience along material aspects formed part of the human model. Here, we need to understand that humans are spiritual entities; therefore, the human model will not be complete without the consideration of subtler attributes such as consciousness and conscience. Third, and most importantly, the physical model should provide the essential feature for its operation. Here, the concept of 6 external stimuli connects a human to the external environment, here the plant or system. These stimuli, even though they connect an operator to the plant, for affecting the action/work, should be further considered to make the human model relatively more complete. Here, the five sense-based organs listed in Bhagavad Gita provided additional ready to use the list that included fine stimuli and sense action organ, viz. Voice/Speech (Mouth/Tongue), Hands (for affecting an action, Legs (for support, movement and for affecting actions) and two other, i.e., Genitals and Anus, which do not directly contribute to human reliability and did not form part of the detailed CQB investigation.

It is a recognized fact that engineering plant operations, such as nuclear, aviation, railway processes and chemical plants, are human interaction-intensive ecosystems. Particularly, during emergency or accident scenarios, human actions form a vital component. Of course, this fact has been highlighted by the records of accidents. Even if the performance of the SSCs is characterized in a quantitative manner, it is also required that day-to-day operation of the plant is often based on the qualitative input. This aspect is further endorsed by the fact that reliability data are often not available for arriving at the estimates such that the uncertainty associated with the final estimates is acceptable. Uncertainty beyond a certain level makes the estimates not adequate entities.

This problem was addressed at two levels. At the quantitative level, a new approach to address the relative stress levels from many sources involving a stress–strength model was employed to create the foundation for obtaining the nonperformance probability or failure probability in CQB. Here, the stress ratio obtained from the stress–strength model, as well as the degree of overlapping of the stress and strength model, provides an effective and elegant mechanism of failure probability estimation.

The fuzzy logic approach is very effective in treating qualitative and imprecise information toward providing estimates of possibility and not the probability of failure. Accordingly, all the stress-causing parameters and provision of plant systems that represent plant strength were modeled to provide the estimates of failure probability in line with the interference model in statistics and probability. and be at individual or organizational level were modeling qualitative inputs were model.

It is expected that this chapter provides an improved framework through an extended CQB model, considering the human model, consciousness and conscience and many other aspects, as discussed above, to reduce the uncertainty in estimates. Furthermore, the approach that this book advocates, risk-conscious operations management, is expected to reduce the chances of error and has the potential to create a balanced and risk-conscious culture that enhances plant reliability on the one hand and reduces the risk levels on the other.

References

Bailey P (2020) Paolabailey.com. [Online]. Available: www.blog/paolabailey.com. [Accessed 2020]

Bendix A (2019) Chernobyl was the world's worst nuclear power plant accident. Here's how it compares to Fukushima and Three Mile Island. Businenessinsider.in, Mumbai

Berry J (2018) Medical news today, endorphins: effects and how to increases levels, 6 Feb 2018. [Online]. Available: medicalnewstoday.com/articles/3208. [Accessed 16 Jan 2022]

Endsley MR, Bolstad CA (1994) Individual differences in pilot situational awareness. Int J Aviation Psychol 4:241–264

Hameroff R, Penrose R (2014) Consciousness in the universe—A review of the 'Orch OR theory. Phys Life Rev 11:39–78

Hoffman M (2021) Webmd: picture of the brain—human anatomy. 21 June 2021. [Online]. Available: https://www.webmd.com/brain/picture-of-the-brain. [Accessed 12 Aug 2021]

International Atomic Energy Agency (1989) Human error classification and data collection. IAEA, Vienna

International Atomic Energy Agency (1995) Control room systems design for Nuclear Power Plants. IAEA, Vienna
Kitamura M, Fujita Y, Yoshikawa H (2005) Review of international standards related to the design for control rooms on nuclear power plants. J Nucl Sci Technol 42(4):406–417
NEI (2022) Oct 2019. [Online]. Available: nei.org/resource. [Accessed 21 Jan 2022]
Neurophysiology and Rehabilitions (2022) Edelweiss Publication, [Online]. Available: https://edelweisspublications.com/keyword/43/90/Central-nervous-system. [Accessed 15 Jan 2022]
Prabhpada Interpreter-Translater AC, Bhagatam Geeta
Savarese G, Lund LH (2017) Global public health burden of heart failure. Epidemiology 3(1):7–11
Shutterstock (2022) [Online]. Available: shutterstock.com. [Accessed 21 Jan 2022]
LRS SM (2021) LRS, [Online]. Available: http://lifetime-reliability.com/tutorial/human_error_rate_table_insights. [Accessed 29 Dec 2021]
Stanford Encyclopedia of Philosophy (2021) Stanford University, 19 May 2017. [Online]. Available: https://plato.stanford.edu/entries/consciousness-unity/#Hist. [Accessed 10 Dec 2021]
Synapse (2021) 24 Nov 2021. [Online]. Available: https://en.wikipedia.org/wiki/Synapse. [Accessed 12 Sept 2021]
The India Forum (2021) The India Forum, 2021. [Online]. Available: www.theindiaforum.com. [Accessed 3 Nov 2021]
The National Diet of Japan (2012) The accident at Fukushima Daiichi. The National Diet of Japan
The Six sigma Story: Mumbai Dabbawalas in 2021 (Updated) (2021). [Online]. Available: https://www.henryharvin.com. [Accessed 2021]
Varde PV, Pecht MG (2018) Chapter 10 Human Reliability. In: Risk-based engineering—An Integrated approach to complex engineering systems: A special regerence to nuclear plants. Singapore, p 565
VedAB-S (1996) Vyasa translation: Prabhupada, Bhagavad-Gita—As it is, Mumbai: the Bhaktivedanta book trust
Verma AK, Srividya A, Gaonkar RP (2009) Fuzzy-reliability engineering. Narosa, New Delhi
WHO (2020) World Health Organization, [Online]. Available: https://www.who.int/news-room/fact-sheets/detail/cardiovascular-diseases-(cvds). [Accessed 6 Jan 2022]

Chapter 8
Operational Risk Management

8.1 Introduction

Operational risk management enables plant or system operators to either minimize or mitigate operational risk while optimizing plant performance (Operational Safety Management 2022). Conscious efforts to ensure that all risk identification, reduction and correction measures that mainly include surveillance and monitoring, inspections, maintenance/repairs and performance tracking and compliance follow the highest standards of safety. The final responsibility of plant safety rests with the operating organization (International Atomic Energy Agency 1999a). Since operations are a human interaction-intensive ecosystem, human and organizational consciousness are critical factors in the risk management function (Varde 2021). Specifically, the major functions of plant management are plant operation, often 24/7, to meet the production objective, maintenance function to affect maintenance and surveillance programs, planning that involves design and engineering services, training and qualification, quality assurance, in-service inspection and performance and safety evaluation. The documentation development, review and update and maintenance activities form part of the planning and training and qualification program due to nature of the documentation activity.

The duly approved plant technical specification prescribes the safety limits, safety system settings, limiting condition for operations that the plant staff must comply during all the modes and states of operations. The training and qualification function is responsible for training fresh human resources and complies with the conditions laid out in the training and qualification manual. The periodic requalification and competency checks also form part of the training and qualification program.

The technical specification has an organization management chart and the roles and responsibilities of the higher, middle, operations, maintenance and services engineer. Apart from this, the technical specifications also prescribe criteria for incident categorization into event, safety-significant event, etc., and the surveillance schedule for the safety-significant SSCs.

P. V. Varde, *Risk-Conscious Operations Management*, Risk, Reliability and Safety Engineering, https://doi.org/10.1007/978-981-19-9334-3_8

The operating organization's approved chart also identifies the function of communication with the regulatory body. This function, along with the conditions for reporting criteria to the regulatory body, is also prescribed in the plant technical specification. There are special functions to identify typical hazards specific to plant hazard potential; for example, for nuclear plants, the radiation and health services wing has responsibility for radiation monitoring and other advisory services in support of operational and maintenance activities.

There is a plant level safety committee. All the safety cases are reviewed by this committee, and the cases of safety significance, as prescribes in technical a specification, are reported to the regulatory committee. There is an arrangement of three-tier regulatory reviews. The lowest level is the unit level committee, followed by the plant-specific safety committee, and the highest is the Apex Regulatory Safety Committee. The proposals, depending on their safety significance and regulatory manual, are subjected to these committees. In many countries, the application of PRA study as part of a risk-informed approach is used in support of regulatory review. Safety is the overriding factor in all the above functions. The quality assurance function ensures the quality in all the aspects of operations, including planning, control, supervision, plant modification, training and qualification, documentation, safety committee and particularly regulatory compliance. In this sense, the quality assurance function essentially follows a safety management activity for the plant.

8.2 Integrated Operations Risk—A Perspective

Operational risk for the purposes of safety critical systems can be defined as the '*the increase in likelihood or consequences or both due to failed process, people and system due to internal or external initiating events or inadvertent or intentional actions*'. This definition also conforms to the mathematical expression of risk at the same time provide a perspective that is in line with other definitions of operational risk, such as the one for banking/financial systems (Basel Committee—Operational risk 2022). Unlike the financial system, the potential consequences are radiation and chemicals, posing threats to the loss of life, the environment and property, including financial losses.

Complex system operational safety considerations must go beyond its predominant risk attribute to ensure that the exposure of risk to the public, property, economics and occupational members is targeted to zero risk or to a minimal acceptable level. For example, for nuclear plants, the PRA and deterministic approach model the potential for 'radiation risk' exposure during normal as well as accident conditions largely to members of the public. This is a vital and critical part of the risk evaluation. However, the operation of the plant should consider the industrial risk, security risk including cyber threats, loss due to plant reliability (unintended trips), unavailability (plant shutdown) due to unavailability of services that lead to loss of credibility of

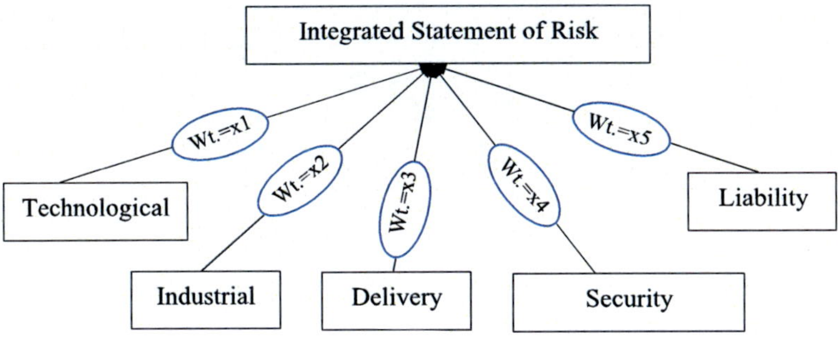

Fig. 8.1 Integrated operational risk representation

the services to customer, operational profit-risk, liability and so on. Of course, radiation risk remains the dominant risk; however, the other components also need to be considered as part of a holistic approach to risk.

Figure 8.1 presents the concept of integrated risk, where the parameter weight factor (Wt = xi) has been introduced to arrive at an integrated statement of risk.

The accidental release of radiation is considered the major component for risk modeling for nuclear plants; for processes or chemical plants, the hazard potential plant could be the accidental release of chemicals; or for aviation and transport systems, the accident involving fatalities or injury to the passengers and operating crew becomes the major subject of modeling.

Often, the different risk categories are interrelated or affected. For example, an accident involving radiation consequences will also lead to long-term loss of services, availability and revenue and might include industrial consequences. A reactor trip due to genuine conditions or spurious reasons introduces a transient in the normal operation of the reactor and puts demand for a safety system for reactor shutdown, decay heat removal systems (if the trip results in tripping of the main coolant system) and so on. These transient conditions call for the start of the safety system to meet the safety functions and often some human actions. Any deviation from the set shutdown plant configuration, for example, one of the shutdown cooling pumps failed to start and operate on demand, calls for interlock logic to automatically actuate the standby pump and some operator action, if required so on. Here, there is a fear of the transient taking the plant from a normal shutdown state where manual error or nonperformance of the redundant system poses a threat to safety function capabilities. Finally, the reactor trip might cause a loss of unavailability or a threat to the shutdown cooling function, which would lead to a loss of services.

The liability issues come into account when certain accident conditions, be radiation or industrial accidents, attract legal and insurance aspects. The operating organization has the primary responsibility of any liability that is attracted as a legal provision to ensure suitable compensation for potential fatality or loss of property caused by accidents to the members of the public. Generally, the operating organization maintains insurance against this potential liability, and thereby, the insurance company is responsible for the compensation as per the contract conditions that specify, limiting amount and so on.

This section is concerned with the integrated operational safety performance indicator index that can be used with proper weightage factor to evaluate the integrated statement of risk. To avoid the whole exercise becoming too subjective, some national or international models of safety performance indicators can often be adopted, for example, the IAEA safety performance indicator system (International Atomic Energy Agency 2000a). These indicators conform to this attribute of the 'harmonized' procedure and have some common basis when two plants are compared. In the IAEA system, both deterministic and probabilistic attributes have been used.

8.2.1 Technological Accident Risk

The risk potential from an engineering facility operation depends upon the major hazard potential associated with the respective technology. For example, the risk potential from a nuclear plant has at its core the radiation as hazard, a chemical plant deals with chemical hazard, transport systems deal with vehicle or rolling stocks induced accidents, space system deals with loss of mission, loss of crew or even loss of science, or air born accident induced by aircraft/aviation systems, industrial hazard deals with health hazard to the occupational workers. Often, there are overlapping hazard potentials; for example, the nuclear industry has to address apart from nuclear radiation-related aspects and industrial and chemical as well as transport system-induced hazards. Every technology/industry must ensure that the system operation should reduce the hazard to occupational workers in general and the public, particularly to the minimum acceptable level.

Since nuclear safety is at the core of this book, the subject, i.e., 'operations risk management', provides fundamental guidelines and frameworks for operational risk management for nuclear systems, which, of course, can be adopted for other technologies with few systems technological specific requirements (International Atomic Energy Agency 1999b). The major deterministic components include maintaining the highest availability of safety systems, implementation of good practice, periodic audit, monitoring safety performance indicators along with use of moderns, etc. Furthermore, the development of risk-conscious culture, along with other applications of PRA, such as the use of optimized technical specifications, risk-based maintenance practices, the deployment of PRA-based operational support systems (Varde et al. 1998), aging risk assessment and management as well as regulatory review and life extension (Varde et al. 2011), is some of the vital aspects that need to be considered. Radiation risk assessment through the evaluation of the likelihood and consequences of potential accidents has been discussed at length in this book. Both the PRA and deterministic insights have been used as part of risk-based engineering. Hence, the scope of the section is to discuss operational risks associated with complex systems, which can be adopted for other safety critical systems.

For this section, the PRA model of the plant provides insights into the risk associated with various accident sequences. This, in turn, provides information on the combination of components, human action or software or systems at a higher level, the failure of which results in accidents. It is important to note that in many nuclear plants, risk monitors are used by the shift operating staff for managing operations, keeping an eye on the change in risk levels. This subject will be discussed in the application section of this chapter.

Apart from this, the availability of a simulator plays a central role in training, retraining and evaluating the human factor for normal operation as well as for accident conditions. It is accepted that the availability of a simulator is vital for reducing human error in real-time conditions.

8.2.2 Industrial Hazard Risk

As per the IAEA guideline, the industrial safety system should be an integral part of the overall safety philosophy, which includes ensuring excellence in industrial safety practices (International Atomic Energy Agency 2018). Furthermore, the IAEA safety series 2/2 (International Atomic Energy Agency 2016) provides that 'the operating organization shall establish and implement a program to ensure that safety-related risks associated with nonradiation-related hazards to personnel involved in activities at the plant are kept as low as reasonably achievable'. The nonradiation-related safety program shall include arrangements for the same elements as radiation safety, viz. planning, identification, implementation, monitoring and review of the relevant preventive and protective measures, and it shall be integrated with the nuclear and radiation safety program. The only difference between radiation safety and industrial safety is that the hazard of industrial safety is distributed locally and very in nature, e.g., chemical, fire, height, etc., unlike the nuclear hazard where the hazard potential is located at core, pool, etc.

The occupational safety and health (OSH) standards and guidelines at the national and international levels often provide an adequate knowledge base to implement the system to ensure that industrial risk is at the lowest in a complex system.

However, in the context of the risk-conscious framework, it is required that the human and organization factor aspects are revisited. This requires risk-conscious culture development and implementation in line with what has been discussed about the radiation risk model. The major steps are (a) there should be a top-down drive to sensitize the occupational staff about risk aspects derived from a structured approach, (b) it should be emphasized that at root, it is human factors that are responsible for the industrial risk, (c) work on strength and weakness of general and accepted attitude-related aspects and identify the areas for improvement of organizational and human attitude, (d) a system of review of industrial aspects and corrective actions and monitoring and documentation and (e) assess the safety compliance attitude of the organization and ensure that a robust and quality system exists to ensure affecting compliance to the comments of safety committees.

8.2.3 Plant Operational Unreliability and Unavailability Risk

The risk-conscious culture has the fundamental principle that risk reduction in a measurable way is an overriding priority, and the decision should be made considering the overall benefits of operation keeping risk as low as minimum. In real-time scenarios, the situation in the control room is often dominated by a complex scenario where choosing plant operation may be a beneficial option while enacting the measures through additional surveillance, administrative and technical provisions and so on. There is also consideration of introducing a transient in the system by tripping or shutdown of the plant while the risk appears to be apparent or notional. Furthermore, every situation has its own subjective/objective evaluation of losses associated with disruption by introducing a transient tripping or shutdown of the operation of a nuclear power plant and interruption in the electrical supply and its societal consequences in a nuclear plant. Uncertainty is isotope delivery to support health care systems.

However, in this era of globalization and a competitive market economy, nuclear energy must compete with other energy systems. The current safety philosophy provides an alternative option of risk-informed principles and methodology to address the traditional prescriptive and conservative approach in the regulatory domain.

In this context, even though complex and to a large extent subjective, the operating organization has to have a mechanism to assess the risk-associated loss of reliability, i.e., interruption-free operation of the plant as well as the loss of availability deliverables of operations. This requires an algorithm to be developed; however, the challenge here is that we are dealing with too many subjective parameters and conditions where an organizational approach is required to evaluate the risk associated with operational loss.

The trends of safety performance indicators along with plants and other specific indicators are used to evaluate the risk associated with operational loss. A suitable numerical weightage factor supports the estimation of the integrated risk component.

8.2.4 Security Risk

Unlike safety, which focuses on unintended conditions or events leading to radiological consequences from authorized activities, security focuses on the intentional misuse of nuclear or other radioactive material by nonstate actors to cause harm. There are two types of threats, internal and external; however, security risks normally focus on external threats (International Atomic Energy Agency 2010a).

There are two types of security risk to a nuclear plant: (a) external object-induced radiation hazard and (b) risk due to cyber attach (Park and Lee 2017a). Making a risk-based framework with its major elements, like (a) use an integrated approach of probabilistic and deterministic approach, role and evaluation of human factor, (c) monitoring and surveillance provision, (d) uncertainty characterization and (e) comparison with the goals and criteria. The deterministic approach evaluates the level of defense-in-depth provision, level of security protection in terms of physical protection, fencing, intruder detection system that includes CCTV camera, lighting, sensors, etc., and the human layer of security, the potential hazard or threat perception from the plant. The probabilistic model evaluates the threat conditional probability, adequacy of hardware, software and human layer of protection. Finally, this evaluation produces the strength and weakness of the security system and provides risk estimates of security risk.

The specific challenges in security risk assessment are modeling of software failure in digital systems of the plant that has the potential to block the system for its successful functioning and monitoring. Normally, these assessments are performed by employing the condition probability that the event has been initiated and the response of the security provision is evaluated.

In this section, the objective is to evaluate the security risk based on the security performance criteria and their trends moderated by the weighting factors. The IAEA safety performance shaping factors need to be expanded to include the security performance shaping factors as overlapping safety and safety aspects that require an interface to verify and validate and implement the intended objectives of the corresponding functions (International Atomic Energy Agency 2010b).

8.2.5 Liability Risk

Even though the probability of accidents for nuclear plants with the potential of radiation series consequences in the public domain, as the record of the nuclear industry shows, is very low compared to other sources of energy production. The reason is the multiple barriers for protection in nuclear plants. However, the public interest needs to be protected; however, the risk to members of the public is low. For example, the An Indian Parliamentary Act provides for civil liability for nuclear damage and prompt compensation to the victims of a nuclear incident through a no-fault liability regime channeling liability to the operator, appointment of Claims Commissioner, establishment of Nuclear Damage Claims Commission and for matters connected therewith or incidental (Government of India 2010, International Atomic Energy Agency 1964).

As mentioned, the operating organization is primarily responsible for civil liability. The implementation of this involves the availability of insurance against damage toward channeling the operator's liability for damage in the public domain, which provides adequate protection to stakeholders while ensuring the public interest. The insurance company might consider many plant-specific, socioeconomic aspects and potential risks from the operation of nuclear plants or any industry to workout insurance conditions, including premiums. In many countries, the core damage frequency, or release frequency from containment or the final statement of risk in the public domain, forms one of the metrics in support of decisions. The objective of this section is to have a preliminary discussion that should be indicative of liability as one of element of integrated risk, and the details do not form the scope of the discussion.

8.2.6 Integrated Risk Formulation

This section addresses the formulation of an integrated risk statement based on the five risk categories, viz. postulation radiation risk, industrial hazard, risk due to loss in reliability and unavailability and liability risk, as discussed in the preceding sections. The mathematical expression for integrated risk assessment based on the model given in Fig. 8.2 can be given as

$$R(t) = \sum_i r_i(t) \tag{8.1}$$

$R(t)$ is the time-dependent integrated risk from the plant, and r_i represents the individual component of the risk. The individual component is expressed as follows:

$$r_i(t) = w_i \ \mathrm{Li} \ * \ \mathrm{Ci} \tag{8.2}$$

For the we are discussing, the value of i ranges from 1 to 5, that represent nuclear, industrial, deliver, security and liability risk, respectively. Since these expressions carry subjectivity and imprecise expressions for variables, a fuzzy treatment of this model can provide an effective approach for integrated risk evaluation. The reason is that even though the statement of risk from nuclear radiation accidents is derived from a PRA, the other risk characterization will be largely based on expert elicitation.

Given this background, a fuzzy ANN model, as shown below, with five input nodes, 3 middle layer nodes and one output node can be considered. The ANN representation for obtaining the integrated risk statement is as follows:

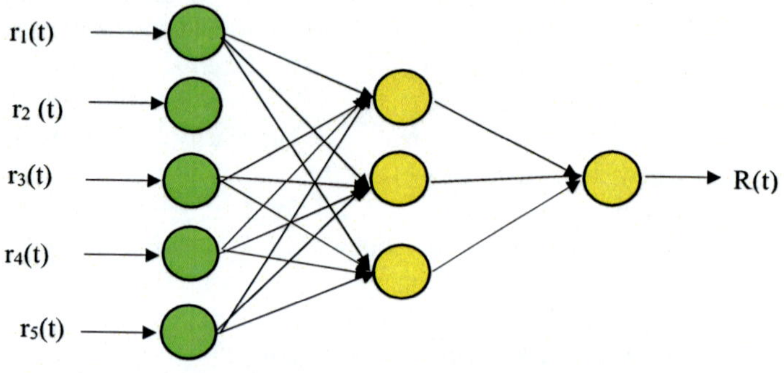

Fig. 8.2 Fuzzy ANN model for evaluation of integrated risk from the plant

The procedure to develop the input vectors for the fuzzy ANN simulation is as follows:

1. Chose the risk categories (here, it is out of five) required for integrated assessment
2. Normalize the rise statements for harmonizing required for integration modeling
3. Assign the weightage factor for each risk category based on the risk-significance inside and outside of the plant
4. Develop a fuzzy model
5. Develop a transfer function employing an ANN architecture with risk categories as input and metrics for integrated risk at the output layer
6. Perform training of the ANN employing 80% of the identified vectors
7. Perform the recall test using the remaining 20% vectors.

After satisfactory trials in plants or simulators, the decision is made to implement the system as part of a risk monitor. The output of the risk monitor developed using Level 1 PRA as well as probabilistic security risk assessment along with industrial, operational (plant unavailability induced risk for internal as well as external socioeconomic impacts) and liability risk the plant has improved insights into risk-based decisions in the control room. There is a conscious effort not to go into subtler and specific aspects on risk modeling as (a) the nature of this application is very subjective, (b) even though the fuzzy ANN approach has been employed for discussion, plant can use other mathematical or intelligent approach for modeling and (c) the procedure for selecting weights for each of the risk categories is a matter of judgment as also the normalization procedure.

8.3 Risk-Based/Risk-Informed and Risk-Conscious Approach

At this point, it is vital to understand the distinguishing features between risk-informed or risk-based approaches and risk-conscious philosophy with respect to operations management. The following are these features:

1. The risk-informed approach has its origin in the premise of supporting regulatory applications using PRA insight in support of decisions, while risk-conscious methodology has been designed for application in operations ecosystems.
2. The risk-conscious approach is an extension of risk-informed/risk-based methodology where the fundamental human existence is where the focus is on reaching human roots for reaching the basic human root causes such that an effective corrective mechanism can be implemented toward eliminating or reducing human errors.
3. The development and growth of risk-conscious culture is key to risk-conscious culture.
4. The objective of risk-conscious culture is to develop a culture where risk-based thinking becomes integral to all the functions of operations.

5. Risk-conscious culture is a way of life that does not remain confined to the physical limits of plants, but it is part of the consciousness of the operational staff.
6. The risk-conscious culture enables institutional and effective considerations of not only safety but also security through the attributes of conscience.
7. In risk-conscious culture, the emphasis is on operating the fundamental human model while performing precursor analysis, human root cause analysis and other elements, such as the treatment of imprecise qualitative input typical to the operational environment, while making this input to the risk model quantified estimates.

In summary, the risk-conscious approach is tailored for operational ecosystems while providing an effective mechanism for regulatory compliance. Figure 8.6 shows the evolution of consciousness from information to providing a risk-conscious culture. This figure underlines the core distinguishing feature of risk-informed to risk-conscious approaches to address operational requirements.

In other words, this figure underlines that information and knowledge are not adequate until they become part of conscious thinking and powers human cognition for improved performance.

8.4 Operational Safety Performance Indicators Approach

The development and application of a safety performance indicator approach provide an effective and integrated methodology that, apart from recording the status of indicators, provides an approach for trending these indicators. IAEA-TECDOC on operational safety performance indicators provides an effective assessment and management of operational safety (International Atomic Energy Agency 2000a). The IAEA indicators can either be directly used or can be adopted while maintaining plant characteristics and requirements of safety. There has been active interest in applying these standards. Figure 8.3 depicts the structural hierarchy of the IAEA adopted in the RCOM approach. In this figure, the text in white depicts the IAEA approach, while the text in black shows the RCOM approach and how it overlaps with the IAEA model of operational safety performance indicators (Fig. 8.4).

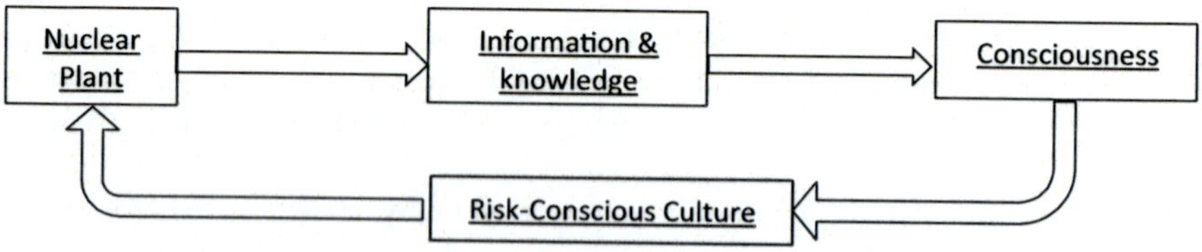

Fig. 8.3 Importance of consciousness vis-à-vis information

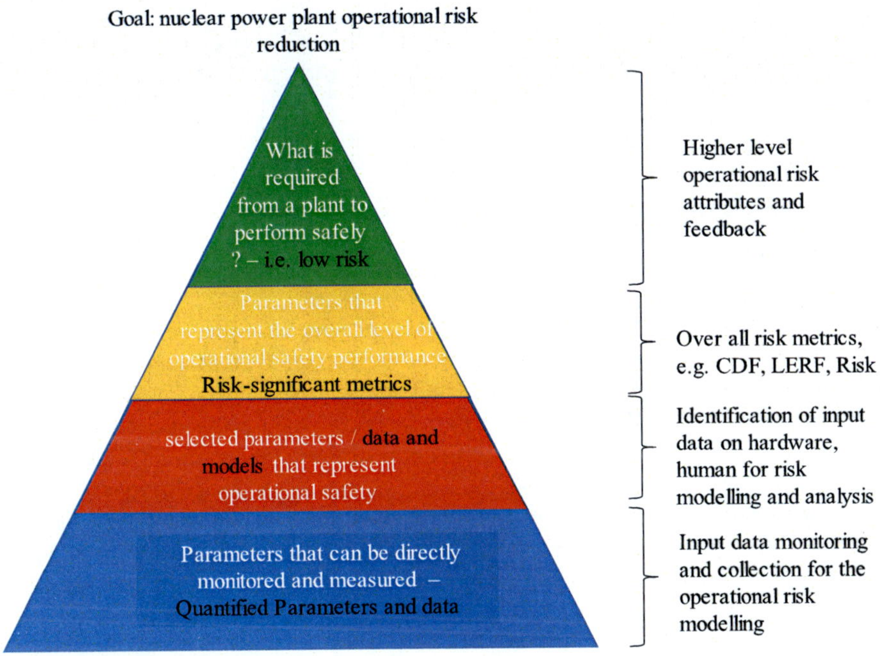

Fig. 8.4 IAEA framework adopted for the RCOM approach for operational risk management

IAEA implements three key elements, viz. (a) the plant operates smoothly—plant normal operation, (b) the plant operates with low risk—NPPs emergency condition and (c) the plant operates with a positive safety attitude—attitude of plant worker for safety. These elements have been amply addressed directly with risk reduction as the major driver in place of safety in line with IAEA model quantified attributes and metrics required for operational safety improvement. Since PRA is at the core of RCOM, the quantification aspects bet it as input or output in terms of plant initiating event frequency for process systems, system unavailability at safety systems and support system level and core damage frequency at plant level. Apart from this PRA, it provides an elegant framework for risk-significant prioritization for SSCs and parameters. It may be noted that IAEA metrics also provide PRA results, as discussed above.

Even the IAEA approach for operational safety indicators intuitively or naturally aims to quantify attributes; however, provision exists in terms of ranking the parameters and tracking and monitoring the trends in parameters. However, the PRA approach in RCOM provides a mechanism for monitoring hardware, human and software failure trends and prioritization for in-service inspection, maintenance management and the development of operator support systems.

Furthermore, the third key element directly or indirectly seeks to have a safety culture such that the attitude of the workers toward, procedure, work, inspection, surveillance, maintenance, etc., is positive. In RCOM, the risk-conscious culture is a common thread that runs through all the not only activities but also awareness, thinking and alertness as an attitude to have the highest level of dedication for the plant.

8.5 Integrated Operational Risk Assessment Management Framework

PRA in general and Level 1 PRA in particular provide an effective framework for operational risk assessment and management. In fact, the RCOM approach utilizes, apart from the management of issues on hand, long-term strategies for risk reduction to or eve below the acceptance limit, and the strategy will be discussed in the following section. Therefore, the risk assessment aspect being the subject of this book further elaboration is not considered necessary.

Figure 8.5 shows the major elements of operational risk management, viz. identify, evaluate, measure, impact assessment, prioritize, correction implementation, control, ensure monitoring and quality control of the modified system and the last but most important the documentation.

There are two ways to identify the risk potential from a plant: (a) by developing insights from a Level 1 PSA and plant accident analysis study and (b) real-time data on incidents, failures and near miss events and performing accident precursor analysis again by mapping the potential of these events employing the available Level 1 PRA risk potential.

Fig. 8.5 Major elements of operational risk management

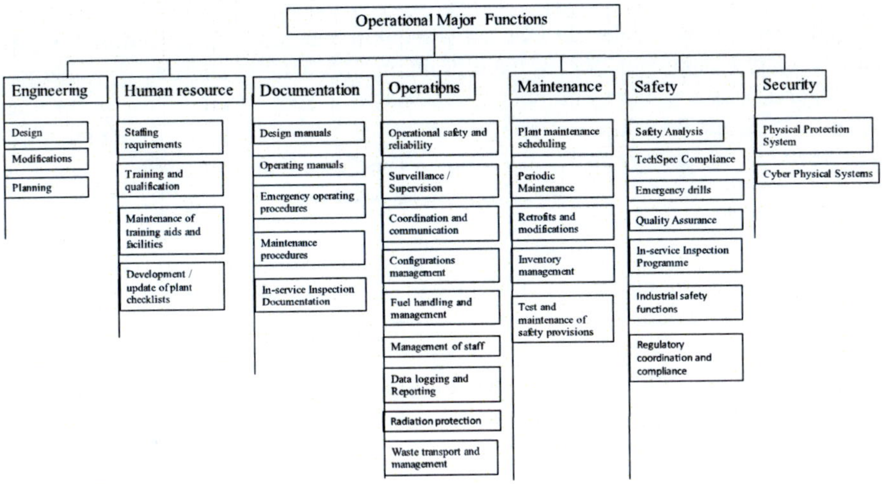

Fig. 8.6 Typical operation functions

8.5.1 *Identification*

The operation ecosystem in terms of major functions can be structured and presented as shown in Fig. 8.6. It may be noted that this depiction is not exhaustive, and many more ways can be used to define operational activities. However, for the purpose of this book, we have considered this representation. To create a risk-conscious ecosystem, we need to identify or select the area or activities as given in Fig. 8.6 for upgrades, modification or policy change. The term 'identify' also refers to the identification of external and internal hazards and the preparation of the list of applicable initiating events. Further identify the plant defenses, i.e., safety functions that include safety systems and human actions to bring the plant deviation to normal. Identify the resources required as well as the logistics and the scope and objective of the subject project. It is also required that the approach required and the decision making criterion be spelled out. For example, if a plant does not have a Level 1 PRA model of the plant, then the first activity should be safety analysis. This will drive one to take up probabilistic risk assessment (PRA) with at least limited scope Level 1 (International Atomic Energy Agency 1988). If it is decided to use a deterministic approach, the feasibility of the approach should be checked or verified.

8.5.2 *Evaluation*

Based on the available qualitative observation and experience and, if possible, historical data as well as associated performance trends, it evaluate the gap area, i.e., the available limitation of the subject system or proposal and the expected performance of the system particularly related to the assessment of expected risk reduction. Formulating a method/approach for evaluation among the options and decision metrics and criteria forms a major component of evaluation. For example, the evaluation process might include the development of an integrated approach employing, on the one hand, the deterministic/operational assessment and event tree for accident sequence while fault tree methodology for system modeling, on the other hand, toward obtaining the list of the minimal cut-set for establishing a qualitative basis for risk assessment. Furthermore, the scope of evaluation might include a comparison with a similar system elsewhere. This insight provides insights into the evaluation in terms of the strength and limitation of the evaluation options. The experience also suggests that plant data from records and simulator experiments that involve the design of thermal hydraulic simulation and operator response efficiency for a given situation, e.g., limiting temperature and other process constraints, provide robust ground for preparing requirements for quantitative risk assessment.

8.5.3 *Quantitative Assessment*

This is a critical step in risk-conscious operations management and deals with the development of quantitative and measurable metrics in support of support decision making. This includes, based on the insights and information from the evaluation process, the development and quantification of the PRA model by plugging in hardware, human and software reliability failure probability toward assessing the system unavailability and further core damage frequency as a measure of the safety level of the plant. Here, some aspects that change the course of the decision should be, as far as possible, to use the model that reflects the built and operating state of the plant. These aspects are consideration of common cause failure, assessment of uncertainty, assessment of human factors as applicable to plant operations and emergency procedures. Human factor review is also required for the subject change that is being retrofitted or incorporated in the plant. The system level unavailability estimates quantified list of accident sequences and the statement of core damage frequency generally form the core insights in assessment. It is a good practice to map human factors on the modified system and based on these provisions, such as requirement of automation—if the time window for human action is a limiting factor, review of emergency operating procedure—toward confirming that plant provisions are meeting the requirement of emergency operating procedures, and last the reevaluation of emergency operating procedure for establishing its capability in terms of meeting the characteristics of the postulated accident conditions, e.g., operational and environmental stresses, spare part availability feasibility to plant staff in time, transport of material to and from the site, etc.

8.5.4 Prioritization

The fourth step, i.e., 'prioritize', addresses the ranking of hardware and human action, particularly the common cause failure aspects, based on their respective risk-significance. *Quantitative measure is also referred to as the importance measure.* The importance measure of an SSC can be determined in success space considering the impact of an increase in the reliability of the system due to an increase in the reliability of individual components. Similarly, the importance measure can also be determined in the failure space by considering the impact of an increase in the failure probability of an individual component on the overall system failure probability. Since the objective, here, is to use the minimal-cut-set obtained from PRA to evaluate the risk-importance measure, the scope of the discussion will be limited to risk-importance measures. In principle, the risk importance of the component is a function of component unavailability and the location of the component in the system. The risk importance of components in parallel, redundant or standby configuration will be less compared to the series configuration. Even for redundant configurations, the common cause group (CCG) comprising parallel, redundant or standby units will always carry higher importance. When we look at the list of cut-sets, most of the time, the CCFG tops the list when arranged according to risk-importance measures. Before, we go ahead with discussion on risk-importance evaluation, we must understand that the cut-set comprised of single, or more than one element. Any one element failing in the cut-set does not result in an accident scenario. All elements of the accident sequence should fail then only the accident occur. The second aspect is the analysis level, i.e., the system level or plant level. The system level analysis produces a list of minimal cuts comprised of safety function unavailability, and the safety function includes component failure, human error and common cause group that lead to system failure. Therefore, we are dealing in the unavailability zone.

When the analysis is performed at the plant level, the minimal cut-set list comprises an initiating event and the safety system or safety functions. The initiating event frequency and the safety function unavailability together lead to the minimal cut-set or accident sequence frequency. Hence, it can be summed up that at the system level, the minimal cut-set is unavailable, and at the plant level, the accident sequence has frequency as its unit.

A general mathematical model of the risk-importance measure at the system level can be formulated as follows:

$$I_i(t) = f(\Delta U_i(t), \Delta U_S(t)) \tag{8.3}$$

where $I_i(t)$ is the risk-importance measure of component i, $\Delta U_i$is the change in component i unavailability, and $\Delta U_S\ (t)$ is the change in system unavailability due to the location of component i in the system configuration (series/parallel). The above can be applicable at the system level in PRA, where the fault tree (for a complex system) or reliability block diagram (for a relatively simple system) techniques are used for modeling.

We will discuss here the Fussell-Vesely (FV) Importance, Risk Reduction Worth Importance (RRW) and Risk Achievement Worth (RAW) Importance measures as follows:

8.5.4.1 Fussell-Vesely Measure of Risk-Importance

At the system level, the Fussell-Vesely importance measure for component i is given as:

$$I_i^F(t) = \frac{F_i[U(t)]}{F_S[U(t)]} \tag{8.4}$$

where $F_i[U(t)]$ is the contribution of component I to the overall system unavailability, and $F_S[U(t)]$ is the reference system unavailability. This model facilitates the estimation of component risk importance, and once components are arranged according to their importance, we have a list prioritized according to their risk-significance or risk impact that can be used to prioritize operational, maintenance and test activities.

As mentioned, the list of cut-sets, i.e., accident sequences, along with their predicted frequencies, is one of the major outputs of a Level 1 PRA. The accident sequences when seen in conjunction with the deterministic accident analysis provide information on the level or severity of consequences. Here, we have the information on the frequency or likelihood and consequences associated with each accident sequence, i.e., the risk associated with the accident sequence.

To make a practical use of this vital information, we list the cut-set along with its frequency of occurrence to arrive at the risk importance of each cut-set and to prioritize the cut-sets such that the resources can be optimized considering the risk-importance measure.

The risk-importance models, viz. Fussell-Vesely, are given as follows:

$$I_i^F = \frac{I_i(t)}{I_S(t)} \tag{8.5}$$

$I_i(t)$ and $I_S(t)$ are the frequency of the subject cut-set (i) for which the importance measure is being evaluated and the system summation of the failure frequency of all the cut-sets in the system, respectively.

8.5.4.2 Risk Reduction Worth Measure of Risk Importance

The results of PRA at an intermediate stage are expressed by a list of risk-significant cut-set lists. The Risk Reduction Worth (RRW) importance measure also works with the cut-set list (at the system or plant level) for arriving at the risk importance of an element of cut-set, viz. initiating event, component, common cause group or human action. The RRW assessment addresses obtaining the plant level accident sequence or minimal cut-set frequency by assuming that the subject element of the accident sequence is 'perfect' by setting the unavailability of this element to 0 in all the minimal cut-sets having this subject element. Obviously, the minimal cut-sets having this element, be it component failure, initiating event, human error will be eliminated from the reference list of plant level minimal cut-set or accident sequence.

Now, we have two frequencies, the reference accident sequence frequency, which also includes the accident sequence having the subject component, which is the summation of all the accident sequences obtained from the Level 1 PRA. This is the core damage frequency estimate. Second, we have the core damage frequency obtained by excluding the minimal cut-sets having the subject element. The Risk Reduction Worth and Risk-achievement Worth enable evaluation of the risk-importance level from the minimal cut-set or accident sequence list as follows:

The RRW is expressed in two ways, ratio and difference, as follows:

The RRW ratio model is as follows:

$$I_i^{Rw} = \frac{F_P[\mathrm{CDF}(t)]}{F_P[\mathrm{CDF}(t) I U_i(t) = 0]} \tag{8.6}$$

F_P[CDF] is the frequency of plant CDF.

$F_P[\mathrm{CDF}|U_i(t) = 0]$ is the CDF frequency obtained after setting the element i unavailability to 0, i.e., the component is highly reliable.

Accordingly, RRW provides a measure for risk reduction by increasing the subject element reliability such that operational and maintenance resources can be optimized. This enables the identification of elements/components for reliability improvement to effectively implement the risk reduction program.

The RRW difference model is as follows:

$$I_i^{\mathrm{RW}} = F_P[\mathrm{CDF}(t)] - F_P[\mathrm{CDF}(t) U_i = 0] \tag{8.7}$$

8.5.4.3 Risk Achievement Worth Measure of Importance

The Risk Achievement Worth (RAW) measure of importance is an effective measure where the objective is to achieve a set risk goal. For example, the current core damage frequency, say, is 2.0×10^{-5}/year, and the regulatory target has been set to 1.0×10^{-6}/year. Therefore, the objective is to achieve the target CDF by identifying the components sensitive to an increase in unavailability. This RAW is a useful measure for this type of situation. The procedure involves setting the element I unavailability to 1.0 (opposite of what we did in RRW).

Similarly, the RRW RAW can be measured by ratio as well as difference techniques. The.

Ratio model:

$$I_i^{RA} = \frac{\text{CDF}_P[U(t)U_i = 1]}{\text{CDF}_P[U(t)]} \tag{8.8}$$

Difference model:

$$I_i^{RA} = \text{CDF}_P[U(t)U_i = 1] - CDF_P[U(t)] \tag{8.9}$$

As seen, the effect of risk increase by setting the system SSC unavailability to 1.0 will result in risk increase, thereby enabling identification of CCS by analyzing the risk sensitivity to increase in unavailability. A plant often comprises over tens of 1000's of SSC, and a computerized program can easily formulate a list of the component's importance for system risk prioritization.

8.5.5 *Impact Assessment*

Impact assessment is either performed as part of risk assessment or during risk management. Making assumptions related to data, models and uncertainty in knowledge forms part of risk assessment. Apart from this, there are situations when the uncertainty associated with the data are relatively high to provide required confidence in the analysis. One of the approaches to impact assessment is performing *sensitivity analysis*. For example, for one component, say, rupture disk a passive component that provides passive barrier between two interfacing system locations. Rupture disk provides a positive and often passive barrier between two interfacing system, say main coolant recirculation system and the emergency core cooling (ECCS). This rupture disk is supposed to rupture during a loss of coolant accident (LOCA) scenario for injection of coolant into the main coolant recirculation system, while during normal operation of the plant, it should remain healthy (leak free). For such components, it is a challenge to obtain reliability data, particularly for the required specifications from a generic source. Normally, a limited number of tests are carried out to obtain confidence. However, these data are not adequate to have a robust probability of failure data. The analyst assumes a failure probability of 1×10^{-3} during the modeling and analysis. However, this was one assumption, and its impact needs to be analyzed. Therefore, a sensitivity analysis is performed by varying the probability of failure data from 10^{-2} to 10^{-4} and how it affects the core damage frequency results in PRA. If the results are within the permissible range, then this assumption stands validated. Similarly, if it is felt by analysts that the uncertainty band is relatively higher for a data point, a separate sensitivity analysis is performed to ensure that this much uncertainty is acceptable in some situations.

8.5.6 Corrective Actions

One of the major outcomes of the PRA study is 'recommendations' of the study related to identifying weakness or limitation in plant configurations, safety management and operations policy aspects, overall human factor requirements, test and maintenance requirements aging issues, etc. Further, specific tasks might include a review of the testing frequency of SSCs, upgradation/retrofit/requirements or incorporation of modification in the existing system, changing the operation policy related to allowable outage time for maintenance, identifying SSCs that are not part of the testing schedule or even identifying some sections of the safety system that, if affected in an unsafe manner, will remain latent until the demand is placed on the system.

Regulatory processing and compliance are integral components of corrective programs. Therefore, a detailed safety analysis is subjected to an independent review before initiating or submitting the application for regulatory review. Risk impact assessment, e.g., through a sensitivity analysis or by simulating the sessions in a plant simulator for ensuring the reliability and capability of the modifications, upgraded or modifications and this aspect, i.e., internal review, is vital to ensure that the final objective of risk reduction is met. These studies, at times, are performed on a living PRA model or through a risk monitor.

In many cases, the corrective action program involves replacing the existing SSC with new SSC in the system. The initial assessment can be carried out with available plant-specific data on other systems or the data available from generic systems. As the components are subjected to increasing testing and operation, the data are built, and reliability estimates can be generated for the subject component.

8.5.7 Quality Control and Quality Assurance

Once the corrective actions program is implemented, planning is required to identify the performance trend monitoring in the affected system, services and policies. A reporting mechanism that includes meeting regulatory oversight requirements and associated compliance modalities should also be developed. The quality checklist developed for the PRA application analysis should form part of regulatory submission such that adequacy of the change can be discussed. The quality attributes for respective applications in available national and international standards can be the reference guidance document for this purpose. The aim here is to ensure that the risk reduction objectives have been achieved. A quality assurance program should be developed for hardware, software, human factor aspects and environmental factors to achieve the goal of risk reduction. A retraining and refresher program is required to ensure that human factors such as system performance in terms of failure-free operation, availability and safety depend on human factors such as operation, maintenance and services.

8.5.8 *Documentation*

The change should be affected in all the related drawings, manuals, procedures and checklists and window forms for the dissemination of knowledge throughout the plant operating agencies. Regulatory communications and compliance documents should also form part of the training and qualification program.

8.6 Precursor Analysis

This section presents the precursor analysis framework that enables proactive approach accident prediction due to SSC failure human or organizational factors. The risk-conscious approach to accident precursor analysis assumes that human failure is a common denominator for any failure.

8.6.1 *General*

Accident precursor analysis (APA) in its present form is relatively new compared to the traditional approach to operational safety that involves routine/periodic plant surveillance, walkdowns, clock plant/field area supervision and further incident investigation, safety performance review identification, documentation, etc., toward looking for deviation from normal operations or anomalies and identification of latent faults such that enhanced monitoring, implement immediate and long-term proactive or corrective actions such that anomalies do not pose a threat of failures.

However, the operating experience in general and the accident investigation or root cause analysis of failures that resulted into the accidents revealed that the symptoms of incipient failures were available well in advance in the form of deviation from normal operation or anomalies. The record of the accidents in the nuclear industry shows that the indications of latent anomalies are often revealed when incident investigations are carried out after the event becomes a reality.

8.6.2 *Review of Major Accidents and Events and APA Requirements*

TMI 1976—The PRA study showed that the pressurizer tank relief valve failing to close mode will lead to the system inventory loss and might lead to loss of coolant from the system. The operator error that led to blocking of ECCS can be traced back to procedural or training limitations (United States Nuclear Regulatory Commission

1980; Phimister et al. 2004). In Chernobyl 1986, factors such as specific characteristics of the reactor and design of the control rod and associated enabling of the control rod and procedural violation were directly related to plant safety culture. In fact, two earlier accidents at RBMK reactors had already indicated weaknesses in the characteristics and operations of RBMK units, and the accident in one of the two of these RBMKs (Leningrad Unit-1) is considered to be by some as a precursor to Chernobyl (International Atomic Energy Agency 1992). In the case of Fukushima in 2011, the operator-regulatory compliance protocol and associated compliance culture along with the management of emergency scenarios, particularly decision making in the public domain, led to the final conclusion that this disaster was manmade (Diet of Japan 2011). The NEA report on precursor events has highlighted the nature of the precursor and one overall conclusion that all three major nuclear power accidents (TMI, Chernobyl and Fukushima), as well as several of the lesser-known close calls, had precursors in the previous, although often or in the same country (Nuclear Energy Agency 2014). Furthermore, in the Devis Basse plant, many operational observations provided evidence of an off-normal phenomenon, boric acid-induced degradation of the carbon steel nozzle location and an indication in the ventilation system regarding the change in air quality as a precursor for a passive phenomenon that was revealed as nozzle degradation due to chemically induced corrosion (Report: DAVIS-BESSE Reactor Vessel Head Degradation 2022; NASA Office of Safety and Mission Assurence 2006). In the area of space transport systems operations, the anomalies whose failure mechanisms were integral to the two major losses or major accidents, viz. Columbia and Challenger occurred prior to those accidents. The accident investigation reports found that the processes in place at the time did not respond to the prior anomalies in a way that shed light on their true risk implications (NASA 2011).

8.6.3 Identification of Gap Areas

Even though operations activities such as periodic reviews, data logging, surveillance, supervision of specific observations during maintenance and plant walkdown and regulatory reviews have the potential to capture symptoms or anomalies that deviate from normal conditions/operations and are further analyzed for their potential for being candidate precursors to accidents, the general experience, as we have discussed in the preceding paragraph, is that they often may not always form a candidate for serious considerations such that the consequences of failure or accident are reduced or avoided.

From the above discussions, it can be seen that accident precursor analysis and management form part of the accident investigation, i.e., after the accident has already occurred, and to that extent, this is a reactive approach for affecting the corrective or mitigating action program (Phimister et al. 2004). However, the practical requirement is the implementation of a proactive approach to the accident precursor analysis framework. The weakness in the systems identified through the detailed analysis and corrected before the accident or failure or the consequences actually occur.

Recent literature shows that there is a growing interest in the development and application of precursor analysis models and methods, particularly toward identifying anomalies, analyzing anomalies for their potential for being precursors to incipient failures and accidents and eliminating or avoiding failures or accidents, proactively. Here, along with the traditional deterministic operational management approach, the PRA approach provides an effective framework for mapping operational data on a PRA model for accident potential analysis. It can be argued that, keeping in the lessons learnt, the international atomic energy agency has been actively pursuing and encouraging the deployment of resources for developing and integrating the APA program in nuclear power plant operations member states (International Atomic Energy Agency 2004). The PRA-based approach introduced a framework that is more systematic and rational based and further provides a quantitative basis for the evaluation and prioritization of precursor events. The system unavailability and conditional core damage probability provide the required metrics for decision making and prioritization of the safety issues. The available literature also shows that even space industries such as NASA have a framework for APA (NASA 2011). The space system is primarily the mission critical in nature of the APA approach, even though at the core, it remains the same; however, the treatment of APA elements is system specific. For example, in NASA APA analysis, the loss of crew, loss of mission, loss of vehicle or loss of science form the decision metrics. The major elements for space systems are building a case load from operational data, identification and generalization, evidence gathering and grading of anomaly conditions. In both, i.e., nuclear as well as space methodology analysis of human action along with organizational factors forms an integral part of the analysis.

8.6.4 Risk-Conscious APA Framework

The discussion thus far has established the requirement of the APA program in an operations ecosystem and upgraded APA to have risk as the driving factor for nuclear plants, where the PRA model and insights are central to risk-conscious initiatives (International Atomic Energy Agency 2001, 2004). It has been established in every sense that precursors to accidents were there in most of the accidents; however, their risk potential was underestimated. The second observation is that for this lapse or error, human or organization elements were essentially responsible. This fact was most of the time not found to require attention, as even in many instances, human or organization error or limitation was identified as a direct cause of the accidents; however, in other cases, SSC or software failures were identified as a direct cause, and a detailed and in-depth analysis would have pointed to human or organizational limitations for these so-called SSC or software failures. The review of methodologies further pointed out that there is a need for identifying, viz. the case in point is NASA's approach that the mechanism of failure is also critical; therefore, for a robust APA framework, failure mode and mechanism and effect analysis (FMMEA) or an improved science-based approach is required to understand the deeper failure mechanisms at the material and physics levels. The major requirement for an effective APA framework is an approach for evaluating the time taken for developing an anomaly into a failure. Here, the risk-conscious approach proposes a physics-of-failure or data-driven approach for developing prognostics and health management techniques such that corrective action can be initiated before the onset of failure. The last availability of at least a Level 1 PRA is a must for assessment of the risk potential of a precursor and further general management of the precursor program.

Figure 8.7 shows that the risk-conscious APA framework, of course developed for nuclear plants, can be adopted for other safety critical systems. This section will discuss the major elements of this framework and associated rationales.

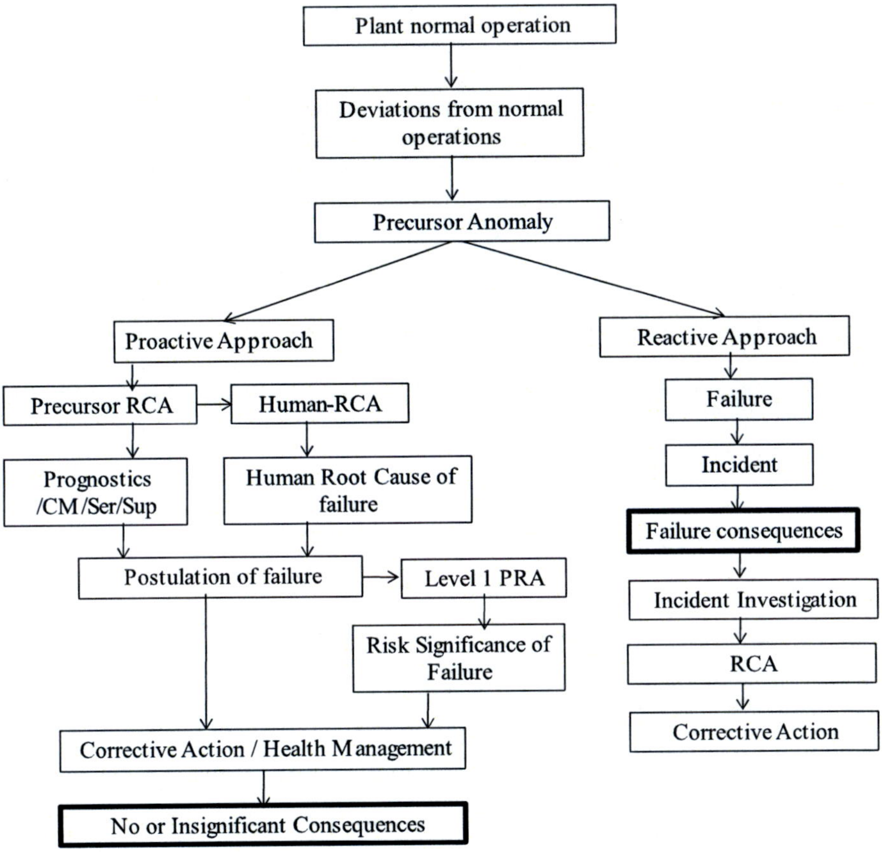

Fig. 8.7 Risk-conscious approach to the accident precursor analysis program

8.6.4.1 Plant Operation

In the context of this discussion, in line with the definition of a precursor, any deviation from normal operation is referred to as an anomaly. In addition, the subject here is most of the accidents before it becomes a reality which provides the symptoms (or may remain dormant/latent) in the form of either as SSC performance or change in physical or chemical design or operations characteristics, or a definite trend in human or organizational behavior or culture that there is an incipient accident or failure, or consequences are in the offing. Only it is matter of time when the right opportunity or additional event combinations of conditions will make the undesired condition a reality. Here, normal plant operation refers to power operation, low-power operation and shutdown for schedule maintenance. The deviation from normal operation refers to anomalies of plant performance, SSCs and human behavioral symptoms that might need attention for detailed analysis.

8.6.4.2 Precursor Anomaly Recording and Management

In the RCOM framework, the identification of anomalies follows a two-pronged approach. First, the identification through established deterministic plant activities, such as a conscious approach to surveillance, walkdown, supervision, maintenance, tests, tracking for human and organizational aspects, without any prejudice such that the input or observation does not receive the notice of a system of anomaly recording. This aspect can be left to a module called screening, where a conscious decision is made about the risk potential of an anomaly by employing typical PRA precursor nodes, where the input is subject to processing through initiating event or system unavailability potential. Establishing the operations anomaly case log and processing includes a procedure for dispensing each case by mapping it through a deliberative and a quantitative screening procedure. An anomaly is subjected to a) operational identification and review and second through the PRA procedure for screening against the established PRA metrics.

8.6.4.3 Binning for Potential Safety Significance

Even though the operating plants do have provisions for capturing anomaly detection, and most of them have the potential to exhibit the characteristics of a proactive approach, for example, periodic supervision of plant systems, surveillance programs, quality audits, etc., the history of accidents and events suggest that additional or alternate and more formal, improved and structured approaches with feedback and corrective action and tracking provisions are required that are effective in mapping the anomaly to failure such that risk reduction can be achieved and evaluated through quantified metrics. Second, all the anomalies do not require elaborate resources, as they do not pose a challenge for the risk of accidents. This points to a graded approach to precursor management. There can be many levels; however, Fig. 8.7 shows the two levels of the proactive approach and the traditional more or less reactive approach. Even though screening is an elaborate procedure, for the purpose of two-level binning, some examples of major guidelines/rules are formulated as follows:

1. An anomaly that has the potential for common cause failure should be given the highest priority.
2. The preliminary mapping of anomalies on the PRA model, where the anomaly could be a precursor to an existing initiating anomaly, should be put under an anomaly of higher significance.
3. An anomaly with the potential for catastrophic failure after a critical point should be considered a threat event and may be considered a priority.
4. Any behavioral trend at the individual or collectively at the organizational level should be given the highest priority, particularly because it has potential safety and security aspects.
5. General attributes that point to human attitudes, such as alertness, awareness, integrity, honesty, commitment to duty, promptness and the ability to work in a team culture, should also form guidelines in view of short- and long-term safety and security goals and accordingly be treated for binning.
6. The screening approach should be based on apart from deliberative, PRA-based mapping against the established quantitative criteria/metrics.
7. Certain events/anomalies might require detailed analysis for evaluating their threat potential; on a conservative basis, these events can be binned for proactive approaches.

The guidelines, as above, provide the basis for binning these events either for proactive evaluation or reactive evaluation or can even be ignored for future decisions.

8.6.4.4 Proactive Approach to Precursor Analysis

As shown in Fig. 8.8, the main modules of the proactive approach that includes root cause analysis, if required to the depth such that it can identify human root causes; prognostics for assessing remaining time for the anomaly to mature as failure, in fact even failure mode mechanism and effect analysis can also be used to assess the terminal consequences; the outcome of this step is understanding the elements like nature, mode, severity and time available for management of the consequences of the failure; the mapping of the postulated failure on Level 1 PRA model of the plant could be a parallel activity where the final evaluation can be made on the basis of information that has been consolidated about the elements discussed in the previous step; the final iteration or mapping of the failure scenario/failure on PRA model provides the risk-significance of the event, and accordingly, the health management or corrective action program can be implemented along with the trending and monitoring the performance of the corrected system. The following section briefly presents the core elements of each step as part of proactive monitoring.

Generally, RCA is performed after an accident or failure has occurred as part of an incident investigation to understand the underlying causes of failure that contribute to

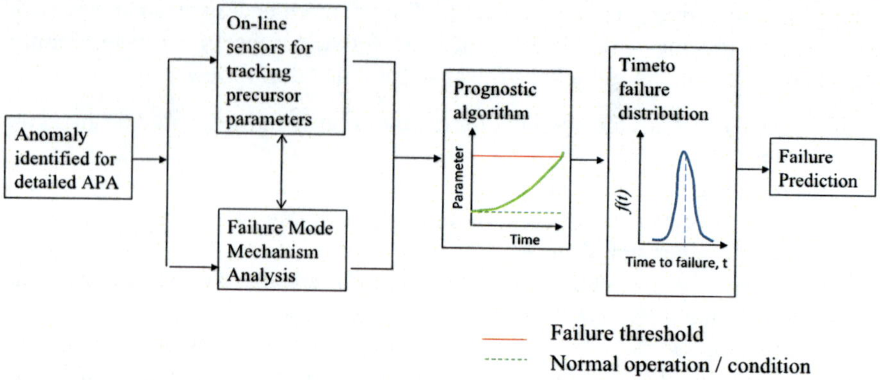

Fig. 8.8 Simplified representation of prognostics as part APA implementation

the occurrence of undesired events, as shown as an element of the reactive approach. Precursor RCA requires a mechanistic or physics-of-failure approach to model the factors responsible for deviation from normal operation/condition, further trending and prediction of the instant of failure, which can be achieved through science-based physics-of failure or data-driven prognostic algorithms (Varde and Pecht 2012). The challenge for precursor RCA is to model for predicting degradation trend and second the evaluation of failure threshold. This is needed, as apart from the probability of failure, we are also looking for its instant of failure with acceptable uncertainty such that corrective or health management schedules can be decided such that the consequences of failure are either avoided or reduced.

Once the anomaly has been identified, there is a need to make a decision about whether the mechanical or structural analysis will be performed based on data driven, and then the available data will be subjected to analysis to map the level of degradation. Alternatively, if the requirement is to employ physics-of-failure or mechanics of failure, then the prediction will follow the science-based approach and where environmental testing will form integral to modeling. If the approach is a failure mode or the prognostic approach is a resource consuming science-based option, as shown in Fig. 8.8, it is directly applicable to SSCs. The other approaches include conditioning monitoring and periodic in-service inspection, surveillance testing of safety systems, operational audits, etc. If the anomaly is related to human or organizational aspects, then the approach to track and correct the deficiency will require another approach through simulator training and experiments, motivational exercises, administrative changes or even spiritual roots. This aspect will be discussed later as part of human root cause analysis in the following section. However, whether it is SSC failure or human or organizational deficiency, the aim in risk-conscious operations management is to reach the human roots responsible for the deviation and eventual failure potential or failure itself.

Apart from trend monitoring, the following step involves the development of a distribution of time to failure such that time to failure can be predicted with reasonable accuracy. This aspect depends on the availability of data either obtained through experiments or using plant-specific experience or generic data from similar facilities.

The prediction accuracy as discussed depends upon the parameter, set of sensors, accuracy and dependability of the physics-of-failure (Pof) model and eventually the probabilistic estimation modeling and accuracy of estimation along with uncertainty bounds.

8.7 Consideration of Human Factors in the RCOM

In risk assessment, the reliability data are categorized as hardware and software failures, human errors and common cause failures. Even though it is not practical and may not be required to arrive at the fundamental cause, i.e., human error, the RCOM approach takes a principle position that even hardware and software failure have its root in human failure, be it design, material selection or manufacturing, operational, organizational, environmental and quality control issues (Varde 2021). In the operational environment, human failure also needs to be considered from beyond the man–machine interaction failure perspective and requires an open window that enables the identification of organizational issues that might have led to failure. Based on the experience and insights from root cause failure analysis of incidents and accidents, these considerations highlight that human error generally forms the root cause(s) of failure. For example, failure of a motor operated valve could be due to failure of motor to start due to overload, and this overload was caused by jamming of shaft with the housing—a mechanical failure that could not be identified by higher current meter reading available locally in the previous maintenance schedule. Here, even though the cause that is included in PRA modeling is failed to open—a hardware electrical failure, but the RCA, if performed, would indicate that actually it was human error to identify and correct the fault.

There are situations when the design fault remains latent and reveal only when the component is integrated into the plant or during a condition that was not postulated during the design. Similarly, inadequate quality control during design, fabrication and acceptance testing due to inadequate specification or improper QA procedure reveals during operations phase (Quality Control Techniques—Production Operations Management 2022). At times, one comes across a situation when an ad hoc decision/provision to get over or as an immediate solution for facilitating plant operation for a limited period or action for the next available opportunity does not find attention and remains in a compromised position until the failure occurs. These failures could be attributed to material selection, manufacturing or quality control organizational issues.

What follows is recognizing that along with the direct cause of the failure, the root cause should also be analyzed to avoid or reduce the recurrence of the fault. The most important insight from the above discussion is that 'an organizational consciousness that every failure has a root of human error' goes a long way in sensitizing the staff to remain alert and aware about the job and consequences of failure.

Second. The existing approach to risk analysis starts from the component or human action type and mode of failure as the basic event along with the associated failure probability. This procedure works in support of significant component safety prioritization. In risk conscious culture, the only fundamental roots or basic causes for failures should be used as much as possible. It is recognized that it is not possible to reach the human root causes responsible for failure, and an attempt should be made to perform the analysis until the identified intermediate root has the potential to eliminate or reduce the chances or recurrence of the subject failure. In risk-conscious culture, the objective is to understand human behavior and use this information to improve the risk management system. Therefore, we can say that in risk-conscious cultures, human and organizational aspects acquire higher priority. Let us discuss the implementation aspect of the above approach. Let us assume that in the plant, there were subsequent failures.

8.8 Root Cause Analysis with Special Interest on Identifying the Human Roots

There is an intrinsic assumption in the root cause analysis procedure in RCOM that for most of the failures, be they hardware, software, organizational and management related, the fundamental root cause is human individual or organizational error (Worldcat 2017). The second assumption is that to address the issues related to human failure, one must go down to the level of human specific causes fundamental causes of failure. This means we need to have a human model and the failures should be related, failure of consciousness, cognition, conscience or any physiological aspects such that recurrence of a type of human weakness in terms of these human attributes such that not a specific error or typical omission, commission and so on, but in terms of fundamental roots and the organization can focus on these attitudinal aspects to avoid recurrence of the failure.

As shown in Fig. 8.9, the root cause analysis comprised three levels. Level 1 is preliminary to understand the type of failure, whether it is design, manufacturing/construction/installation/commissioning, quality, O&M, training and qualification, or organizational or management-related failure. Generally, the incidence/failures that form part of Level 1 analysis have very low risk-significance.

Level 2 analysis is performed in more detail, and the idea is to understand the type of failure in terms of conventional taxonomy, viz. error of commission, error of omission, mistake, lapse, etc. This requires more data and information to support the categorization.

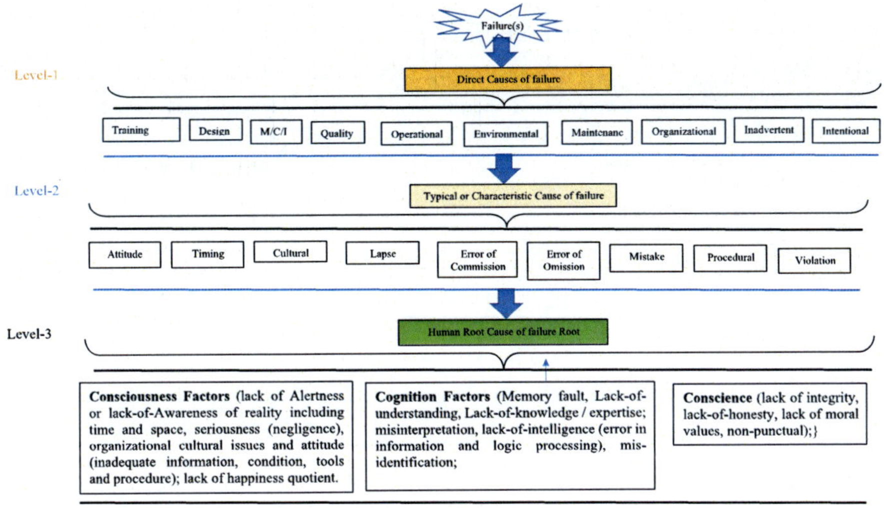

Fig. 8.9 RCOM approach to characterize the causes of human error

8.9 Risk Metrics for Operational Risk Management

The safety philosophy of the plant ensures an adequate safety margin between the operating regime and accident conditions to ensure a higher level of safety. Table 8.1 shows the major elements of risk-conscious operations management metrics. The idea is to implement conditions such that the accident probability is very low. For example, the core damage frequency of the plant is ~ 1.0×10^{-5}/year, which means that the plant might not attract core damage. On the other hand, we have an operating regime where safety-significant events at times threaten the operating regime boundary. There are safety aspects that limit plant reliability and availability. Risk-conscious operations management considers a metric that is graded in terms of risk such that the metrics can be used to obtain risk insights from risk scenarios ranging right from the operating regime to the risk regime, such as core damage frequency, by employing deterministic and probabilistic criteria. The advantage is that these intermediate criteria act as precursors to core damage or system unavailability and provide early indications of plant risk-conscious culture.

The major probabilistic indicators or metrics for operational risk management include the trends of plant core damage frequency, safety systems, engineering safety features and safety support system unavailability, frequency of initiating event frequencies, common cause failure events and their comparison with the national goals and criteria. The evaluation of uncertainty associated with these estimates should form an input in operational decision making. Any estimate with an unacceptable level of uncertainty needs not form part of the decision making. In terms of log normal distribution, an error factor of < 3.0 can be considered acceptable. Keeping

Table 8.1 Risk-conscious operations management metrics

S. No	Metrics	Target value mean/median	Remarks
1.	Level 1 PRA: core damage frequency (per year) Old plants (built before eighties) New plants (built after eighties)	$< 1 \times 10^{-4}$ $< 1 \times 10^{-5}$	PRA insights: These core damage frequencies are applicable for full-scope Level 1 PRA that includes full power, low power and shutdown states and other internal hazards (internal fire and flood; and external events, that includes, seismic, external flood, aircraft crash (Uncertainty bounds have not been shown)
2.	Accident sequence that has potential to core damage	$< 5 \times 10^{-4}$	This is a precursor condition or state of core damage That means core damage has not occurred but the accident sequence in question has threat perception to the plant. For example, even though 3/3 is the failure criteria for shutdown cooling pumps, but the 2 pumps have remained unavailable for a short time during a shutdown
3.	Violation of technical specification condition, either safety limit, limiting condition for operation, safety system setting, surveillance program	Operating regime boundary safety limit/limiting condition for operation is violated	Deterministic indicator filing the gap between normal operating regime and threat perception
4.	Increasing trends in common cause failure of safety systems	More than 1	Deterministic indicator filing the gap between normal operating regime and threat perception
5.	Increase in human error as root has been identified leading to significant events	More than 3	Deterministic indicator filing the gap between normal operating regime and threat perception
6.	Major plant fire/flood/seismic event	2×10^{-1}	Deterministic indicator filing the gap between normal operating regime and threat perception

(continued)

Table 8.1 (continued)

S. No	Metrics	Target value mean/median	Remarks
7.	Individual shutdown system unavailability (considering that there are two independent shutdown systems)	$\leq 1 \times 10^{-3}$	If there is only one shutdown system, then the unavailability should not be $< 1.0 \times 10^{-5}$ ($\leq$ SIL-4 limit) IEC (International Atomic Energy Agency 2010a)
8.	Decay heat removal system unavailability	$\leq 1 \times 10^{-4}$	PRA insights
9.	Emergency core cooling system	$\leq 1 \times 10^{-3}$	PRA insights
10.	Emergency power supply system (Class III)	$\leq 1 \times 10^{-3}$	PRA insights
11.	Class II power supply system	$\leq 1 \times 10^{-4}$	PRA insights: For a given mission time
12.	Class II power supply system	$< 1 \times 10^{-5}$	PRA insight: For a given mission time

in mind the available literature and regulatory practices, the achievable targets for the various metrics are as follows:

This is one way to even connect a Level 1 PRA study to intimately connect with the plant. The use of plant-specific data, human factor interaction modeling, mapping technical specification conditions and incidences as precursors are some of the ways in which a PRA reflects not only plant design but also operations. The other extreme but undesirable condition is that PRA analysts perform PRA studies based on generic data, design configurations, accident analysis reports, emergency operating procedures, etc., without or with minimal involvement of operating plant staff. This is the reason often you come across PRA study where there are very few recommendations for risk reduction. Even if couple of recommendations are there one can see that these can be formulated even without doing PRA.

The PRA modeling and the results and insight should be sensitive to plant risk-culture.

As seen, the deterministic as well as probabilistic criteria form the operations risk metrics. Generally, core damage is not expected during the life of the plant, but there are incidents and situations that might have threat perceptions, and these events form a risk potential for the plant and provide real-time insight for risk reduction measures.

Items 1 and 7–12 are the risk indicators available directly from the Level 1 PRA. Indicators 2–6 belong to the category of threat to operating regime and complement the probabilistic criteria (Table 8.2).

Table 8.2 Major activities/functions for risk-conscious applications

Activity	Approach for risk-conscious application	Requirement	Model and methods	Metrics for decision making	Constraints and limitations	Remarks/benefits
Category: engineering						
Design	Risk-based design	Human root cause analysis for design errors	Design analysis considering stress–strength model	Operational and safety-risk reduction and expected uncertainty	Quality of evaluation due to availability, especially aging considerations	Design based on current compliances
Modifications	• Risk-based prioritization of modification and engineering • Applicability of new configurations	Past performance data and compliance to regulatory requirements	Risk-significance evaluation of candidate modifications	A list based on prioritized risk-significance for candidate SSCs	The quality of remaining life estimation may determine the adequacy of the approach	Prioritized list where top 20% compliance tend to meet 80% safety goals
Planning for shutdown and aging management	Identification and prioritization of aging issues and assessment of risk-margins	Information and data on life extension. Aging studies and collection of data with associated uncertainties with performance of SSCs	Deterministic as well as probabilistic analysis—remaining life estimation	Compliance to existing performance and risk reduction measures	Science and technology of material degradation and remaining life evaluation	Life extension of plant complying the regulatory licensing criteria and goals

(continued)

Table 8.2 (continued)

Activity	Approach for risk-conscious application	Requirement	Model and methods	Metrics for decision making	Constraints and limitations	Remarks/benefits
Category: human resource						
Staffing requirements	Incidents operational analysis to evaluate human error where the nonavailability of actions led to incidence or exacerbation of situation Availability of remote operational facility from the local area Requirement of redundancy of operator positions	Risk-significance of the area based on aggregate SSCs component of risk from areas Training model of the plant for area manning	Identification of human nonperformance issues/nonactions that impact safety Technical specification requirements	Contribution of human error to total risk due to nonaction or lack of local monitoring	Initial considerations of personal area monitoring the basis and arguments Requires a management approach to effective solution as also affecting changes to enhanced automation and monitoring	Staffing is one of the major factors and an expensive preposition. An optimized staffing based on risk-based arguments serve long terms goals of safety
Training and qualification	Inclusion of probabilistic along with deterministic knowledge toward characterization of risk-significance of SSCs, areas, layout and routing, human action and O&M aspects	PRA study, Simulator and training aids for components	Risk model of the plant, incident analysis records and insights from root cause analysis	Checklist, written exam, walkdown and interview. Finally, monitoring of performance after induction in real-time on hands on jobs	This is the first step and responsibility of training group to inculcate risk-conscious culture as part of safety culture of the plant	The trainees become aware and alert to risk component in their plant due to random or deliberate actions. This has potential to avoid human failures and related incidents—a challenge to plant safety and security

(continued)

Table 8.2 (continued)

Activity	Approach for risk-conscious application	Requirement	Model and methods	Metrics for decision making	Constraints and limitations	Remarks/benefits
Maintenance of training aids and facilities	Monitoring of staff performance particularly new officers	Quality and adequacy of training depend on updated checklists, updating of simulator and intelligent aids with the matching features of the plant	Periodic evaluation of training methods, aids and simulation approaches for its adequacy	Quality level of training reflected in trainees' performance in the field	Monitoring and corrective mechanism periodicity is critical component	Training is a key to performance level of staff in real-time specially during initially stages in their induction for regular services
Developing and updating of plant checklists	Periodic review for matching with the latest status of design configuration and operational policy	Matching of training with plant configurations and requirements	Periodic training and retraining along with refresher courses, lecture programs	As above	Due to round the clock operation of the plant running courses and lectures covering all the levels poses challenges	Updating of staff knowledge, awareness levels and alertness specially to new aspects can avoid incidents
Documentation						
Design manual	Characterization of risk category of SSCs and related features and related to Engineering Standards, like ASME Section III, Section XI, etc.	Design basis for system and SSCs and its relation to its safety categorization Role of human action and adequate provision as also time window available should be factored into the design	Role of process and safety systems in PRA modeling, risk-significance of individual SSCs, identification of redundancies, diversity and potential CCFs	Contribution of the systems and individual SSCs to the net core damage frequency, risk reduction worth (RRW) for respective SSCs	For SSCs with specific design and configuration, additional description in respect of compensating or sensitivity analysis is required	A design basis PRA is based on the generic reliability data. As the plant gets into operations mode, the model should be updated to reflect as built and operating characteristics Manuals should be updated periodically

(continued)

Table 8.2 (continued)

Activity	Approach for risk-conscious application	Requirement	Model and methods	Metrics for decision making	Constraints and limitations	Remarks/benefits
Operating manual	Applicability and adequacy of human action as also potential human actions that might take the plant from normal operation to deviation from normal operation should be identified	Identification and prioritization of human error and its evaluation	Use of CQB model for human reliability evaluation given the safety performance influencing factors	For any operating procedure accumulated human error should not be more than the acceptable limit, say 10^{-3}	Availability of human error data could be prohibitive for this evaluation	Once critical human errors are identified the chances of plant entering into deviation from normal operation state is reduced that results into reduction of safety and reliability risks
Emergency operating procedure	Human reliability evaluation to ensure applicability and adequacy of human action as also potential human actions that might take the plant from emergency scenario to accident condition should be identified	Identification and prioritization of human error given the time window available for actions and special challenges posed by emergency scenario, like access to the area, operational and environmental stresses, etc.	CQB model with applicable stress-factors, context, operator's attitude should be considered for evaluation of the adequacy of the emergency operating procedures	The accumulated human error probability in an EOP should not be more than the stipulated limit, say 10^{-2}	Emergency scenario at times becomes highly uncertain. Therefore, a PRA study should be reviewed to get an understanding of residual risk Computerized and AI-based procedure might be effective in such situations	The PRA-based operator support system approach facilitates implementation of emergency operator procedures to reduce the possibility of human error

(continued)

Table 8.2 (continued)

Activity	Approach for risk-conscious application	Requirement	Model and methods	Metrics for decision making	Constraints and limitations	Remarks/benefits
Maintenance procedures	Risk-based maintenance management based on conditional core damage probability. Availability of maintenance procedures based on identified and prioritized maintenance actions leads to normal startup and operation of SSCs on demand	Reduce the probability of active or latent faults such that the SSCs start and operate on demand toward fulfilling intended safety or reliability criteria	Adoption of PRA study to support estimation of CCDP and optimized maintenance intervals based and allowable outage time for SSCs based on the operating experience on SSCS	Apart from CCDP-based maintenance prioritization, evaluation of frequency and allowable outage time forms the major risk metrics	Availability of risk monitor and evaluation of CCDP, maintenance intervals and allowable outage time for safety-significant SSCs is a resource consuming option	A well-designed risk-based maintenance management system is key to overall risk reduction
In-service inspection	Risk-based in-service inspection with consideration of human factor	Availability of LOCA*ECCS accident sequences and ECCS emergency operating procedures	Identification of accident sequences and evaluation of conditional core damage frequency contribution for identified piping segments. Identification of nature of inspection method and associated frequency through a risk metrics	Risk metrics with likelihood of failure of a segment and consequences (CCDP) as two vectors toward identification and prioritization	The coverage of ISI, adequacy of inspection, human factors associated the ISI might induce uncertainty in results of ISI Revision of document periodically based on new insights and data	ISI is one of the elements of defense in depth toward ensuring structural integrity

(continued)

Table 8.2 (continued)

Activity	Approach for risk-conscious application	Requirement	Model and methods	Metrics for decision making	Constraints and limitations	Remarks/benefits
Core operations activity						
Operational safety and reliability	Plant system configurations based on single failure criteria, redundancy and diversity ensure required resilience and management through human interventions. As operations, management is a human action interactive ecosystem, human factors like right awareness, alertness for required response, cognitive capability for analysis of transient and corrective actions are critical to safety and reliability. Availability of plant risk monitor has potential to empower staff in decision making	An efficient human–machine interface, operator aids, like procedures, information system identification and diagnostic support systems. Audio-visual display and static and dynamic mimics to facilitate the human–machine interface	Highly trained staff to ensure system reliability along with ensuring technical specification stipulations. Risk-conscious decisions can be permitted through only policy decisions related AOT and STI for process or safety systems The risk monitor can address the issue of overconservatism by evaluating the risk levels for various equipment outages more than plant technical specification stipulation. Surveillance and test intervals based on PRA analysis might provide optimum test interval and maintenance periodicity	The decision criteria based on instantaneous as also change in CDF, CCDP as part of risk monitor aided decision making provides an elegant mechanism for risk-conscious decisions	The scope of risk monitor in terms of plant and system coverages might prove restrictive in decision making. Additionally, quantification of and acceptable level of uncertainty might pose interpretation issues	Keeping in the advantages of incorporating risk monitor, many NPPs have risk monitor in support of decision making

(continued)

Table 8.2 (continued)

Activity	Approach for risk-conscious application	Requirement	Model and methods	Metrics for decision making	Constraints and limitations	Remarks/benefits
Surveillance and supervision	Alertness, awareness, cognitive reflexes, stability of mind during emergency are key to human performance	Operator training in simulator environment along with actual scenario conditioning is the core requirement for correct responses, especially during emergencies	Application of risk-conscious culture ensures knowledge of risk mitigation resilience development through CQB approach which is the vital factors for human performance	Monitoring and tracking the performance of staff specially for emergency or stressful situations should be part of human resource management	Availability of database of human performance to provide the needed guidance and support, if needed	Human reliability methods—specially employing stress inducing factors are proving to be useful for operational ecosystems
Coordination and communication (for accident management)	Risk-conscious management, particularly that involve thinking for in support of coordination and communication	Highly vigilant, well-aware and alert staff about the plant status particularly about potential risk levels while dealing with plant staff, other agencies	Accident sequence evaluation through risk monitor for instantaneous CDF value and postulated scenario	CCDP, CDF, affected system and SSCs unavailability. Risk metrics for decision making	Even though in every PRA model, there is a special consideration for ensuring 'completeness' for coverage of the analysis, however, a due to highly complex dynamics achieving 100% guarantee may not be always possible	With risk-conscious culture in place and effective, risk component due to lack of coordination and communication can be reduced significantly

(continued)

Table 8.2 (continued)

Activity	Approach for risk-conscious application	Requirement	Model and methods	Metrics for decision making	Constraints and limitations	Remarks/benefits
Configuration management	*Same as operational safety and reliability*					
Fuel handling and management	Provision of operator aids to facilitate his focus, concentration, high awareness and alertness about the handling activities	Fueling is highly human–machine centric activity therefore specially trained crew, use of high-quality component during repair, maintenance and an exhaustive machine qualification after maintenance or repair	The fueling checklist should be subjected to human reliability analysis Deployment of intelligent operator support system Continuous learning from operating experience and a perpetual program for upgrades and improvement	High fuel handling system reliability and availability Number of incidents related to fueling	Fuel handling system requires a just approach to automation for two reasons. The reliability of automated should be very high on one hand; on the other, there should be adequate time and access window for corrective action. This is very subjective and tends to be restrictive	The potential risk-significance of fueling can have dependence on mode of fueling, i.e., on-line (reactor operating mode) and off-line (in reactor shutdown mode)

(continued)

Table 8.2 (continued)

Activity	Approach for risk-conscious application	Requirement	Model and methods	Metrics for decision making	Constraints and limitations	Remarks/benefits
Management of staff	*Same as training and qualification*					
Data logging and reporting	Risk-conscious approach plant performance monitoring and on-line or periodic data logging	The performance data are central prediction of plant reliability, safety and in terms of assessing performance trends plant dynamic status evaluation	The existing approach to monitoring depending on safety significance of the parameters categories into safety related, performance related, requiring timely action or providing the change in status which is required for plant reliability. The risk-significance of the parameters which instead of having major distinct categories provides continuous rating using numerical risk-significance of the parameter	Data availability and quality are crucial for periodic safety performance evaluation. As also for future analysis of system performance in general and degradation	There is no empirical model as to predict the parameter for categorization based on safety significance. The optimization or parametric study based on safety significance to be used as part of risk-conscious culture	Performance monitoring, root cause analysis and investigation of incidents often require historical data from the plant data repositories

(continued)

Table 8.2 (continued)

Activity	Approach for risk-conscious application	Requirement	Model and methods	Metrics for decision making	Constraints and limitations	Remarks/benefits
Radiation protection	Nuclear industry has huge operation experience on using the metrics that include evaluation of potential risk (loss of reliability, safety (fatalities, health effects due to random human failures) security (deliberate human actions) and the minimum dose required) such that benefits outweigh the dose consumption first during normal operation and later based on the accident histories at international level for accident conditions	Research on using risk-based approach to provide an improved alternative to as low as reasonably achievable (ALARA) as ALARA is considered subjective (Department of Energy 1997) while RP activities have issues that need be based on ethics and requires human conscious considerations (Yim 2013)	Optimization considering the benefits vs consumption	Applicable dose (numerical value) for typical jobs in the plant	ALARA is subjective and a empirical model based on data, experience and ethical principle require further research	Even though ALRA is subjective, it has worked well for the growth of nuclear industry

(continued)

Table 8.2 (continued)

Activity	Approach for risk-conscious application	Requirement	Model and methods	Metrics for decision making	Constraints and limitations	Remarks/benefits
Radioactive material transport and management	Risk-based/informed principles that includes security risk should be adopted to translate to plant risk-consciousness	During radioactive substance, transport and waste management to be evaluated and adequately designed against postulated safety and security risk	Radioactive transport risk assessment studies to be adopted for probabilistic evaluation	Acceptably low metrics approved by regulatory body should be set for the plant for compliance	Safeguard and related issues do not form part of this metrics	The description is limited to technical aspects, and safeguard regulation is not within the scope of this approach
Maintenance						
Plant maintenance scheduling	Risk-based maintenance management (planning along with industrial safety requirements)	• Safety grad activities should comply technical specification stipulations • Availability of qualification procedure for parts, activities, workmanship, etc. • Bench marking of typical maintenance activities in terms of time taken from time of starting to time of completion marked by handover to operations, quality of jobs and material/spare part and consumables	• Risk or reliability-based maintenance management • Stipulated periodicity ± time provisions • Outage time ± • Required documentation preferably in computerized maintenance management procedure	• The maintenance activities should not induce faults/failures • Completion time • Quality level of work	Possible constraints, required time, resources, like spares, human resource, logistics like jigs fixtures	Maintenance is vital to plant safety and reliability

(continued)

Table 8.2 (continued)

Activity	Approach for risk-conscious application	Requirement	Model and methods	Metrics for decision making	Constraints and limitations	Remarks/benefits
Periodic maintenance (field repair and replacement work)	• Risk-based maintenance management—(field work and supervisions ensuring industrial safety precautions mainly fire, flooding, falling of object from heights)	• Maintenance authorized through work permits system • Adherence to approved maintenance procedure • Changes in intended procedure, parts specifications and quality should be authorized and compensated adequately, and the same should be recorded and reported • Identification of crew responsible for maintenance activity, including field supervisor • Normalization activities should comply operations and maintenance protocols for handing over of SSCs	• Reliability/risk-based maintenance management model • Reporting of feedbacks and observation into computerized maintenance management system • Performance tracking at least for initial periods	Fault tree operation of the equipment till next maintenance schedule	–	Maintenance is vital to plant safety and reliability
Inventory management—spare parts, consumables	Operations research inventory management model ensures just-in time supply chain based on SSCs failure rates	Availability of spares and consumables as and when needed	Operations research inventory management model	How far the inventory management satisfies just-in-time receipt as also availability of spare on perpetual basis	Organizing financial resources/administrative approvals might pose challenges. Operation of risk-significant SSCs should, in no case, affect the availability of the subject SSC	Availability of spare parts is key to efficient and effective management of maintenance program

(continued)

Table 8.2 (continued)

Activity	Approach for risk-conscious application	Requirement	Model and methods	Metrics for decision making	Constraints and limitations	Remarks/benefits
Periodic testing of safety systems	Risk-based approach to surveillance test interval (STI) and allowable outage time (AOT) estimation	A Level 1 PRA study for risk-simulation Studies and risk monitor is preferred	STI and AOT estimation modeling and analysis for optimizing these subject STI and AOT	Satisfying the acceptable CDF and CCDP acceptance criteria	Adequacy of data and requirement of safety function and associated failure criteria	It is preferred that STI and AOT are evaluated as part of technical specification restructuring as part of risk-informed regulation using risk criteria
Safety						
Safety analysis	Integrated approach involving deterministic analysis along with probabilistic risk assessment	The regulatory requirement is deterministic safety analysis along with probabilistic analysis For implementing risk-conscious culture, the integrated approach to risk analysis is desired	Integrated risk-based engineering model for prioritization of SSCs, evaluation of operating procedures, deployment of risk monitor and PRA-based operator support system as operator's, technical specification optimization, etc.	• Availability of approved safety analysis documents in plant updated periodically • For plant applications, the change in CDF and CCDP and risk-importance ranking	Plant SSCs failure data, CCF and human factor models and uncertainty characterization approach. Development of dynamic PRA is still in developing stages	Integrated approach tends to handle the long-term limitation of built in overconservatism with not very sound and adequately proven rationales and basis

(continued)

Table 8.2 (continued)

Activity	Approach for risk-conscious application	Requirement	Model and methods	Metrics for decision making	Constraints and limitations	Remarks/benefits
Technical specification compliance	Risk-informed approach for regulatory compliance	Approved at least Level 1 PRA for cases where internal configurational aspects of the plant, like configurational changes, incident review, violation of technical specification cases, etc.	Applications based on PRA model meeting technical specification stipulations or in support of regulatory waiver	Technical specification stipulations, regulatory approval, risk-based criteria like system unavailability, CDF, and CDP	Same as above	Same as above
Emergency preparedness (through periodic emergency drills)	Based on the risk insights on expected risk levels derived from Level 1, 2 and 3 PRA	Reassessment of requirements and periodicity of emergency drills and evaluation response and thereby feedback for improvements, if any	There are three levels drills envisaged, they are a) plant emergency, b) site emergency and c) off-site emergency (public domain). A decision is based on the postulated risk level in the plant, in plant site and in public domain obtained by Level 1, 2 and Level 3 PRA	Major metrics is as follows: Plant emergency is based on Level 1 results, i.e., CDF/year estimate Site emergency is based on Level 2 and 3 PRA results that shows the consequences are limited to site only based on LERF and release and consequence model and Level 3 PRA results if the consequences are envisaged in public domain	Nonavailability of Level 2 and 3 PRA can be compensated by deterministic insights	Emergency drill frequency

(continued)

Table 8.2 (continued)

Activity	Approach for risk-conscious application	Requirement	Model and methods	Metrics for decision making	Constraints and limitations	Remarks/benefits
Quality assurance	Risk-based audit and quality-conscious policy in engineering inspection and quality control activities	Quality control and assurance in all engineering, maintenance, operations and safety and security services activities including dimension and transportation	A philosophy of 'quality is fundamental and prerequisite to reliability and safety assurance'. A model for risk-conscious identification and prioritization of activities	Prioritization based on risk-significance and further modification employing consciousness (full awareness) and conscience (highest ethical standards) principles	The attribute 'quality' is intrinsically subjective which is controlled by specifications; however, a risk-conscious approach can reinforce the prevailing quality assurance systems	
In-service inspection	Risk-based in-service Inspection	Level 1 full scope PRA, expertise and a resolution at highest level in plant to implement risk-based inspection	Use of one among available standard approaches with modification to reflect plant characteristics for improved results	Likelihood vs consequences risk metrics -consequences are represented by CCDP or potential degradation and likelihood by failure frequency of the subject segment	Requirements of plant-specific data. The quality of results depends on quality of data	

(continued)

Table 8.2 (continued)

Activity	Approach for risk-conscious application	Requirement	Model and methods	Metrics for decision making	Constraints and limitations	Remarks/benefits
Industrial safety	Risk-based management approach systematically the elements Identification of initiating causes, analysis, prioritization, monitoring and feedback and corrective actions	Reduce health hazards and fatalities	Plant data collection or use data from similar plants with similar characteristics, create an integrated PRA model toward postulating plant safety provisions to arrive at dominant areas of concern, for upgrades and corrective action	Risk to occupational worker and trend monitoring	Application of PRA modeling is resource consuming, as also creating an industrial risk-consciousness is a challenge	In nuclear industry, highest standard of nuclear safety are maintained evident from the operating experience of 2 major accidents in 18,500 accumulated, i.e., a mean frequency of 1.08×10^{-4} r – y (World Nuclear Association 2010). That's where the industrial safety and overexposure chances are occupying much of the focus
Regulatory compliance	RB approach to supervision and compliance (Reserve Bank of India 2014)	Prioritization for management of safety issue	A qualitative or quantitative—PRA model for the management of supervisory functions	Risk-index and higher coverage of risk-compliance with 80:20 philosophy, i.e., 20 percent high risk cases addresses 80% of risk-concerns	Design of a robust mutually agreed facility-regulatory operational framework	Smooth communication between facility and regulatory authority

(continued)

Table 8.2 (continued)

Activity	Approach for risk-conscious application	Requirement	Model and methods	Metrics for decision making	Constraints and limitations	Remarks/benefits
Security						
Physical protection system (PPS)	Integrated probabilistic security risk assessment framework for operation and training of staff	Training of security and related agencies on plant services of physical protection systems Apart from design basis threat approach application (International Atomic Energy Agency 2013) of vulnerability based probabilistic approach (Teagarden et al. 2004)	Risk-conscious awareness-based identification and prioritization as also implementation of computerized intelligent surveillance involving vision and sound and deviation notification. Higher alertness and awareness through lectures, training and conduct of courses	Higher alertness and improvement in human response during surveillance	Some of the qualitative parameters were handled through fuzzy approach to quantification	Modern methods based on advanced technology with higher consciousness and conscience attribute are vital for improved approach to PPS
Cyber physical systems	Vulnerability assessment against security threat	Protection of complex network- based CPS and Internet of things (IoT) with intelligent components for embedded critical infrastructures. (Peng et al. 2013)	Integrated approach risk assessment for vulnerability assessment	Robustness and resilience against the postulated threats	Security risk assessment of a system based on embedded (hardware and software operations) technology, CPS, IoT and complex network	Often the separation of internal only (intranet) and internal (internal and external) along with firewalls, audit-based approaches are employed

References

Atomic Energy Regulatory Board (2011) Security of radioactive sources in radiation facilities, AERB/RF-RS/SG-1. AERB, Mumbai, India

Basel Committee—Operational risk, BASEL [Online]. Available https://www.bis.org/list/bcbs/tid_28/index.htm. Accessed 23 Mar 2022

Department of Energy (1997) Applying The ALARA Process For Radiation Protection of the Public and Environmental Compliance with 10 CFR PART 834 and DOE 5400.5, DOE-STD-ALARA—DOE STANDARD (Draft), DOE, Washington DC, 1997+

Diet of Japan (2011) Fukushima accident—a manmade disaster. Diet of Japan, Japan

FATF (2019) Financial action task force—terrorist financing risk assessment [Online]. Available www.fatf-gafi.org/publications/methodsand trends/document/Terrorist-Financing-Task-Force-Risk-Assessment-Guidance.html. Accessed 2019

Gupta SK (2009) Management by consciousness, India: Sri Aurobindo Institute of Research in Social Sciences. Sri Aurobindo Society, Pondicherry

Government of India (2010) The civil liability for nuclear damage act. Government of India, New Delhi

International Atomic Energy Agency (1964) Civil liability for nuclear damage—official records, international conference, Legel Series No. 2. International Atomic Energy Agency, Vienna

International Atomic Energy Agency (1988) Probabilistic safety assessment of nuclear power Plants (Level 1), IAEA PRA Guide P-4. IAEA, Vienna

International Atomic Energy Agency (1992) INSAG-7: The Chernobyl accident : updating of INSAG-1. IAEA, Vienna

International Atomic Energy Agency (1999a) Strategies for competitive nuclear power plants, IAEA-TECDOC-1123. International Atomic Energy Agency, Vienna

International Atomic Energy Agency (1999b) Management of operational safety in nuclear plants, INSAG-13. IAEA, Vienna

International Atomic Energy Agency (2000a) Operational safety performance indicator, IAEA-TECDOC-1141. International Atomic Energy Agency, Vienna

International Atomic Energy Agency (2000b) Operational safety performance indicators for nuclear power plants, IAEA-TECDOC-1141. IAEA, Vienna

Internal Atomic Energy Agency (2001) Application of Probabilitsic Safety Assessment for Nuclear Power Plants, IAEA-TECDOC-1200. IAEA, Vienna

International Atomic Energy Agency (2004) Precursor analyses—the use of deterministic and PSA based methods in the event investigation process at nuclear power plants, IAEA-TECDOC-1417. IAEA, Vienna

International Atomic Energy Agency (2010a) Objective and essential elements of a states's nuclear security regime, Nuclear Security Series No. 20. International Atomic Energy Agency, Vienna

International Atomic Energy Agency (2010b) The interface between safety and security at nuclear power plants, INSAG-24. International Atomic Energy Agency, Vienna

International Atomic Energy Agency (2010c) Objective and essential elements of a state's nuclear security regime, IAEA Nuclear Security Series No. 20. IAEA, Vienna

International Atomic Energy Agency (2013) Objective and essential elements of a State's nuclear security regime—IAEA Nuclear Security Series No. 20. IAEA, Vienna

International Atomic Energy Agency (2016) Safety of nuclear power plants : Commissioning and operations, Safety Standard Series No. 2/2 (Rev. 1). International Atomic Energy Agency, Vienna

International Atomic Energy Agency (2017) Managing the financial risk associated with the financing of new nuclear power plant projects, IAEA Nuclear Energy Series No. NG-T-4.6, IAEA, Vienna

International Atomic Energy Agency (2018a) Industrial safety guidelines for nuclear facilities, Nuclear Energy Series No. NP-T-3/3. International Atomic Energy Agency, Vienna

International Atomic Energy Agency (2018b) Industrial safety guidelines for nuclear facilities, IAEA Nuclear Energy Series No. NP-T-3.3. IAEA, Vienna

International Atomic Energy Agency (2019) Human factor engineering in the design of nuclear power plants, Specific Safety Guide o. SSG-51. IAEA, Vienna

International Electrotechnical Commission (2010) Functional safety of electrical/electronic/programmable electronic safety-related. IEA

Jones RB (2012) Risk-based management: a reliability-centered approach—practical, cost effective methods for managing and reducing risk. London

Joshi NS, Varde PV (2019) Root cause analysis. In: International Conference on Reliability, Safety and Hazard—2019, Lulea, Sweden

Mordecai Y, Dori D (2012) Risk oriented systems engineering: integrating risk into systems modelling with OPM. In: 3rd International Engineering Symposium, CESUN 2012, Delft University of Technology

NASA Office of Safety and Mission Assurence (2006) The Devis-Besse close call—system failure case study. NASA, Washington D.C.

NASA (2011) NASA accident precursor analysis handbook. NASA/SP-2011–3423, Version 1.0. National Aeronautics and Space Administration, Washington D.C.

Nuclear Energy Agency (2014) Working group on operating experience report on Fukushima Daiichi NPP precursor event. Nuclear Energy Agnecy, Paris

On managing people (2011) Harvard Business School [Online]. Available harwardschool.ac.in

Operational Safety Management (2022) Industrial Software [Online]. Available https://industrial-software.com/solutions/aveva-operational-safety-management/. Accessed 4 Feb 2022

Park JW, Lee SJ (2017a) Probabilistic risk evaluation of cyber attack on a nuclear power plant safety. In: NPIC & HMIT 2017, California

Park JW, Lee SJ (2017b) Probabilistic risk evaluation of cyber-attach on a nuclear power plant safety. In NPIC&HMIT 2017, San Francisco, CA

Peng Y, Lu T, Liu J, Gao Y, Xie F (2013) Cyber-physical System risk assessment. In: 2013 ninth international conference on intelligent information hiding and multimedia signal processing

Pearson S (2022) What is operations management—definition and core concepts, tallyfy.com [Online]. Available https://tallyfy.com/operational-risk-management/. Accessed 14 Mar 2022

Phimister JR, Bier VM, Kunreuther C (2004) Accident precursor analysis and management: reducing technological risk through diligence. National Academies Press

Quality Control Techniques—Production & Operations Management [Online]. Available https://www.managementstudyguide.com/quality-control-techniques.htm. Accessed July 14 2022

Report: Davis-Besse Reactor Vessel Head Degradation, United States Nuclear Regulatory Commission [Online]. Available https://www.nrc.gov/reactors/operating/ops-experience/vessel-head-degradation/lessons-learned/lessons-learned-files/lltf-rpt-ml022760172.pdf. Accessed 28 Mar 2022

Reserve Bank of India (2014) Compliance functions in banks. Reserve Bank of India [Online]. Available https://rbidocs.rbi.org.in/rdocs/Notification/PDFs/CFBNF12032015.pdf. Accessed 21 Mar 2022

Souse S, Nunes E, Lopes I (2015) Measuring and managing operational risk in industrial processes. FME Trans 43:202–295

Taleb NN (2004) Fooled by randomness—the hidden role of chance in life and in the markets, 2nd edn. Penguin Books

Teagarden G, True D, Burns E (2004) Probabilistic consequence analysis of security threats—a prototype vulnerability assessment process for nuclear power plants. Electric Power Research Institute, Campbell

United States Nuclear Regulatory Commission (1980) Three Mile Island—a report to the commissioners and to the public—NUREG/CR-1250. NRC, Washington D.C.

Varde PV (2021) Risk-conscious operations management in nuclear plants. In: American nuclear society—probabilistic risk assessment—risk-informed technology (PSA-21), Virtual Meeting

Varde PV, Pecht MG (2012) Role of prognostics in support of integrated risk-based engineering in nuclear power plant safety. Int J Prognostics Health Manag 8:2153–2648

Varde PV, Sankar S, Verma AK (1998) An operator support system for research reactor operations and fault diagnosis through a connectionist approach and PSA based knowledge based systems. Reliab Eng Syst Saf 60:53–69

Varde PV, Joshi NS, Mishra V, Rai KK, Satheesh C (2019) Development of risk-informed inspection for BARC research reactors—Dhruva. IAEA Meeting, Vienna

Varde PV, Joshi NS, Agarwal M, Srivastava A, Bandi LN, Varde PV, Joshi NS, Shrivastava A, Bandi LN, Kohli AK (2011) Development of an integrated methodology in support of remaining life assessment of a Co-60 based gamma irradiation facilities. Nucl Eng Des 241

Wall IB, Haugh JJ, Worleg DH,Wall IB, Haugh JJ, Worlege DH (2001) Recent application of PSA for managing nuclear power plant safety. Prog Nucl Eng 39(3–4):367–425

Wikipedia (2019) Wikipedia [Online]. Available https://en.wikipedia.org/operations-management/. Accessed 20 Dec 2019

Worldcat (2017) Human error as the root cause of severe accidents at nuclear reactors. Bezpecnost Jaderne Energie 25(7–8):197–217, 25(7–8):197–217

World Nuclear Association (2022) Safety-of-nuclear-power-reactors, World Nuclear Association [Online]. Available https://world-nuclear.org/information-library/safety-and-security/safety-of-plants/safety-of-nuclear-power-reactors.aspx. Accessed 21 Mar 2022

Yim MS (2013) Ethical Issues in ALARA. In: First Asian Workshop on ethical dimension of the radiological protection, Daejeon

Chapter 9
Artificial Intelligence Based Approach for Operator Support System

9.1 Introduction

The reactor operation ecosystem, from the point of man–machine interaction characteristics, can broadly be divided into two major categories, viz., normal operations and emergency conditions. The immediate response of reactor systems to any undesired initiating event manifests as a transient condition, and depending on the adequacy initial automatic and latter combined automatic and manual action bring back the normal operation of the plant or if not controlled or the mitigative actions are not adequate, then might lead to accident conditions.

Even the reactor control design philosophy stipulates no operator action during the initial period into the transient, which could be in the range of 15–30 min in existing or older reactor systems to approximately 3 h for advanced systems (Atomic Energy Regulatory Board 2015). For example, for the case of the Loss of Coolant Accident scenario, the plant design provides passive safety features, a higher coping time of the plant in terms of reserve of resources, such as emergency power supply, extended cooling capability through higher reserve inventory or provision of effective leak collection of circulation inventory.

However, the challenge that the operator experiences is higher cognitive load due to a sudden burst of cognitive stresses associated with decision making each time and information overload in a relatively short time, at least during the initial period into the transient. The situation in the control room and affected areas requires comprehending the actual scenario while analyzing the plant status for further action toward ensuring the safety of the plant.

Therefore, the human–machine interaction is critical to the safety and reliability of the plant. Over 70% of accidents are attributed to human error and organizational factors. Often, the reason could be that the cognitive load on the operator, particularly during emergency scenarios, is expected to be high, which might lead to erroneous decisions. In fact, it is generally accepted that 80–90% of accidents are due to human error. For example, approximately 70% of aircraft accidents have been attributed to human error. In a Finnish study, human errors were involved in 84% of serious

P. V. Varde, *Risk-Conscious Operations Management*, Risk, Reliability and Safety Engineering, https://doi.org/10.1007/978-981-19-9334-3_9

accidents and in 94% of fatal accidents (Human Error 2022). Apart from direct man–machine interaction, the root cause of failures has often been found to be due to inadequate automation in design and ergonomics (Haight 2007). However, there is a consensus among the safety experts that during the emergency scenario, there is severe demand for operator attention and awareness to diagnose the scenario and develop a correct perception of the plant state for making the correct decision. Here, a large paradox is that while on the one hand, during emergency evaluation, the cognitive capability of the operator is expected to be at its best, while the high stress level associated with the emergency scenario, the cognitive capability tends to drop on the other, to a lower level and might reach a level below 20% of normal conditions, which is highly undesirable from safety considerations. These stresses are in response to flooding of the control room with information at least during the initial period into the accident where the scenario is evolving dynamically, and there is always a worry that one should not miss a critical piece of information that is at the root of the evolving scenario.

Another and more important aspect is that when we look at the history of accidents, it will be clear that all accidents are not initiated due to catastrophic faults or failures. Often, the seed of the accidents lies in some anomalies or precursors, e.g., decision and action during normal operation, that were not even remotely considered to form a source of accident initiation. For example, in TMI, the blockage in the resin line of the feedwater system and the action to clear the blockage by blowing compressed air led to air entry into the system, and the subsequent stuck-open condition of the ECCS tank coupled with correction comprehension of the evolving scenario led to the accident (Jang et al. 2017). In Chernobyl, the initial condition was assessing the turbine flow coast down behavior to assess how long the power remained available to the cooling pumps. This is a normal scenario. What led to the initiation of the accident was operating the reactor in a prohibitive zone, i.e., less than 600 MW (thermal).

Keeping in view the affirmation of the Vedic literature that consciousness is fundamental to human existence as well as the investigation in this regard, the RCOM postulates that level of consciousness has bearing on human response in general and cognition in particular. If we extend this argument to the context of the operator support system, then human intelligence, awareness, alertness, commitment, dedication level, morality and ethics, etc., are directly relevant to operations management, and the operator support system should aid these capabilities for identifying the deviation, diagnosis and associated corrective actions and particularly for identifying the plant conditions and, if needed, corrective or mitigative actions during emergency scenarios. Another human attribute is conscience, which is responsible for moral values, integrity, commitment to work, credibility of behavior and the perception of right and wrong–which is critical for evaluating security-related aspects (Giubilini 2021).

There is consensus that any application in an NPP should be capable of addressing risk-based requirements that include safety and security, i.e., be it due to unintended or random or intended actions that increase the risk of accidents/incidents (International Atomic Energy Agency 2010). The scope of the approach being discussed here has three major components, transient identification–to identify the event, diagnostics–to identify the component failure and prognostics–to postulate the trend of scenario development in time at the component level and last the advisory for health management. For normal operation, it is only the prognostic module or condition monitoring module that trends the critical parameter and identifies any deviation from normal operations. The rate of change will determine how much time is available with the operator for acting before the consequences of failure become a reality.

9.2 Historical Perspective and Literature Review

The nuclear industry has traditionally been conservative and has always been conscious of the application of new technology. One of them is the application of digital technology, and the other is artificial intelligence or machine learning technology. In fact, the experience suggests that for deployment AI systems is one of the desired conditions that plant should preferably have digital control instead of analog, due to ease of multiplicity of signals and adoption, interfacing the algorithms and interface design. Digital systems are being deployed as a replacement for traditional analog systems either as part of a refurbishment program or as integral design philosophy in new and advanced systems. New plant designs generally employ digital control and protection logic. On the other hand, the progress in the application of AI has been slow. In the existing plant, computer-based systems or AI systems are being developed as part of an information system or safety parameter display system. The available publication shows that active interest exists in the real-time use of AI for transient identification and fault diagnosis and many other applications (Kang and Lee 2022). Additionally, AI has been used in reactor physics areas such as power prediction to overcome modeling for complex computations (Myung-Sub et al. 1993).

In fact, the IAEA made efforts to encourage consideration for the application of AI-based operator support systems in nuclear power plants (International Atomic Energy Agency 1988). Wu et al. all provided a review of the application of the Probabilistic Risk Assessment Technique (PSA) for the development of an operator support system (Wu et al. 1989). The reason was that PRA provides an integrated, graphical and cause-effect framework for knowledge representation. This approach provided an effective framework for the development of diagnostic logic–critical for operator support systems. In fact, probabilistic information at each node and terminal points in the Fault Tree and event tree in PRA provides an elegant mechanism in conflict resolution in support of decision making. Therefore, the knowledge representation using fault trees and event trees provides an effective framework for diagnostics. This knowledge can be translated to develop rule-based expert systems. In fact, Varde P.V. at all have discussed an integrated approach for the development of operator support systems for transient identification and rule-based approaches for diagnostic modules where PRA knowledge has been used for knowledge representation (Varde et al. 1998). It has been shown that the expert system approach may not be effective for transient identification, as the production rule processing follows a sequential procedure and may not be effective and efficient in predicting/identifying the reactor status. Therefore, the artificial neural network approach is being followed among the many intelligent approaches for transient identification.

The safe and secure operation of nuclear power operations is a key to a low carbon energy supply for Industry 4, i.e., 4th Industrial Revolution, where the existing challenges include the increasing complexity of systems, even though there is a very low likelihood, the probability of accidents and the impact of emergency stresses on human factors pose challenges. The application of intelligent systems, the Internet of Things (IoT) and other advanced technologies, such as smart sensors, can further enhance the safety of NPPs (Lu et al. 2020).

9.3 Approaches to Address Human Factors in Operations

9.3.1 Application of Defense-In-Depth

Safety function maintenance in nuclear plant operation is based on the philosophy of barrier protection. This involves features/provisions/actions for maintaining the barrier integrity from the fuel element to the reactor coolant system pressure boundary to the containment envelope. If a safety function cannot be maintained, e.g., due to a failed system or/and unsuccessful procedures, the safety of the environment can still be ensured by maintaining the next level of barrier by means of other safety functions.

As part of this traditional safety approach, the application of single failure criteria, diversity, redundancy, fail-safe operation, quality assurance and conservative design are the major cornerstones in the design, operations and regulation of nuclear plants. This approach makes the reactor system resilient against hardware failure as well as human failure. The human–machine interface design considering ergonomic consideration, training and qualification procedures, approved organizational hierarchy and minimum staff requirements in control room and plant areas, etc., ensuring the availability of safety documents, such as plant emergency operating procedures, a culture of authorization jobs through checklists and strict compliance with technical specifications, are the major requirements to strengthen human elements at the individual and organizational levels. The design also ensured that during any transient, the system should automatically reconfigure to achieve basic safety functions, i.e., ensuring automatic safe shutdown, accomplishment of decay heat removal function and containment isolation, such that no human action is required during 30 min into the transient. Along with this, the application of a quality assurance program forms part of ensuring high quality in hardware as well as highly trained staff in the operations of plants through strict training and qualification programs. This approach proved effective and time tested; however, the insight gained from the history of nuclear accidents suggested that further strengthening of safety design is needed, as seen in the following sections.

9.3.2 *Inherently Safe and Passive Design*

The availability of advanced technology, materials and computational facilities and an improved understanding of core reactor physics, thermal hydraulic processes and structural material properties enabled the development of concepts such as inherently safe and passive system design. For example, ensuring by designing the negative void and temperature coefficient of reactivity ensures that core and fuel integrity is maintained against the accidental increase in reactivity and temperature. The design ensured that during the accident condition, the plant had a higher copying time in terms of ensuring core cooling for an extended period, say, for more than 72 h without operator intervention. Most of the modern plants being built are embarrassing this modern approach in design.

9.3.3 *Optimized Automation in Support of Decision Making*

Automation in principle is required to reduce human efforts to operate and control a process or system such that the reliability and safety objectives are achieved effectively. Generally, the design philosophy of nuclear plants following an experience-based approach has been used in relation to the decision on optimization systems, considering the postulated scenario for normal operation as well as emergency requirements. How much automation? is a question, as the traditional approach to automation tends to be arbitrary. It is not based on a formal scientific-based evaluation. The reason is that the problem is having multiple objective functions on the one hand, while there is not adequate clarity on the constraint. For example, human behavior, particularly during emergency conditions, poses a challenge. The following diagram shows, in principle, the elements to understand automation requirements.

As seen in Fig. 9.1, the operator aids the online sensor as well as operator feed input in an open loop architecture, i.e., no automatic action is affected in the plant through the advisory system. It is interesting to note that the plant design procedure ensures that dependence on the operator is reduced during the on-set of emergencies; however, in case there is failure of automation, then only the operator role shifts to the manual implementation of correction. It must be noted that the high level of redundancy or parallel configuration and strict quality control reduces the chances of operator intervention during the initial period into the event. Of course, the manual action forms part of emergency procedure after initial 10–30 min and in modern plant with the passive and inherent safety features to more than 24 h. as the design intent. Here, one can note that unlike other mission and safety critical systems, nuclear design philosophy does not struggle with the question of 'how much automation is

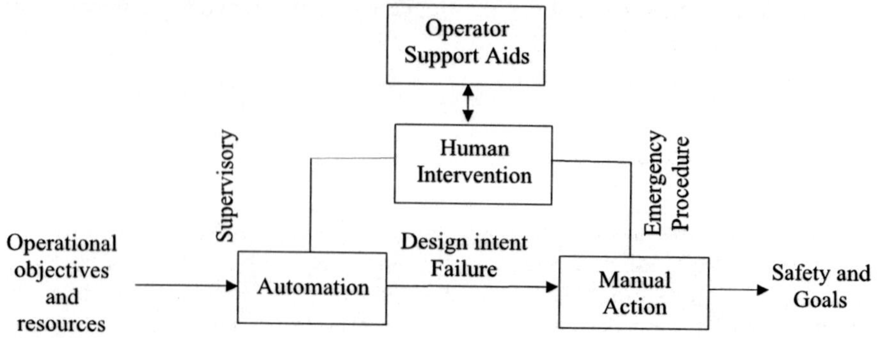

Fig. 9.1 Automation and plant perspective

adequate' to ensure an emergency scenario. Furthermore, questions related to plant reliability and availability might require a thorough review of the requirements of automation.

9.3.4 Training and Plant Simulators

The training program prepares the fresh graduate engineering and science candidates in design and operational aspects at the components, systems and plant levels. At the core, the emphasis is on safety aspects of operation. Apart from hands-on training through lecture programs on plant design and operational aspects, visits and working in control rooms and other plant areas along with the completion of checklists is a prerequisite for licensing and authorization. Plant staff are trained specifically in emergency operational procedures and compliance with technical specifications. The availability of plant simulators has become a regulatory requirement for strengthening training programs. The simulator facility provides an improved opportunity for training staff through real-time interaction with the simulator man–machine interface. This aspect also strengthens the performance evaluation of trainees, as the simulator provides a better degree of freedom and opportunity for in-depth training based on the response of the simulator in real-time. In the existing training programs, the simulator training and evaluation form an integral part of training.

As part of the licensing and authorization process, the trainee must undergo a series of exams, which mainly include the successful completion of checklists, followed by walkdown tests and interviews by the expert committee, before the candidate is authorized to take charge in the direction operation of the plant.

9.3.5 *Control Room Features*

9.3.5.1 General

Ensuring safety at all operational states of a nuclear reactor is the prime objective. Safe operation of the reactor implies maintaining the integrity of the core and containment building to prevent release of radioactivity to the environment. The plant can be said to be in a safe state if certain conditions are true both during normal and abnormal operating states. The conditions necessary to maintain the plant in a safe state are referred to as critical safety functions, i.e., being able to (a) control/terminate the fission chain reaction; (b) remove the fission/decay heat from the core; and (c) prevent the release of radioactivity to the environment. Defense-in-depth philosophy is implemented by engineering the process and safety systems in such a way that there is more than one success path to accomplish each safety function.

Normal operation of the reactor implies safe operation, i.e., plant parameters are well within the operating limits. Deviation from normal operation implies violation of these limits. Provisions exist in the system to cater to all anticipated operational occurrences. However, if not controlled, deviation from operational occurrences may lead to accident conditions. Restoration of normal operation and ensuring safety are the basic goals when the operation is abnormal. However, in general, the action(s) required to achieve these two goals need not coincide. During the time spent analyzing and diagnosing a malfunction state and restoring operations, the plant condition may degrade to such an extent that safety is compromised.

9.3.5.2 Man–Machine Interface–Plant and System Mimic

The hard panels spread across the control room, displaying the plant system mimics, are vital as a static man–machine interface that provides the status of not only the reactor and associated parameters but also the process and safety as well as safety support systems. These systems are also critical for operation, as the switches, buttons and knobs located on the mimic panels also enable the operator to manage the manure system status as per the requirements of the plant/system. The color codes represent a system, while visual on/off indications and measuring instruments provide the required information about the process status and status of parameters.

In modern or refurbished control rooms, the static panels take the place of digital systems as well as the dynamic visual display units, which provide a user-friendly operator-friendly interface that helps in multitasking, interacting with multiple systems on the operator station itself. Often, the replacement of traditional devices/sensors with intelligent sensors allows autonomous diagnostic and efficient communication compared to traditional instrumentation. In new nuclear plants, the control room design is based on digital logic and a VDU-based man–machine interface. In fact, in many old plants that have undergone life extension, the replacement of analog to digital conversion is common, except in protection logics where the decision is based on exhaustive safety review. Here, how to address the uncertainty associated with software reliability of software in digital systems poses challenges.

9.3.5.3 Proactive Approach to Precursor Analysis

The operator needs to be able to do three things: (a) understand what the normal conditions are in all the systems relevant to the current overall status of the plant; (b) recognize, with the help of available data (e.g., alarms, indications, recorders, etc.) the safety significance of the abnormal condition; and (c) to know how to respond correctly to restore the normal operation without compromising safety. The concept of safety function maintenance introduces a systematic approach for mitigating the consequences of an event based on a hierarchy of protective actions. To accomplish the safety functions, the operator does not need to know precisely what event has occurred. However, he does need to know what safety function must be accomplished, what the success paths are available, and the current plant state. By monitoring the selected sets of parameters that identify the status of the safety function being out of range or by certain alarm conditions, the threat to safety maintenance can rather easily be ascertained. Once threats to safety functions are identified, the safety systems designed to maintain the threatened safety functions are initiated automatically or manually.

9.3.6 Major Operator Aids

9.3.6.1 Hardcopy/Computerized Emergency Operating Procedures

Emergency procedures exist in the plant to manage the anticipated unusual conditions as well as accident conditions in the form of hard copy sequence steps. However, in real-time situations, the operator often must make decisions based on his or her mental model of the plant due to time constraints. During abnormal conditions, the initial approach to analyze the problem is event-based. However, on failing to reach any conclusion, as may happen during multiple failures, the emphasis shifts to safety function maintenance.

9.3.6.2 Safety Parameter Display Systems

Post-TMI accident and its safety reviews pointed to deployment operator aid in the control room of the nuclear plant that presents a consolidated and abstract display of safety parameters of the plant, and this eventually led to developing a subset as operator aid called the safety parameter display system (SPDS) (Joseph and George 1983). Interest in the deployment of safety parameter and display systems (SPDS) by nuclear plant operating organizations was growing and peaked during the late 1980s and early 1990s.

However, in 1986, based on a survey evaluating the performance of safety parameter display systems (SPDSs) with respect to status and quality in operating nuclear reactors, the USNRC issued a notice to utilities to review the performance and applicability of the SPDS in their plants and consider actions, if needed, to preclude the issues identified in this survey (USNRC 1986). The problem identified in this survey ranged from Lack of SPDS availability because of gross system malfunctions, Display of unreliable or invalid data and alarms, Poor acceptance of SPDS by operators because of reliability problems, Failure of management to integrate SPDS into the operational environment, Changes and interruption of SPDS display from outside the control room, Inadequate documentation of SPDS and failure to control system testing and modifications and Slow SPDS response to some operator commands.

9.3.6.3 Computerized Operator Information System

In the later period, digital technology slowly replaced the old analog systems in the nuclear plant. The operator aids also transformed into a digital operator information system to more advanced aids, such as risk monitoring and operator support systems with transient identification and diagnostic capabilities. The computerized operator information systems were less complex and similar to the SPDS as a man–machine interface system compared to dynamic operator support systems.

9.3.6.4 Risk Monitor

PRA technology was found to be a mature and promising technology for addressing real-time issues in the late nineties. The popularity of the development and deployment of risk monitoring grew later in the early two thousand years. In many nuclear plants across the world, the risk monitor has been deployed to support the operator in decision making in normal operation to evaluate the impact of a decision on risk, e.g., evaluation of the effect of an increase in allowable outage time and assessment of options available in support of the decision to address the safety case for the plant. Risk monitoring and related aspects have been discussed in Chap. 6 on risk simulation.

As seen, the nuclear industry was aware of the challenges involved in addressing the human factor and associated issues in the control room of the nuclear plant. This may be the reason the safety culture in the nuclear industry grew over time, and the claim that the nuclear industry maintains the highest standard of safety is endorsed and finds widespread acceptance. However, what is more important is that the nuclear industry also accepts the fact that sustained efforts can only where the risk of accidents becomes acceptably low is still a goal where we say the 'work is in progress' reflects part of nuclear culture.

It seems that these reviews provided vital learning lessons to nuclear plants as safety parameter display systems either as part of any emergency support aid, such as intelligent operator advisory systems, or in standalone mode are still being actively used in nuclear plants.

9.4 Role of Machine Learning in Operations

Traditionally referred to as 'intelligent systems', e.g., artificial neural networks (ANNs), fuzzy rule-based expert systems, genetic algorithms, support vector machines along with data analytics and data mining, robotics and the Internet of things (IoT) in recent times, are put under the class of 'machine learning' systems. These systems are actively being developed for application in nuclear plants in general and safety and security in particular to support the development of advanced technologies. Some of the examples are smart sensors and transmitters, prognostics and health management technologies for failure prediction, operator advisory systems in the control room, imaging identification for routine as well as strategic applications, robotics for automatic response and for application in radiation areas to reduce the radiation dose consumption which at times becomes prohibitive during normal as well as emergency conditions, are some of the examples of deployment of machine learning and data mining in nuclear applications.

The reason why these technologies are growing and replacing human activity is that they are all inspired by human physical and cognitive functioning, as these technologies have demonstrated better performance than the human itself. Of course, some caution is needed before these techniques are adopted for regular application and induction for safety critical systems. This is because any new technology brings along with it new issues and modes of failure. Furthermore, how much autonomy can be given to this system is also a big question, as this system might supersede humans in some domains, but the major question is how to treat these systems; they do not have consciousness and conscience or the capability to distinguish ethically right or wrong action, like humans.

9.4.1 Human Model as Inspiration for Machine Learning

The machine learning in this chapter refers to artificial neural networks and fuzzy rule-based systems for transient identification and fault diagnosis, respectively, for the development of an operator support system. ANN algorithms are inspired by the functioning of the brain. In the human brain, the connection of neurons, called axons, along with chemical processes in the cell body and axons provides the fundamental mechanism for cognitive processing. This processing provides two major attributes, viz., information processing and large memory, in support of all human activities,

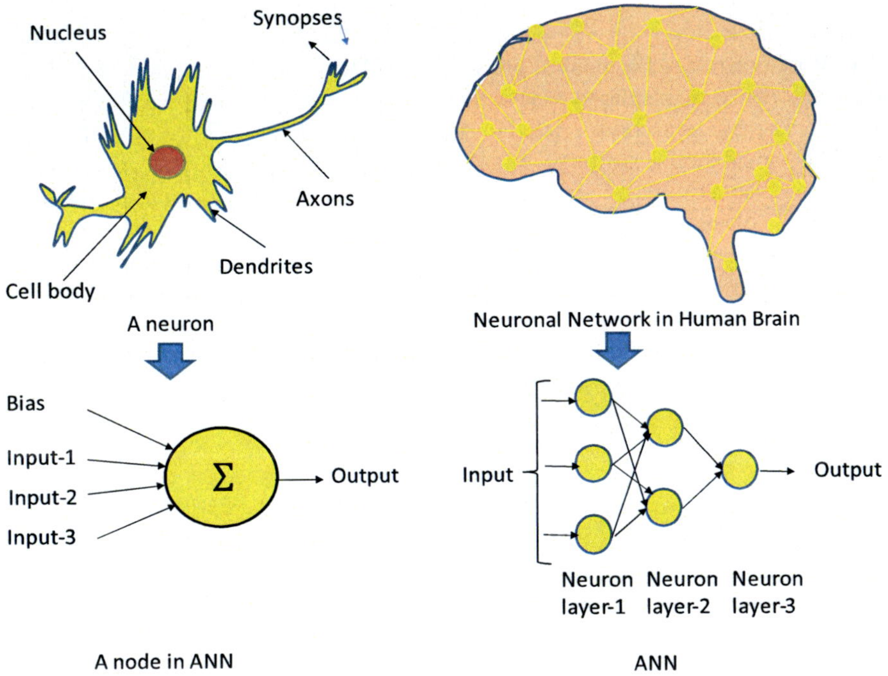

Fig. 9.2 How the human brain inspires machine learning–ANN modeling

including establishing a connection with the external and internal world in general and decision making.

Figure 9.2 shows a schematic of the brain, where it depicts the activation areas among the neurons although axons–the connecting elements. The synapse location at dendrite tips between two neurons provides the junction for passing neural signals to the connected neurons. The construction features of a typical neuron that inspires the formulation of a node in ANN, as seen on the left side of the figure, provide the basic building block of machine learning through the implementation of an ANN. The only difference is that the activation of the nodes is the driving mechanism in ANN, while in Brain, the connection of neurons facilitates memory and information processing functions.

9.4.2 *Can Machine Learning Exhibit Artificial Consciousness Potential*

Artificial consciousness is one of the areas of research that is in its initial stages of enquiry. The major reason is that the field of robotics, as well as autonomous operations such as self-driving automobiles, has already entered societal applications. Robots are used for many applications that include service sectors. With these advances, there is a considered view 'whether robots or AI in general can be conscious systems. There are questions related to AI agents capable of handling ethical issues, emotions, or subjective attributes, such as humans.

There are some fundamental requirements that need to be addressed. In the field of artificial consciousness, there are two major considerations. The first deals with the functional aspects of consciousness, and the second deals with the phenomenal aspect. The functional aspect addresses the functional aspects of cognitive-conscious operations in the brain with respect to rational thinking, evaluation, reasoning, learning, decision, etc. There is certain confidence in the AC community that, in a limited sense, computers or AI agents can exhibit functional consciousness. However, there is a need to test the hypothesis that gives confidence in attributing functional consciousness.

The second, i.e., phenomenal consciousness that addresses subjective attributes, such as the capability of an AI agent, to exhibit the quality of experiencing emotions, sensory qualities, such as smell, judgment of right or wrong. There is not much success in AC with respect to phenomenal consciousness.

The popular Turing test, which is extensively used to qualify the AI application, might not work with respect to qualifying AI agents, as consciousness poses challenges in handling subjective qualities, referred to as the problem area of consciousness itself. Therefore, any AI agent to qualify for possessing consciousness capabilities must be tested and demonstrated reliable operation and meeting the risk acceptance levels.

In this chapter, AI agents are employed for operator support systems that essentially must demonstrate their capability with respect to the following consciousness attributes:

Memory: Memory is a fundamental component required to process the acquired information/data in hard or soft or biological elements that involve encoding, storage, retrieval and modification based on the new input for further calls from the address locations. This information is used in many life or engineering processes in support of learning, reasoning, diagnostics, analysis, pattern matching and decision making either in isolation or along with other inputs and experiences. Human memory plays a vital role in building, modifying and storing conscious experiences. In the same vein for AI systems, memory can also be a pseudo part of the conscious system.

Learning: Learning is a core conscious experience in the sense that the rate of learning in humans is a result of the intensity of conscious stimuli, e.g., seeing, reading, smelling and touching. Generally, the learning quotient is defined as ease in terms of time, accuracy, memorizing the aspects of trade, knowledge, skill, or rules of the subject. In this context, AI can be considered conscious as tools such as artificial neural networks or machine learning algorithms have been performing learning based on past data presented to the learning algorithm, such as machine learning or artificial neural network algorithms; unsupervised, classifying and categorizing common attributes of the data called clusters, e.g., support vector machine; or reinforced learning algorithms that modify the trend of learning based on the new inputs available during learning. Accordingly, it can be argued that AI has the capability to learn consciousness; in fact, in some respects, it is better than humans, as chances of error on leaned patters are less and at times inferior to humans when new data or some external input is required in support of decisions.

Reasoning: Generally, referred to as logical processing that involves using the existing information that includes knowledge and/or data, is an integral part of the human thinking involved, particularly analysis in general and diagnostic, particularly for arriving at the goal or solution to a complex problem. Consciousness required for cognitive processing of new information in conjunction with the existing database of information. The level consciousness of brain faculties is directly responsible for the precision and accuracy of the reasoning. The human capability or capacity of reasoning is adversely affected during emergency conditions in support of decision making. In this sense, computers might play a critical supporting role in decision making.

From the above, in the context of AI application to operator support systems, considering the advantages available to computerize processing, it can be argued that computer-based operator support systems demonstrate the capability of being conscious or posing consciousness or exhibiting conscious attributes.

9.4.3 The Performance Metrics for an Intelligent System

In the artificial intelligence parlance, the above system is referred to as the online intelligent operator advisory (IOA) agent. These systems required bench marking its performance against the performance metrics by performing the recall tests using a reactor signature having a coverage factor that provides the required confidence for putting to end use, i.e., for control room application.

The performing metrics for the nuclear plant operations advisory system are as follows:

1. Reliability of transient prediction (for trained scenario) >99%
2. Online learning capability from the new scenario
3. Recovery capability from wrong advisory or no-goal found situation
4. Online diagnostic capability: 90%
5. Time taken for transient identification: a factor of 0.01 for the human operator
6. Diagnostic time: a factor of 0.01 for human operators
7. Training convergence error (RSMS): 0.001
8. Dynamic capability: 100%

These metrics are given only for guidance purposes and can be expanded or modified or expanded based on the domain-specific requirements. For example, the identification and diagnostic module should exhibit dynamic capability in terms of facilitating reinforcing new insight or input during the session such that the probability of correct decisions can be improved to such an extent that the management has confidence in the deployment of the advisory system for real-time application in the control room of the plant.

The understanding of the domain and the task environment is critical to the success of the application. In fact, one of the profound opinions and observations for the very limited growth of artificial intelligent system applications was that (a) even on a small scale or as a prototype level, the application could be demonstrated successfully; however, when the size of the model increased for real-time systems, the systems did not satisfy the performance requirements, and (b) a lack of data, experience and intimate knowledge of the system and an inadequate set of hypotheses to check the performance of the system led to failure of AI systems during real-time applications. This is where the importance of big data, data mining, and data analytics found ground in support of AI development.

9.4.3.1 Observability to the Agent

What it means for an operator support system placed in the control room of the plant? The instrument sensors and the associated telemetry along with human sense bases, such as the eye, ear, nose, skin and mind, enable an individual operator and his team to form a perception of a situation or requirement and decision making. In the nuclear plant, the control rooms are designed to provide and present coverage of the plant that is adequate to support him in ensuring safety and reliability. There are some areas in the plant, in addition to the control room, which are called local areas, and these areas are again manned by technical staff. The critical signals from these areas are communicated to the control room; however, certain equipment and systems for which the information is not sent to the control room are supervised locally. The local area technicians have the responsibility of control and supervision of the equipment and systems and remain in touch with control room staff for smooth operations of local systems and equipment. Therefore, the overall operational ecosystem has distributed architecture with the main control room as the central processing and action centers.

With respect to the operator support system, the observability coverage can be adequate in terms of the sensor network along with other audio and visual systems. Even if it is felt that some information is not coming to the control room online, there will be provisions to input this information interactively manually. However, this applies to the diagnostic module in OSS, while the transient identification module that is basically configured using ANN will have 100% observability as the transient characteristics, say fast evolution of scenario, does not allow any manual interaction.

9.4.3.2 Nature of Data Deterministic and Probabilistic

Even though most of the data received online and sampled periodically in the range of 10 milli sec interval milliseconds can be considered as crisp data, particularly the trip and alarms that remain either 'on' or 'off' level have deterministic characteristics. However, the kind of uncertainties associated with the collection of these signals and using it for transient processing poses challenges such as registration timings, the requirement of correct analysis even for missing and fuzzy signals, and the assumption of randomness in input trip and alarm signals. For analog or measured continuous parameters or the interpretation of operational inputs, fuzzy and stochastic considerations should be effective assumptions.

9.4.3.3 Training of Staff

The operational requirements in a nuclear plant have all three major categories or levels of training. The first category addresses skill-based training and qualification. In this category, recruitment is done at the higher secondary school level. Here, the emphasis is placed on increasing awareness of different systems and components, keeping in mind the manning of local plant areas, including control room jobs, to support senior staff. The second level is the lead-hand or supervisory level, where the qualification requirement is a higher secondary level with minimum experience criteria. The supervisory is an important link between the shift in charge and other operating staff. The plant is managed in the round the clock by shifting two trained and qualified engineers. The senior shift engineer is responsible for overall supervisory and decision making activities. He is responsible for the safety of the plant in the round-the-clock shift. He reports to plant management.

9.4.3.4 Multiobjective Environment

Even though in most of the conditions and functions for intelligent operator advisory systems, safety is an overriding attribute, there is also a need to build and demonstrate in the intelligent modules to ensure that spurious warnings are not generated that might reduce plant reliability and availability. Here, the preprocessing of the singles such that only validated singles only form input to the intelligent module. The preprocessing involves 2-out-of-3 voting redundancy confirmation, testing the modules for noisy and uncertain input, etc. Choosing the right algorithm or system for a given application. For example, for multiobjective problems, the genetic algorithm is considered the right candidate.

9.4.3.5 Domain Specific Attributes

The literature shows that one of the critical conditions for the success of any machine learning application is coverage and considerations of domain-specific requirements and expertise. For example, the intelligent operators aid designed to cater to automated emergency operator support systems should possess adequate capability for knowledge representation and processing, inferencing keeping in view that safety is an overriding factor. These tools should be capable of processing not only static rules and data but also capturing and processing dynamic scenarios in a manner that is more effective and efficient than the human operator. The system might come across a situation where it enters an unknown territory in terms of rules and data in such situations, and the system should track back or reach a safe state such that consequences due to wrong decisions can be avoided. Furthermore, some applications might require hardening of the system due to conditions created by the emergency itself; for example, the processing circuit of the hardware of an intelligent system might be affected by the increase in radiation levels during an emergency in a nuclear plant. Furthermore, the operating synthesis should validate at every state that the safety goals are met through special metrics. For example, for a nuclear plant ensuring core cooling, fast response to shut down and subcriticality, levels of ionizing radiation consequences should be the basic building principles for processing the information during normal as well as emergency conditions.

More factors, such as the capability to process the input in parallel as well as in sequential mode, are key requirements. For example, in an intelligent operator support system, the parallel processing requirement for identifying plant emergencies employing an ANN-based approach and further sequential processing through a production rule-based diagnostic approach might provide an effective strategy for an operator support system.

9.4.3.6 Major Postulations/Assumptions

It is recognized that operations management has many facets and many dimensions, and the subject 'operations research' specifically caters to most of the situations in operations management, such as identification, prioritization, diagnosis, supply chain modeling, queuing process and modeling, software reliability aspects, warranty and market evaluation. However, considering the scope of safety critical systems, the machine learning aspects important to this family of systems, such as nuclear facility operations, process and chemical systems and transport systems such as space, railway and aviation, this chapter has been tailored to address the requirements of these systems. Accordingly, the major postulates are as follows:

1. The human model in this approach is based on considerations of consciousness, cognition and conscience along with the six senses for interaction with the outside environment.

2. The operating crew is trained and qualified in various aspects of design, operations, maintenance and services necessary for smooth operation and handling emergency conditions.
3. The control room along with the process console for performing various activities directly for the plant, such as power maneuvering, start-up, shutdown, is central to the operator support system.
4. It is postulated that intelligent processing features of the operator support system include conscious factors, such as awareness, alertness, cognitive factors, such as memory, rationalizing, diagnostics, correlating information for identification, inferencing and decision making; and the conscience factors such as integrity, dedication, sincerity, support the operator's decision making capability.
5. The operator support system envisaged here acquires the signals/input related to various processes and manually fed queries generated by the system. The system operators are in an open loop in the sense that the decisions and actions are performed by the operator only. Therefore, even though the machine learning paradigms have been developed and tested rigorously, the responsibility of actions is with the operator only, and the system plays only an advisory role.

Apart from the above postulates, there are many assumptions that can be discussed as and when the development activities are discussed in the next section.

9.5 Development of a Machine Learning-Based Operator Support System for Nuclear Plants

9.5.1 General

The design approach involved the development of the Intelligent Operator Advisory (IOA) system, even though it uses the state of the art in the design of artificial intelligence in general and machine learning in particular along with the experience and lessons learned in the development and testing of the operator support system developed by the author to comply with the performance metrics set out for IOA (Varde et al. 1996).

Design Objective: Development of an intelligent operator support system for a nuclear plant to increase human cognitive and action reliability toward reducing risk associated with human error, particularly during emergency conditions.

Scope: The scope of this system is to model (a) the model accident precursor based on the information available in the control room and (b) the dynamic aspects of the emergency operating procedure, (c) the risk reduction as the overriding priority by using the PRA model of the plant and (d) where PRA knowledge is inadequate to address a situation model-based and operating experience from the basis.

9.5.2 Generic Requirements of AI for OSS

Nuclear plants represent an example of highly complex engineering systems. Given that we are talking about a system that generates energy through a neutronic chain reaction that needs to be controlled effectively by just maintaining the system to be critical in terms of neutron population generation and loss balance, decay heat keeps producing even when the system is shut down, and third and most important contain the radiation in the reactor boundary. Furthermore, the interconnection and dependence between hardware, software and human interactions make the operations demands highly skilled human resources. Even though automation in nuclear plants caters to many challenges, a higher demand for reliability and safety of the nuclear plant requires higher autonomy during operation. There is an opinion that the operator's role is vital for enhancing reliability on the one hand and risk reduction on the other, as human cognition is critical in support of choosing a course of action and decisions. Higher autonomy with the availability of intelligent AIDS might work to reduce cognitive load on operators during emergencies (Basher and Neal 2003).

Figure 9.3 shows a broader perspective of nuclear plant operation, the role of operators and the critical role of operator support systems in operations ecosystems. The process and electronics from analog and digital monitoring systems and sensors of instrumentation channels communicate the states of parameters to control room on status of systems, on neutronics, thermal hydraulics, temperature, pressure, flows, structural properties, chemical parameters, parameter values, etc. The signal also carries the status of the reactor, systems and equipment that inform the operator operational status of the reactor and safety systems.

The control rooms are designed such that the operator can effectively and efficiently oversee the reactor status, manipulate, reactor, equipment and components remotely from the control room. The availability of information in the control room ensures that the operator obtains the required information regarding the status of the reactor and associated systems. The reactor console in the control room enables the operator to determine the manure reactor status, viz., start-up, power raising and lowering, shutdown and monitor, and assess the performance of critical systems such as reactor shutdown, reactor neutronic systems, trip and alarm logging.

The operator and the support staff are licensed staff in the control room and are specially trained, licensed and authorized at various levels. The number, qualification levels, and availability of staff in the control room are governed by plant technical specifications. Any noncompliance attracts regulatory review. The technical specification also prescribes the duties and responsibilities of the operation, maintenance and services staff.

The availability of an operator support system is not a regulatory requirement; however, the plant operator information is becoming a regular feature of the control room. This system informs the operator status of safety systems and other important parameters often named safety parameter display systems. This feature often includes a trip and alarm system that enables the operator to understand the reactor condition.

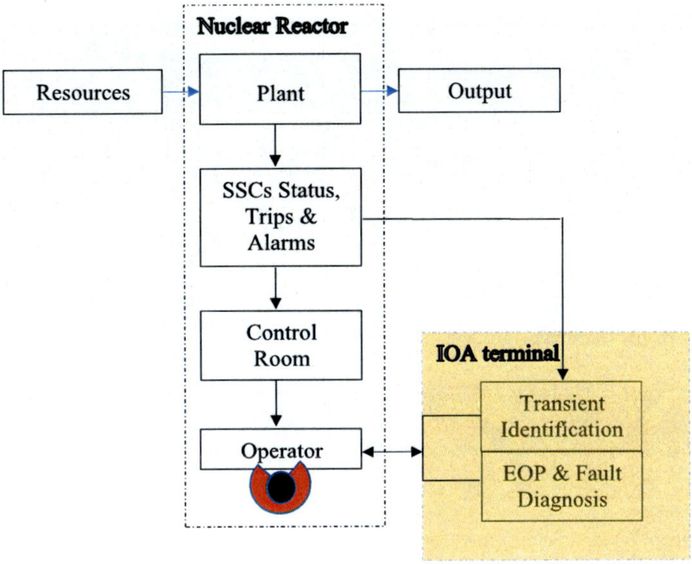

Fig. 9.3 Role of intelligent operator advisory system in nuclear plant control room

The subject of this chapter is the development of an intelligent transient identification and diagnostic system for normal and emergency conditions, which is shown in yellow on the right side of Fig. 9.4. As seen, this is an open loop advisory system, and its use and further action based on the advice received from IOS is the operator's choice. This system does not work in closed feedback loops like many other automatics systems, as the domain of supervisory activities such as transient identification and diagnosis is stochastic or probabilistic in nature. Hence, it supports operators in decision making based on the input IOS receiving online. Of course, the development of the transient identification module poses challenges as the scenario is dynamic and has to deal with fuzzy, incomplete, and at times noisy data. However, the advantage is that the module is subjected to rigorous tests before it is considered for real-time use. Similar complexity is involved with the diagnostic module, which works on fuzzy rule-based expert systems. In operations, communications are often made in a linguistic manner that carries with it an expression about the level and complexity of information. For such scenarios, the use of fuzzy logic for the development of production rules in the diagnostic module provides an effective strategy to address operations ecosystems.

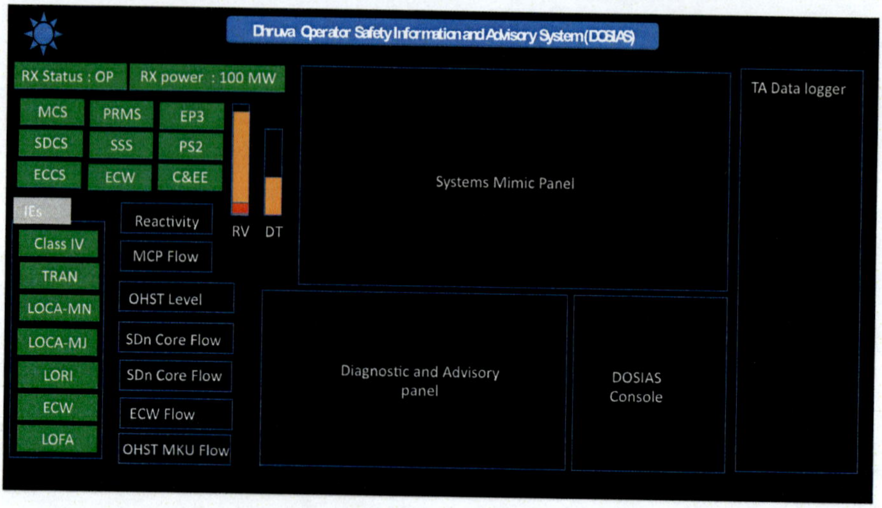

Fig. 9.4 Typical layout operator advisory system

9.5.3 Artificial Neural Network for Transient Identification

Generally, the plant transients in any engineering are characterized by three major factors: (a) they change the operating status of the plants, e.g., from normal operational state to safety shutdown state, (b) the applicable process status as represented by plant control and logic, e.g., trips, alarms, sudden change on process parameters; change their respective status in relatively short time say ranging from millisecond to seconds and may continue until minute when the plant condition stabilizes in a mode, as mentioned to say shutdown state, (c) each registration pattern is unique to a given state or condition, say the transient loss of power– whether the initial condition was operating state, number of feeders tripping, normal shutdown state during the transient, etc., and last and most important (d) even though operators are trained for taking decisions during transients, due to flooding of signals the transient scenario demands higher consciousness and cognitive requirements in support of decisions.

Based on the research and literature review, the artificial neural network (ANN) approach, which provides a classification strategy through a pattern matching approach, provides an elegant mechanism for transient identification. The second and major advantage is that the ANN approach provides the solution in a very short time compared to the operator generally takes for comprehending the status of the plant. Therefore, in the operator advisory system, the ANN approach has been deployed for transient identification. The output of the ANN triggers the knowledge-based system for initiating fault diagnosis, which is the next required module for the operator support system. The ANN module consists of m transients that need to be modeled in a plant; therefore, a cluster architecture has been discussed here, as shown in Fig. 9.5.

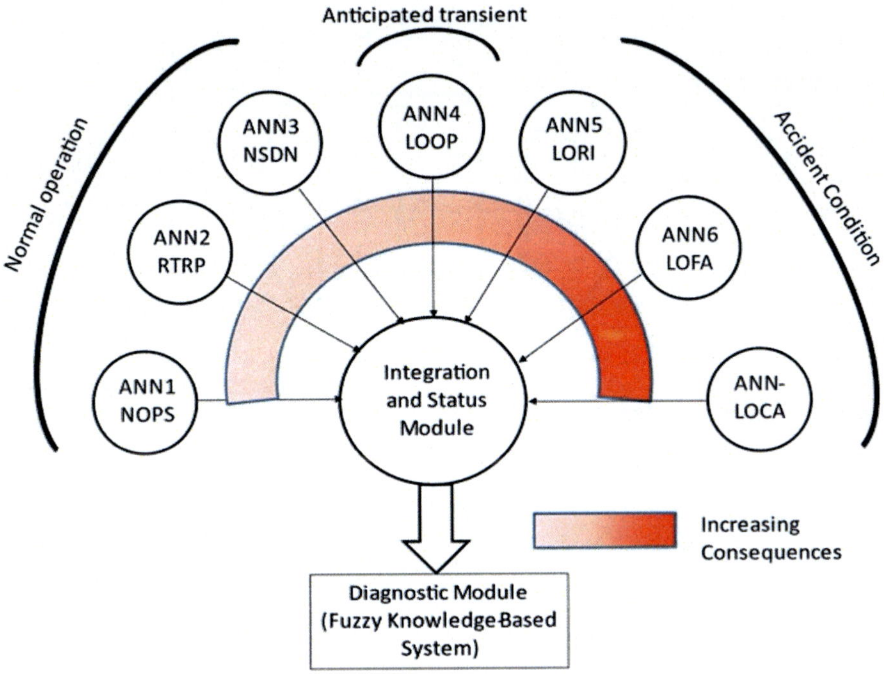

Fig. 9.5 Representation of transients in the ANN module

9.5.3.1 Implementation of Machine Learning Through ANN Framework

A. **Basic Concept**

A node and the connection between successive nodes form the basic building block of the ANN. Figure 9.6 shows a typical node showing the input of the processing algorithm through an activation function–in this case, a sigmoid function and the results presented through the output node. The weighted input is presented to the node as seen in the function. The improvement of weights at nodes is a measure of the significance of that node as the ANN learns. Mathematically, this processing can be represented by a mathematical function as follows:

$$O = C_o + \sum_{i=1}^{N} C_i \emptyset_i (x_1, x_2, x_3 \ldots . x_n), \tag{9.1}$$

Here, O represents the output at a node, which is a function of inputs $x1, x2, x3 \cdots x_n$; C_i *is a constant;* and Φ represents a polynomial function. One of the major strengths of an ANN algorithm is that it works well even when the input–output relation exhibits a nonlinear characteristic, which is the case for transient identification in an

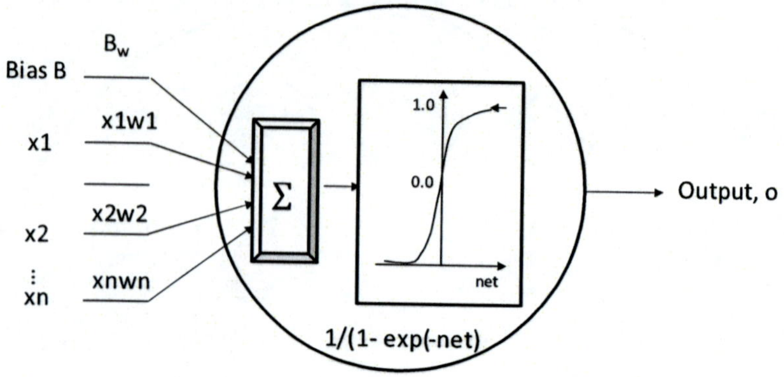

Fig. 9.6 A typical node in ANN

Fig. 9.7 Typical 3-layer ANN structure

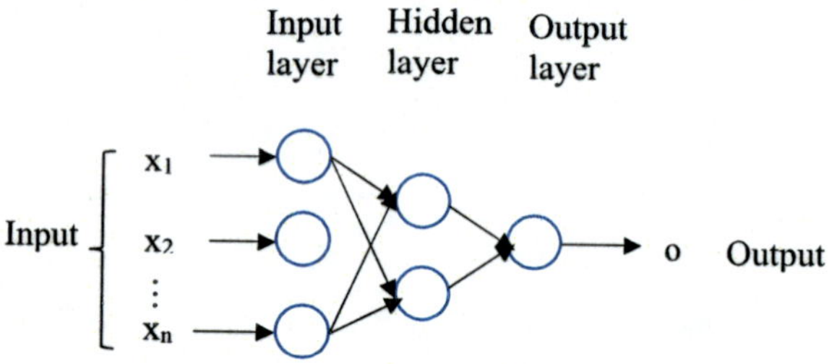

automated engineering environment. Accordingly, the sigmoid function represents a processing at node as follows:

Furthermore, Fig. 9.7 shows the essential features of an ANN. Typically, at the structural level, ANN has three layers, viz., one input layer, one or more than one hidden layer and one output layer, to present the result of ANN computation.

The number of neurons in the input layer is governed by the number of inputs in an input vector, while the number of neurons in the output layers is determined by the requirements of presenting the results, while the number of hidden layers and the number of neurons in the hidden layer is still a topic of debate, and often expert judgment or iterative experiments are performed to determine the number of hidden layers and neurons in each layer.

At this stage, it is necessary to define the function 'net'. Before that, it is also necessary to understand the requirement of a bias node in ANN computation. Let us first see what makes a bias voltage in an electronic circuit. Of course, it provides the reference minimum voltage required for the satisfactory operation of a control card. On similar lines, the bias node carries the reference bias intensity for the functioning, and since it is a software algorithm, it provides a reference value around which the circuit should function. Now, let us express the mathematical expression for 'net' for any node j in each layer k for input coming from xi nodes in the 'k-1' layer (previous layer) modified by weights 'w' of connecting nodes i-j as follows:

$$\text{net}_j = \sum_i w_{ij}^k x_i^{(k-1)} + b_j^{(k)} \tag{9.2}$$

As discussed above, w_{ij}^k is the connection weight between the ith node in the k-1 layer and the jth node in the kth layer. $x_i^{(k-1)}$ is the output from the ith node in layer k-1, and b_j^k is the bias node in the kth layer. The output from x_j^k is estimated here employing the sigmoid function of *net* as follows:

$$x_j^k = \frac{1}{1 + \exp\left(-\text{net}_j^k\right)} \tag{9.3}$$

9.5.3.2 Training Algorithm

The fundamental relationship between the input layer and its corresponding metrics forms the training set. The table shows the input training sets for the loss of off-site power full power operations and the shutdown condition. The feed-forward and backpropagation network (BPN) learning algorithm is one of the learning algorithms that has been used in training the ANN. The BPN algorithm works in forward pass that initiates from the input layer to the hidden layer and further from the hidden layer to the output layer. At the output nodes, the desired and actual output is compared and provides the network at each node with an error. An average error is also evaluated at the layer level. The average error at the output in each iteration is compared with the set goal until the goal or set error is achieved. This process of iterative error estimation is called convergence. In the ANN designed here, a root mean square (rms) error estimation was used. The rms error formulation implemented in this network is expressed as:

$$E_{rms} = \left[\frac{\sum_{i=1}^{p} \sum_{j=1}^{n^k} \left(d_{pj} - x_{pj}^k\right)^2}{p * n^k}\right]^{0.5} \tag{9.4}$$

where

E_{rms} *rms* error
p Number of patterns in training set
L Number of layers
d Desired output
x Actual output.

Training involves an iterative process of weight modification and convergence until the RMS error reaches the set goal. The successive modification of modes in each iteration is also referred to as batch training. As discussed above and the

Table 9.1 ANN configuration in the transient identification module

Parameter no	Description	Metrics
1.	Number of neurons in input layer	26
2.	Number of neurons in the output layer	1
3.	Number of hidden layers	1
4.	Number of nodes in the hidden layer	15
5.	Learning rate	System assigned dynamic
6.	Momentum rate	System assigned dynamic
7.	Target average rms error	0.001
8.	Target individual error at nodes	0.0001
9.	Maximum number of iterations	1000
10.	Time for convergence/training	3 min

information given in tables about input and output vectors, the training configuration of the ANN can be given as shown in Table 9.1

Training Vectors

Formulation of input pattern

During a transient that is change of status, say from normal operations to automatic shutdown happens nearly in a short time say in a range of milliseconds to a minute depending upon the specific characteristics of a given parameter in terms of its time constant. This means that all the parameters do not achieve the new state at once but at different times. For example, the reactor trip registers in milliseconds, while the coolant flow might take 50 to more time depending on the system inertia. This applies to most of the parameters. What it means for designing an ANN algorithm is that it should exhibit dynamic characteristics in the temporal domain. This requirement was met by having a dedicated ANN for each category of transients. In this chapter, a 'loss of off-site power' (LOOP) transient from full power operation has been considered for demonstration of the ANN approach for transient identification.

A. **Identification of the parameters**

Two broad categories of transients are postulated in the design basis of the nuclear reactor, viz., anticipated transients and design basis accidents. Of course, the third category includes beyond-design basis accidents. However, for the purpose of demonstrating the transient identification approach, we will focus on anticipated transients–the transients that are expected to occur in the life of the plant more than once. For example, the subject here is loss of off-site power (LOOP). Since LOOP events occur at least once in a year and there exists good experience in responding to this situation, the station blackout situation should be avoided. Generally, there is a pattern of registration of process and safety parameters in the control room data logger. The second reason for the choice of LOOP event is that it can be easily simulated on a plant simulator as part of training as well as performing the recall tests. The 37 parameters selected for the ANN training are acquired by a data acquisition module either from the simulator or from the plant as per requirements. The time constant for the data acquisition is 1 cps.

Table 9.2 shows the pattern that includes (a) reactor trips that follow binary pattern 0 or 1, i.e., the parameter is cleared state (normal state) or registered state (deviation from normal), respectively. Reactor trips automatically shut down the reactor without operator intervention. (b) Process alarms, such as reactor trip alarms, also follow binary logic 0 or 1. (c) Process parameter value of the process parameter on the instrument scale, which might range from minimum to maximum, but it is not a binary value.

The binary parameter occupies values of 0 and 1, as mentioned above, but for the purpose of ANN simulation, the chosen value for 0 is 0.1, and for 1, it is 0.9. The reason is that the values o and 1 make the convergence at times prohibitive, as it takes a long time to arrive at the corresponding rms error during training, while there is no distinct advantage by going below 0.1 or above 0.9.

Furthermore, the process parameters are normalized between 0 and 1 such that the ANN input vector is consistent, and all the input nodes can be represented between 0.1 and 0.9. The procedure for normalization is as follows. Assume the actual PCW pump flow is 28,000 lpm, while the instrument range is 40,000 lpm. The normalized value of the pump flow is 28,000/40,000 = 0.7. The input pattern for the training of ANN is represented by 37 × 17 metrics, where 37 are parameters represented as row in row and 17 are the epoch, i.e., time stamping for which the ANN is trained dynamically for transient–normal automatic shutdown due to off-site power failure, as shown in Table 9.2. The first and second columns show the serial number and nomenclature of the 37 parameters identified for this transient. For each parameter, the remaining 17 rows show the parameter status or normalized value (applicable for instrument reading) that forms a vector for training.

9.5.3.3 ANN Training

The simulation and iterative process is integral to optimizing the parameters for ANN. The training configuration is shown in Table 9.1. The objective is to achieve the convergence of the network to the acceptable aggregate or overall rms error. The ANN code was written in the SCILAB environment. The simulations are performed with identified input vectors by performing parametric studies to arrive at the number

Table 9.2 (a) TR-1: Loss of Off-Site Power during Power Operations—Normal Shutdown. (b) TR-2: Loss of Off-site Power–during routine refueling shutdown condition (system depressurized)

(a)

S. No	Symptom/Parameter	Epoch of time																
		1	2	3	4	5	6	7	8	9	10	11	12	13	14	15	16	17
1.	Time (Min.)	0	0.01	0.1	1	2	5	15	60	120	180	240	300	360	420	480	540	600
2.	N. power	1	0.3	0.1	0.1	0.1	0.1	0.1	0.1	0.1	0.1	0.1	0.1	0.1	0.1	0.1	0.1	0.1
3.	MPSF-Ch-A	0	1	1	1	1	1	1	1	1	1	1	1	1	1	1	1	1
4.	MPSF-Ch-B	0	1	1	1	1	1	1	1	1	1	1	1	1	1	1	1	1
5.	MPSF-Ch-C	0	1	1	1	1	1	1	1	1	1	1	1	1	1	1	1	1
6.	PCP-1	1	0.8	0.3	0.1	0	0	0	0	0	0	0	0	0	0	0	0	0
7.	PCP-2	1	0.8	0.3	0.1	0	0	0	0	0	0	0	0	0	0	0	0	0
8.	PCP-3	1	0.8	0.3	0.1	0	0	0	0	0	0	0	0	0	0	0	0	0
9.	PCW flow	1	0.5	0.1	0	0	0	0	0	0	0	0	0	0	0	0	0	0
10.	PCW PR	1	0.2	0.1	0	0	0	0	0	0	0	0	0	0	0	0	0	0
11.	DP	0.7	0.2	0.1	0	0	0	0	0	0	0	0	0	0	0	0	0	0
12.	SDP-1	0	1	1	1	1	1	1	1	1	1	1	1	1	1	1	1	1

(continued)

Table 9.2 (continued)

(a)

S. No	Symptom/Parameter	Epoch of time																
		1	2	3	4	5	6	7	8	9	10	11	12	13	14	15	16	17
13.	SDP-2	0	1	1	1	1	1	1	1	1	1	1	1	1	1	1	1	1
14.	SDP-3	0	1	1	1	1	1	1	1	1	1	1	1	1	1	1	1	1
15.	SDP-1 FL	0	0.1	0.3	0.5	0.5	0.5	0.5	0.5	0.5	0.5	0.5	0.5	0.5	0.5	0.5	0.5	0.5
16.	SDP-2 FL	0	0.1	0.3	0.5	0.5	0.5	0.5	0.5	0.5	0.5	0.5	0.5	0.5	0.5	0.5	0.5	0.5
17.	SDP-3 FL	0	0.1	0.3	0.5	0.5	0.5	0.5	0.5	0.5	0.5	0.5	0.5	0.5	0.5	0.5	0.5	0.5
18.	ARNU	0	1	1	1	1	1	1	1	1	1	1	1	1	1	1	1	1
19.	CVDV	0.5	0	0	0	0	0	0	0	0	0	0	0	0	0	0	0	0
20.	EP3-Bus-J-UV	0	1	0	0	0	0	0	0	0	0	0	0	0	0	0	0	0
21.	EP3-Bus-k-UV	0	1	0	0	0	0	0	0	0	0	0	0	0	0	0	0	0
22.	EP3-Bus-L-UV	0	1	0	0	0	0	0	0	0	0	0	0	0	0	0	0	0
23.	EP3-Bus-J-ALT	0	0	1	1	1	1	1	1	1	1	1	1	1	1	1	1	1
24.	EP3-Bus-K-ALT	0	0	1	1	1	1	1	1	1	1	1	1	1	1	1	1	1
25.	EP3-Bus-L-ALT	0	0	1	1	1	1	1	1	1	1	1	1	1	1	1	1	1

(continued)

Table 9.2 (continued)

(a)

S. No	Symptom/Parameter	Epoch of time																
		1	2	3	4	5	6	7	8	9	10	11	12	13	14	15	16	17
26.	EP2-Bus-O-UV	0	0	0	0	0	0	0	0	0	0	0	0	0	0	0	0	0
27.	EP2-Bus-P-UV	0	0	0	0	0	0	0	0	0	0	0	0	0	0	0	0	0
28.	EP2-Bus-Q-UV	0	0	0	0	0	0	0	0	0	0	0	0	0	0	0	0	0
29.	EP2-Bus-R-UV	0	0	0	0	0	0	0	0	0	0	0	0	0	0	0	0	0
30.	EP3-Bus-S-UV	0	0	0	0	0	0	0	0	0	0	0	0	0	0	0	0	0
31.	EP3-Bus-T-UV	0	0	0	0	0	0	0	0	0	0	0	0	0	0	0	0	0
32.	Plant CC-Overhead Level	0	0	0	0	0	0	0	0	0	0	0	0	0	0	0	0	0
33.	Plant CC-SOC (State-of-charge)	0	0	0	0	0	0	0	0	0	0	0	0	0	0	0	0	0
34.	Tertiary cooling	0	1	1	1	1	1	1	1	1	1	1	1	1	1	1	1	1
35.	Fire water Inj	1	0	0	0	0	0	0	0	0	0	0	0	0	0	0	0	0
36.	Plant emergency usable BT-line up Reserve-1 availability	1	0	0	0	0	0	0	0	0	0	0	0	0	0	0	0	0
37.	Plant emergency Usable SFSB reserve-2 availability	1	0	0	0	0	0	0	0	0	0	0	0	0	0	0	0	0

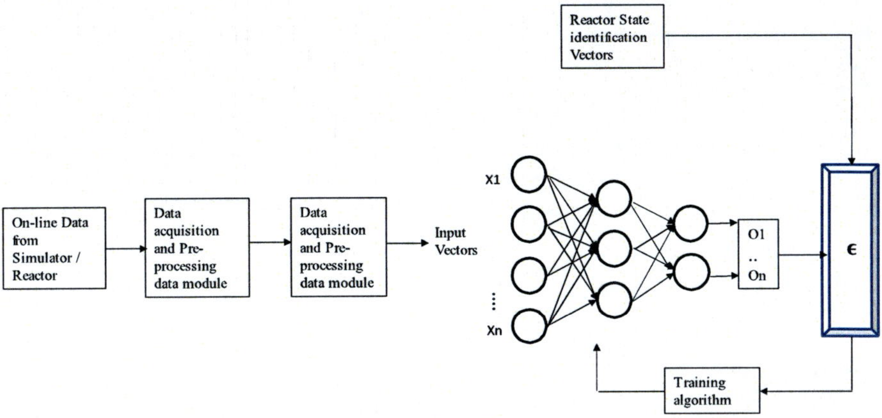

Fig. 9.8 Training algorithm for the ANN Varde et al. (1998)

of hidden layers and the number of neurons in the hidden layer while targeting the network performance. This sets the stages for performing the recall tests. The ANN employs the sigma function as the activation function. The training is performed in batch mode, i.e., each of the 17 patters is fed to the ANN sequentially. During each forward processing, the change in weights is computed, and during backpropagation, safety is affected on the output and hidden layers. Finally, as expected, the rate of convergence is relatively fast, and toward the end of training, the rms error saturates. The parameters that were used for this final convergence were a single hidden layer and 25 nodes in the hidden layer. It is worthwhile to note that appropriating 0.1 for 0 and 0.9 for 1 helped achieve convergence faster. This aspect was checked while performing the recall tests on the network with the trained or learned as well as new patterns with anomalies that can affect performance during the real-time condition of the plant. The training was terminated by the network when the goal, i.e., a rms error of 0.001, was achieved.

Figure 9.8 shows the training algorithm. The first three steps are online data acquisition, identification of patterns and preprocessing such that ANN can present these inputs, as shown in Table 9.2. The output during each forward movement is compared with the desired output that leads to modifying weights for backpropagation. Backpropagation continues until convergence is achieved. Figure 9.8 shows the convergence of the RMS error with respect to the number of iterations.

9.5.3.4 Testing and Validation

Even though the training vectors were based on the data that form the pattern, ANN testing is required to check its robustness and prediction accuracy for real-time conditions encountered in the plant. For example, 2-out-of-3 majority voting logic in designs and control such that one sensor registration failure does not make

Table 9.3 Events and variables

Event	Parameter/Status	Fuzzy variable
A	Lub-oil circulating line leakage	Low/Medium/High
A	Lub-oil pump operation fault	Low/Medium/High
C	Heat exchanger tube leakage	Low/Medium/High
D	Tube fouling (causing increased resistance to oil flow)	Low/Medium/High
E	Pressure anomalies at the downstream side of the oil filter	Low/Medium/High

the parameter invalid. Similarly, noise might form part of the data coming to the network. At times, the data received in the plant are such that additional inputs are required to comprehend the evolving scenario. Input in the plant might be used for safety parameters.

Therefore, recall tests are performed on the ANN module as follows:

a. Input with trained pattern
b. Deviation in the input patterns due to missing data
c. Patterns with ambiguous data
d. Pattern with sensor failure data.

Table 9.3 presents the results of the recall tests. The criterion to predict the success level of ANN was that the network should behave the way a trained operator behaves. The RMS error provides confidence in the prediction. For example, one sensor failure is anticipated in the plant, but this represents the state of the plant where there is concern and needs immediate attention. The degradation in RMS error should be expected, but it should not exhibit cause for serious concern. Noisy data filtering is required in any application; here, the network should be robust and accurate against the background notices up to the accepted level. Network should also be robust to the extent; any tangentially new pattern is provided then it should try to match the nearest matching pattern but with higher rms error.

One plant transient is identified, and the next step is to initiate the diagnostic module to ensure that the response of the plant is as intended in terms of safety equipment functions as well as process systems for the desired status.

(b)

Symptom/Parameter	Epoch of time																
	1	2	3	4	5	6	7	8	9	10	11	12	13	14	15	16	17
Time (Min.)	0	0.01	0.1	1	2	5	15	60	120	180	240	300	360	420	480	540	600
Neutronic power	0.1	0.1	0.1	0.1	0.1	0.1	0.1	0.1	0.1	0.1	0.1	0.1	0.1	0.1	0.1	0.1	0.1
MPSF-Ch-A	0	1	1	1	1	1	1	1	1	1	1	1	1	1	1	1	1
MPSF-Ch-B	0	1	1	1	1	1	1	1	1	1	1	1	1	1	1	1	1
MPSF-Ch-C	0	1	1	1	1	1	1	1	1	1	1	1	1	1	1	1	1
PCP-1	0	0	0	0	0	0	0	0	0	0	0	0	0	0	0	0	0
PCP-2	0	0	0	0	0	0	0	0	0	0	0	0	0	0	0	0	0
PCP-3	0	0	0	0	0	0	0	0	0	0	0	0	0	0	0	0	0
PCW flow	0	0	0	0	0	0	0	0	0	0	0	0	0	0	0	0	0
PCW PR	0.1	0.1	0.1	0.1	0.1	0.1	0.1	0.1	0.1	0.1	0.1	0.1	0.1	0.1	0.1	0.1	0.1
DP	0	0	0	0	0	0	0	0	0	0	0	0	0	0	0	0	0
SDP-1	1	1	1	1	1	1	1	1	1	1	1	1	1	1	1	1	1
SDP-2	1	1	1	1	1	1	1	1	1	1	1	1	1	1	1	1	1
SDP-3	1	1	1	1	1	1	1	1	1	1	1	1	1	1	1	1	1
SDP-1 FL	0.5	0.5	0.5	0.5	0.5	0.5	0.5	0.5	0.5	0.5	0.5	0.5	0.5	0.5	0.5	0.5	0.5
SDP-2 FL	0.5	0.5	0.5	0.5	0.5	0.5	0.5	0.5	0.5	0.5	0.5	0.5	0.5	0.5	0.5	0.5	0.5
SDP-3 FL	0.4	0.4	0.4	0.4	0.4	0.4	0.4	0.4	0.4	0.4	0.4	0.4	0.4	0.4	0.4	0.4	0.4
ARNU	1	1	1	1	1	1	1	1	1	1	1	1	1	1	1	1	1
CVDV	0	0	0	0	0	0	0	0	0	0	0	0	0	0	0	0	0
EP3-Bus-J-UV	0	1	0	0	0	0	0	0	0	0	0	0	0	0	0	0	0

(continued)

(continued)

(b)

Symptom/Parameter	Epoch of time																
	1	2	3	4	5	6	7	8	9	10	11	12	13	14	15	16	17
EP3-Bus-k-UV	0	1	0	0	0	0	0	0	0	0	0	0	0	0	0	0	0
EP3-Bus-L-UV	0	1	0	0	0	0	0	0	0	0	0	0	0	0	0	0	0
EP3-Bus-J-ALT	0	1	1	1	1	1	1	1	1	1	1	1	1	1	1	1	1
EP3-Bus-K-ALT	0	1	1	1	1	1	1	1	1	1	1	1	1	1	1	1	1
EP3-Bus-L-ALT	0	1	1	1	1	1	1	1	1	1	1	1	1	1	1	1	1
EP2-Bus-O-UV	0	0	0	0	0	0	0	0	0	0	0	0	0	0	0	0	0
EP2-Bus-P-UV	0	0	0	0	0	0	0	0	0	0	0	0	0	0	0	0	0
EP2-Bus-Q-UV	0	0	0	0	0	0	0	0	0	0	0	0	0	0	0	0	0
EP2-Bus-R-UV	0	0	0	0	0	0	0	0	0	0	0	0	0	0	0	0	0
EP3-Bus-S-UV	0	0	0	0	0	0	0	0	0	0	0	0	0	0	0	0	0
EP3-Bus-T-UV	0	0	0	0	0	0	0	0	0	0	0	0	0	0	0	0	0
Plant CC-Overhead level	0	0	0	0	0	0	0	0	0	0	0	0	0	0	0	0	0
Plant CC-SOC (State-of-charge)	0	0	0	0	0	0	0	0	0	0	0	0	0	0	0	0	0
Tertiary cooling	1	1	1	1	1	1	1	1	1	1	1	1	1	1	1	1	1
Fire water Inj	1	0	0	0	0	0	0	0	0	0	0	0	0	0	0	0	0
Usable BT-line up reserve-1 availability	1	0	0	0	0	0	0	0	0	0	0	0	0	0	0	0	0
Plant emergency usable SFSB reserve-2 availability	1	0	0	0	0	0	0	0	0	0	0	0	0	0	0	0	0

9.6 Diagnostic Module Fuzzy Knowledge-Based Expert System for Diagnosis

9.6.1 General

This section presents a comprehensive methodology for fault diagnosis that enhances the existing condition-based maintenance capability of engineering systems and reduces the dependence on human experts for fault diagnosis using a fuzzy rule-based approach (Nain 2011). This approach can be used in a complex system where failure data are either not available or scarcely available, which makes use of extensive data-driven techniques prohibitive. The methodology involves deployment of the Level PRA model for knowledge representation and further development of diagnostics. Even though FMMEA forms an integral to this approach, in this chapter, the results of FMEA/FMECA have been used to identify the failure of the SSCs at the basic nodes in the fault tree. The qualitative and quantitative power of the fault tree is employed for structuring the diagnostic algorithm and understanding its contribution in terms of quantitative estimates or probability at the system and component levels and risk at the plant level in support of operations and maintenance management. This knowledge, most of the time in the form of 'if-and-then' representation, is then utilized for making fuzzy rules in a fuzzy rule-based diagnostics system. This mathematical model gives the healthy baseline and failure definition of various failures, and this information is used in the membership function for developing a fuzzy rule-based system. Finally, a case study is carried out on a diesel generator used in a nuclear plant (Nain 2011).

9.6.2 Development Approach

This framework has been developed for online mode for interfacing the system with a plant simulator for training and evaluation and for real-time operation in the control room of the plant as an Operator Advisory System. The PRA technique for knowledge representation enables integrating risk-conscious attributes in the system. For example, the identification of transient and diagnostic capabilities is expected to enhance operator awareness and alertness and thereby the appropriate response-taking capability of the advisory system. Even though the data acquisition module is interfaced with a plant computerized information system in an online mode, the feedback or corrective action implementation depends on the operator's decision; in this sense, this system works in an open loop mode.

Figure 9.9 shows the flow chart depicting the functional architecture of the operator advisory system. The operational part of the system has been discussed considering loss of off-site power (LOOP), an anticipated transient that is expected to occur more than a couple of times depending upon the off-site grid reliability. We have already discussed the application of the ANN approach for transient identification.

Therefore, one LOOP event is identified, and the control passes to the knowledge-based diagnostic system. Based on the data available, the diagnosis is initiated. Depending on the condition of the plant, the diagnosis is terminated, confirmed by the 'goal found' function in the code. For example, if the response of the plant, in terms of status of the equipment status, safety parameters and other operator and feedback from the operator, during the LOOP event is as intended, then the system operation is terminated, and once the operator acknowledges and resets the system to make it ready and available for online surveillance.

If the plant response is not as intended as sensed by the advisory system, then by default, the symptom-based module is automatically selected. The process for detecting the failed component such as a Diesel Generator or a breaker to operate is identified based on the online signal coming to the advisory system along with operator input. The detailed diagnosis is initiated by initiating a fuzzy rule-based system to diagnose the cause of failure. Furthermore, system operation continues to perform detailed diagnostic, as discussed employing risk-based approach as also the corrective action as an advisory for removing or isolating the fault for restarting the failed system. If the threat is averted such that safety functions such as plant cooling capability are restored, then the system operator, as above acknowledges, resets the

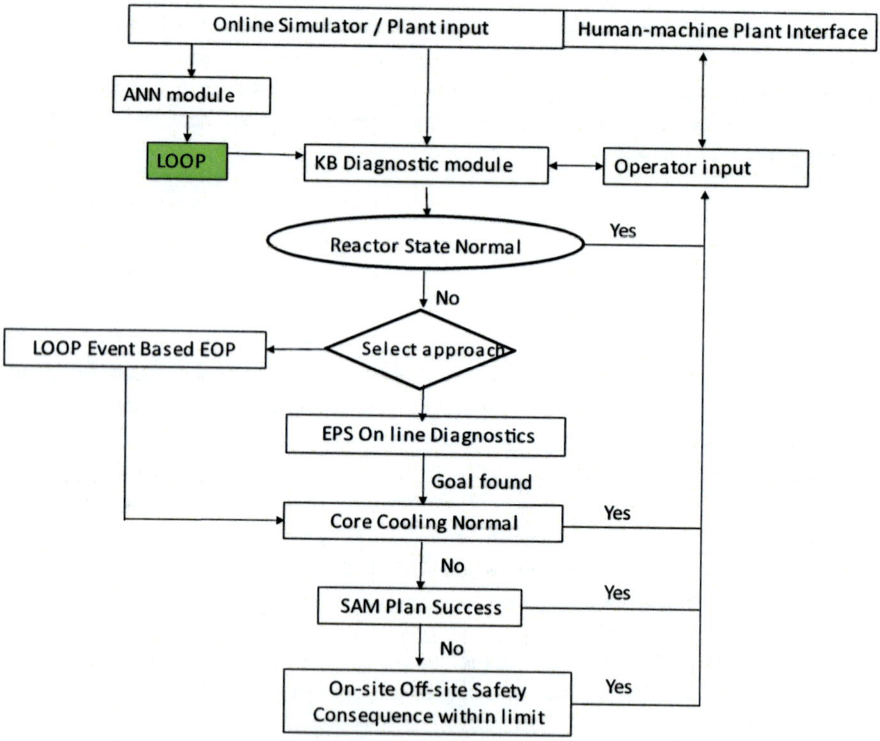

Fig. 9.9 Functional structure of the diagnostic module

system. Here, if the required operator can initiate a plant EOP, the operator advisory system is not able to mitigate the system failure conditions.

Similarly, if the threat to safety cannot be averted, then the plant severe accident management plan is activated to mitigate, avoid and reduce the consequences in the on-site or off-site domain. However, the scope of this section does not cover the implementation of SAMP plant as (a) its probability is very low and (b) the diagnostic algorithm is same as the EOP which is discussed here, only difference is that some additional data and diagnostic and plan actions are needed.

9.6.3 PRA Knowledge Representation and Rule Extraction

The risk-conscious approach requires extraction of production rules and data at three levels, viz., at plants where the risk metrics core damage frequency derived from the accident sequences, second, safety system or safety support systems with unavailability, and at the component level, the failure of hardware, software or human actions that has the potential to introduce transients at the system and further at the plant level. These conditions pose a threat to safety and are referred to here as risk metrics.

9.6.3.1 Knowledge Representation

The production rules in fuzzy knowledge-based systems are represented by employing if-and-then algorithms that have three components: antecedents, logic, or fuzzy operators (membership functions).

Suppose a logic requires an entity A and B and C for enabling the condition D. Then, a typical general structure with crisp and fuzzy logic is shown as follows:

If
Condition A = True AND
Condition B = True AND
Condition C = True
Then
Condition D = True
For Fuzzy logic
If
Condition A fuzzy membership-A (Valid) AND
Condition B fuzzy membership-B (Valid) AND
Condition C fuzzy membership-C (Valid) AND
Then
Inference D = Valid

Often, the knowledge available from

In operations ecosystems, even the crip knowledge is transformed into fuzzy knowledge to enable the conversion of linguistic variables into building knowledge, as the interpretation of PRA data and knowledge requires expert opinion for interpreting the numbers or probabilities. For example, if an operations expert is asked how the performance of the Emergency Power Supply (Class III) system is? Even though the unavailability as shown on the even tree is $5 \times 10^{-}4$/demand, operations staff would like to express his opinion as Unavailability as *'Low', Medium; 'High', and 'Very High'*. Alternatively, even on a membership unavailability scale ranging *(0 for low and 1.0 for very high)* 0, 0.1, 0.2, 0.3 ……0.9, 1.0, the staff would like to give a rank called membership.

For example, the unavailability target for an emergency power supply system (EPS) stipulated by a regulatory design document is say Unavailability $\leq 10^{-3}$ then it requires an intimate understanding as to what is very low, low, medium high and very high. However, the knowledge of numerical estimates weighted by operations staff where they have a choice to put on scale their expert opinion from 0 to 1.0 as membership function when the anchor number 1×10^{-3} is available in support of expressing their opinion. In fact, the fuzzy approach works with crisp parameters such as Success (1)/Failure (0), Operating (1)/Failed (0) and Open (1)/Closed (0) used as part of crisp logic. In fuzzy logic, there is an option to include 0 and 1 as decision variables as part of the membership function.

This is likely why fuzzy expert systems are slowly replacing traditional production-based expert systems.

9.6.3.2 Fuzzy Rule Extraction from Event Tree

Event tree analysis as part of PRA is employed for modeling and analysis of accident scenarios. Figure 9.10 shows a simplified and limited scope event tree for loss of off-site power (LOOP) event tree. Simplified as the objective is to demonstrate the rule extraction and development of risk perception with respect to each accident sequence or condition in the plant and therefore a simplified event tree. Limited scope as all the systems involved in a given function, viz. core cooling has not been included in the event tree. For example, if the electrical system fails, then the diverse hydraulic power water system that operates a turbine coupled on the same shaft as the electrical motor that drives the shutdown cooling pumps does not form part of event three.

The LOOP event for the purpose of fuzzy rule extraction is based on the postulation of total and instantaneous failure of all the redundant feeders. The confirmed LOOP event initiates an automatic reactor trip through its protection system. Due to the unavailability of the main circulation system, the shutdown system also starts automatically and ensures the decay heat removal function, as shown in the event tree shown in Fig. 9.10. The LOOP is propagated based on the response of safety systems and safety support systems. Sequence nos. 1 and 3 attain a safe state. In Sequence 1, all the safety systems have performed as intended; however, in Sequence 3, even though emergency power has failed, the shutdown cooling is successful due to the successful functioning of a diverse power water system (water from an overhead tank drives a turbine that operates the shutdown cooling pump), and the shutdown core cooling function is accomplished.

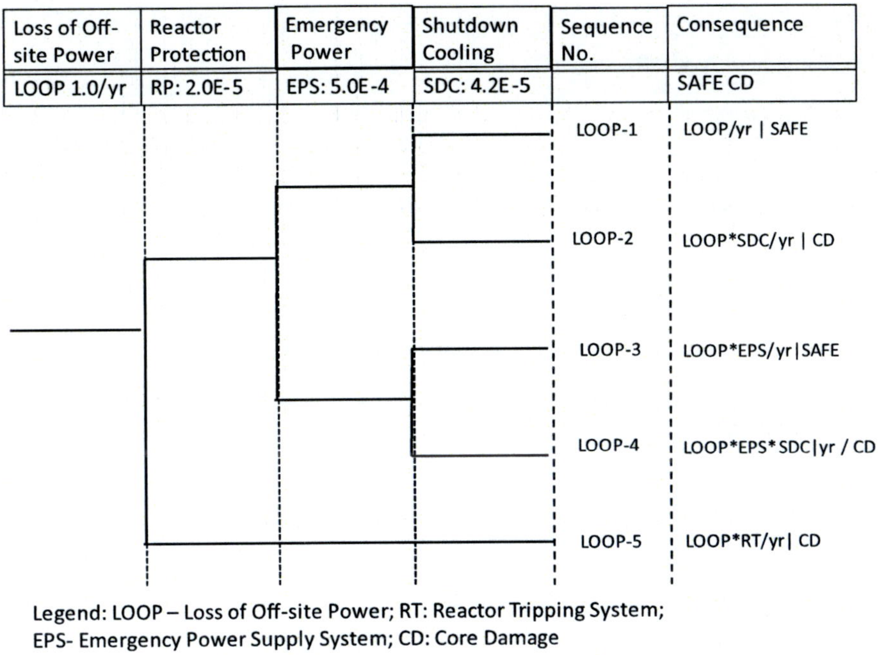

Fig. 9.10 Event tree approach for knowledge representation

The risk insights from the above graphic presentation in an operator advisory system are (a) the graphical representation of event progression to comprehend plant risk levels, (b) rule extraction from the event tree logic representation can be used to activate the applicable emergency operating procedure (EOP) module and (c) the quantitative estimate of risk is vital in support of decisions and thereby helps structure mitigation plans and procedures. For example, the Emergency Power Failure requires implementation of emergency operating procedure for Diesel Power on-site Diesel Generator system failure diagnostics EOP Sequence No 3 and 4 require. Sequence 4 postulates the core damage state, while 3 is the case reduction of the safety margin, as shown in Fig. 9.11. The failure of the reactor protection system alone leads to core damage states; however, the likelihood of these scenarios is negligible, as plants have two redundant and often diverse protection systems.

The ANN confirms the LOOP scenario. This input, along with the status of safety systems, has provisions to mimic the characteristics of a dynamic system by employing conditional probability nodes in the simulation. As per the inference of the Event Tree rules, further diagnostics might be required either for the EPS or SDC cooling system. For the purpose of demonstration, the EPS system has been considered to demonstrate the applicability of the fault approach for diagnostic rule extraction, including failed subsystem analysis.

The example of rule extraction is as follows:

For Sequence -2

If
LOOP = membership function Valid (i.e., 1.0 during real-time scenario) AND
EPS = membership function Valid (i.e., 1.0 during real-time scenario)
Then
EOP-2: SDS Diagnostics

For Sequence –4 (assuming that SDC has failed due to EPS failure)
IF
LOOP = membership function Valid (i.e., 1.0 during real-time scenario) AND
EPS = membership function Valid (i.e., 1.0 during real-time scenario) AND
SDC = membership function Valid (i.e., 1.0 during real-time scenario)
Then
EOP-4: EPS Diagnostics (the fault tree in the next section initiates diagnostics for EPS)

Fig. 9.11 Extraction of fuzzy rules from event tree

9.6.3.3 Fault Tree Approach for Fuzzy Rule Extraction

The EPS comprised two redundant Diesel Generators, Viz DG-A and DG-B, of the same rating, and each DG was sufficient to take plant essential loads. Apart from this, the EPS has power supply buses, breakers (Breaker-A and Breaker-B) and support for each DG, as shown in Fig. 9.12. The failure criterion for the EPS is failure of supply from both DGs to various essential loads, either independently or due to common cause failure (CCF). This situation, if not mitigated, might lead to station blackout conditions. EPS failure has been assumed to occur due to failure of any of its subsystems, such as control and protection, lub-oil, fuel oil and jacket cooling systems. The Bus Fault and DG Breakers failure were also modeled as independent events.

Figure 9.13 shows the fault tree for the EPS. As seen, the electrical bus fault that facilitates the power supply available to the essential load, DGs independent and CCF are the major causes for EPS failure. Only one DG, either DG-A or DG-B failure, does not result in EPS failure, as each DG can cater to essential loads in the plant. Here, it may be noted that only two CCFs have been identified in the EPS fault tree. The human error due to operations and maintenance accounts for procedural error, human–machine interface error, calibration error, maintenance error related to reverting back the DG units, Breakers and other support system normalization procedure after testing that have potential due to common crew, gadgets and tools, raw material or spare and others related leaving certain material unattended. A beta-factor of 0.07 has been used to account for CCF, as the system has been assumed to be in service and has seen over 200 demands that are adequate to reveal design

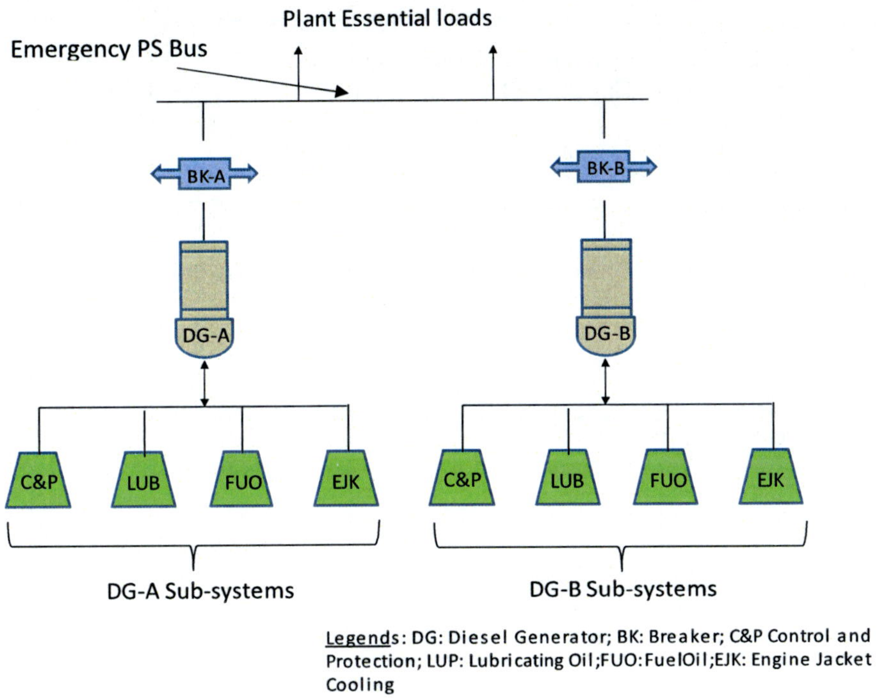

Fig. 9.12 Simplified emergency power supply system

and manufacturing faults. Even though there is physical separation between DG sets looking at the fire potential in the DG area, conservative fire has been considered one of the CCFs.

The online data collected from the DG instrumentation sensors are undervoltage on the common EPS bus, i.e., both DGs have not started on Loop demand. The online signal indicated that DG-1 has not started due to a trip *'Low lub-oil pressure'* during priming itself, and even though DG-2 has started, its breaker has not closed on auto. The rule extraction for diagnosis and inference for identifying the faulty system is as follows (Fig. 9.14):

The next step is to initiate the diagnostic procedure for identifying the direct cause of DG-1 and DG-2 failure. We will deal with the lub-oil pressure diagnostic only as the causes of breaker failure often found to be straight forward, either breaker control supply failure or breaker mechanical problems. The low lub-oil diagnostics are performed as follows:

Process- and component level diagnostics often require, apart from available online signals, the human–machine interaction to arrive at the direct cause of the anomaly or component failure. There are two parameters for which the audio/visual trip/alarm is available, viz. the DG trip audio/visual trip for low lub-oil pressure when

If
DG-1 = membership function = ≥ 0.9 AND
DG-2 = membership function = ≥ 0.9
Then
Emergency DG-1Check
Emergency DG-2 Check

If
Emergency DG-1 Check = Low lub Pressure ≤ 1.0 kg/cm membership ≥0.8
Then
Lub-oil Diagnostics

If
Emergency DG-2 Check = DG Breaker Status 'Not Closed membership ≥0.8
Then
DG-2 Breaker Diagnostics

Fig. 9.13 Representation of fuzzy production rules

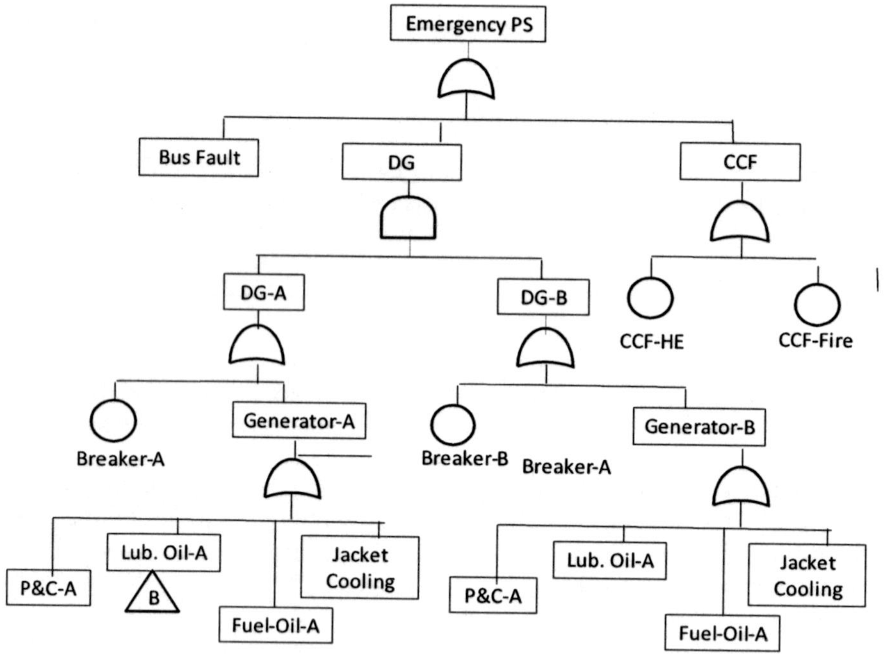

Fig. 9.14 Fault tree for fuzzy rule extraction

pressure drops below or <1.0 kg/cm, and the lub-oil temperature was high when the lub-oil temperature reached 90 degrees Celsius.

For low lub-oil pressure, the reason could be (a) leakage from the lub-oil circulation line, (b) failure of the lub-oil pump to start, (c) heat exchanger tube failure where the leaked-out oil finds its way into cooling water on the secondary side, (d) tube oil-side inside fouling that leads to drop in -oil pressure and (e) choking of oil filter.

Similarly, the rule for high lub-oil temperature can be developed. Here, it can be noted that the lack of secondary side cooling, which reduces the service water flow to the lub-oil heat exchanger, is one of the major causes. This can be due to engine bearing or other component increases in mechanical friction or higher than normal exhaust air temperature along with internal fouling of the tube (oil-side), which leads to oil flow reduction through the tube, clogging of the flow path on the secondary side, i.e., outside the heat exchanger tube, and choking of the filter, which can be the major reason for higher oil pressure. However, it can be noted that a drop in lub-oil pressure is one of the major parameters that is sensitive to the health of the lub-oil system as well as DG health along with Jacket cooling water temperature being higher than normal.

9.6.3.4 Fuzzy Inferencing in Diagnostic Systems

The fuzzy expert system approach has been derived from the production rule structure in the form of the IF-AND-THEN representation of logic processing. The If part is generally called the 'Antecedent', and the THEN component provides the 'Consequence'. The only difference is that the fuzzy production rules are allotted the membership function that occupies values in the continuum from 0 to 1.0, unlike production rules that generally follow binary logic (0/1, to reflect true and False) to check the validity of a given condition. A fuzzy set An in X is expressed as follows:

$$A = \{(x, \mu_A(x)) | x \in X\} \tag{9.5}$$

where $\mu_A(x)$ is the membership function of the fuzzy set An in x. Figure 9.15 shows the major steps involved in the generation of fuzzy inference. The first and major step is the collection of domain data and information. This requires expertise and experience in the design, operation, maintenance and generation of diagnostics. Many intelligent expert systems have been found not to provide the intended performance due to lack of expertise on domain data knowledge and experience. The input, which is either in the form of crisp data and expert opinion, is transformed by employing a function; it can be triangular, trapezoidal, etc., depending on the requirement. The fuzzy rule base developed a priori is subjected to an inference algorithm based on the actual plant online or user supplied inputs. The fuzzy inference is subject to processing in the defuzzification module. Figure 9.16, which implements DG diagnostics, illustrates the complete procedure and finally produces the crisp output that is based on the initial input based on crisp or expert opinion.

Fig. 9.15 Fault tree for lubrication oil system anomalies/failure

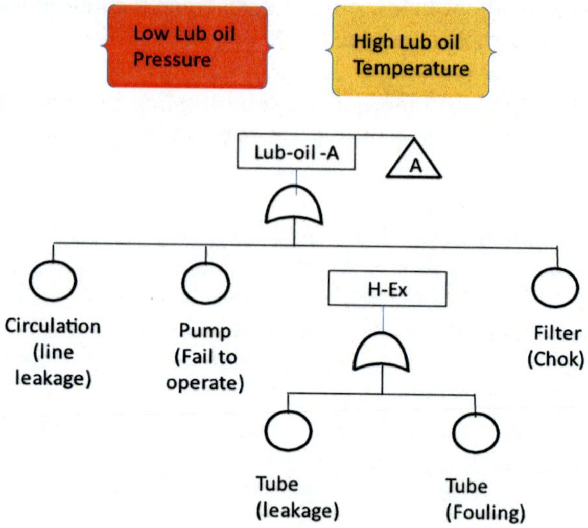

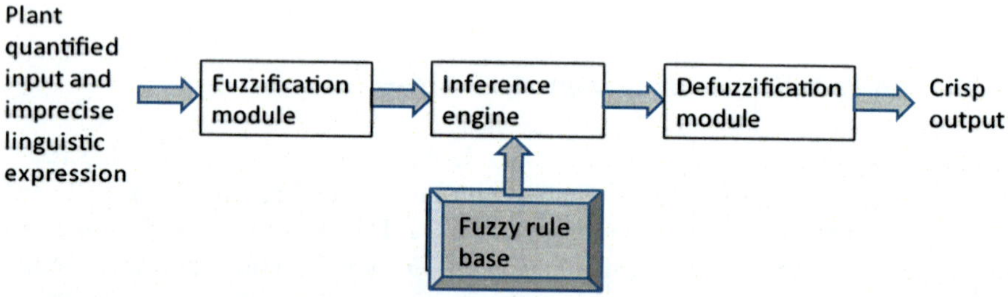

Fig. 9.16 Architecture of the diagnostic module in the operator advisory system

The diagnostic tool has been developed for the lubricating oil system of the DG. The first step is to assimilate as much knowledge that includes system normal operation parameter values, trip/alarm settings on the instruments, and expert opinion in the form of linguistic variables about the system performance and operation. Reference to the design and operation of the system, maintenance and operations manuals, maintenance data cards and PRA model of the system, etc. Particular contribution of subsystems to failure of DG and maintenance and experience.

Here, any parameter when it reaches:

'High' value indicates complete failure; 'Medium' indicates action needed, 'Low': indicates fault intensity is low.

The idea is to assess the risk potential of the system in support of the development of a fuzzy expert system such that a reasonable diagnostic coverage can be obtained. The reason is that in a risk-conscious system, the deployment of resources should be commensurate with the risk potential of the entity that is being modeled.

Suppose the top even, i.e., the DG failure, is represented by '*T*; then, as seen in Fig. 9.13 and the reference to Table 9.3, the cut-set equation of the fault tree can be given as:

$$T = AUBUCUDUDUE \tag{9.6}$$

U: represents the logical symbol for Union (other symbol $\bigcap$ is used for interaction sets).

There are 5 events, and each event can occupy three fuzzy values 3 (low, medium and high). Accordingly, there are $5^3 = 125$ logical combinations of logic events, i.e., rule formulations are possible. This demonstrates the power of the fuzzy logic approach to rule formulation. The traditional fault tree follows a binary (true or false) approach, which facilitates 25 rule formulations. The fuzzy approach enables us to capture the real-time scenario in operational ecosystems. The binary approach often makes the decision process less effective and in some at times prohibitive. The five parameters/status of equipment provides more than one fuzzy rule-based diagnostic. There are two modes of when DG failures *'fail-to-start' (FTS)* and other 'failed-to-run' (FTR) continuously. Table 9.4 illustrates the process of inferencing using the five events. Furthermore, many more rules have been developed.

Except for Vector-1, which formulates the fuzzy rule for DG's standby condition, the other five vectors show formulations related to failure/anomalies in the lub-oil system fault tree that lead to different levels of failure/degradation of DG.

Between two events in the table, a column is the logic operator Union and Interaction, represented by $\bigcup$ and $\bigcap$, respectively. It may be noted that the crisp and qualitative, i.e., the numerical value of the DG/lub-oil parameter and operator-judgment, respectively, are used to formulate; however, these parameters form the basis for the formulation of fuzzy input. However, these crisp parameters do not form in the scope of this discussion, as finally, the whole formulation is based on three states.

The development of a fuzzy rule base derived from plant diagnostics, along with evaluating the condition for their validity when subjected to an inferencing strategy and later to a defuzzification process, is central to the implementation of fuzzy rule-based diagnostics. There are several fuzzy defuzzification techniques given in the open literature; however, considering the nature of the problem used for illustration, estimation of the centroid of the membership function appears to provide an elegant approach to defuzzification. The centroid (C) for a given fuzzy membership function geometry can be given as follows (Nain 2011):

$$C = \frac{\int \mu_A(z)z\mathrm{d}z}{\int \mu_A(z)\mathrm{d}z} \tag{9.7}$$

Figure 9.16 provides an overview of the fuzzy inferencing approach based on the fuzzy input vectors that represent a rule. In fact, this inference can go further down to investigate each element discussed in Table 9.4. For example, for element B that deals with lub-oil pump be it starting pump or DG driven lub-oil recirculating pump.

Table 9.4 Fault tree input for the fuzzy logic inference approach

Vector	A	LO ∪/∩	B	LO ∪/∩	C	LO ∪/∩	D	LO ∪/∩	E	Fuzzy output (Centroid evaluation) (P)	Crip out put	Inference
V-1	0.0/Low	∪	0.0/Low	∪	0.1/Low	∪	0.1/Low	∪	0.1/Low	Low	0.1: Lub-oil parameters are normal	Reference vector: DG standby
V-2	0.1/Low	U	0.7	U	0.2/Low	U	0.1/Low	U	0.1/Low	Medium/B	Lub-oil pressure is dropped from normal of 5.0/kg cm^2 to 3.0 kg/cm^{-2} due to pump bearing issues	DGanomalies due to DG pump anomalies to be corrected
V-3	0.6/Medium	U	Low/0.3	U	0.1/Low	U	0.3/Low	U	0.3/Low	Medium/A	Oil pressure is dropping due to circulating line leakage estimated to be 100 ml per min	Lub-oil circulating line leakage to be corrected

(continued)

Table 9.4 (continued)

Vector	A	LO ⋃/⋂	B	LO ⋃/⋂	C	LO ⋃/⋂	D	LO ⋃/⋂	E	Fuzzy output (Centroid evaluation) (P)	Crip out put	Inference
V-4	0.1/Low	U	Low/0.3	U	0.1/Low	U	0.7	U	0.1/Low	Medium/B	70% Fouling of tube	Drop in oil pressure and flow causing lub-oil temperature high
V-5	0.1/Low	U	0.5	U	0.1/Low	U	0.5	U	0.9/high	High/E	80% chocking of filter	Filter chocking led to low oil pressure–and tripped the DG (presence of metal particles after analysis suspected bearing failure)
V-6	0.9/High	U	Low/0.3	U	0.1/Low	U	0.3/Low	U	0.3/Low	High/A	Oil leakage from the lub-oil system caused DG trip	Lub-oil circulating line leakage to be corrected

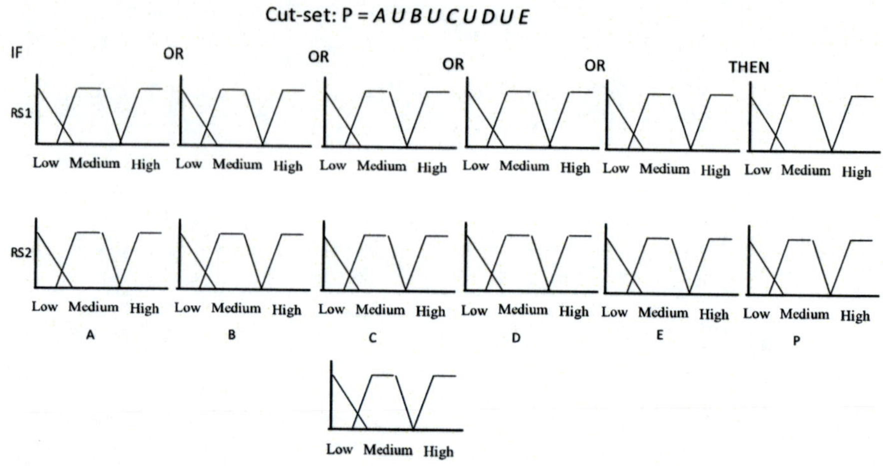

Fig. 9.17 Graphical representation for fuzzification and defuzzification

Further development for DG pump failure provides the causes of DG starting pump failure, i.e., pump failure due to starting logic failure, mechanical problems, or power supply failure (Fig. 9.17).

Assuming that most of the inputs are derived online, the automatic fuzzy diagnostics aid the operator in an automatic mode identification of the cause of the DG problem, i.e., failure of the subsystem and further the cause of subsystem failure.

9.7 Final Remarks

This chapter presents the applicability of the machine learning approach to the development and application of operator support systems. This approach is superior to either a hardcopy-enabled or computerized event-based approach. Even if the operator support systems are developed employing a traditional programming-based approach, they might not meet the requirements of emergency scenarios due to nonlinear characteristics that are further challenged by fuzzy or imprecise input. It is generally accepted that intelligent approaches based on machine learning algorithms, such as ANNs for transient identification and fuzzy rule-based approaches for diagnosis, provide an effective mechanism.

If we look at the scenario during the era of 1980 when the intelligent approaches were actively being developed but suffered many roadblocks and finally only this one decade that it came in new avatars of machine learning algorithm that it found relatively improved acceptance. One of the reasons why it did not see much success was that many at times its domain-specific knowledge and expertise, which are vital for the development of an intelligent system, was not given the importance that it possibly deserved. Therefore, one of the major conclusions of this chapter is that domain experts should be part of machine learning-based projects, particularly the validation and verification components of the project.

References

Atomic Energy Regulatory Board (2015) Design of light water reactor based nuclear power plants, AERB CODE NO. AERB/NPP-LWR/SC/D. AERB, Mumbai, India

Basher H, Neal JS (2003) Autonomous control of nuclear power plants. ORNL/TM-2003a/252. U.S. Department of Energy

Giubilini A (2022) Conscience, Stanford encyclopedia of, 11 Feb 2021. [Online]. Available: https://plato.stanford.edu/entries/conscience/. [Accessed 5 Apr 2022]

Haight JM (2007) Automation versus human intervention: what is the best mix for optimum system performance? A case study. Int J Risk Assess Manage 7(5)

Human Error (2022) OSHwiki, 28 Feb 2022. [Online]. Available: https://oshwiki.eu/wiki/Human_error#:~:text=It%20is%20generally%20accepted%20that,in%2094%25%20of%20fatal%20accidents. [Accessed 4 Apr 2022]

International Atomic Energy Agency (1988) Use of expert systems in nuclear safety. IAEA, Vienna

International Atomic Energy Agency (2010) Interface between safety and security for nuclear power plants, INSAG-24. IAEA, Vienna

Jang S, Park S, Jae M (2017) Development of an accident sequence precursor methodology and its application to significant accident precursors. Nucl Eng Technol 49:313–326

Joseph PJ, George WL (1983) A history and overview of the safety parameter display system. IEEE Trans Nucl Sci 30(1)

Kang JS, Lee SJ (2022) Concept of an intelligent operator support system for initial emergency responses in nuclear power plants. Nuclear Eng Technol

Lu C, Lyu J, Zhang L, Gong A, Fan JYY, Li X (2020) Nuclear power plants with artificial intelligence in industry 4.0 era: top-level design and current applications—a systemic review. IEEE Access 8:194315-194332

Myung-Sub R, Cheon WS, Chang SH (1993) Power prediction in nuclear power plants using back-propagation learning neural network. Nucl Technol 94:270–278

Nain A (2011) Diagnostic and prognostics using reliability based methods for diesel generator using in nuclear plants. M. Tech. Thesis of HBNI, Mumbai

USNRC (1986) Information notice No. 86–10: safety parameter display system malfunctions. United States NUclear Regulatory Commission, Washington D.C.

Varde PV, Sankar S, Verma AK (1998) An operator support system for research reactor operations and fault diagnosis through a connectionist framework and PSA based knowledge based systems. Reliab Eng Syst Saf 60:53–69

Varde PV, Sankar S, Verma AK, Prakash P (1996) OPAD—An expert system for research reactor operations and fault diagnosis. In: Proceedings of ASME/JSME's 4th international conference on nuclear engineering (ICONE-4). New Orleans, USA

Wikipedia, "Wikipedia," Consciousness, [Online]. Available: https://en.wikipedia.org/wiki/Artificial_consciousness

Wu JS, Apostolakis GE, Okrent D (1989) Probabilistic Risk Assessment and intelligent support systems. Nucl Eng Des 113:269–279

Chapter 10
Risk-Conscious Maintenance Management

10.1 Introduction

In this era of globalization, the importance of maintenance cannot be emphasized more than ever due to one of the otherwise contrasting multiobjective goals that essentially deal with plant reliability-risk, availability-risk, financial-risk, reputation and liability-risk and most important safety-risk. This aspect is directly relevant to all domains, such as nuclear, space, aviation, defense, chemical, and transport, including finance and societal. The competitive environment stresses operating organizations to address performance issues while meeting risk reduction goals while chasing regulatory compliance on the one hand and the business bottom line on the other. This translates into every additional, e.g., a 1% increase in production capacity can serve not only the business bottom line but also work toward enhancing resources for proactive regulatory compliance while chasing safety goals. It can be argued that an effective maintenance management program and strategy that ensures multiobjective functions with safety as an overriding priority is critical to the optimum utilization of plants or assets to meet business objectives.

The traditional preventive maintenance (PM) approach that has served well for over six decades is under review due to its inherent limitations (International Atomic Energy Agency 2018). The first and foremost is the periodicity, which generally remains constant irrespective of real-time performance in terms of failure frequency and causes of failure. PM suffers from drawbacks such as no sound basis for frequency and type of maintenance. Maintenance planning generally depends on either expert opinion or supplier recommendation, which tends to be, at times, conservative and results in over maintenance. There is consensus that new and advanced strategies are required to strengthen not only maintaining plans but also improving system reliability and delivery. To this end, reliability centered maintenance (RCM) (Vishnu and Regikumar 2016), condition-based maintenance (CBM) (Computers

P. V. Varde, *Risk-Conscious Operations Management*, Risk, Reliability and Safety Engineering, https://doi.org/10.1007/978-981-19-9334-3_10

and Industrial Engineering 2012), prognostics and health management (IEEE Standard Association 1856; International Atomic Energy Agency 2013), predictive maintenance (PRM) and In-service Inspection-based approach to maintenance (International Atomic Energy Agency 2006), and few cases of risk-based maintenance management (RBMM) (Khan and Haddara 2003; Arunraj and Maity 2007) are also being considered by industries. Except for breakdown maintenance (Poor et al. 2019), which is reactive, i.e., responds only after the SSCs have failed, the other strategies are proactive in nature. Furthermore, in recent times, the management component, where advanced, planning, scheduling, evaluation, assessment of modes and mechanism of failure on a regular basis, forms an integral part of maintenance management. Often, professional software, referred to as computerized maintenance management systems, is used to make all maintenance management activities more systematic and effective (Labib et al. 2008).

The principal elements of defense-in-depth as applied to risk-critical systems are (a) prevention of deviation from normal operation, (b) corrective action to recover from deviation, (c) application of emergency operating procedures failure, (d) severe accident management and (e) protection of the public from the hazard (International Atomic Energy Agency 1996). Maintenance management has a critical role to play either planned or unplanned all through these five stages of reactor states. During normal operation that includes normal reactor operation and shutdown conditions, maintenance management addresses the regular repair, replacement, testing and qualification program of the plant. During deviation from normal operation, maintenance might require repair or replacement of the failed component. When the plant enters emergency conditions, maintenance actions become critical to mitigate emergency scenarios to reduce the risk due to potential hazards. When the plant enters the severe accident management phase, the emergency procedures are called off, and the severe accident management plan is initiated, which requires actions to protect containment by ensuring integrity and affecting planned actions to avert consequences in the public domain. The last phase that deals with eliminating or reducing accident management in the public domain requires maintenance services, apart from affecting ad hoc as also planned repair to providing services to emergency management teams that deal with logistics and services for affecting the emergency plan.

There is also consensus that human factors are central to the quality of maintenance or success of any maintenance strategy (Nkosi et al. 2020). Whatever may be the strategy if human factor is not addressed properly then success of maintenance planning cannot be guaranteed or ensured. Therefore, it is required that consciousness attributed forms part of maintenance management (Chrleslandry.com 2021). Even though a risk-informed approach to nuclear plants is increasingly being used for regulatory review applications (International Atomic Energy Agency 2005), for safety critical operations ecosystems in general, an approach is required that goes beyond the information attribute that works for developing culture by integrating human risk-consciousness evolve into organizational consciousness and further national and international consciousness (Consultdss.com 2018). This chapter discusses the risk-conscious management approach to complex systems, and there is an attempt to demonstrate that this approach serves well in addressing the various risk-objective functions while ensuring the business interest of the organization (Varde 2021).

10.2 Equipment Life Cycle Aspects

The equipment life cycle considerations form the basis for modern maintenance approaches, as this concept provides a sound basis to see the context of maintenance with respect to the risk and reliability aspects of complex engineering systems. As this concept covers the complete life cycle reliability description right since installation and commissioning of the SSCs in the plant, its behavior all through the useful life period and in later stages, i.e., the aging phase, when the SSCs enter the aging period. Figure 10.1 shows the typical bathtub curve for the hardware component.

The life and failure rate are depicted on the *x*-axis and *y*-axis, respectively. There are three major regions in the classical bathtub curve: the first on the left side shows the initial or infant mortality period, the middle one is a useful period directly relevant to operations and maintenance management, and the rightmost is the aging period of the component life cycle. It is very important to understand how bathtub curves are generated for a specific component in an industrial reliability testing scenario and a complex safety critical environment, as this is directly relevant to its interpretation to operational maintenance management. Let us consider an example of a nonrepairable component. Consider an industry manufacturing electric bulbs and want to generate reliability data or say MTTF estimate of its 500 W light bulbs. The design of the experiment comprised testing 100 bulbs. The time to failure forms the test terminating criteria, also referred to as the Type I test. All 100 bulbs are tested assuming that

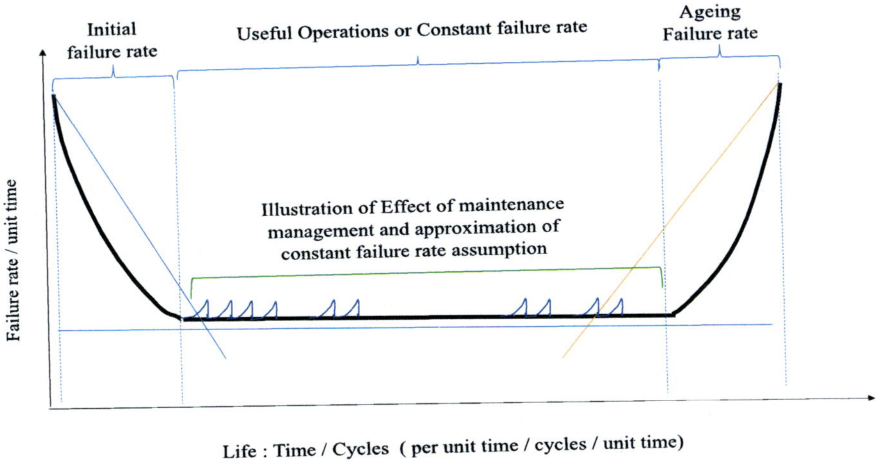

Fig. 10.1 Illustration of the component complete life cycle through the Bathtub curve

there are no voltage fluctuations and the only stress acting on the bulbs at exactly the same levels. In short, we assume that there are no circuits or power supplies, environments, etc., specific variations.

Once the test started for all the bulbs at a single instant of time, i.e., with one switch, all the bulb power supply was placed ON. The initial observation on first five was that bulbs started failing and by the end of the day 5% population failed in two days. For the remaining 95% of the population, the failure rate was constant, i.e., 1 bulb per 5th day until the test was reached, as shown in the aging period, when the failure rate again started picking up and was reached, as indicated by the second and third regions of the bathtub curve. The test was concluded when the designated number of failures was reached.

The root cause analysis reveals that the failures during the initial or the first phase, also referred to as the infant mortality period, were attributed mostly to quality-related issues. Therefore, when the test crossed the infant mortality period and entered phase II, i.e., the useful operations period, the components with inferior quality were screened out, and only the healthy component entered the race during the second phase. The failure rate, as mentioned above, remained constant until the test entered Phase III, where the aging phenomenon manifested. In the aging region, the root cause of failure was attributed to burning out of filament of the bulb.

This test provided two major inputs, viz., root causes of the failures, i.e., quality-related issues during the first phase and filament durability in the third phase, and the estimate of the mean time to failure of the population, both related to the reliability of the component. The reliability community is always blamed for the assumption that the failure data follow an exponential distribution. Whether it is PRA modeling or reliability analysis, it is very common to use the exponential distribution. The first reason why an exponential distribution is considered for risk and reliability modeling is that this is a very simple distribution and enables practicability of modeling the failure phenomenon. However, let us discuss why the exponential distribution does serve the purpose of its applicability for the constant failure rate region, i.e., phase II of the bathtub curve. Let us discuss how the maintenance strategies in general and periodic maintenance or preventive maintenance activity strategy justify the use of exponential distribution not in mathematical sense but to address the applied aspects. You can see small triangle-like objects lined up on the constant failure rate line. These blocks draw our attention to the effect of maintenance. For example, maintenance activities are performed on the component and bring its failure rate to the reference point at the bathtub curve. However, as soon as the component is put under service, the degradation starts to increase the component failure rate, and this increase will occur until the next maintenance is performed to lower the failure rate back to the reference minimum value. The line passing through the average point of the failure rate can be argued to be a constant failure rate line. This observation provides a basis for considering the exponential distribution as a practical approximate solution.

The bathtub curve approach also provides an effective framework for life cycle modeling at the plant level in support of the assessment of plant conditions in an integrated manner. The only difference is that in plant level modeling, the data are not the instantaneous failure rate by plant availability. This particular discussion is

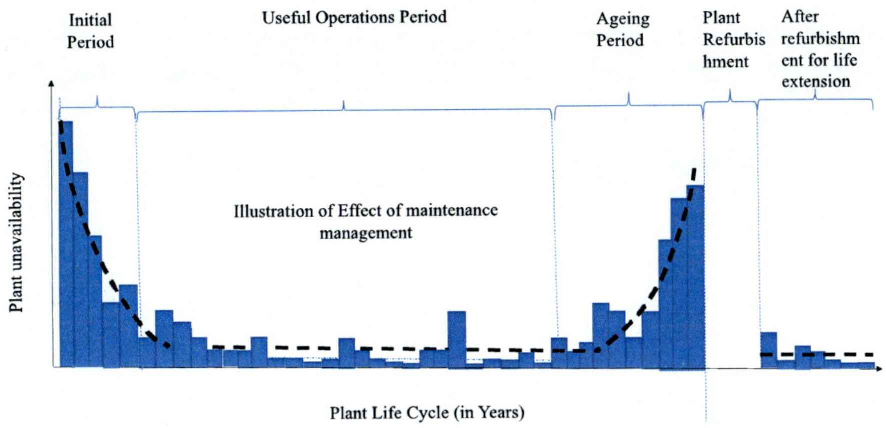

Fig. 10.2 Bathtub curve depicting the plant's integrated life cycle, including the effect of refurbishment (extended maintenance) on life extension

based on insights into a 40 MW_{th} research reactor. At the outset, it is made clear that the data shown in this illustration are not plant-specific but assumed data, as the objective here is to show the feasibility and applicability of the approach at the plant level (Fig. 10.2).

10.3 A Brief Overview of the Evolution of Maintenance Management

The concept of maintenance management has evolved since the inception of human civilization. The concept of maintenance can be found in ancient books such as Veda's that were written over 10,000 years ago, where the major focus was the maintenance of social systems, structures and systems (Prabhupada 1998). Later, maintenance management was part of ground and road, water and transportation until late 1800 AD. However, in a formal manner, it was introduced as part of operations management of mission critical systems in general and military, life sciences, engineering and aviation systems during the industrial revolution starting in the eighteenth century. The importance of quality assurance, reliability and maintenance was further endorsed during World War 1 (1914–1918) and 2 (1939–1945). The relatively recent developments with respect to the Industry 4 revolution, where maintenance management has been one of the major focuses (Garcia and Garcia 2019).

The major types of maintenance that started with breakdown maintenance to preventive, need-based, condition-based and, in more recent times, predictive and prognostic health management techniques have evolved over the year. The other major activity in support of operating management is surveillance in-service inspection and testing. ISI and testing of SSCs is performed in support of structural health assessment and ensuring high availability factors of the standby safety systems on demand. Often, the ISI and testing also form part of maintenance management. The following sections will discuss the major maintenance practices that have evolved over the years toward strengthening maintenance management as a strategy in support of plant operations management.

The bathtub curve approach for system reliability and availability is relevant at the system and plant levels and does not remain relevant at the component level. For example, there is a nuclear plant that logged into an operating period of over 40 years.

10.3.1 Breakdown Maintenance

As mentioned, breakdown maintenance is the reactive maintenance strategy, where the attention of maintenance management essentially remains on the repair and replacement of failed components (Poor et al. 2019). This strategy is employed where the failure of SSCs is considered to not adversely affect the business bottom line or system reliability or lead to risk. Breakdown or corrective maintenance can be either planned or run-to-failure. The essentially addresses the prioritization of maintenance activities, assessing and managing the resources, which include spare, logistics, human and training. Among the listed maintenance strategies, this strategy is relatively simple in implementation and employed where risk levels are negligible, and the system can be made available without much penalty, be it financial or bottom line.

The IAEA glossary defines corrective or breakdown maintenance as '*Actions that restore, by repair, overhaul, or replacement, the capability of failed SSC to function within acceptable criteria*' (International Atomic Energy Agency 2018). In summary, the failure(s) dictate the corrective maintenance strategy against the proactive strategies, such as preventive maintenance.

10.3.2 Periodic Maintenance

Periodic maintenance, also referred to as time-based maintenance, involves structuring a maintenance program with fixed periodicity, for example, daily, weekly, monthly or annually depending upon the experience of the performance of the subject devices, criticality of its services and business bottom lin. Often, the availability of resources might dictate the level of periodicity of the maintenance program. maintenance IAEA glossary defines the periodic maintenance strategy as '*a preventive maintenance program consisting of servicing, parts replacement, surveillance testing at predetermined intervals of calendar time, operating time, or number of cycles*' (International Atomic Energy Agency 2018).

Periodic maintenance suffers from drawbacks; for example, the frequency and type of maintenance are determined based on expert judgment and often tend to be arbitrary in nature. There is not much scope for a systematic review of the system performance toward directing resources such that failure insights form to decide upon the requirements of the type and frequency of maintenance. Periodic maintenance might often result in over maintenance, which in turn leads to chances of human-induced failure.

10.3.3 Condition-Based or Predictive Maintenance

This is an exclusive approach to maintenance where the condition of the SSCs is monitored online or offline to track the degradation of critical parameters of the system (Computers and Industrial Engineering 2012). Often, online diagnosis also forms part of the condition-based maintenance. The prediction of degradation or failure in support of prediction is central to the condition-based approach. Deployment of an online system that includes right from the sensor on the equipment, data acquisition system, and processing of information and, if needed, the diagnostic module is a fundamental requirement for implementation of condition-based maintenance. Special SSCs, such as core cooling pumps and cover gas compressors, are provided with bearing/winding temperature, seal leakage and flow sensors to track degradation, if any, such that action can be taken and the onset of failure can be avoided.

10.3.4 Preventive Maintenance

The preventive maintenance (PM) approach is the most widely employed strategy for maintenance management for engineering systems in general and safety critical systems. The IAEA glossary defines PM as *'Actions that detect, preclude, predict, and mitigate degradation of SSCs to sustain or extend its useful life by controlling degradation and failures to an acceptable level'* (International Atomic Energy Agency 2018). The PM is proactive in nature in the sense that unlike breakdown maintenance, it does not wait for the SSCs to fail corrective actions. Additionally, PM is generally planned and periodic in nature, i.e., maintenance actions are performed at a predetermined interval.

10.3.5 Reliability Centered Maintenance

Reliability centered maintenance (RCM) is a relatively new strategy for maintenance. Even though the aviation industry perhaps was the first to implement an

RCM strategy for aircraft maintenance management, it is slowly being used in many other industries, including the process and chemical industries (Nowlan and Heap 1078). The RCM process is intended to ensure that maintenance tasks are performed based on the systematic analysis of requirements in terms of likelihood and consequences of failure. SSCs with a higher likelihood or consequences of failure are the priority items for consideration. Furthermore, maintenance actions on the SSCs are determined through the analysis of the cause of failure. This strategy ensures that maintenance intervals and maintenance actions are performed and reviewed periodically to ensure higher reliability, availability and safety of assets while meeting the business bottom line.

10.3.6 Prognostics and Health Management

The prognostics and health management (PHM) strategy is the most advanced and promising package for ensuring asset reliability and health management. However, for structural health assessment, the PHM has been used in some form in the sense that online signals are obtained from a network of sensors and the data are processed in an integrated module to provide insight into the degradation of mega structures such as dams and bridges and to ensure failures. Apart from this PHM, it has been extensively applied in the aviation industry for structural health monitoring (Rodrigues et al. 2012). One area that is the focus of PHM is electronics, where considerable R&D has been performed and the results are very encouraging (Pecht 2008). Even though the nuclear industry is traditionally conservative and adopts generally proven technology with required confidence, the available publications show that active interest exists in the nuclear sector for the use of PHM as a tool to address the issue of the aging fleet of nuclear plants (Coble et al. 2015; Prognostics and Health Management 2012).

10.3.7 Risk-Based Maintenance Management

Safety is an overriding priority in engineering systems in general and safety critical systems such as the nuclear, aviation, space & aviation, chemical & process, rail & road sectors (Varde and Pecht 2018). Currently, either PM or RCM approaches are employed; however, considering the objective function, as above, the risk-based approach provides an effective framework for maintenance management. For example, the availability of the Level 1 PRA model in most nuclear power plants and the availability of related codes and standards provide an added impetus and inspire the nuclear industry to adopt a risk-based approach to nuclear plants (Varde and Pecht 2018). As such, risk-based in-service inspection methodology is extensively being used in the nuclear and process industries (Failed 2015; Institute et al. 2016). Furthermore, technical specification optimization, where two major attributes, i.e., surveillance test interval and allowable outage time estimation, are encouraged for nuclear power plants, is encouraged (International Atomic Energy Agency 1993). In fact, there is growing interest and developmental activities in advanced stages for the implementation of risk-informed in-service inspection for nuclear research reactors (Khairmode et al. 2016). A similar argument applies to other safety critical sectors and motivates management to use a risk-based approach for test and maintenance programs.

If we compare and evaluate the above approaches for their efficiency, effectiveness and the availability of the state of the art for its application, we find that every approach has its positive and limitations, and the final message is that further improvements are needed. Here, a need for a maintenance management strategy is required that addresses the requirements of safety critical systems. For example, the breakdown maintenance approach is generally not being used, as SSC failure is not allowed and industries have resorted to either preventive or time-based proactive approaches. Therefore, the PM approach with fixed time intervals is common in most critical safety industries with the use of specific, condition-based approaches for special or high-risk applications. Often, testing and in-service inspection (ISI) are considered part of maintenance management programs, as at the system level, both of these functions are coordinated by the operating crew; on-site, they are performed by essentially a subset or the maintenance group only. Therefore, for the purpose of this chapter, both these aspects will be discussed suitably as and when considered needed.

Maintenance management is a human interaction-sensitive activity; therefore, human consideration and related aspects, such as training and qualification, maintenance procedures, maintenance aids, computerized maintenance management tools, root cause analysis frameworks, availability of computerized aid to identify and prioritize risk-based priority and evaluation systems such as risk monitoring, a system to track regulatory compliance checks, and, most importantly, a system to track the wellbeing of the staff, are critical requirements to ensure a quality maintenance management system in place.

10.4 Major Challenges

Even though there is a reasonable realization that human error is one of the major contributors to the accident and if detailed RCA is performed for incidents and equipment failure, it can be expected that human error contributes to many direct maintenance and associated management activities. Generating an awareness about this aspect not only at staff directly involved in maintenance activities but also among supervisory, middle management and apex management and decision making bodies is at the core of risk-conscious maintenance management. This requires a structured and integrated approach to maintenance management. The greatest challenge is developing a proposal and seeing it through across all management levels. The reason is that it requires a plant policy, resources, development of goals and criteria with respect to management functions. For example, a quality policy and bench marking and a system root cause analysis where the analysis will strive to reach the human root for any software, hardware and direct human errors. Second, implementation of systematic and dedicated precursor analysis framework should be in place–which is possible only when each and every member of the plant is having higher consciousness for any deviation from normal and second a system to evaluate the potential of these deviations to get transformed as a catastrophic failure.

The RCMM at the core has a systematic and rational-based risk model derived from the plant PRA that facilitates the identification and prioritization of SSCs as also areas for risk reduction, which might include training and qualification, deployment of modern computerized and often intelligent software with an objective to reduce human error. For example, the maintenance procedures, which are for the most part available in hardcopy form, can be on custom made intelligent maintenance diagnostic and support systems that guide the staff and provide an audio/visual signal for errors. Here, the chances of reducing human-induced common cause failure should also be given high priority. For example, the calibration error, threat due to poor oil quality for lubrication and cooling purposes, or errors while reverting back the system after completion of maintenance to normal operation where the error remains latent until a demand is there for the safety function.

It cannot be emphasized more that at times the maintenance function goes beyond normal power operation and the shutdown condition to emergency conditions. Here, the challenge is that the time available is often a prohibitive factor, while the stress levels are higher than normal, and other issues such as the availability of the right spares, staff and expertise for execution of the procedure if the timing is beyond office hours. Here, the silver lining is that the PRA document has records of most of the postulated accident sequences or scenarios, and the estimation of the likelihood and consequences of a given sequence involves the assessment of the available time window and the actual time required for a job as one of the factors that forms part of the accident consequence evaluation. What is required is a revisit of the PRA model with respect to most of the human interaction that is expected during emergency conditions and the associated requirements for the successful completion of a maintenance function during an accident situation. Furthermore, even though it is highly unlikely, the plant might require mitigative actions, and if the situation is aggravated, then the severe accident management plan of the plant needs to be validated with the PRA model to ensure that there is coherence in PRA accident postulation and severe accident management guidelines (SAMG) available with the plant. It is expected that in challenging times, such as emergency conditions, the maintenance staff will face higher stresses, and this is where the training module, even for maintenance, should prepare the staff for facing emergencies.

The availability of a system for tracking the performance of maintenance management actions goes a long way in the learning and development of proactive attributes. For example, identification at anomaly stages itself (before the failure occurs) such that failures can be avoided, meetings and discussions to identify plant technical and management weaknesses and a teamwork approach for problem solving. This can be achieved by developing performance metrics that identify the gap between what is there and what is required from risk reduction considerations, where goals and targets are written on the wall for everyone to see and work on it. For example, as for management, the regulatory compliance requirements status can be one of the major metrics. This might include implementation of plant modifications its deadline, improvement in training program, and refresher courses identified for the staff, number of technical specification violations and the contribution to these due to maintenance related aspects, status of root cause analysis number of pending cases. As seen above, a plant's general consciousness, where the 'proactive' attribute along with highly aware, involved and dedicated staff, can make the difference. Here, ethical and moral attributes play a critical role.

Provision of brain storming sessions that are all inclusive in characteristics such as staff is encouraged to put their observation, questions and the challenges that are available and how it can be addressed. Staff's emotional connection with the plant is a key factor, and the pride and happiness the individual takes in asserting that he belongs to this organization is a factor that will determine other human attributes.

Management should make all efforts to ensure that the happiness quotient in the organization is higher. Some established practices proven over ages, such as Meditation, Yoga, Motivational Seminars and Social outreach that even connect family members have the potential to transform organizations.

10.5 Risk-Conscious Maintenance management—The Framework

For safety critical systems, the existing procedure considers safety as an overriding priority. This is achieved through adherence to basic safety principles that include defense-in-depth and compliance with safety criteria and goals. Technical specifications are adhered to in words and spirit, plant procedures are followed, regulatory compliance is ensured, safety aspects are adhered to in decision making, and highly trained and qualified human resources are posted for various plant jobs. Goal and not the means to reach this goal, it is established that it is not safety considerations at the qualitative level. In this chapter, a new concept of maintenance called risk-conscious maintenance management (RCMM) is introduced. The word conscious reflects the importance of human factors in maintenance. Second, along with the many other approaches, the management also forms part of the RCMM strategy. The following section addresses in detail the RCMM.

The framework, as shown in Fig. 10.3, essentially comprised three major modules, Viz, technical, management and risk & reliability modules, as shown on the left, center and right sides, respectively, in the figure. These three modules work in an integrated manner, i.e., share data and information between the modules for the implementation of field optimized activities related to maintenance management while tracking the performance of the approach. The evaluation of performance indicators provides insights into the adequacy and effectiveness of the risk-conscious maintenance management system. In this case, there is a need for improvement

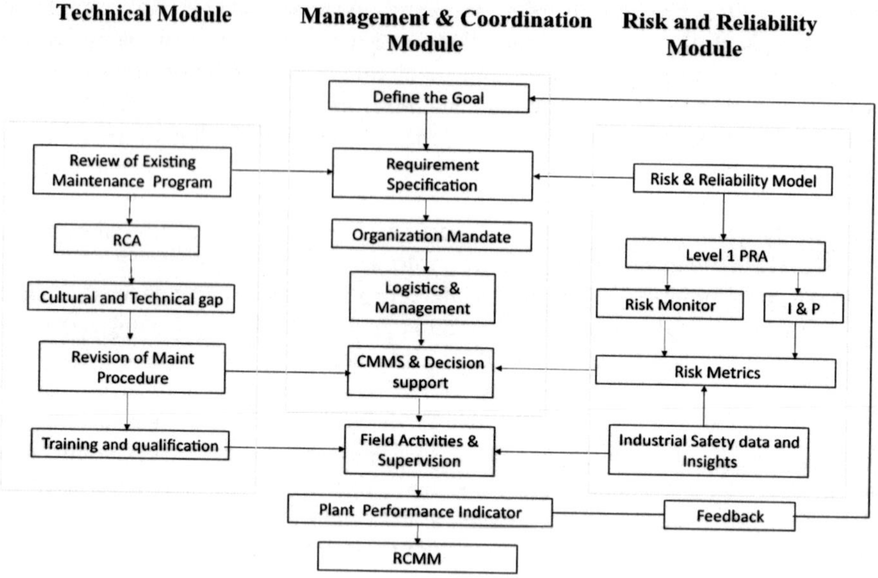

Fig. 10.3 Framework for the RCMM

feedback, and the process is further fine-tuned until the desired results satisfy the objective and aim of the RCMM system as set out in the plant management system.

We briefly discuss how this framework functions in an integrated manner. Before we go to develop a risk-conscious maintenance management program, we should define the goal in the form of performance as well as system expectations.

Once we have a clarity about our goals, it becomes necessary to develop requirement specification, i.e., the requirement specification document involving team members from operations, maintenance, services and experts from PRA, logistics and training, and quality assurance and any other member whom the management feels necessary. This document will include input from the review and requirements module as well as from the risk and reliability module to understand the strength and capability and requirement of the PRA model of the plant. The organizational mandate reporting hierarchy, allocation of resources and logistics, and assignment of responsibility are the few requirements that are critical to the success of the RCMM project.

The RCMM center, which has fully functional computerized maintenance management software along with a risk monitor platform operated by the trained staff, is required to efficiently manage RCMM coordination, updating, reporting and management activities. The input from cultural and ongoing updates on the technical gap between what is going on and what is actually required based on the root cause analysis about the type of corrective actions to be performed and the information from risk metrics developed for evaluating the maintenance/surveillance frequency on a dynamic basis is required to keep the maintenance program as 'living RCMM'. The maintenance management and shutdown schedule together with input on industrial safety insight determines the actual field activity in real-time. The feedback from the maintenance activities is directed to fine tune the program modules.

10.6 Implementation

In line with the RCMM framework, this section has been organized into three main modules, viz, management and coordination, technical, risk & reliability evaluation module, as follows.

10.6.1 Management and Coordination

As shown in Fig. 10.3, this module addresses management and administrative functions in RCMM. Additionally, this module acts as an interface between the technical and risk and reliability modules to ensure and from information exchange between the two modules to achieve the objective function, i.e., the development and implementation of the RCMM. This module has seven submodules. These modules and associated functions have been discussed as follows:

10.6.1.1 Policy

The RCMM is a new approach to maintenance management where the 'risk and associated parameters' are not just overriding provisions but the core driver for maintenance management. Therefore, the organizational goal should reflect this aspect. Second, the RCMM inherently extends its scope to the unavailability and reliability aspects. Furthermore, the scope of the RCCM, apart from maintenance activities, extends to surveillance programs such as testing and in-service inspection and further if the plant management decides then to aging management. Furthermore, even though the RCMM structure enables focused efforts on reducing and managing risk, the resources required should be considered before making an RCCM policy. Accordingly, management should adopt the RCMM by having an RCCM as one of the operational elements of operational policy. The RCMM approach and policy and associated administrative, technical and risk & reliability might form part of technical specifications documents.

Even though organizations can formulate the policy and the elements that form the part of the policy statement, they should address the following aspects.

1. A commitment to risk reduction and management is the core value for the organization.

2. Even though the regulatory oversee element is vital considering the dynamics of maintenance management, an element of flexibility should be available to the plant- and grassroots-level staff.
3. Commitment of support and resources from management includes the plant hierarchy, reporting and feedback.
4. The approved PRA study and an explicit approval for the application of RCMM.

10.6.1.2 Requirement Specifications

The availability of requirement specifications is critical to the successful development of the implementation of the RCMM program. The major motivation for the development specification is a policy statement. The policy statement is the endorsement and acceptance of management support for the RCMM program.

The lessons learned from the existing maintenance management program and an organizational consensus to improve the existing maintenance management program with respect to the performance goals and the way to achieve this, including the availability of logistics support expertise, resources and administrative setup, have bearing on the sound and robust RCMM. The performance goal can primarily relate to achievable targets on risk reduction, enhancement of systems quality, reliability and availability toward supporting the risk reduction and management program. Furthermore, there is detail about the quality attributes that define the overall capability and adequacy of the available PRA study and, more importantly, its regulatory approval for RCMM. The PRA model of the plant is central to the implementation of the RCMM. The procedure to be used for benchmarking PRA study and its acceptance criteria provides required confidence in dealing with logical and analytical aspects in plant safety management. The employment of risk prioritization schemes for SSCs in the plant and the capability of maintenance test interval estimation approaches and the provision of integration of qualitative attributes to accommodate expert opinions in support of decision should be detailed in requirements specification.

Requirement specifications should be checked and reviewed for their adequacy, applicability and capability and potential to systematically meet the set performance indicators and RCMM policy top goals. A level 1 regulatory review and approval should be a must, as RRCM implementation is a policy safety matter. This approval can be termed first state approval to go for RRCM. The second state approval is required on the complete RRCM document and its implementation, while a third stage approval will be when the performance of the RRCM in terms of its performance goals after a brief experience, say of one to two years, is reviewed by the regulatory agency.

This description is not exhaustive, and further, it is the purpose of this section to provide the exact format and scope of requirements specifications. Hence, the plant and consultant can develop the specifications based on the policy and requirements and other factors that might affect the objective and scope.

10.6.1.3 Organizational Mandate

For the successful implementation of RCMM, it is required that there is a directive from top management for implementation of the RCMM program. A project report should be prepared that provides aims, objective, scope, requirement of resources, in terms of human, finance, logistic and administrative requirements and operating policy and the new elements of the approach that require attention and approval of organizational top leadership. There should be explicit discussion on the benefits, in terms of equipment and plant risk reduction and management, increase in reliability and availability and possible cost savings that is expected compared to the existing approach to maintenance management in the organization. Of course, the requirements specification forms part of the project report, say as annexure to the main document. The implementation approach should also indicate different phases in terms of the time and scope of the areas for implementation for better clarity. An approved document is vital for the smooth implementation of RCMM.

10.6.1.4 Logistics and Management

Implementation of the RCMM program requires right from alteration in maintenance setup, creation of a team for overall project execution and management along with reporting hierarchy, availability of a Level 1 PRA study document, safety analysis document, establishment of a central maintenance management control room, availability of an updated risk monitor local area teams, Computerized Operations and Maintenance Risk Management (COMRM) station, trained crew to manage the routines, communication protocol and facilities with plant control room and different plant areas, etc. Spare inventory management system, a tool crib and a quality assurance and management system. Facility for training and qualification of staff in respective trades and positions. In the RCMM approach, only qualified and trained staff meeting the CQB attributes are deputed for the work functions.

10.6.1.5 COMRMS and Decision Support

Two major maintenance management software programs, viz., COMRMS (Computerized Operations and Maintenance Risk Management) and Risk Monitor software, are central to the maintenance and surveillance test management program. The COMRMS is a modified version of CMMS (computerized maintenance management) software systems. In COMRMS, there are additional provisions, in terms of risk ranking, safe and unsafe failure data recording and management, managing the input and output from root cause analysis as well as human root cause analysis. The information from the Living Risk monitor of the plant on the safety significance of systems and components, failure probability of components, and the results of sessions performed in the risk monitor form part of the decision making. The human root cause analysis root cause analysis framework enables the scope of the root cause analysis program to reach the human roots beyond the typical lapse, memory, mistake, error of commission and error of omission, to behavioral issues and further down to corresponding specific root human faculty and associated environmental or related responsible for the human failure.

The two major responsibilities of the COMRM and decision support module are to a) develop and issue the refueling shutdown agenda and the day-to-day management of the maintenance task, which include coordination with the plant control room, health physics, and various maintenance and surveillance testing teams spread over the plant and b) coordinate the maintenance jobs during normal operation of the plant, ensuring that it does not impact plant reliability and safety in operation and other jobs related to industrial safety and associated modifications and upgrades.

10.6.1.6 Tracking of Plant Performance Indicators

Setting the performance indicator and tracking these indicators shows the effectiveness of the RCMM program. Examples of some major indicators at the component and equipment levels are the contribution of the individual unit to the system and plant level risk and reliability, availability, and incidences. At the process and safety system levels, the major indicators are incidences of reactor trip due to process system failure and the overall contribution of systems to system unavailability. Safety systems' major performance indicators are the availability of the system on demand toward bringing the reactor to a safe state or mitigating the consequences of the accident condition. Human performance during maintenance is the key indicator. Furthermore, the dominant causes of human failure arrived through the human root cause analysis and the corrective action, and its effectiveness is an over-encompassing indicator.

10.6.1.7 Assessment and Feedback

The last organizational and management step is to assess the effectiveness of the RCMM; otherwise, revisit the complete system and subject it to the review and reconsiderations.

10.6.2 Technical Module

This module seeks to deal with risk-cultural function and evaluate the gap areas by reviewing the existing maintenance program and the potential benefits that can be achieved due to new technical aspects of maintenance management.

10.6.2.1 Review of Existing Maintenance Program

This is a crucial and the first step for the consideration of an alternate approach that includes the consideration of major features that need to be realized through RCMM. It starts with review of the existing maintenance approach that include the principal features if it is preventive fixed time-based then how for it remain time-based, is it that there is a significant load of breakdown or corrective maintenance, is the plan availability getting adversely affected and more important is it that maintenance failures are significant contributors to incidents/accidents or precursor to accidents. All these aspects will come to surface if a review of the maintenance program is performed by a team comprising maintenance, operations, logistic and risk management experts. The objective is to understand the system maintenance requirements and the limitations and potential of the current maintenance program. The insights developed through this step will be major guidelines for the preparation of the requirements specification of RCMM. In fact, this step can also help in determining the strategy and framework of the proposed maintenance management program. For example, it could be an integrated approach that captures essential features of more than one approach, such as RCMM, preventive maintenance, breakdown, condition-based, prognostic and health management, depending upon the requirements of the individual system or area, equipment characteristics and availability of resources, such as funding position, human resources, facilities and expertise. In fact, root cause analysis of events, failures and incidents in the plant can provide valuable insights into understanding the limitation of the existing maintenance program, as root cause analysis also helps formulate the new maintenance strategy.

10.6.2.2 Root Cause Analysis

Root cause analysis (RCA) can be a powerful tool to understand the fundamental and generic causes and limitations adversely affecting an organization's performance indicators. The problem with RCS is that it is very resource consuming and an expensive proposition it is carried out in the true spirit of reaching the fundamental roots. Often, the root cause analysis program, even for serious accidents, takes the form of a couple of brain storming sessions, and in this sense, RCA may not produce its full potential and may be abused rather than used. RCA often forms part of regulatory compliance; here, it also becomes a ritual to arrive at certain conclusions quickly rather than understanding the fundamental causes such that the event does not occur. Second, there is a trend to point to one or more than one human failure as the root causes that facilitate the event becoming a reality. The ensuing procedure to fix the responsibility to an individual is often unfortunate. As there were not adequate efforts to understand the situation, for example, organizational factors, complexity of the system that were beyond the cognitive capability given the stresses associated with the scenario, personal attributes and motivation and causes associated with the lack of motivation, physiological and psychological factors and causes responsible for these factors, limitation of training program. If it is a design or hardware failure, then often a detailed and systematic engineering analysis is needed, and brain storming might not help much. For example, if a part has failed or did not perform its intended function, the reason could be design, fabrication, quality, or operational procedures. In such cases, detailed structural, thermal hydraulic, chemical, aging (mechanical, electrical) and quality-related analysis is necessary to reach the fundamental roots.

However, as mentioned above, root analysis is a resource consuming option; the other fact is that it is not feasible to perform RCA for all events and failures. This calls for a strategy to prioritize and categorize events and incidents based on risk-based criteria and use this strategy to allocate the available resources for understanding maintenance management requirements. The proposed approach involves the assessment of likelihood, i.e., the frequency and consequences associated with the undesired event, and arrives at the risk factors for each event as follows:

$$\text{Risk} = \text{likelihood} * \text{consequence} \tag{10.1}$$

The parameter likelihood or frequency of event and associated consequences provide the risk factor that can be used to allocate resources in terms of time, finance, manpower, logistics that go into performing the RCA. The availability of risk factors for the events helps in the prioritization of the events. It is generally postulated that the top 20% risk-significant events provide coverage for 80% risk reduction in the plant. Then, the RCA program involves categorizing these ~ 20% events as events of higher safety significance; among the remaining events, they can be assigned the second, third and fourth categories depending on their risk-significance. Furthermore, the detailing and depth that goes into performing an RCA is directly proportional to resource allocation. According, three levels, viz., Level 1, Level 2 and Level 3 RCA, can be categorized in RCMM. Level 1 RCA is performed for selected high-risk-significant events/failures, while Level 2 and 3 RCAs are performed for medium- and low-risk-significant events/failures. The risk metrics for management of the RCA program are as follows (Fig. 10.4).

Categorization of three levels facilitates risk-conscious approach implementation, enabling an elegant mechanism and a practical solution to the constraints posed by

Likelihood / Consequences	High	Medium	Low	Insignificant
High	Level 1 ($10 \geq RF \geq 8$)	Level 1 ($10 \geq RF \geq 8$)	Level 2 ($7 \geq RF \geq 4$)	Level 2 ($7 \geq RF \geq 4$)
Medium	Level 1 ($10 \geq RF \geq 8$)	Level 2 ($7 \geq RF \geq 4$)	Level 2 ($7 \geq RF \geq 4$)	Level 2 ($7 \geq RF \geq 4$)
Low	Level 2 ($7 \geq RF \geq 4$)	Level 2 ($7 \geq RF \geq 4$)	Level 3 ($3 \geq RF \geq 0$)	Level 3 ($3 \geq RF \geq 0$)
Negligible	Level 2 ($7 \geq RF \geq 4$)	Level 3 ($3 \geq RF \geq 0$)	Level 3 ($3 \geq RF \geq 0$)	Level 3 ($3 \geq RF \geq 0$)

Fig. 10.4 Risk metrics for RCA program management. *Note* The risk factors that form the basis applicable level of RCA are shown in bracketed numerals in the risk metrics

limited resources on the one hand and a considered approach that addresses the challenges of managing risk by employing risk priority considerations. The three levels of RCA are categorized in Table 10.1. The events do not pose any series risk factors or issues, are categorized as Level 3 RCA and do not require significant resources and require a typical time of less than week. For the events, having medium risk factors is performed in more detail than Level 3 but less than Level 1 events. Accordingly, they require modest resources and can be completed in less than 2 months. Events for which risk factors are higher, as evident from the PRA precursor analysis, ranging from 10 to 8, are performed in detail and can take up to ~ 6 months. In a year, Level 1 events are not more than one event a year on average. In IAEA parlance, Level 1 events are also referred to as safety significant events, while Level 2 events can be related to the 'event' category, while Level 3 events can be approximated to the incident or equipment failure category. Level 1 events are those events that violate the technical specification clauses and require regulatory clearances for the resumption of plant operations.

Table 10.1 Salient features of three levels of RCA

Attributes	Level 1	Level 2	Level 3
Qualitative category characteristics	A qualitative combination of frequency and consequence as follows: High frequency and high consequences High frequency and medium consequences High consequences and medium frequency	A qualitative combination of frequency and consequence as follows High frequency and low consequence Medium frequency and medium consequence Medium frequency and low consequence Low frequency and high consequences Low frequency and medium consequence Insignificant frequency and high consequence Insignificant frequency and medium consequence; high frequency and negligible consequence	A qualitative combination of frequency and consequence as follows: High frequency and Negligible Consequence; Medium frequency and negligible consequence Low frequency and low consequence Low frequency and negligible consequence Insignificant frequency and high consequence
Quantified category characteristics: (risk factor (RF)	**$10 \geq RF \geq 8$** (High Risk Significance)	*($7 \geq RF \geq 4$)* Medium Risk Significance	*($3 \geq RF \geq 0$)* Low Risk Significance
Typical time required after the event	3–6 months (in case of major accidents it might take more time also)	1–2 months	3–10 days

(continued)

Table 10.1 (continued)

Attributes	Level 1	Level 2	Level 3
Typical features of analysis	First information report, data collection, interview with the technical and managerial staff, consultation with the experts, sample collection, preparing chronological development and associated details, recovering the audio and video recording related to the incidents, logical employing fishbone, logic tree, cause and effect, etc., analysis, if required detailed, failure analysis that might involve thermal hydraulic, structural, probabilistic, physics modeling, aging assessment through corrosion, wear and other modes of degradation assessment. Analysis and identification of human and organizational roots. Preparation of RCA report	First information report, data collection, interview with the technical staff, collection of additional evidence, sample analysis in the specialized labs, brain storming sessions, chronological development and associated details, recovering the audio and video recording related to the incidents. Possibility of this event propagating to a Level 2 or Level 3 with additional condition should be indicated in the report	First information report, brain storming sessions to understand the direct causes of the event. Action taken to avoid recurrence of the event. A brief report providing the description of event, a chronology of occurrences and a brief analysis should be submitted to plant. Possibility of this event propagating to a Level 2 or Level 3 with additional condition should be indicated in the report
Availability/access to laboratories for detailed sample analysis	Detailed sample characterization, analysis and testing & simulation might form the integral to understand the basic causes of failures or error. Simulator experiments to understand the root causes of human failure integral to Level 1 procedure	A limited sample analysis to understand the direct and first level roots of failure	May not be needed
Detailed failure analysis	Elaborated program	Yes	No
Detailed risk impact consequences analysis	Probabilistic risk assessment insights: impact on unavailability and CDF		Not required
Common cause failure potential of the event	Is the event a CCF or a precursor to CCF event if yes, the analysis should be presented in RCA; otherwise		Nil
Expected depth of the analysis	Identification of human error and responsible factors	Identification of responsible causes	Report the findings along with possible causes
Condition monitoring, prognostic or operational feedback	Requirements/need for trend monitoring employing a suitable technique should be addressed such that recurrence of event due to degradation of components can be avoided	Monitoring and feedback requirements need to be analyzed	Monitoring and feedback provision may not be needed

(continued)

Table 10.1 (continued)

Attributes	Level 1	Level 2	Level 3
Reporting	A detailed report with procedure, acceptance criteria used, technical specification violation, if any applicable quality assurance criteria, requirements/need of modifications, compensating mechanism, level of confidence on nonrecurrence, identification of direct, root and human specific causes. Regulatory reporting criterion should be complied with. The RCA report should be circulated to all and peers and forms part of the plant event database	A detailed report with procedure, acceptance criteria used, technical specification violation, if any applicable quality assurance criteria, requirements/need of modifications, compensating mechanism, level of confidence on nonrecurrence, identification of direct, root and human specific causes A plant/regulatory reporting criteria should be complied with. The RCA report should be circulated to all and peers and forms part of the plant event database	A brief report for operational record should be prepared and form part of incident database
Risk and availability impact on plant	PRA mapping to assess the impact of the event is performed. Effect on plant operation in terms plant availability and risk management to explicitly be brought out	PRA mapping is required to understand whether the event could be a precursor to accidents	These events have local impacts, say at component level only and risk impact is ruled out
Administrative/training requirements	Development of a new procedure for the task and whether retraining/refresher/modification in organizational aspects, e.g.	Requirement of retraining and may not be needed	Not applicable
Quality aspects	If any quality-related noncompliance is recorded and identified, then the actions to mitigate a quality-related issue should be	Potential of event to act as precursor accident is Mapped. In this context review of quality aspects forms part of RCA	An investigation to ensure that the failure is not due to quality issues. A debate on whether the failure poses a generic issue is useful
Peer review requirements	A detailed report to be submitted to management and peer groups for review and comments		
Regulatory oversight and review	Submission to regulator for review forms part of the safety management Status of compliance with regulatory recommendation	– Submission of the RCA if required as part of compliance	– Regulatory role is not anticipated unless it is recurrent issue and has potential to impact safety

10.6.2.3 Identification of Cultural Gap

The RCA event and SSCs failure database along with industrial safety incidents, etc., along with regulatory review findings forms major input for identifying cultural issues important to safety and availability. Human resource management and simulator section insights go a long way along with plant walkdown experiences is critical to hands-on work experiences. The record of schedule health monitoring could through lights on the latent human aspects along with the wellbeing quotient of employees, especially in control room and plant areas, not directly; however, subjected in-depth analysis can shed light on the individual and organizational level consciousness. Often, it is required to conduct surveys under the specialist's guidance to understand the human satisfaction level.

The parameters on which the gap areas can be identified include, e.g., trend of incidents, typical equipment downtimes compared to bench marked standards, plants and area general tidiness and upkeep, maintenance of records, time taken to acquire a specific information, happiness quotient and feeling of being part of the organization. Regulatory overview findings, etc. Availability of safety, maintenance and operational documents.

The insights into the assessment of the culture gap enable us to determine the issues and adequacy of the prevailing atmosphere that defines the risk attitude and production capability. Major aspects might include related to work culture, administration and management, interpersonal interaction communication, work ethics, dedication level, how far the members of the organization feel connected with the organization and among themselves and finally reflect on the overall organizational consciousness. For example, employing the happiness quotient is one of the major indicators that might help identify many issues that need attention.

10.6.2.4 Revision of Maintenance Procedure Program

The RCA, particularly carried out to the human root cause analysis (HRCA) level at times, requires modification in terms of change in technique or methodology or even focus of maintenance in terms of jobs to be carried out; if the human error was the cause of the incident, then the change in training and retraining could be an important insight from HRCA, as the evaluation of the existing frequency of maintenance and testing, etc., are some of the important aspects that should be discussed at the plant and regulatory levels with a view to eliminate the possibility of the recurrence of the event.

The modification in change in techniques and methodology or procedure rests with operating organization and is normally managed internally by plant authority. To address human error related to issues at the individual or organizational level, simulator experiments along with consultation with experts are often needed. These issues at times are very complex, as they might require some actions at the plant and organization levels and additional resources. It is up to the plant how these aspects can be addressed in a phased manner. Changing the frequency of testing and maintenance requires revisiting the probabilistic risk assessment and iterative simulation procedure to arrive at the optimum frequency. Tracking the improvements due to the changes/modifications implemented in the context of subject incidents should be part of plant culture to ensure the continuous reduction of risk factors.

10.6.2.5 Training and Qualification

The second important aspect of insights from the root cause analysis in general and human roots in particular is to provide input on modification or changes in the training and qualification of the program. The RCA/HRCA insights may be related to the five major aspects of the training and qualifications program: (a) checklist procedure, (b) design and operations manual and documents, (c) in plant or hands-on training, (d) safety documentation, (e) plant modifications and (f) regulatory or internal safety compliance status.

Any limitations revealed in the RCA training and qualification program are related to organizational factors, as above. For example, the checklist can be reviewed to ensure that apart from declarative, procedural and diagnostic knowledge, there is an adequate reflection of risk aspects. For example, trainees should have adequate knowledge of the risk implications of the maintenance and testing jobs performed during routine as well as periodic testing tasks. The objective should be to ensure that the process equipment and systems should not be allowed to fail in an unsafe manner, and human error in maintenance should not lead to initiating events. For standby systems and equipment, the maintenance procedure should ensure that the system and equipment should be successfully pressed into operation, and no maintenance actions or procedures are blocking procedures used should fail the systems when demand is pressed. In fact, trainees should be aware of the risk importance of various systems and equipment based on qualitative or quantitative risk ranking to understand the importance of the maintenance action and normalization procedures performed after maintenance. The risk insights should be used to use to underline the design and operational aspects, such as mode of operations, failure criteria, mechanism of failures and the vital element of procedural elements that protect the component failure. All the design and operation manuals and drawing should have a high light risk reduction mechanism or protection built into the system and procedure that are effective in recovering the failed equipment to operational states. These aspects should be taught through testing and simulation kits and facilities designed for specific components.

The RCA that points to lack of knowledge, lapses and mistakes error of omission or commission should be developed to the basic human roots to address human specific aspects, such as lack of motivation, attention, focus or organizational aspects that leads to distraction, inadequate reward and punishment system and if required further down to physiological or psychological aspects as the factor affecting these issues.

All the safety significant event root cause analyses should be mapped on the probabilistic risk model of the plant to systematically analyze the causes of error and failure and the requirement of actions to be performed to avoid recurrence of these events. In fact, all the checklists should conform to PRA knowledge and insights. Root cause analysis should use the PRA or risk monitor simulation to assess the factors responsible for making the event happen and evaluate the impact of the failure in terms of consequences at the local, i.e., component, at the system and finally at the plant level.

All plant modifications suggested by the RCA should be evaluated for their impact on risk reduction through risk simulation, and finally, the procedural steps should be validated on a simulator or equipment facility. The effect on core damage frequency and system availability should be analyzed as in support of RCA as well as validating the adequacy of the modification.

Depending on the level of applicability, all regulatory applications or compliance should be verified for benefits in terms of risk reduction or availability considerations. The PRA provides an effective and elegant framework to map events on accident sequence descriptions or system level cut-sets to evaluate the adequacy of regulatory compliance.

10.6.3 Risk and Reliability Module

Risk and reliability have qualitative as well as quantitative notions. The availability of data governs the choice of this option, i.e., if the quantified reliability data are available or the organization decides to develop its reliability database, then quantitative reliability and risk models are created, and this insight is used to develop risk-conscious applications. However, if access to quantified data is not available and the organization feels that qualitative modeling works for them, then qualitative risk and reliability analysis is performed, and the insight from these qualitative rankings is used to develop risk-conscious applications. For example, most nuclear plants have probabilistic risk assessment studies that enable them to use PRA for developing operational and maintenance applications. Of course, a qualitative PRA can also work for a plant if the reliability database is not available for developing a quantified version of PRA. Other qualitative methods include Failure Model and Effect Analysis, where qualitative gradings of frequency and consequences are used to evaluate the risk and reliability of systems. The application of the reliability block diagram, fault tree, and event tree methods are some of the other methods that enable risk and reliability modeling at the qualitative and quantitative levels.

10.6.3.1 Probabilistic Risk Assessment

In this book, a conscious decision is made to use the full-scope Level 1 PRA model of the plant to develop risk-conscious applications. The full-scope PRA normally includes the analysis of internal and external applicable events, full power, low power and shutdown conditions and the reactor core as the source of radioactivity. The scope of external events includes seismic, flood, impact of external objects, fire, etc. The statement of the level and mode of core damage conditions provides the statement of consequence, while the associated frequency provides the statement of likelihood. The major output of the Level 1 PRA is the estimate of the core damage frequency, which provides the statement of risk at the system level.

To ensure that the PRA qualifies for risk-conscious operations management, it should meet the criteria of (a) it meets the quality attributes required for PRA application-related decision making, technical specification optimization, and maintenance management, and plant-specific data requirements, (b) full-scope PRA application, if it is not full-scope due to nonavailability of either internal or other than full power operation, then the PRA team should provide an adequate set of limitations in terms of application areas and (c) frequency of feedback in terms of re-evaluation of human reliability considerations should be performed adequately, keeping in that normally human reliability data come with relatively more uncertainty compared to hardware components.

For effective use of a PRA, it is important that the PRA should be updated periodically with a frequency of at least once in a year such that it reflects the latest model of the plant. Second, an identification and prioritization (I&P) program should be place that reflects the risk importance of the SSCs. In fact, it is advisable that a risk monitor is available in support of decision making.

10.6.3.2 Risk Monitor

The availability of a qualified and approved PRA provides for the development of a risk monitor, or in the context of risk-conscious application, this tool can be termed a risk-conscious operations management system. A risk monitor is an advisory system that should be accessible to control room staff in support of decision making. Risk monitoring should be tested for various application scenarios that might include evaluation of the increase/decrease in unavailability or CDF for change in system configuration, allowable outage time and frequency of individual components and, most importantly, the uncertainty in the results. It should be ensured that the results meet the acceptability criteria and are in line with the current technical specification conditions. If not, these aspects should be reviewed, and technical specifications should reflect allowances for flexibility along with its basis and conditions. Finally, the technical specifications should be approved by the regulatory body.

There should be provision in the plant to interpret the quantified output of risk monitoring such that the rationale for the decisions at the qualitative or quantitative level is recorded in support of the decision.

10.6.3.3 Identification and Prioritization

This is the major and superior aspect that can directly be derived from a PRA study. When SSCs in the plant are identified and ranked based on their importance measures derived from PRA, it becomes convenient to use this input for plant management activities in general and dynamic maintenance management programs. The technical specification document of the plant provides a schedule for surveillance and testing of safety systems; however, this schedule is based on experience or expert opinion, including recommendations of the vendor/supplier, which tend to be conservative. This leads to excessive wear and tear as well as an increase in human error contribution to maintenance. There is a need for the identification and optimization of the optimum test and maintenance schedules as well as the type of maintenance activities that should form part of the maintenance procedure based on the systematic and rational-based framework. The PRA provides an elegant framework for identifying and prioritizing the schedule of maintenance. Furthermore, the PRA-based technical optimization approach provides for the evaluation of the optimum frequency of testing and maintenance, as the evaluation of allowable outage time is just a few of the major activities.

For process equipment, optimization occurs based on the assessment of initiating event frequency and their importance for plant risk and reliability or availability. Traditionally, qualitative judgment and lessons leant from other older plants and associated insights are used for time-based preventive maintenance schedules. However, the PRA provides for a dynamic maintenance schedule for process systems based on their risk and reliability importance.

The subject of PRA-based test and maintenance frequency and allowable outage evaluation will be separately discussed in the section on specific topics in maintenance management.

10.6.3.4 Industrial Safety

The modeling and analysis of industrial events in the plant forms part of the risk-conscious culture and is treated with due importance along with nuclear safety. The insights derived from the RCA of industrial events should be treated in line with nuclear events, e.g., the ranking of the events, analysis of the causes of the events and evaluation of respective consequences in terms of injuries, disabilities, fatalities, and the frequency of occurrence of these events should be analyzed in detail. The priority of the issues should provide the basis for corrective actions and feedback programs. Certain aspects of events, such as fire and floods that form part of nuclear risk assessment should be analyzed as part of industrial accidents along with other industrial events. It is important that the analysis, provision for detection, control, protection and management of industrial events comply with the national and international industrial codes and standards. The periodic surveillance of these systems should also follow the procedure adopted for nuclear safety events. Depending on the regulatory directive and availability of resources, preferably quantitative metrics should be developed and employed for industrial safety management.

10.7 Application of RCMM—Surveillance Test Maintenance Interval Optimization

10.7.1 General Background

This section presents the surveillance test interval optimization for an emergency core cooling system (ECCS) of a reference nuclear plant (Chowdhury and Varde 2010). The ECCS is a safety system or, more precisely, the Engineered Safety Feature (ESF) of the plant for ensuring plant safety or mitigating features for accident conditions involving an unlikely scenario of loss of breach in the primary coolant system boundary and loss of coolant from the system.

Plant management feels that the emergency core cooling system feels that the present testing frequency of ECCS is greater, and this needs to be reduced. The qualitative arguments are as follows: (a) the reference reactor is a low-pressure and low-enthalpy system, and the current configurations and automatic safety provision are adequate to ensure plant safety even if the unlikely event involving a breach in the pressure boundary may not lead to nuclear consequences; (b) past testing over a period of 10 years shows that the system is reliable enough to start and perform on demand and cater to safety requirements without operator intervention, at least during the initial period of at least 30 min; (c) frequent testing is resource consuming and involves dose consumption; (d) frequent testing is inducing aging due to the wearing phenomenon; and (e) the human intervention involved as part of the test and maintenance program on the system and component has the potential for human error might adversely affect system unavailability. However, these arguments require an integrated approach to present a clear picture linking the above arguments with adverse effects on plant risk and unavailability.

The Level 1 PRA model along with the availability of an experimental risk monitor enable modeling and simulation required to demonstrate the safety case of the plant. The first requirements for optimization are availability of the plant-specific failure data of the component and trend of degradation during service or in standby condition and its location of component in the system configuration. The second, the facility number of crew and station for maintenance/test.

10.7.2 Objective Functions

Keeping in mind the major concerns of the plant, the optimization scheme involved, radiation exposure by considering Man-rem consumption. It was required to create these models as follows.

10.7.2.1 Unavailability Model

The component unavailability U is a function of the intrinsic component unavailability, testing time contribution to unavailability, unavailability introduced due to human error and unavailability due to aging that includes wear and degradation. The unavailability equation at the component level is as follows:

$$U = f(U_T + U_t + U_H + U_W + U_D) \tag{10.2}$$

Accordingly, unavailability is related to the function of all these components taken together as an additive function:

$$U = \frac{1}{2}\lambda T + \frac{t}{T} + F_H \theta_0 N + f_W \lambda \left(\frac{\tau_o}{\tau_A}\right)\left(\frac{8760}{T}\right) + f_D \lambda \left(\frac{\tau_{lt}}{\tau_{av}}\right) T \tag{10.3}$$

10.7.2.2 Cost Model

The resources required are proportional to the number of surveillance tests; for example, the human resources required for hands-on jobs, supervision, coordination, tools and facilities and related costs increase with increasing testing and maintenance. Even though the weightage of cost is much lower than risk factors, an increase in risk due to safety system unavailability or system failure without completing the mission, resource consideration forms part of optimization with an appropriate weighting factor. However, surveillance tests and maintenance are required to detect degradation or passive, potential or incipient failures in time such that the performance of the system is acceptable when the system is called upon to perform the safety function.

For the sake of simplicity, resource optimization has been performed through the cost function. That is, cost associated with making the required resource available forms one of the objective functions. Accordingly, the cost function is formulated as follows:

$$\text{Cost} = C_t\left(\frac{L}{T}\right) + \sum\left[\left(\frac{T}{l}\right)r\right] \tag{10.4}$$

where:

C_t: Cost per testing
L: Expected life of the plant
l: Average life of each component
r: Replacement cost associated with each component.

The first term provides the estimate of the lifecycle cost of testing. The second term provides the cost of replacement of the failed component. If we assume that testing the standby system with a plant in normal operation, then the cost per testing is given as:

$$C_t = C_m N_m \tau_t \tag{10.5}$$

C_t: Total cost per human
C_m: Average cost per human
N_m: Human resources required for testing in numbers
τ_t: Time required in hours for each test.

In this case, the plant is shut down during testing (if stipulated by technical specifications requirements), resulting in loss through put or loss of deliverables that can be computed in terms of revenue. Then, this cost will also be added to the total cost of testing. Here, it is assumed that the risk implications during safety system shutdown can be ignored.

10.7.2.3 Dose Consumption Model

In this model, it is conservatively assumed that dose consumption is proportional to the number of persons involved in surveillance testing and maintenance. Second, the dose to the individual or a team is directly related to the general location background radiation field and second due to a hotspot. However, in this model based on experience/expert opinion, a judgment was used to arrive at average dose consumption during the testing. No hot spot has been encountered in the subject area. The model is as follows:

$$Dt = R_t(L/T) \tag{10.6}$$

D_t: Dose consumption per test (in millimeters).

Dose consumption increases with the number of humans involved in each testing and further number of testingings during the life of the plant. An experience suggests that the dose consumption is relatively small during surveillance testing. Hence, a judgment can be used to include this factor in optimization.

The traditional or classical approach to solve a multiobjective function problem involves assigning weights w_i to each objective function based on the importance or sensitivity of the respective function to the final solution. The approach is simple and elegant, as the multiobjective function problem can now be modeled as a single objective function, as given in the following equation:

$$\text{Min} z = w_1 z_1(x) + w_2 z_2(x) + \cdots w_k z_k(x) \tag{10.7}$$

The limitation of this approach is that it gives only a solution and cannot find a solution from a population uniformly distributed over a nonconvex trade-off surface.

For complex multiobjective functions, genetic algorithms (GAs) have often been found to be efficient.

In the following section, an approach that focuses on a single objective function, i.e., 'risk' optimization, is presented. The operating organization feels that system risk is the dominating attribute of concern. However, later on modeling can be done to understand the sensitivity of the parameters like cost and dose consumption,

10.7.3 Risk-Based Approach to Surveillance Test and Maintenance Interval Optimization

10.7.3.1 System and Plant Level Modeling

In this approach, a risk-based methodology that creates a basic platform for ST&M interval estimation methodology was developed for surveillance testing in interval optimization for emergency core cooling systems (ECCS) of a reference research reactor. ECCS is an engineered safety feature (ESF) that remains standby during normal conditions and starts automatically in response to loss of coolant accident (LOCA) in the event of uncontrolled loss of inventory from the system. The ECCS injection at the inlet and outlet of the reactor core such that irrespective of the location of the leak (at inlet/outlet), the coolant system makes up cooling are maintained. The ECCS has four major basic functions: (a) monitoring the integrity of the coolant system boundary, (b) sensing the LOCA scenario and location, (c) generating the demand for ECCS and (d) collecting the leaked-out inventory or stored water inventory into the system, thereby ensuring reactor core cooling. Furthermore, the procedure to rectify the leak and mitigation measures, if needed, involves manual procedures. Figure 10.5 shows the major features of ECCS that are relevant for this study.

Figure 10.6 shows the fault tree for the ECCS for core injection at the outlet of the core. The ECCS is tested based on its surveillance frequency stipulated by the technical specification. It may be noted that this surveillance test interval is traditionally decided based on expert opinion that is based on the operating experience of this or other similar plant in the beginning and revised if required based on the plant-specific experience. Since the STI is based on expert opinion, there is subjectivity that comes with this decision, and at times, it even becomes too conservative. The numerical approach discussed here is based on the integrated reliability model of the plant and governed by the quantified estimates of failure rate data and the system configurations. To demonstrate the relationship between the surveillance time intervals on the fault tree unavailability, one example is presented to demonstrate the methodology with the following assumptions:

The optimization is to be done for a single objective function, i.e., minimization of system unavailability, where the major consideration is loss of availability during standby or dormant mode due to latent catastrophic or degradation-related issues.

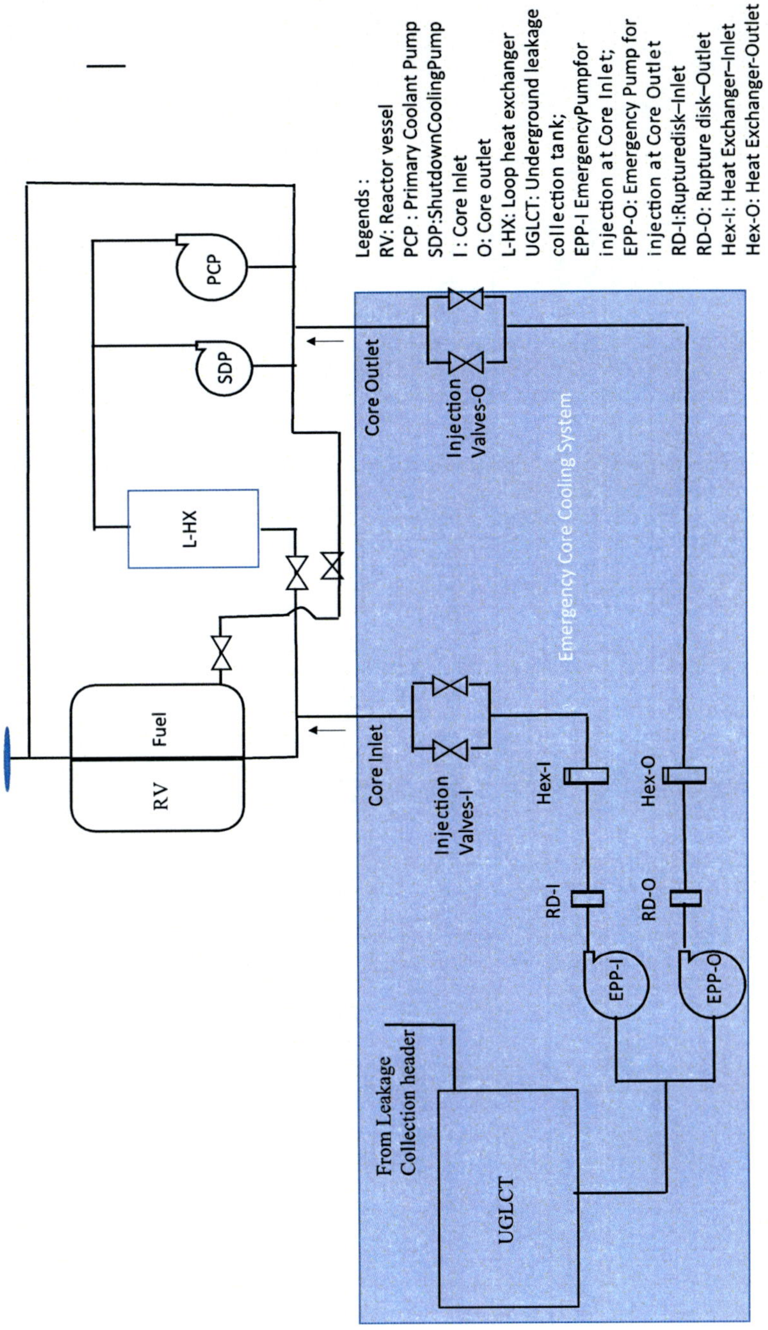

Fig. 10.5 Emergency core cooling system for reactor cooling function during LOCA scenario

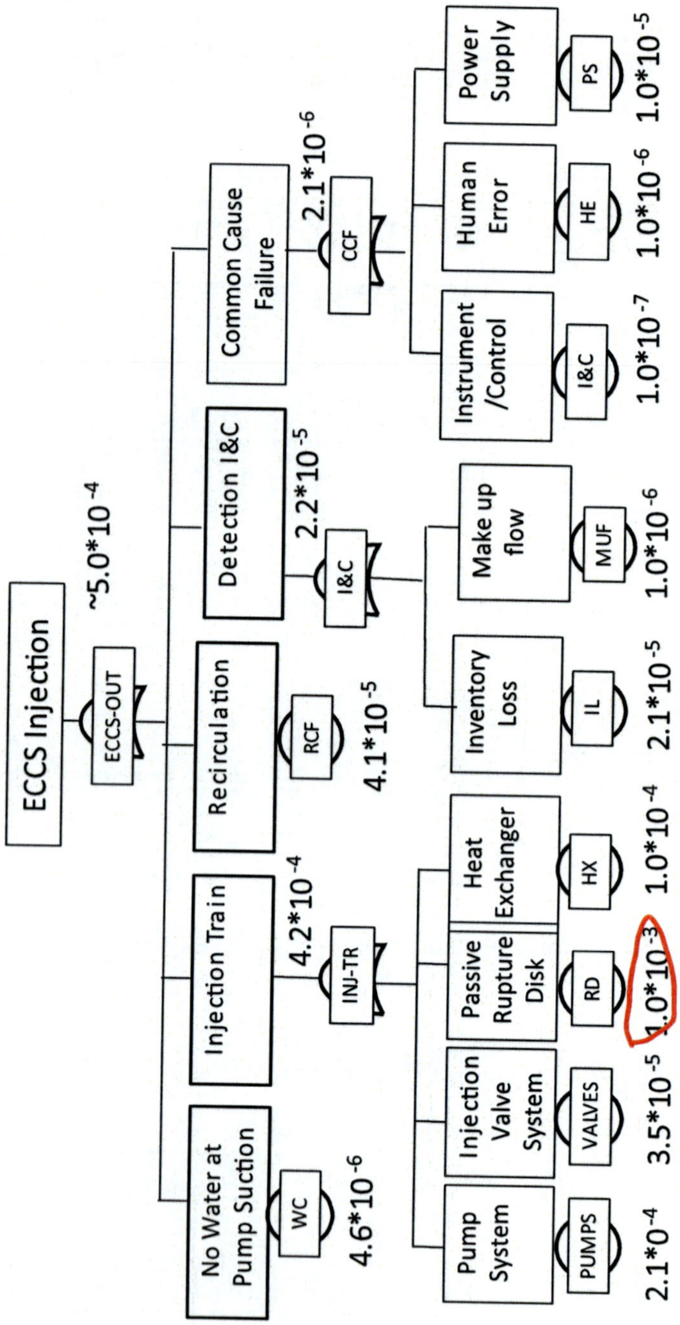

Fig. 10.6 ECCS simplified and modified fault tree for demonstration

Often, the test duration component also forms part of the unavailability. However, if it is justified that during plant shutdown, the ECCS system unavailability contribution to the plant risk is low, a decision can be made consciously to ignore the contribution due to test duration component optimization of the test interval. The plant-specific data used in the study reduce uncertainty to an acceptable level. For demonstration a simplified plant schematic along with fault tree are adequate to demonstrate the advantages and limitation of this approach. Use of terminal probability to simplify the model input reliability data. This case study is for demonstrating the PRA approach to STI evaluation. Therefore, some assumptions have been made to effectively demonstrate the power and limitation of this approach. The case study addresses a reference plant and not for any specific reactor or system, as the idea, as mentioned, is to demonstrate the approach. Even though a typical ECCS system in a plant could be more complex in terms of the design of the I&C system, the number of hardware and software components might involve human actions for mitigating the emergency scenario. However, in this example, we have a simplified ECCS system model and accordingly a simplified model of fault tree and event tree. A rare approximation approach has been used to model the failure phenomenon and evaluate the standby system failure probability. The reference value of ECCS unavailability refers to the estimate if unavailability with the current ST&MI and this just for comparison purpose to any change that is being discussed here. Furthermore, in the fault tree, it can be observed that the injection train branch is a major contributor to the ECCS reliability, while other intermediate states in the fault tree contribution are insignificant to the top event probability. Accordingly, only injection train has been considered to represent ECCS failure probability, and other branches, such as water crisis, recirculation, CCF and I&C branch, have not been considered for evaluation of ECCS and CDF estimates.

A simplified model of the event tree for the LOCA is shown in Fig. 10.7. Here, it has been assumed that accident sequence (AS) Nos. 1, 3 and 5 do not result in core damage. Remaining AS 2, 4, 6 and 8 safe risk implications. It can be seen that AS No. 7 does not involve ECCS performance; nevertheless, LOCA being an initiating event and reactor safe shutdown in response to an initiating is a must to ensure safety. This event has been categorized as resulting in CD conditions. In this simulation, the purpose is to estimate the frequency from this point of view; only AS 2 is relevant, as other ASs have very low frequencies. Hence, for CDF frequency estimation, only this AS, i.e., No. 2 has been considered. Table 10.3 shows the estimates obtained from the parametric study where the CDF was obtained by varying the ST&M from 730 h (1 month) to 8760 h (12 months) (Table 10.2).

10.7.3.2 Results of the Parametric Study

Table 10.3 shows the estimates obtained from the parametric study where the ECCS system was unavailable and the CDF frequency was further varied at the plant level by varying the surveillance ST&M interval from 730 h (1 month) to 8760 h (12 months/1 year). This range is based on the operating experience and resources

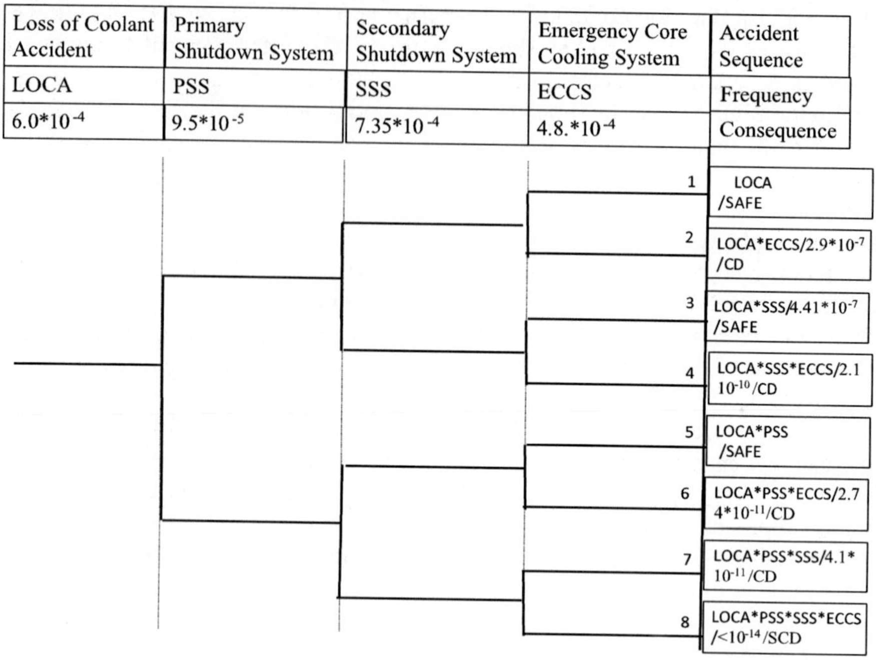

Fig. 10.7 A simplified event tree for demonstrating arriving at a statement of risk

Table 10.2 Failure data for fault tree analysis

S. no.	Component	Component boundary	Failure mode	Failure rate/probability	Remark/data
1	Centrifugal pump system	Pump, motor, Connection between inlet and outlet valves. breaker, Control supply including the actuating relay	Fail to start and fail to run	$3.0 * 10^{-4}$	Generic and plant-specific
2	Valve (motor operated) system	Valve, Motor, Terminal connection. Control actuating relays	Fail to open the Valve	$1.37 * 10^{-4}$	"
3	Rupture disk–isolation	Rupture disk, and flange connections	Failed to rupture	$1 * 10^{-4}$	"

(continued)

Table 10.2 (continued)

S. no.	Component	Component boundary	Failure mode	Failure rate/probability	Remark/data
4	Heat exchanger–plate type	Heat exchanger, connection on primary and secondary side up to isolation valves	Cooling failure	$1 * 10^{-5}$	
Terminal/intermediate node probabilities postulated/assumed					
5	Water crisis at pump suction	Tank and associate piping and valves and strainer	CCF-Unsafe	$4.6 * 10^{-6}$	”
6	Recirculation failure	Pumps, forward and return line and associated components, viz, valves, heat exchanger, etc	Unsafe	$1.7 * 10^{-6}$	”
7	CCF	Leading to unsafe failure (I&C, Power Supply and Control)	Unsafe	$2.1 * 10^{-6}$	”

needed, as well as its impact on reactor availability and last the overriding and deciding parameters, i.e., postulated risk estimates. At the outset, there are few considerations that need to be discussed. The source of the data, particularly for vital components, such as pumps and valves, includes injection, check or nonreturn, isolation and their motive modes, such as fluid force, motor, air operated and heat exchanger available, available from plant-specific sources based on operating experience of over 20 years. For many components, the data were used from generic sources. Human error probability was estimated in the detailed model essentially from the plant-specific sources, while available human reliability models were used to evaluate the applicability of these estimates as well as the applicability of generic models themselves. Furthermore, the experience has been that most of the time, it is the common cause of failures (CCFs), especially human error, as potential CCFs need special attention. For any risk study, considerations of uncertainty may be needed. The example discussed here avoids further complexity due to uncertainty considerations. However, while making a decision, this aspect must be considered. The plant management is alert to the extent any if deficiency is noted the implementation of modifications be it additional redundancy, operational policy changes, change in maintenance practices, etc., are affected in a time bound manner with highest priority. The availability of a plant simulator facilitates risk simulation studies by

Table 10.3 Results of the parametric study to evaluate CDF as a function of ECCS unavailability for different ST&MI

ST&MI in months/hours (International Atomic Energy Agency 2018)	ECCS unavailability (only random failures) (Vishnu and Regikumar 2016)	CDF (due to only random failures (Computers and Industrial Engineering 2012)	ECCS of unavailability due to testing (IEEE Standard Association 1856)	ECCS unavailability (due to random failure and testing) (International Atomic Energy Agency 2013)	CDF (due to random failure and testing in ECCSs) (International Atomic Energy Agency 2006)
1/730	$1.84 * 10^{-4}$	$1.1 * 10^{-7}$	$3.3 * 10^{-2}$	$3.3 * 10^{-2}$	$1.2 * 10^{-5}$
2/1460	$3.64 * 10^{-4}$	$2.2 * 10^{-7}$	$1.6 * 10^{-2}$	$1.6 * 10^{-2}$	$9.7 * 10^{-6}$
3/2190	$5.74 * 10^{-4}$	$3.3 * 10^{-7}$	$1.1 * 10^{-2}$	$1.1 * 10^{-2}$lePara>	$6.3 * 10^{-6}$
4/2920	$7.37 * 10^{-4}$	$4.4 * 10^{-7}$	$8.2 * 10^{-3}$	$8.9 * 10^{-3}$	$4.9 * 10^{-6}$
5/3650	$9.17 * 10^{-4}$	$5.5 * 10^{-7}$	$6.6 * 10^{-3}$	$7.5 * 10^{-3}$	$3.9 * 10^{-6}$
6/4380	$1.1 * 10^{-3}$	$6.6 * 10^{-7}$	$5.5 * 10^{-3}$	$6.6 * 10^{-3}$	$3.9 * 10^{-6}$
7/5110	$1.3 * 10^{-3}$	$7.7 * 10^{-7}$	$4.7 * 10^{-3}$	$6.0 * 10^{-3}$	$2.7 * 10^{-6}$
8/5840	$1.5 * 10^{-3}$	$8.8 * 10^{-7}$	$4.1 * 10^{-3}$	$5.6 * 10^{-3}$	$2.4 * 10^{-6}$
9/6570	$1.64 * 10^{-3}$	$9.9 * 10^{-7}$	$3.6 * 10^{-3}$	$5.2 * 10^{-3}$	$2.1 * 10^{-6}$
10/7300	$1.83 * 10^{-3}$	$1.01 * 10^{-6}$	$3.3 * 10^{-3}$	$5.1 * 10^{-3}$	$1.8 * 10^{-6}$
11/8030	$2.0 * 10^{-3}$	$1.2 * 10^{-6}$	$3.0 * 10^{-3}$	$5.0 * 10^{-3}$	$1.8 * 10^{-6}$
12/8760	$2.15 * 10^{-3}$	$1.29 * 10^{-6}$	$2.7 * 10^{-3}$	$4.8 * 10^{-3}$	$1.62 * 10^{-6}$

running sessions by blocking a set of SSCs or even operator responses such that the overall impact of these deviations can be studied, and proper action can be initiated.

For the simulation of the plant PRA model for postulated ECCS performance and its impact, increasing the STI&M interval from 2190 h to 4380 shows the change in unavailability for only random failure and net change in CDF; unavailability due to (time required for) testing alone; and change in unavailability due to combined effect of random failure and testing and the net effect on CDF.

The above estimation was used to evaluate the impact of testing time on the overall results of the ECCS optimization. It is seen from Table 10.3 that columns 2 and 4 provide estimates of ECCS unavailability based on random failure and testing contribution alone, respectively. It can be noted that during the initial period, i.e., for three months, the ECCS unavailability is dominated by contribution due to testing, and later, the random failure contribution starts increasing overall unavailability, and finally, it is the random failure that becomes a significant component of overall unavailability. Columns 3 and 6 show the impact of system unavailability on CDF median estimates.

10.7.3.3 Decision Criteria

The available codes and standard require stipulates the CDF unavailability target of $\leq 10^{-3}$ per demand. Furthermore, for safety demonstration of the CDF value for old and new plants, the accepted figures are $\leq 10^{-4}$ and $\leq 10^{-5}$/yr, respectively. It may be noted that these figures applicable for Full-scope Level 1 PRA, i.e., conservation of contribution due to internal and external events with the reactor core and spent fuel in storage as the source of radiation, for a complete range from full power, low power and shutdown operation. External events mainly include seismic, flood, fire and the impact of an external object. The security considerations do not form part of the above CDF statement. However, it is desirable that insights from probabilistic security risk assessment (PSRA) results should be part of decision making.

In summary, the above example was for the demonstration of a simplified procedure and forms an important input in support of decision making. However, the unavailability of 95% confidence as part of uncertainty characterization, use of generic data, no consideration of other factors, such as dose consumption and financial implications and accelerating effect of aging on performance of equipment due to excessive testing, availability of full-scope PRA and human reliability considerations, aging status of the SSCs that also includes safety as process systems, and a comprehensive consideration of plant risk-consciousness are critical to decision making.

To overcome some of the limitations of the above approach, which are lacking in multiobjective optimization, further studies were performed by employing a genetic algorithm approach for ECCS test interval optimization (Chowdhury and Varde 2010).

10.7.4 Genetic Algorithm—An Intelligent Approach to Optimization

10.7.4.1 Background

Genetic algorithms (GAs) are very powerful tools for optimization problems (Deb 2001). In computer science and operations research, a genetic algorithm (GA) is a metaheuristic inspired by the process of natural selection that belongs to the larger class of evolutionary algorithms (EA). Genetic algorithms are commonly used to generate high-quality solutions to optimization and search problems by relying on biologically inspired operators such as mutation, crossover and selection (Krishnanand 2009). The surveillance test interval (T) optimization ideally requires, particularly when the search space is very large where it is important, local minima to be avoided (Deb 2001). Additionally, when the optimization problem addresses multi-objective functions, genetic algorithms have been shown to be an effective approach. Intuitively, the GA starts with a random activation of solution points in the solution space and, while performing three stages, namely, selection, crossover and mutation, a typical phenomenon inherent to biological and genetic sciences and arrives at the final solution space.

10.7.4.2 The GA with Real Parameters for ST&MI Optimization

The essential feature of the GA is that it is an evolutionary operation in line with the biological processes involved in the development of offspring to parent. Chowdhury and Varde have developed a novel approach where the real-parameter GA has been applied for a multiobjective optimization problem, as this approach makes the algorithm simple by not having the overload of coding and decoding of the solution (Deb 2001). The chromosome is a fundamental unit, and the GA works in a population of chromosomes that is subjected to selection criteria through the fitness function and operation of crossover and mutation. The candidate solution space or chromosome is characterized through bit strings (e.g., 1,010,101). Accordingly, there are two basic requirements for the GA operation: the genetic representation of the solution domain and a function for fitness computation.

The GA procedure starts with the identification of the problem in terms of the identification of the initial solution space, often randomly. The first major step is the selection of parent chromosomes for reproduction. The fitter chromosomes, i.e., the one with higher probabilities are selected. The crossover operator mimics the biological mimics biological recombination. The operator randomly chooses a locus and exchanges strings after the identified locus before and after between two chromosomes and creates two children, e.g., two chromosomes 1,100,101 and 1,011,111 with the identified locus at three produce new children crossed over strings 1,101,111 and 1,010,101. Next and third steps where the mutation operator flips one bit in the child chromosome, e.g., 1,101,111 could be mutated in the 7th position to 1,101,110.

The above procedure of selection, crossover and mutation is very common but not unique, and many techniques have been used by researchers keeping in view the requirements of the problem in hand that deal with varying complexity of specific domains.

The above procedure was continued upward until termination criteria were fulfilled. Depending on the nature of the problem and the requirements of resources and deliverables, some of the list of criteria, of course, is not exhaustive and the desired level of optimum conditions are achieved. A decision is made based on a review of the combination of conditions during the operation of the algorithm, max. no generations, time, computational time, etc.

10.7.4.3 Modeling the GA Algorithm for Multiobjective Optimization Problems

The objective of this section is to create an effective model for evaluating the surveillance test frequency for an emergency core cooling system (ECCS) for a reference reactor. The ECCS is a standby safety system or an engineered safety feature in the plant and remains in standby mode. ECCS fault tree modeling forms part of creating the PRA model of the plant. The ECCS unavailability and core damage frequency form the risk metrics at the system level and plant level, respectively. The optimization of the surveillance test intervals is carried out to minimize the objective functions, unavailability, cost function and dose consumption.

As this is a multiobjective function problem, Pareto Dominance was used in this project. The search space remains between the maximum and minimum applicable range that has been derived from the domain specific operations and maintenance records and design documents. Therefore, a real-parameter model of the genetic algorithm has been adopted for this application in place of binary strings. Other specific aspects/assumptions deal with the use of both Stochastic Remaining Roulette Wheel Selection and Ranking-Based tournament selection along with crowding distance as a tie breaker parameter have been used separately along with BLX-α Crossover and Nonuniform mutation. The flowchart that depicts the implementation of the GA procedure is shown in Fig. 10.8.

The present assignment can be classified as a search problem where GA has been used to optimize multiple objective functions. For multiobjective optimization, Pareto's dominance concept has been used (Krishnanand 2009). In this section, some concepts regarding Pareto Dominance are introduced, and then how they have been incorporated in this work is shown.

There are three objective functions: Unavailability, Cost and Man-rem consumption. The search space spawns between maximum and minimum STI values. Both Stochastic Remain- der Roulette Wheel Selection and Ranking-based tournament selection along with crowding distance as tie breaking parameter have been used separately along with BLX-α Crossover and Nonuniform Mutation. The details are given in subsequent subsections. The whole process is represented in a flowchart in Fig. 10.8.

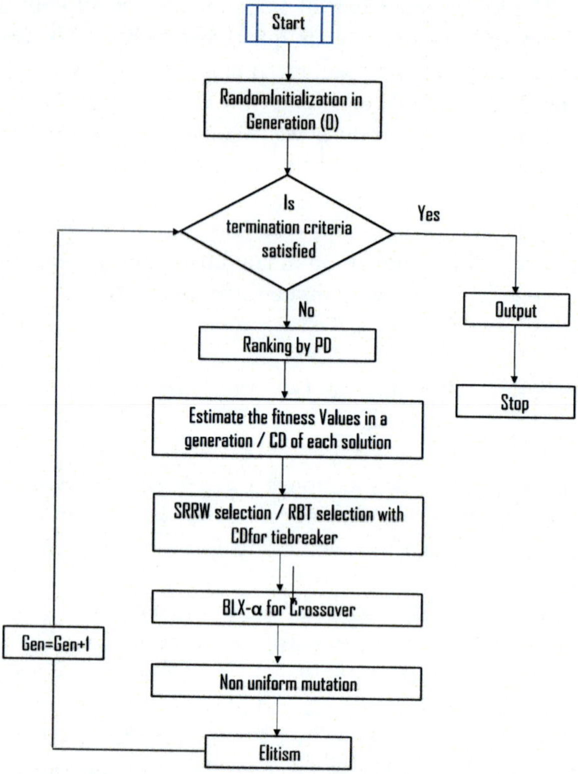

Fig. 10.8 Flow chart of the optimization algorithm

A. Selection

Selection is a very important operator in GA. Here, the objective functions, as mentioned earlier, are Unavailability, Cost and Dose Consumption. The classical approach to solve the multiobjective function approach involves assigning weight w_i to each normalized objective function $z_i(x)$ to facilitate conversion of the problem to a single parameter optimization problem. As shown in the following equation.

$$Minz = w_1 z_1(x) + w_2 z_2(x) \cdots w_k z_k(x) \tag{10.8}$$

where k is the number of objective functions

Since a single objective is used, it is also computationally efficient. However, it has a serious disadvantage. It gives only one solution and hence cannot find solutions uniformly distributed over a nonconvex trade-off surface (Zitzler et al. 2000). Therefore, Pareto's dominance14 has been used, which will be demonstrated with an example. Suppose we have two solutions T1 and T2. Now, according to Pareto's Dominance, T1 will Dominate T2 if and only if

Unavailability $(\underline{T}\ 1) \leq$ *Unavailability (T1) and Cost* $(\underline{T}\ 1) \leq$ *Cost (T2) and Man rem consumption (T1)* $\leq$ *Man rem consumption (T2)*

Or
Unavailability ($\underline{T}$ *1*) $\leq$ *Unavailability (T1) and Cost* ($\underline{T}$ *1*) $\leq$ *Cost (T2) and Man rem consumption (T1)* $\leq$ *Man rem consumption (T2)*
Or
Unavailability ($\underline{T}$ *1*) $\leq$ *Unavailability (T1) and Cost* ($\underline{T}$ *1*) $\leq$ *Cost (T2) and Man rem consumption (T1)* < *Man rem consumption (T2)*

A set $A \subseteq Y$ (where Y denotes the entire decision space) is called a Pareto optimal set iff $\forall a \in A$: there does not exist $x \in Y$: x dominates a. In this case, all the solutions in A will be called nondominated solutions because there are no other solutions that can dominate those solutions in terms of Pareto Dominance. Now, for selection from a generation, we first remove the Pareto optimal set of nondominated solutions and rank them as 1. Then, again from the rest of the solutions of the generation, we remove the nondominated solutions and rank them as 2. This procedure continues until all the solutions of the generation have been ranked 15. Now, we can very well see that those solutions having lower ranks are more likely to be the optimized value. With this understanding, a fitness function f(STI) has been defined with Eq.

$$f(T) = 1/\text{rank}(T) \tag{10.9}$$

As much randomness is inherent in genetic algorithms, the results often vary among different runs of the program. However, because of the inherent stochasticity of the Roulette Wheel selection process, many more perturbations may creep into end results. Therefore, from the fitness values of solutions and the average fitness value of the generation, the expectation for each solution has been calculated.

$$E_T = f(T)/T_{\text{avg}} \tag{10.10}$$

Now, we deterministically place the integer number of copies of ESTI for each solution STI in the mating pool to participate in crossover and mutation. Then, from the remaining fractional part, we perform roulette wheel selection. This ensures that at least one or more copies of more fit solutions will go to the mating pool. Otherwise, the inherent stochasticity of Roulette Wheel selection may not allow the fitter solutions to be absorbed in the mating pool. This whole process of selection is known as Stochastic Remainder Roulette Wheel selection.

Along with the stochastic remainder roulette wheel approach, a crowding distance-based approach has also been attempted, which aims to obtain a uniform spread of solutions along the best-known Pareto front without using a fitness sharing parameter. This process was used in NSGA-II (Chowdhury and Varde 2011). This procedure is demonstrated below.

Step 1. Rank the population and identify nondominated fronts F_1, F_2, …, F_R for each front $j = 1, \ldots, R$ repeat steps 2 and 3.

Step 2. For each objective function k, sort the solutions in F_j in ascending order. Let $l = |F_j|$ and $T_{[I,k]}$ represent the ith solution in the sorted list with respect to the objective function k. Assign $\mathrm{cd}_k(T_{[1,k]}) = \infty$ and $\mathrm{cd}_k(T_{[l,k]}) = \infty$, and for $i = 2, \ldots, l-1$ assign with the following

$$\mathrm{Cd}_k\left(T_{[i,k]}\right) = \frac{(z_k(T_{[i+1,k]}) - (z_k(T_{[i-1,k]}))}{(z_{k,\max} - z_{k,\min})} \tag{10.11}$$

Step 3. To find the total crowding distance cd(T) of a solution T, sum the solution crowding distances with respect to each objective, given by equation, i.e.,

$$\mathrm{cd}(T) = \sum_k \mathrm{cd}_k(T) \tag{10.12}$$

This approach has an advantage over other approaches that use niche counts for diversification. Here, we do not have to specify any user defined parameter such as a share or the k-th closest neighbor (Rudolph 1998). This crowding distance measure has been used for tiebreakers in a selection technique called the crowded tournament selection operator: Randomly select two solutions $T_1$and T_2; if the solutions are in the same nondominated front, the solution with a higher crowding distance win. Otherwise, the solution with the lowest rank is selected.

Crossover

In our current approach, simple or naive crossover cannot be used, as the solutions consist of only one component. Therefore, for the current work, Blend Crossover has been used, which is also known as BLX-α crossover (Zitzler et al. 2000). For two parent solutions $STI_{1,\mathrm{gen}}$ and $STI_{2,\mathrm{gen}}$ (assuming $STI_{1,\mathrm{gen}} < STI_{2,\mathrm{gen}}$), the *BLX-α* randomly picks a solution in the range

$$[T_{1,\mathrm{gen}} - \alpha\left(T_{2,\mathrm{gen}} - T_{1,\mathrm{gen}}\right), T_{2,\mathrm{gen}} + \alpha\left(T_{2,\mathrm{gen}} - T_{1,\mathrm{gen}}\right) \tag{10.13}$$

$$\left[\mathrm{STI}_{1,\mathrm{gen}} - \alpha\left(\mathrm{STI}_{2,\mathrm{gen}} - \mathrm{STI}_{1,\mathrm{gen}}\right),\ \mathrm{STI}_{2,\mathrm{gen}} + \alpha(\mathrm{STI2},\mathrm{gen} - \mathrm{STI1},\mathrm{gen})\right] \tag{10.14}$$

Thus, if u is a random number between 0 and 1, the following is an offspring given by the equation:

$$T_{1,\mathrm{gen+1}} = (1-\gamma)T_{1,\mathrm{gen}} + \gamma T_{2.\mathrm{gen}} \tag{10.15}$$

where $\gamma = (1+2\alpha)u - \alpha$

Generally, *BLX-α* performs better when $\alpha = 0.5$, and the same value of α has been used for our work. However, BLX-α has an interesting property that the location of the offspring depends on the difference in parent solutions. This will be clear from the equation:

$$T_{1,\mathrm{gen+1}} - T_{1,\mathrm{gen}} = \gamma(T_{2,\mathrm{gen}} - T_{1,\mathrm{gen}}) \tag{10.16}$$

If the differences between the parent solutions are small, the difference between the offspring and parent solution is also small. This property of the search operator allows us to constitute an adaptive search. If the diversity in the parent solutions is large, an offspring population with a large diversity is expected, and vice versa. Thus, such an operator allows us to search the entire space early on (when a random population over the entire space is initialized) and concentrate the search more on the later stages when the population tends to converge in some region of the search space.

Mutation

Mutation is an important operator in the GA, as it attempts to introduce additional variability in the population so that other areas of the search space can be explored. There are many mutation operators available for real-parameter GA. However, for mutation in the subject problem, we employed *a nonuniform mutation operator* (Zitzler et al. 2000) because of its adaptive property with a slight modification. This operator takes a single parent (T_1, gen) and produces a single offspring ($T_{1,\text{ gen}}$ + !) given by equations:

$$T_{1,\text{gen}+1} = T_{1,\text{gen}} + \Delta\left(\text{gen}, T_{\max} - T_{1,\text{gen}}\right) \text{ if a random bit is 0} \tag{10.17}$$

$$T_{1,\text{gen}+1} = T_{1,\text{gen}} - \Delta\left(\text{gen}, T_{1,\text{gen}} - T_{\min}\right) \text{ if random bit is 1;} \tag{10.18}$$

The function $\Delta(t, y)$ returns a value in the range [0, y] such that the probability of $\Delta(t, y)$ being close to increases as t increases (here, we have used t as the generation number). This property causes the operator to search the space uniformly when the generation number is small and then very locally in the later stages. The following functional form has been used given by equation

$$\Delta(t, y) = \text{yr}\left[1 - \left(\frac{-t}{t_{\max}}\right)^3\right] \tag{10.19}$$

where r is a random no in the range [0, 1] and $t_{\max}$ is the maximum value of the parameter t.

Elitism

Elitism is the process of preserving the elite solutions from each generation. Sometimes, we may lose the elite solution of a generation because of crossover and mutation. Therefore, we may decide to transfer them directly to the next generation without participating in any crossover or mutation. However, in contrast to being closer to the optimal value, they are quite useful in pulling the populations toward the optimal value through crossover and mutation. Therefore, a trade off has been taken between these two contradictory requirements by sending 10 percent of the elite solutions directly into the next generation and allowing the remaining 90 percent of the elite solutions to participate in crossover and mutation.

10.7.4.4 Case Study–Optimization of STI&M

In Sect. 10.7.3, a risk-based approach with a single objective function that addresses only system unavailability optimization or parametric study is discussed. The formalism of the earlier procedure is not tailored for including the other two parameters, viz., resource cost and dose consumption, in one simulation. As mentioned earlier, the GA approach is very effective for multiobjective function optimization. This power of the GA is derived from the random functions approach, which is integral to its core elements, viz., selection, mutation and crossover. The major objective of this case study was to demonstrate the power of GA with real-parameter considerations for a similar work on the nuclear plant Dhruva at BARC Mumbai. In fact, in this case study, the authors have developed models for unavailability, which is expected to be effective for unavailability considerations. The only difference compared to the previous case was that here, the objective was to demonstrate results for multiobjective function considerations, so the unavailability model was not extended to the plant level, i.e., evaluation of core damage frequency evaluation. Of course, at the system level, the statement of unavailability itself is a metric for risk.

Here, the consideration of random failure as testing or system outage also forms part of the study. The advantage is that optimized testing time enables estimation of Allowable Outage Time (AOT) as part of technical specification optimization. Therefore, in this chapter, the scope of evaluation does not remain limited to surveillance test/maintenance interval optimization but also includes 'allowable outage time' optimization, although it has not been explicitly cited in this chapter thus far. Now, the original considerations are that for this case, testing is also carried out during reactor shutdown with the system in a depressurized state. Therefore, the question is whether the emergency core cooling system is required or the system can be reverted back to normal standby mode to meet the design safety objective, i.e., injection in emergency mode either immediately for major LOCA or at least 30 min into the initiation of loss of inventory stipulated for minor LOCA conditions. The reference testing time or outage time for the test was assumed to be 24 h. However, with plants in depressurized conditions, the possibility of major LOCAs is highly improbable. Furthermore, the insight on man-rem consumption for the identified scenario as also resources required (along with aging contribution due to excessive testing) can be additional input in support of decisions.

10.7.4.5 Results and Discussion

Using the procedure, the three parameters unavailability, man-rem consumption, and cost have been optimized with respect to STI&M. All the failure probability data for the system components along with crossover and mutation probabilities of 0.6 and 0.01, respectively; a population size of 30 is used as input. The GA algorithm was able to find the optimized value of STI&M uniformly over the Pareto front. With the enumeration method, it has been that cost minimizes at STI&M value of 9490. Minimum man-rem consumption occurs at a possible value of 10,000 h. On the other hand, if outage time is included per STI as part of the unavailability calculations where it is assumed that the plant is under normal operation during testing, then unavailability minimizes at STI value 7685 with unavailability at 5.35E−03, whereas if the testing time/STI term is excluded from the unavailability calculation assuming that the plant is in the shutdown state during testing, then the minimum unavailability at STI value 231 is 6.73E−5. The results of this analysis considering cases 1 and 2 are shown in Table 10.4. The results for cases 1 and 2 have also been shown graphically to obtain a visual feel of the three objective functions. The results are shown in Figs. 10.9, 10.10, 10.11 and 10.12. In each graph, the man-rem consumption and unavailability are shown on the x-axis and y-axis, respectively. Each unit on the x-axis represents 40 milli-rem, while cost is represented by the intensity of color; the greater the cost is, the higher the intensity of the representative data point.

The results are given in terms of cost, availability and man-rem consumption. The algorithm is designed to find a solution uniformly over the Pareto front employing ranking base tournament selection with diversification based on the crowding distance approach, whereas stochastic remainder roulette gives clustering in some region of the objective function space.

For visualization purposes, the proposed solutions are represented graphically in the objective function space. All the points searched by the algorithm are shown in the objective function space along with the proposed solutions highlighted in blue. The results using both stochastic remainder roulette wheel selection and ranking-based tournament selection with diversification based on crowding distance of the solutions have been shown separately. Figure 10.9 shows the results for case 1 with ranking-based tournament selection with diversification based on the crowding distance of the solution. Figure 10.10 shows the solution for case 1 of Table 10.4 employing stochastic remainder roulette wheel selection. Figures 10.11 and 10.12 show the results for case II employing ranking-based tournament selection with diversification-based crowding distance and stochastic remainder roulette wheel selection. The STI&M values where cost maximizes and minimizes along with the proposed optimized STI values have also been shown in the graph.

It may be noted that simulation was performed with crowding distance-based tournament selection and stochastic remainder roulette wheel selection for excluding testing time and including testing time along with random unavailability, as shown in Figs. 10.9 and 10.10, respectively.

Table 10.4 Proposed STI values for cases 1 and 2

Case proposed STI values using crowding distance-based tournament selection
1. 9662, 237, 237, 259, 259, 274, 374, 308, 308, 2561, 1795, 538, 538, 797, 3652, 630, 630, 736, 410, 410, 1123, 6569, 8516, 8516, 5052, 4122, 1284, 5851, 5851
2. 9491, 9008, 9933, 9972, 9972, 9972, 8503, 8503, 8845, 8845, 9228, 7652, 7652, 7791, 912, 8101, 8101, 8873, 8647, 8647, 9594, 9594, 9156, 9781, 9781, 9781, 9206, 9158, 9046, 9046

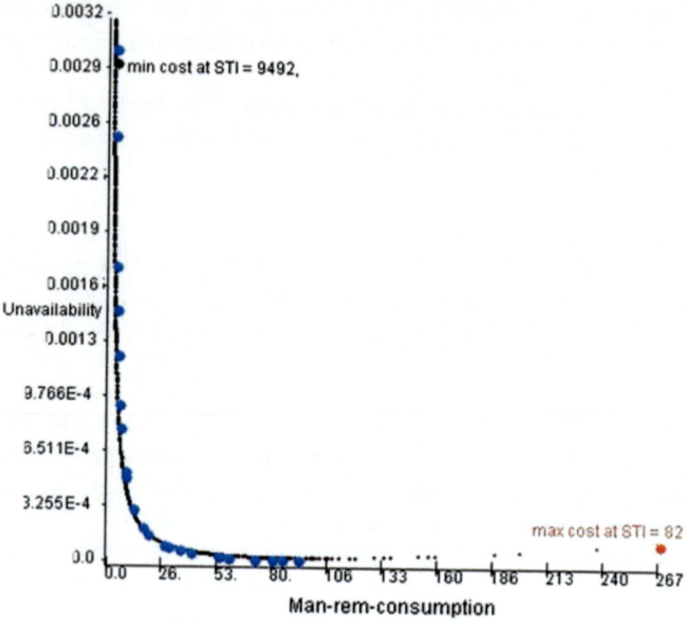

Fig. 10.9 Case 1 (excluding testing time/STI) with crowding distance-based tournament selection

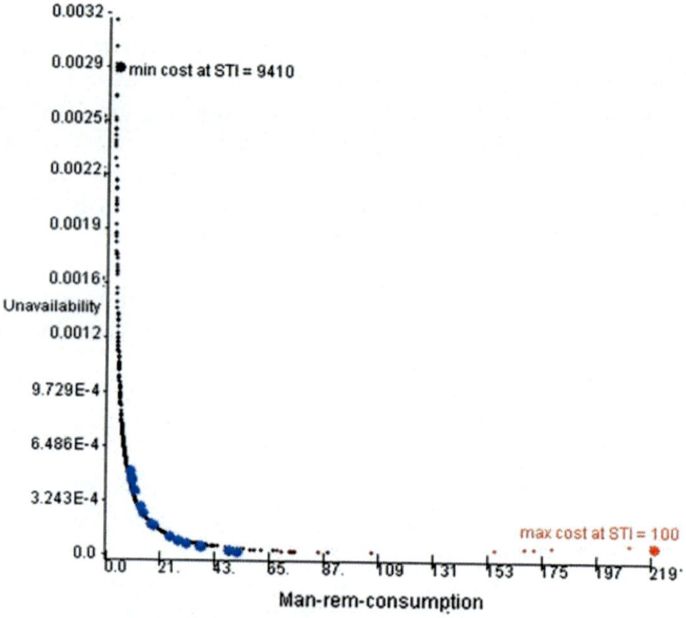

Fig. 10.10 Case 1 (excluding testing time/STI) with stochastic remainder roulette wheel selection

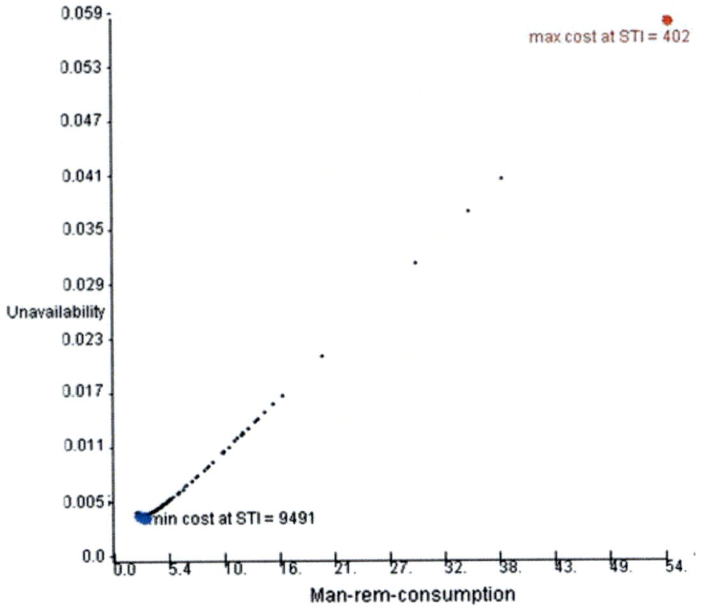

Fig. 10.11 Case 1 (including testing time/STI) with crowding distance-based tournament selection

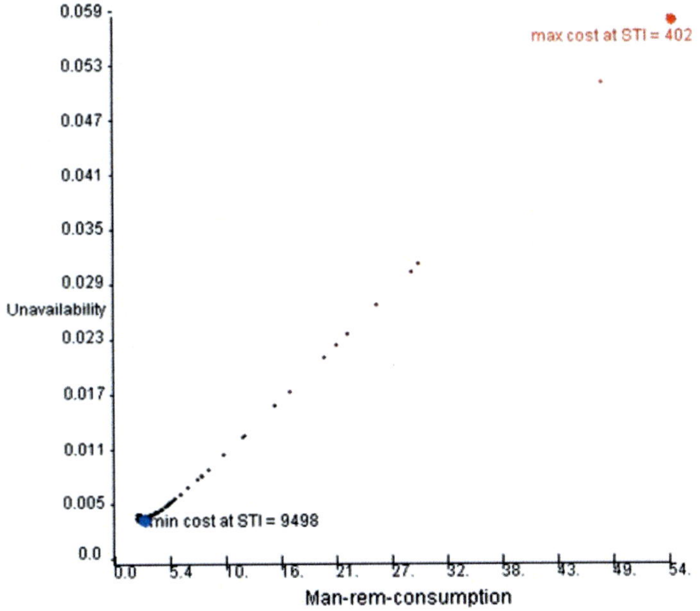

Fig. 10.12 Case 2 (including testing time/STI) with stochastic remainder roulette wheel selection

10.8 Critical Aspects of RCCM

Based on the above discussions in this chapter, the major factors that directly impact the risk-conscious maintenance management program are summarized as follows.

10.8.1 An Integrated Maintenance Management

As discussed in Sect. 10.3, there are various approaches to maintenance management; however, there is a growing realization that for effective maintenance management, an integrated approach where it is possible to have one major approach, such as preventive maintenance, while other approaches, such as condition-based, predictive-based, prognostics and health management approaches, can also form part of the overall maintenance strategy. As the general observation in any industry is that by addressing the 20% higher risk SSCs, 80% of the risk coverage through maintenance programs is possible. As such, every SSC that supports the management objective is important; however, there is some 5–10% SSC that can afford breakdown, but immediate action is required to correct the deficiency. For such SSCs, breakdown maintenance could be an affordable strategy in place of preventive maintenance. The time and effort saved can be allocated to SSCs with higher risk importance. Only an integrated approach organized by employing risk-significant metrics that enables the prioritization of SSCs can effectively generate the risk-consciousness of maintenance staff as well as in its management. For highly safety critical components, for example, risk-critical structural systems, risk-based inspections and prognostics can be implemented such that early warning signals/feedback are available for implementing corrective action programs. For integration, the organizational consciousness should be developed through interactions, deliberations, supervision, lectures and on subjects related to plant maintenance findings, initiatives, the deployment of new tools and methods and, most importantly, the types of maintenance errors, especially common cause failures that are encountered, and the provision of training programs to keep the maintenance staff highly alert, aware and dedicated in their duty.

10.8.2 Human Factor Considerations

Human error is the major contributing factor to maintenance- and testing-related errors. This observation can be argued to be not so. However, if root cause analysis is performed many times, a scene will emerge in which many failures could be related to organizational culture, such as lack of training and qualification programs, quality assurance in spare, hardware, inadequate supervisory services, improper or inadequate maintenance issues, ignoring the deviation or precursor to faults or failures and regulatory compliance, which are also part of human error. Recognizing this aspect that human error is a critical component of maintenance management will benefit the organization in terms of proactive rather than reactive approaches to maintenance management. The staff and management's high level of consciousness with respect to operating machinery SSCs health, issues of aging, training requirements for staff, and having a system of monitoring and feedback might to a great extent reduce the weak pockets in maintenance management. Training of staff in probabilistic risk assessment (PRA) models, methods and data will enable them to think in terms of the risk importance of 'what they do'. The implementation of a human root cause analysis program and online feedback will help manage what is lacking at the individual and organizational levels. Furthermore, the implementation of a systematic precursor analysis program through a structured framework, i.e., PRA supported, and through general consciousness during plant walkdowns, site visits and organizational provision to track and manage this program has significant potential for higher risk reduction. This is because this program will increase the level of awareness and alertness of staff regarding any deviation or near miss that eventually has the potential to become a fault or failure with consequences that can be avoided, or the level of severity can be brought down.

10.8.3 Common Cause Failure

However, safety critical systems design and operations are based on the defense-in-depth philosophy, where the provision of redundancy, diversity, single failure criteria, etc., forms fundament to ensure safety and business objectives. It must be understood that for CCF to become a reality, a coupling mechanism is needed, apart from the fault itself. The coupling factor could be an increase in humidity in the area, flooding, contaminated lubricant oil added to redundant units and human error that disabled all the redundant units, etc. However, common cause failure (CCF) occupies a critical dimension, as CCF can knock off the provisions of redundancy and at times even diversity due to the common effect fail more than one similar redundant provision and disable the safety function. The analysis of risk modeling provides in variable one insight that CCF for a group of components in any safety system contributes in terms of high-risk ranking. Often, human error has been found to contribute to CCF. A robust and validated design that postulates all possible common cause contributing factors works to reduce the CCF during operation. Furthermore, strong surveillance and monitoring and management programs can reduce the chances of CCF. Common cause contributors, such as calibration error, maintenance errors committed by the same crew or the same tools, equipment and quality-related common issues in spare, equipment and services, are some of the CCF contributors. Even communication or documentation aspects might lead to CCF. At a certain level of flooding of an area, seismic or environmental impacts in an area, corrosion, material degradation issues and aging might also appear as CCF.

The design provision to have defense against CCF is separation of equipment so that fault does not propagate to nearby redundant system, location, i.e., locating equipment such that flooding of area does not knock of redundant provisions. In maintenance, more than staff performing work on safety critical systems, the probability of lapse or mistake by having two levels can be reduced. Lack of knowledge in isolating and putting back in-service might lead to CCF due to blocking the services of a system or redundant equipment. Having provision of audio and visual alarms/trips such that unintended fault can reveal in advance, particularly the passive system status provision, can save many CCF events. One strong management tool is having, like safety analysis, a document on common cause failure that is revised based on the plant operations history and feedback on design, precursor potentials. Design upgrades in the form of the diversity of the same service from more than one source is a recognized approach to avoid CCF.

10.8.4 Dynamic Quotient in Maintenance Management

Generally, the maintenance management program follows the static characteristics in terms of surveillance maintenance and corrective action activities that have traditionally worked well. These were often based on previous experience or postulating the occurrence of certain modes and mechanisms and their qualitative frequencies based on experience and adopted as metrics as low, medium, high and very high for the periodicity of maintenance.

Here, it may be noted that there is a growing interest in implementation or risk-based maintenance management approaches that are dynamic in nature. The updated frequencies of the SSCs in the database provide the ground for implementing need-based maintenance. Furthermore, condition-based maintenance and for passive structures, the inspection or surveillance frequencies can be determined based on the risk potential of the safety system and reliability estimates of process systems. The component database provides information about the frequency as well as the modes of failure, and the root cause analysis will provide the need for the failure mechanism. The periodic input on these three aspects makes the program risk-based or for process systems reliability based maintenance management programs. The implementation of risk/reliability centered maintenance requires management provisions in terms of policy, authority, procurements and feedback provisions. Apart from this, performance metrics form part of the risk-based maintenance management program. The availability of a PRA is an asset for developing a risk-based management program, as most of the input for identification and prioritization comes from a living level 1 PRA of the plant.

10.8.5 Adoption of Advanced Technology

Digital technology is being deployed in many engineering and industrial systems, particularly safety critical systems, to improve organizational performance. The major features of RCCM are a) availability of a plant simulator in support of training and qualification program to understand the dynamic aspects of the plant, b) availability of a risk monitor in the training section or control room of a plant and c) computerized maintenance management and maintainer advisory system.

Training on plant simulator and equipment crosssection models and an evaluation system is critical to enhance maintenance management effectiveness. The training on equipment isolation requirements, the protocol of equipment taking over and reverting, and the appreciation of operational requirements for any maintenance activity form a routine but important component of plant operational ecosystems. Requirement of communication between the maintenance team through modern aids such as voice and video systems (which also forms a record), use of auto and alignment features such as the use of laser-assisted alignment, and online handheld aid to record observation/data with downloading features should form an integral part of surveillance and monitoring.

Use of automation features in the plant for various activities has been found to be effective in reducing the fault.

References

American Petrolium Institute (2016) Risk-Based Inspection. API 580, API

Arunraj NS, Maity J (2007) Risk-based maintenance—techniques and applications. Hazardous Mater 142:543–661

Computers and Industrial Engineering (2012) An overview of time-based and condition-based maintenance in industrial application. Computers and Industrial Engineering, vol 63, pp 135–149

Consultdss.com (2018) Risk awareness is key to operational discipline [Online]. Available: https://www.consultdss.com/risk-awarness. Accessed 21 Aug 2021

Chrleslandry.com (2021) Risk consciousness and the creation of liveable cities [Online]. Available: Chrleslandry.com/panel/wp-content/uploads/2013/Risk-consciousness.pdf. Accessed 10 Aug 2021

Chowdhury S, Varde PV (2010) Surveillance test interval optimization for nuclear plants using real parameter genetic algorithms. In: 2010 2nd international conference on reliability, safety and hazard—risk-based technologies and physics-of-failure methods (ICRESH), Mumbai

Chowdhury S, Varde PV (2011) Surveillance test interval optimization for nuclear plants using multi objective real parameter genetic algorithms. Int J Qual Reliab Saf Eng 10(2):159–177

Coble J, Ramuhalli P, Bond L, Hines JW, Upadhyaya B (2015) A review of prognostics and health management application in nuclear power plant. Int Jr Prognost. Health Manage. ISSN 2153-2648,008

Deb K (2001) Multi-objective optimization using evolutionary algorithms. Wiley

Fonseca CM, Fleming PJ (1993) Multi-objective genetic algorithms. In: IEEE colloquium on genetic algorithms for control system engineering, London

Garcia SG, Garcia MG (2019) Industry 4.0 implications in production and maintenance management: an overview. Procedia Manuf 41:415–422

IEEE Standard Association (2017) IEEE standard framework for prognostics and health management. IEEE Std 1856. IEEE, New York, USA

International Atomic Energy Agency (1993) Risk-based optimization of technical specification for operations of nuclear power plants, IAEA-TECDOC-729. IAEA, Vienna

International Atomic Energy Agency (1996) Defence in depth in nuclear safety, INSAG-10. IAEA, Vienna
International Atomic Energy Agency (2005) Risk informed regulation of nuclear facilities: overview of the current status, TECDOC-1436. IAEA, Vienna
International Atomic Energy Agency (2006) Maintenance, periodic testing and inspection of research reactors, safety guide series no. NS-G-4.2. IAEA, Vienna
International Atomic Energy Agency (2013) Advanced surveillance, diagnostics and prognostic techniques in monitoring structure, systems and components in nuclear power plants. IAEA, Vienna
International Atomic Energy Agency (2018) IAEA safety glossary, 2018th edn. IAEA, Vienna
Krishnanand KR (2009) Comparative study of five bio-inspired evolutionary optimization techniques. In: World Congress on Nature & Biologically Inspired Computing (NaBIC), ND, USA
Khairmode AU, Joshi NS, Varde PV, Buktar R (2016) PLM integrated risk informed in-service inspection in research reactors. Int J Eng Res Technol (IJERT) 5(12)
Khan I, Haddara M (2003) Risk-based maintenance: a quantattive approach for maintenance/inspection scheduling and planning. Loss Prevent. Process Indus. 16:561–537
Labib A, Kobbacy KH, Murthy D (2008) Computerised maintenance management system. In: Complex System Maintenance Handbook
Nkosi M, Gupta K, Mashinini (2020) Cause and impact of human error in maintenance of mechanical systems. In: MATEC Web of Conference 312,05001
Nowlan FS, Heap HF (1978) Reliability centered maintenance, A066-579. Office of the Assistant Secretary of Defence, United Airlines, USA
Pecht MG (2008) Prognostics and health management of electronics. Wiley (2008)
Poor P, Zenisek D, Basl J (2019) Historical overview of maintenance management strategies: development from breakdown maintenance to predictive maintenance in accordance with industrial revolutions. In: International Conference on Industrial Engineering and Operations Management, Oilsen, Czech Republic (2019)
Prognostics and Health Management (2012) Role of prognostics in support of risk-based engineering in nuclear plants. In: Prognostics and health management. ISSN 2153-2648,008
Prabhupada AB (1998) Selected verses from vedic scriptures (Adi literature). Bhaktivedanta Book Trust, Mumbai (1998)
Rodrigues LR, Yoneyama T, Mascimento CL (2012) How aircraft operations can benefit from PHM techniques. In: IEEE aerospace conference proceedings (2012). https://doi.org/10.1109/AERO.2012.6187376
Rudolph (1998) On a multi–objective evolutionary algorithm and its convergence to the Pareto set. In: Proceedings of the 1998 IEEE international conference on evolutionary computation
Varde P (2021) Risk-consciousness operations management for nuclear plants. In: American Nuclear Society Probabilistic Safety Assessment—PSA 21, 7–12 Nov 2021, Virtual Meeting, Columbus, USA
Varde PV, Pecht MG (2018) Risk-based Engineering—an integrated approach to complex system: special reference to nuclear plants. Springer, Singapore
Vishnu C, Regikumar V (2016) Reliability based maintenance strategy selection in process plants: a case study. In: Procedia technology 25—global colloquium in recent advancement and effectual researches in engineering science and technology RAEREST 2016
Wellington AS, Vasconcelos VD, Rabello EG (2015) Risk-based inspection in the context of nuclear power plants. In: 2015 International Nuclear Atlantic Conference—INAC 2015, Sao Palo, Brazil
Zitzler E, Deb K, Thiele L (2000) Comparision of multi-objective evolutionary algorithms: empirical results. IEEE Trans Evolut Comput 8(2):173–195

Annexure A
Distributions: A-1 Normal, A-2 F and A-3 Chi-square

See Tables A.1, A.2 and A.3.

P. V. Varde, *Risk-Conscious Operations Management*, Risk, Reliability and Safety Engineering, https://doi.org/10.1007/978-981-19-9334-3

Table A.1 Cumulative standard normal distribution

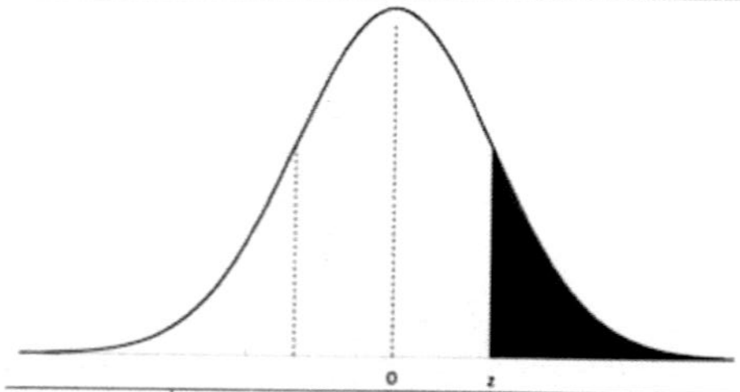

z	0.00	0.01	0.02	0.03	0.04	0.05	0.06	0.07	0.08	0.09
0.0	0.5000	0.4960	0.4920	0.4880	0.4840	0.4801	0.4761	0.4721	0.4681	0.4641
0.1	0.4602	0.4562	0.4522	0.4483	0.4443	0.4404	0.4364	0.4325	0.4286	0.4247
0.2	0.4207	0.4168	0.4129	0.4090	0.4052	0.4013	0.3974	0.3936	0.3897	0.3859
0.3	0.3821	0.3783	0.3745	0.3707	0.3669	0.3632	0.3594	0.3557	0.3520	0.3483
0.4	0.3446	0.3409	0.3372	0.3336	0.3300	0.3264	0.3228	0.3192	0.3156	0.3121
0.5	0.3085	0.3050	0.3015	0.2981	0.2946	0.2912	0.2877	0.2843	0.2810	0.2776
0.6	0.2743	0.2709	0.2676	0.2643	0.2611	0.2578	0.2546	0.2514	0.2483	0.2451
0.7	0.2420	0.2389	0.2358	0.2327	0.2296	0.2266	0.2236	0.2206	0.2177	0.2148
0.8	0.2119	0.2090	0.2061	0.2033	0.2005	0.1977	0.1949	0.1922	0.1894	0.1867
0.9	0.1841	0.1814	0.1788	0.1762	0.1736	0.1711	0.1685	0.1660	0.1635	0.1611
1.0	0.1587	0.1562	0.1539	0.1515	0.1492	0.1469	0.1446	0.1423	0.1401	0.1379
1.1	0.1357	0.1335	0.1314	0.1292	0.1271	0.1251	0.1230	0.1210	0.1190	0.1170
1.2	0.1151	0.1131	0.1112	0.1093	0.1075	0.1056	0.1038	0.1020	0.1003	0.0985
1.3	0.0968	0.0951	0.0934	0.0918	0.0901	0.0885	0.0869	0.0853	0.0838	0.0823
1.4	0.0808	0.0793	0.0778	0.0764	0.0749	0.0735	0.0721	0.0708	0.0694	0.0681
1.5	0.0668	0.0655	0.0643	0.0630	0.0618	0.0606	0.0594	0.0582	0.0571	0.0559
1.6	0.0548	0.0537	0.0526	0.0516	0.0505	0.0495	0.0485	0.0475	0.0465	0.0455
1.7	0.0446	0.0436	0.0427	0.0418	0.0409	0.0401	0.0392	0.0384	0.0375	0.0367

(continued)

Table A.1 (continued)

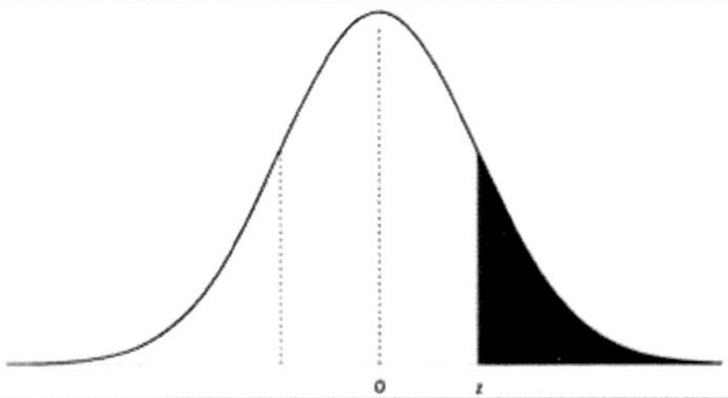

z	0.00	0.01	0.02	0.03	0.04	0.05	0.06	0.07	0.08	0.09
1.8	0.0359	0.0351	0.0344	0.0336	0.0329	0.0322	0.0314	0.0307	0.0301	0.0294
1.9	0.0287	0.0281	0.0274	0.0268	0.0262	0.0256	0.0250	0.0244	0.0239	0.0233
2.0	0.0228	0.0222	0.0217	0.0212	0.0207	0.0202	0.0197	0.0192	0.0188	0.0183
2.1	0.0179	0.0174	0.0170	0.0166	0.0162	0.0158	0.0154	0.0150	0.0146	0.0143
2.2	0.0139	0.0136	0.0132	0.0129	0.0125	0.0122	0.0119	0.0116	0.0113	0.0110
2.3	0.0107	0.0104	0.0102	0.0099	0.0096	0.0094	0.0091	0.0089	0.0087	0.0084
2.4	0.0082	0.0080	0.0078	0.0075	0.0073	0.0071	0.0069	0.0068	0.0066	0.0064
2.5	0.0062	0.0060	0.0059	0.0057	0.0055	0.0054	0.0052	0.0051	0.0049	0.0048
2.6	0.0047	0.0045	0.0044	0.0043	0.0041	0.0040	0.0039	0.0038	0.0037	0.0036
2.7	0.0035	0.0034	0.0033	0.0032	0.0031	0.0030	0.0029	0.0028	0.0027	0.0026
2.8	0.0026	0.0025	0.0024	0.0023	0.0023	0.0022	0.0021	0.0021	0.0020	0.0019
2.9	0.0019	0.0018	0.0018	0.0017	0.0016	0.0016	0.0015	0.0015	0.0014	0.0014
3.0	0.0013	0.0013	0.0013	0.0012	0.0012	0.0011	0.0011	0.0011	0.0010	0.0010
3.1	0.0010	0.0009	0.0009	0.0009	0.0008	0.0008	0.0008	0.0008	0.0007	0.0007
3.2	0.0007	0.0007	0.0006	0.0006	0.0006	0.0006	0.0006	0.0005	0.0005	0.0005
3.3	0.0005	0.0005	0.0005	0.0004	0.0004	0.0004	0.0004	0.0004	0.0004	0.0003
3.4	0.0003	0.0003	0.0003	0.0003	0.0003	0.0003	0.0003	0.0003	0.0003	0.0002
3.5	0.0002	0.0002	0.0002	0.0002	0.0002	0.0002	0.0002	0.0002	0.0002	0.0002

Table A.2 95th Percentile Values for the F Distribution

$f2$	f1									
	1	2	3	4	5	6	7	8	9	10
1	161	200	216	225	230	234	237	239	241	242
2	18.51	19.00	19.16	19.25	19.30	19.33	19.35	19.37	19.38	19.40
3	10.13	9.55	9.28	9.12	9.01	8.94	8.89	8.85	8.81	8.79
4	7.71	6.94	6.59	6.39	6.26	6.16	6.09	6.04	6.00	5.96
5	6.61	5.79	5.41	5.19	5.05	4.95	4.88	4.82	4.77	4.74
6	5.99	5.14	4.76	4.53	4.39	4.28	4.21	4.15	4.10	4.06
7	5.59	4.74	4.35	4.12	3.97	3.87	3.79	3.73	3.68	3.64
8	5.32	4.46	4.07	3.84	3.69	3.58	3.50	3.44	3.39	3.35
9	5.12	4.26	3.86	3.63	3.48	3.37	3.29	3.23	3.18	3.14
10	4.96	4.10	3.71	3.48	3.33	3.22	3.14	3.07	3.02	2.98
11	4.84	3.98	3.59	3.36	3.20	3.09	3.01	2.95	2.90	2.85
12	4.75	3.89	3.49	3.26	3.11	3.00	2.91	2.85	2.80	2.75
13	4.67	3.81	3.41	3.18	3.03	2.92	2.83	2.77	2.71	2.67
14	4.60	3.74	3.34	3.11	2.96	2.85	2.76	2.70	2.65	2.60
15	4.54	3.68	3.29	3.06	2.90	2.79	2.71	2.64	2.59	2.54
16	4.49	3.63	3.24	3.01	2.85	2.74	2.66	2.59	2.54	2.49
17	4.45	3.59	3.20	2.96	2.81	2.70	2.61	2.55	2.49	2.45
18	4.41	3.55	3.16	2.93	2.77	2.66	2.58	2.51	2.46	2.41
19	4.38	3.52	3.13	2.90	2.74	2.63	2.54	2.48	2.42	2.38
20	4.35	3.49	3.10	2.87	2.71	2.60	2.51	2.45	2.39	2.35
21	4.32	3.47	3.07	2.84	2.68	2.57	2.49	2.42	2.37	2.32
22	4.30	3.44	3.05	2.82	2.66	2.55	2.46	2.40	2.34	2.30
23	4.28	3.42	3.03	2.80	2.64	2.53	2.44	2.37	2.32	2.27

(continued)

Table A.2 (continued)

f2	f1									
	1	2	3	4	5	6	7	8	9	10
24	4.26	3.40	3.01	2.78	2.62	2.51	2.42	2.36	2.30	2.25
25	4.24	3.39	2.99	2.76	2.60	2.49	2.40	2.34	2.28	2.24
26	4.23	3.37	2.98	2.74	2.59	2.47	2.39	2.32	2.27	2.22
27	4.21	3.35	2.96	2.73	2.57	2.46	2.37	2.31	2.25	2.20
28	4.20	3.34	2.95	2.71	2.56	2.45	2.36	2.29	2.24	2.19
29	4.18	3.33	2.93	2.70	2.55	2.43	2.35	2.28	2.22	2.18
30	4.17	3.32	2.92	2.69	2.53	2.42	2.33	2.27	2.21	2.16
40	4.08	3.23	2.84	2.61	2.45	2.34	2.25	2.18	2.12	2.08
60	4.00	3.15	2.76	2.53	2.37	2.25	2.17	2.10	2.04	1.99
120	3.92	3.07	2.68	2.45	2.29	2.18	2.09	2.02	1.96	1.91
Inf	3.84	3.00	2.60	2.37	2.21	2.10	2.01	1.94	1.88	1.83

f2	f1						
15	20	24	30	40	60	120	Inf
246	248	249	250	251	252	253	254
19.43	19.45	19.45	19.46	19.47	19.48	19.49	19.50
8.70	8.66	8.64	8.62	8.59	8.57	8.55	8.53
5.86	5.80	5.77	5.75	5.72	5.69	5.66	5.63
4.62	4.56	4.53	4.50	4.46	4.43	4.40	4.36
3.94	3.87	3.84	3.81	3.77	3.74	3.70	3.67
3.51	3.44	3.41	3.38	3.34	3.30	3.27	3.23
3.22	3.15	3.12	3.08	3.04	3.01	2.97	2.93
3.01	2.94	2.90	2.86	2.83	2.79	2.75	2.71
2.85	2.77	2.74	2.70	2.66	2.62	2.58	2.54

(continued)

Table A.2 (continued)

f2	f1						
15	20	24	30	40	60	120	Inf
2.72	2.65	2.61	2.57	2.53	2.49	2.45	2.40
2.62	2.54	2.51	2.47	2.43	2.38	2.34	2.30
2.53	2.46	2.42	2.38	2.34	2.30	2.25	2.21
2.46	2.39	2.35	2.31	2.27	2.22	2.18	2.13
2.40	2.33	2.29	2.25	2.20	2.16	2.11	2.07
2.35	2.28	2.24	2.19	2.15	2.11	2.06	2.01
2.31	2.23	2.19	2.15	2.10	2.06	2.01	1.96
2.27	2.19	2.15	2.11	2.06	2.02	1.97	1.92
2.23	2.16	2.11	2.07	2.03	1.98	1.93	1.88
2.20	2.12	2.08	2.04	1.99	1.95	1.90	1.84
2.18	2.10	2.05	2.01	1.96	1.92	1.87	1.81
2.15	2.07	2.03	1.98	1.94	1.89	1.84	1.78
2.13	2.05	2.01	1.96	1.91	1.86	1.81	1.76
2.11	2.03	1.98	1.94	1.89	1.84	1.79	1.73
2.09	2.01	1.96	1.92	1.87	1.82	1.77	1.71
2.07	1.99	1.95	1.90	1.85	1.80	1.75	1.69
2.06	1.97	1.93	1.88	1.84	1.79	1.73	1.67
2.04	1.96	1.91	1.87	1.82	1.77	1.71	1.65
2.03	1.94	1.90	1.85	1.81	1.75	1.70	1.64
2.01	1.93	1.89	1.84	1.79	1.74	1.68	1.62
1.92	1.84	1.79	1.74	1.69	1.64	1.58	1.51
1.84	1.75	1.70	1.65	1.59	1.53	1.47	1.39
1.75	1.66	1.61	1.55	1.50	1.43	1.35	1.25
1.67	1.57	1.52	1.46	1.39	1.32	1.22	1.00

f1 Numerator degrees of freedom
f2 Denominator degrees of freedom

Table A.3 Chi-square distribution

Chi-square distribution									
Degrees of	α								
Freedom	0.995	0.990	0.975	0.900	0.100	0.050	0.025	0.010	0.005
1	0.0000	0.0002	0.0010	0.0158	2.7055	3.8415	5.0239	6.6349	7.8794
2	0.0100	0.0201	0.0506	0.2107	4.6052	5.9915	7.3778	9.2103	10.5966
3	0.0717	0.1148	0.2158	0.5844	6.2514	7.8147	9.3484	11.3449	12.8382
4	0.2070	0.2971	0.4844	1.0636	7.7794	9.4877	11.1433	13.2767	14.8603
5	0.4117	0.5543	0.8312	1.6103	9.2364	11.0705	12.8325	15.0863	16.7496
6	0.6757	0.8721	1.2373	2.2041	10.6446	12.5916	14.4494	16.8119	18.5476
7	0.9893	1.2390	1.6899	2.8331	12.0170	14.0671	16.0128	18.4753	20.2777
8	1.3444	1.6465	2.1797	3.4895	13.3616	15.5073	17.5345	20.0902	21.9550
9	1.7349	2.0879	2.7004	4.1682	14.6837	16.9190	19.0228	21.6660	23.5894
10	2.1559	2.5582	3.2470	4.8652	15.9872	18.3070	20.4832	23.2093	25.1882
11	2.6032	3.0535	3.8157	5.5778	17.2750	19.6751	21.9200	24.7250	26.7568
12	3.0738	3.5706	4.4038	6.3038	18.5493	21.0261	23.3367	26.2170	28.2995
13	3.5650	4.1069	5.0088	7.0415	19.8119	22.3620	24.7356	27.6882	29.8195
14	4.0747	4.6604	5.6287	7.7895	21.0641	23.6848	26.1189	29.1412	31.3193
15	4.6009	5.2293	6.2621	8.5468	22.3071	24.9958	27.4884	30.5779	32.8013
16	5.1422	5.8122	6.9077	9.3122	23.5418	26.2962	28.8454	31.9999	34.2672
17	5.6972	6.4078	7.5642	10.0852	24.7690	27.5871	30.1910	33.4087	35.7185
18	6.2648	7.0149	8.2307	10.8649	25.9894	28.8693	31.5264	34.8053	37.1565
19	6.8440	7.6327	8.9065	11.6509	27.2036	30.1435	32.8523	36.1909	38.5823
20	7.4338	8.2604	9.5908	12.4426	28.4120	31.4104	34.1696	37.5662	39.9968
21	8.0337	8.8972	10.2829	13.2396	29.6151	32.6706	35.4789	38.9322	41.4011
22	8.6427	9.5425	10.9823	14.0415	30.8133	33.9244	36.7807	40.2894	42.7957
23	9.2604	10.1957	11.6886	14.8480	32.0069	35.1725	38.0756	41.6384	44.1813

(continued)

Table A.3 (continued)

Chi-square distribution

24	9.8862	10.8564	12.4012	15.6587	33.1962	36.4150	39.3641	42.9798	45.5585
25	10.5197	11.5240	13.1197	16.4734	34.3816	37.6525	40.6465	44.3141	46.9279
26	11.1602	12.1981	13.8439	17.2919	35.5632	38.8851	41.9232	45.6417	48.2899
27	11.8076	12.8785	14.5734	18.1139	36.7412	40.1133	43.1945	46.9629	49.6449
28	12.4613	13.5647	15.3079	18.9392	37.9159	41.3371	44.4608	48.2782	50.9934
29	13.1211	14.2565	16.0471	19.7677	39.0875	42.5570	45.7223	49.5879	52.3356
30	13.7867	14.9535	16.7908	20.5992	40.2560	43.7730	46.9792	50.8922	53.6720
31	14.4578	15.6555	17.5387	21.4336	41.4217	44.9853	48.2319	52.1914	55.0027
32	15.1340	16.3622	18.2908	22.2706	42.5847	46.1943	49.4804	53.4858	56.3281
33	15.8153	17.0735	19.0467	23.1102	43.7452	47.3999	50.7251	54.7755	57.6484
34	16.5013	17.7891	19.8063	23.9523	44.9032	48.6024	51.9660	56.0609	58.9639
35	17.1918	18.5089	20.5694	24.7967	46.0588	49.8018	53.2033	57.3421	60.2748
36	17.8867	19.2327	21.3359	25.6433	47.2122	50.9985	54.4373	58.6192	61.5812
37	18.5858	19.9602	22.1056	26.4921	48.3634	52.1923	55.6680	59.8925	62.8833
38	19.2889	20.6914	22.8785	27.3430	49.5126	53.3835	56.8955	61.1621	64.1814
39	19.9959	21.4262	23.6543	28.1958	50.6598	54.5722	58.1201	62.4281	65.4756
40	20.7065	22.1643	24.4330	29.0505	51.8051	55.7585	59.3417	63.6907	66.7660
41	21.4208	22.9056	25.2145	29.9071	52.9485	56.9424	60.5606	64.9501	68.0527
42	22.1385	23.6501	25.9987	30.7654	54.0902	58.1240	61.7768	66.2062	69.3360
43	22.8595	24.3976	26.7854	31.6255	55.2302	59.3035	62.9904	67.4593	70.6159
44	23.5837	25.1480	27.5746	32.4871	56.3685	60.4809	64.2015	68.7095	71.8926
45	24.3110	25.9013	28.3662	33.3504	57.5053	61.6562	65.4102	69.9568	73.1661
46	25.0413	26.6572	29.1601	34.2152	58.6405	62.8296	66.6165	71.2014	74.4365
47	25.7746	27.4158	29.9562	35.0814	59.7743	64.0011	67.8206	72.4433	75.7041
48	26.5106	28.1770	30.7545	35.9491	60.9066	65.1708	69.0226	73.6826	76.9688
49	27.2493	28.9406	31.5549	36.8182	62.0375	66.3386	70.2224	74.9195	78.2307

(continued)

Table A.3 (continued)

Chi-square distribution									
50	27.9907	29.7067	32.3574	37.6886	63.1671	67.5048	71.4202	76.1539	79.4900
51	28.7347	30.4750	33.1618	38.5604	64.2954	68.6693	72.6160	77.3860	80.7467
52	29.4812	31.2457	33.9681	39.4334	65.4224	69.8322	73.8099	78.6158	82.0008
53	30.2300	32.0185	34.7763	40.3076	66.5482	70.9935	75.0019	79.8433	83.2526
54	30.9813	32.7934	35.5863	41.1830	67.6728	72.1532	76.1920	81.0688	84.5019
55	31.7348	33.5705	36.3981	42.0596	68.7962	73.3115	77.3805	82.2921	85.7490
56	32.4905	34.3495	37.2116	42.9373	69.9185	74.4683	78.5672	83.5134	86.9938
57	33.2484	35.1305	38.0267	43.8161	71.0397	75.6237	79.7522	84.7328	88.2364
58	34.0084	35.9135	38.8435	44.6960	72.1598	76.7778	80.9356	85.9502	89.4769
59	34.7704	36.6982	39.6619	45.5770	73.2789	77.9305	82.1174	87.1657	90.7153
60	35.5345	37.4849	40.4817	46.4589	74.3970	79.0819	83.2977	88.3794	91.9517
61	36.3005	38.2732	41.3031	47.3418	75.5141	80.2321	84.4764	89.5913	93.1861
62	37.0684	39.0633	42.1260	48.2257	76.6302	81.3810	85.6537	90.8015	94.4187
63	37.8382	39.8551	42.9503	49.1105	77.7454	82.5287	86.8296	92.0100	95.6493
64	38.6098	40.6486	43.7760	49.9963	78.8596	83.6753	88.0041	93.2169	96.8781
65	39.3831	41.4436	44.6030	50.8829	79.9730	84.8206	89.1771	94.4221	98.1051
66	40.1582	42.2402	45.4314	51.7705	81.0855	85.9649	90.3489	95.6257	99.3304
67	40.9350	43.0384	46.2610	52.6588	82.1971	87.1081	91.5194	96.8278	100.5540
68	41.7135	43.8380	47.0920	53.5481	83.3079	88.2502	92.6885	98.0284	101.7759
69	42.4935	44.6392	47.9242	54.4381	84.4179	89.3912	93.8565	99.2275	102.9962
70	43.2752	45.4417	48.7576	55.3289	85.5270	90.5312	95.0232	100.4252	104.2149
71	44.0584	46.2457	49.5922	56.2206	86.6354	91.6702	96.1887	101.6214	105.4320
72	44.8431	47.0510	50.4279	57.1129	87.7430	92.8083	97.3531	102.8163	106.6476
73	45.6293	47.8577	51.2648	58.0061	88.8499	93.9453	98.5163	104.0098	107.8617
74	46.4170	48.6657	52.1028	58.9000	89.9560	95.0815	99.6783	105.2020	109.0744
75	47.2060	49.4750	52.9419	59.7946	91.0615	96.2167	100.8393	106.3929	110.2856

(continued)

Table A.3 (continued)

Chi-square distribution

76	47.9965	50.2856	53.7821	60.6899	92.1662	97.3510	101.9993	107.5825	111.4954
77	48.7884	51.0974	54.6234	61.5858	93.2702	98.4844	103.1581	108.7709	112.7038
78	49.5816	51.9104	55.4656	62.4825	94.3735	99.6169	104.3159	109.9581	113.9109
79	50.3761	52.7247	56.3089	63.3799	95.4762	100.7486	105.4728	111.1440	115.1166
80	51.1719	53.5401	57.1532	64.2778	96.5782	101.8795	106.6286	112.3288	116.3211
81	51.9690	54.3566	57.9984	65.1765	97.6796	103.0095	107.7834	113.5124	117.5242
82	52.7674	55.1743	58.8446	66.0757	98.7803	104.1387	108.9373	114.6949	118.7261
83	53.5669	55.9931	59.6918	66.9756	99.8805	105.2672	110.0902	115.8763	119.9268
84	54.3677	56.8130	60.5398	67.8761	100.9800	106.3948	111.2423	117.0565	121.1263
85	55.1696	57.6339	61.3888	68.7772	102.0789	107.5217	112.3934	118.2357	122.3246
86	55.9727	58.4559	62.2386	69.6788	103.1773	108.6479	113.5436	119.4139	123.5217
87	56.7769	59.2790	63.0894	70.5810	104.2750	109.7733	114.6929	120.5910	124.7177
88	57.5823	60.1030	63.9409	71.4838	105.3722	110.8980	115.8414	121.7671	125.9125
89	58.3888	60.9281	64.7934	72.3872	106.4689	112.0220	116.9891	122.9422	127.1063
90	59.1963	61.7541	65.6466	73.2911	107.5650	113.1453	118.1359	124.1163	128.2989
91	60.0049	62.5811	66.5007	74.1955	108.6606	114.2679	119.2819	125.2895	129.4905
92	60.8146	63.4090	67.3556	75.1005	109.7556	115.3898	120.4271	126.4617	130.6811
94	62.4370	65.0677	69.0677	76.9119	111.9442	117.6317	122.7151	128.8032	133.0591
95	63.2496	65.8984	69.9249	77.8184	113.0377	118.7516	123.8580	129.9727	134.2465
96	64.0633	66.7299	70.7828	78.7254	114.1307	119.8709	125.0001	131.1412	135.4330
97	64.8780	67.5624	71.6415	79.6329	115.2232	120.9896	126.1414	132.3089	136.6186
98	65.6936	68.3957	72.5009	80.5408	116.3153	122.1077	127.2821	133.4757	137.8032
99	66.5101	69.2299	73.3611	81.4493	117.4069	123.2252	128.4220	134.6416	138.9868
100	67.3276	70.0649	74.2219	82.3581	118.4980	124.3421	129.5612	135.8067	140.1695
101	68.1459	70.9007	75.0835	83.2675	119.5887	125.4584	130.6997	136.9710	141.3513
102	68.9652	71.7374	75.9457	84.1773	120.6789	126.5741	131.8375	138.1345	142.5322

(continued)

Table A.3 (continued)

Chi-square distribution

103	69.7853	72.5748	76.8086	85.0875	121.7686	127.6893	132.9747	139.2971	143.7122
104	70.6064	73.4130	77.6722	85.9982	122.8580	128.8039	134.1112	140.4590	144.8913
105	71.4282	74.2520	78.5364	86.9093	123.9469	129.9180	135.2470	141.6201	146.0696
106	72.2509	75.0918	79.4013	87.8208	125.0354	131.0315	136.3822	142.7804	147.2470
107	73.0745	75.9323	80.2668	88.7327	126.1234	132.1444	137.5167	143.9400	148.4236
108	73.8989	76.7736	81.1329	89.6451	127.2111	133.2569	138.6506	145.0988	149.5994
109	74.7241	77.6156	81.9997	90.5579	128.2983	134.3688	139.7839	146.2569	150.7743
110	75.5500	78.4583	82.8671	91.4710	129.3851	135.4802	140.9166	147.4143	151.9485
111	76.3768	79.3017	83.7350	92.3846	130.4716	136.5911	142.0486	148.5710	153.1218
112	77.2044	80.1459	84.6036	93.2986	131.5576	137.7015	143.1801	149.7269	154.2944
113	78.0327	80.9907	85.4728	94.2129	132.6433	138.8114	144.3110	150.8822	155.4662
114	78.8618	81.8362	86.3425	95.1276	133.7286	139.9208	145.4413	152.0367	156.6373
115	79.6916	82.6824	87.2128	96.0427	134.8135	141.0297	146.5711	153.1906	157.8076
116	80.5221	83.5293	88.0837	96.9582	135.8980	142.1382	147.7002	154.3438	158.9771
117	81.3534	84.3768	88.9551	97.8740	136.9822	143.2461	148.8288	155.4964	160.1460
118	82.1854	85.2250	89.8271	98.7902	138.0660	144.3537	149.9569	156.6483	161.3141
119	83.0182	86.0738	90.6996	99.7067	139.1495	145.4607	151.0844	157.7995	162.4815
120	83.8516	86.9233	91.5726	100.6236	140.2326	146.5674	152.2114	158.9502	163.6482

Annexure B
Probability Plotting

Probability plotting is a graphical way of determining if a given data follows certain distribution or not. This is done by plotting the data on a probability paper of the distribution to be tested. Each distribution has its own probability graph paper.

Y-axis on the probability plot represents CDF.

Procedure for Probability Plotting

- Arrange the data in ascending order.
- Determine the cumulative distribution function using relation:

$$\mathrm{cdf} = \frac{R - 0.375}{n + 0.25}$$

  where n is the size of sample space, i.e., the number of data points.
- Plot the available data points against the CDF on the probability graph.
- Fit a straight line through the points on the graph. Perform regression analysis to determine suitability of the fitted line.

Normal Distribution

Consider following data available from some observations of a normally distributed random parameter.

$$n = 25$$

P. V. Varde, *Risk-Conscious Operations Management*, Risk, Reliability and Safety Engineering, https://doi.org/10.1007/978-981-19-9334-3

Data	Arranged data	Rank (R)	CDF
− 0.2298	− 2.24933	1	0.024752
− 0.57922	− 1.46891	2	0.064356
0.480481	− 0.94645	3	0.10396
− 0.38683	− 0.63206	4	0.143564
0.421616	− 0.57922	5	0.183168
1.087706	− 0.4385	6	0.222772
− 2.24933	− 0.38683	7	0.262376
1.80448	− 0.2298	8	0.30198
− 0.63206	− 0.21463	9	0.341584
1.316458	− 0.0584	10	0.381188
1.551592	0.176938	11	0.420792
− 1.46891	0.330885	12	0.460396
0.176938	0.343221	13	0.5
3.46626	0.421616	14	0.539604
− 0.21463	0.438645	15	0.579208
0.486283	0.480481	16	0.618812
0.330885	0.486283	17	0.658416
1.267902	1.087706	18	0.69802
1.090514	1.090514	19	0.737624
− 0.94645	1.267902	20	0.777228
− 0.4385	1.316458	21	0.816832
0.343221	1.551592	22	0.856436
− 0.0584	1.80448	23	0.89604
2.534975	2.534975	24	0.935644
0.438645	3.46626	25	0.975248

See Fig. B.1.
Linear regression analysis:

Slope	0.781653
Y—intercept	− 0.3601
R^2	0.963079

The above plot is for 25 data points. As the number of data points is increased, the data plot will follow the straight line more closely.

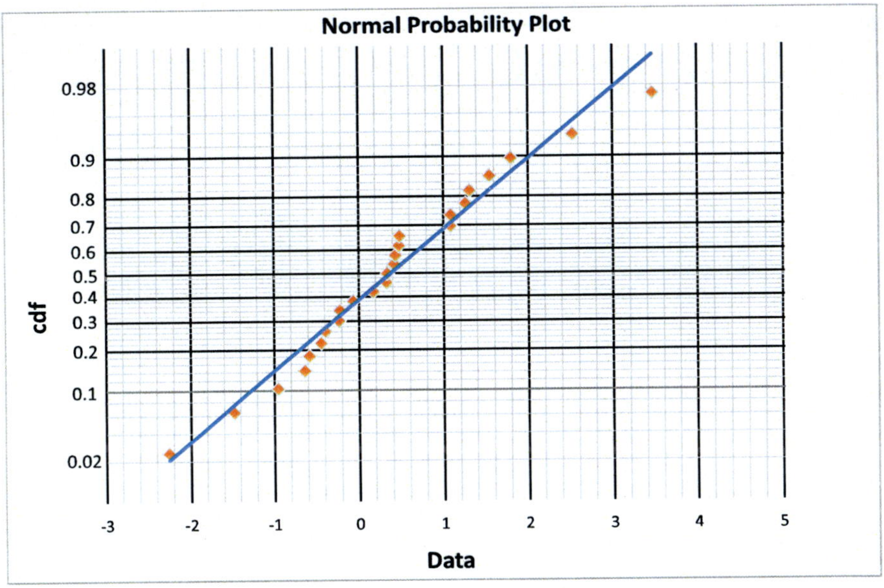

Fig. B.1 Normal probability plot for normally distributed data

Log-Normal Distribution

Consider following data available from some observations of a log-normally distributed random parameter.

$$n = 25$$

Data	Arranged data	Rank(R)	CDF
0.421619	0.225015	1	0.024752
1.08043	0.296972	2	0.064356
0.296972	0.328407	3	0.10396
0.328407	0.336531	4	0.143564
0.993174	0.345908	5	0.183168
4.63034	0.421619	6	0.222772
0.463168	0.463168	7	0.262376
1.449732	0.465404	8	0.30198
0.79805	0.476017	9	0.341584
3.056762	0.540316	10	0.381188
0.336531	0.79805	11	0.420792
1.033093	0.824962	12	0.460396
1.737639	0.993174	13	0.5

(continued)

(continued)

Data	Arranged data	Rank(*R*)	CDF
3.006	1.033093	14	0.539604
4.684278	1.08043	15	0.579208
1.089731	1.089731	16	0.618812
0.225015	1.449732	17	0.658416
0.476017	1.737639	18	0.69802
0.345908	2.112932	19	0.737624
10.49037	2.431748	20	0.777228
0.540316	3.006	21	0.816832
2.112932	3.056762	22	0.856436
0.824962	4.63034	23	0.89604
2.431748	4.684278	24	0.935644
0.465404	10.49037	25	0.975248

See Fig. B.2.

The above plot is for 25 data points. As the number of data points is increased, the data plot will follow the straight line more closely closely (Fig. B.2).

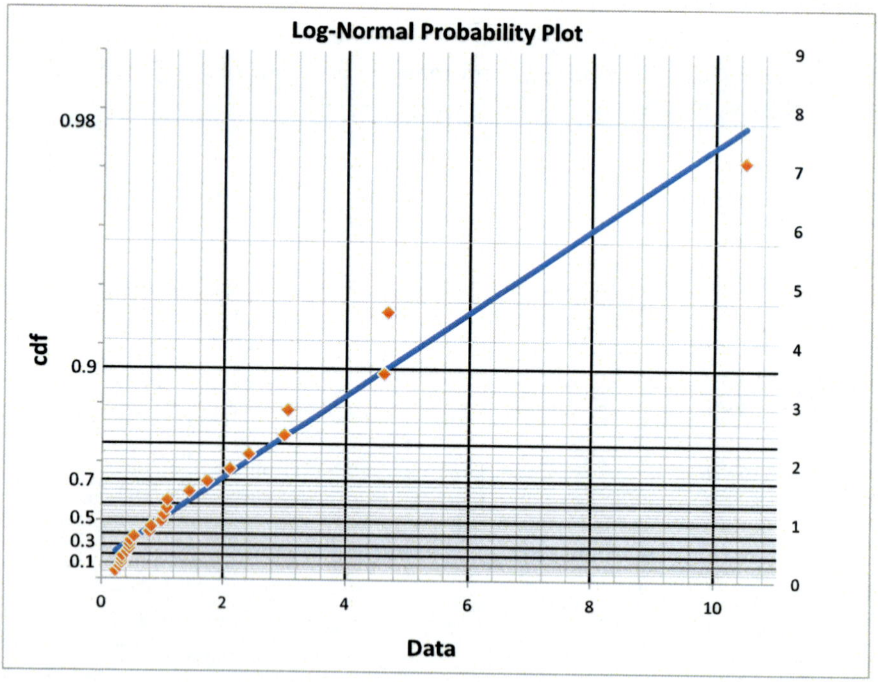

Fig. B.2 Log-normal probability plot for log-normally distributed data

Slope	0.705958
Y—intercept	0.320609
R^2	0.967892

Weibull Distribution

Weibull probability density function is given as:

$$f(x) = \left[\frac{\beta}{\alpha}\left(\frac{x-\gamma}{\alpha-\gamma}\right)^{\beta-1}\right]\exp\left\{-\left(\frac{x-\gamma}{\alpha-\gamma}\right)^{\beta}\right\}$$

where

α is scale parameter and f is shape parameter
β is location parameter
γ location parameter

Starting from zero, location parameter is zero.

$$f(x) = \left[\frac{\beta}{\alpha^{\beta}}x^{\beta-1}\right]\exp\left\{-\left(\frac{x}{\alpha}\right)^{\beta}\right\}$$

We get cumulative distribution function by integration

$$F(x) = 1 - exp\left\{-\left(\frac{x}{\alpha}\right)^{\beta}\right\}$$

Rearrangement of the equation gives:

$$\ln[-\ln(1-F)] = 2.303\,\beta\,\log_{10}(x) - \beta\,\ln(\alpha)$$

Comparing the above equation with the general straight line equation

$$Y = MX + C$$

Shows that the plot between 1 n [— 1 n (1 — F)] and $\log_{10}(x)$ will be a straight line with slope 2. 3 0 3 β. Weibull probability graphs allow to directly plot between F and x.

Weibull probability plot for a normally distributed data is constructed as follows:

Arranged data	Rank(R)	CDF
1.01	1	0.024752
1.790423795	2	0.064356
2.312881288	3	0.10396
2.627271519	4	0.143564
2.680115059	5	0.183168
2.820832755	6	0.222772
2.872499257	7	0.262376
3.029526943	8	0.30198
3.044705174	9	0.341584
3.200928516	10	0.381188
3.436270255	11	0.420792
3.590217018	12	0.460396
3.602552571	13	0.5
3.68094744	14	0.539604
3.697977336	15	0.579208
3.739813076	16	0.618812
3.745614462	17	0.658416
4.347037754	18	0.69802
4.349846104	19	0.737624
4.527233979	20	0.777228
4.575790106	21	0.816832
4.810923957	22	0.856436
5.063812038	23	0.89604
5.794306586	24	0.935644
6.725591813	25	0.975248

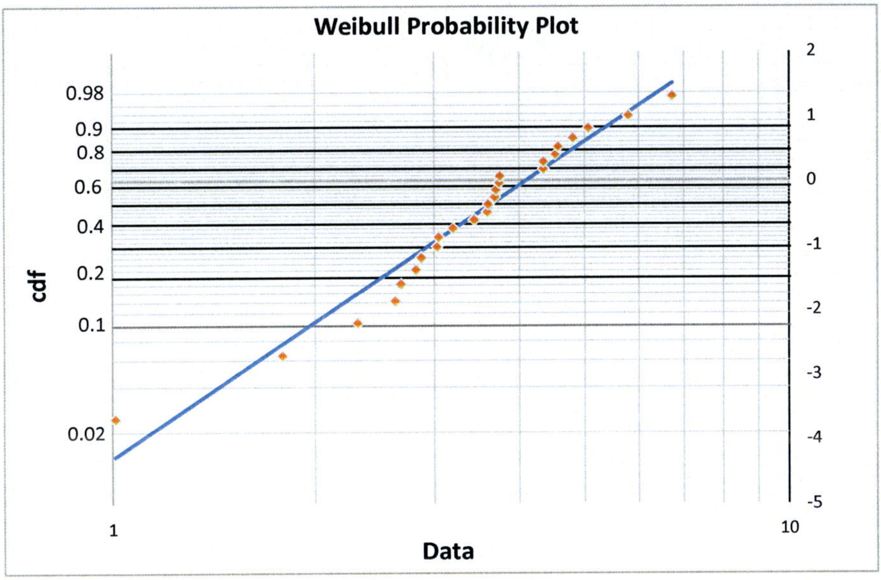

Linear regression analysis:

Slope	7.030298
Y—intercept	− 4.30834
R^2	0.959744

$$\text{shape factor}, \beta = \frac{7.030298}{2.303} = 3.05$$

Weibull probability plot for a normally distributed data is constructed as follows:

Arranged data	Rank (R)	CDF
0.225015	1	0.024752
0.296972	2	0.064356
0.328407	3	0.10396
0.336531	4	0.143564
0.345908	5	0.183168
0.421619	6	0.222772
0.463168	7	0.262376
0.465404	8	0.30198
0.476017	9	0.341584

(continued)

(continued)

Arranged data	Rank (*R*)	CDF
0.540316	10	0.381188
0.79805	11	0.420792
0.824962	12	0.460396
0.993174	13	0.5
1.033093	14	0.539604
1.08043	15	0.579208
1.089731	16	0.618812
1.449732	17	0.658416
1.737639	18	0.69802
2.112932	19	0.737624
2.431748	20	0.777228
3.006	21	0.816832
3.056762	22	0.856436
4.63034	23	0.89604
4.684278	24	0.935644
10.49037	25	0.975248

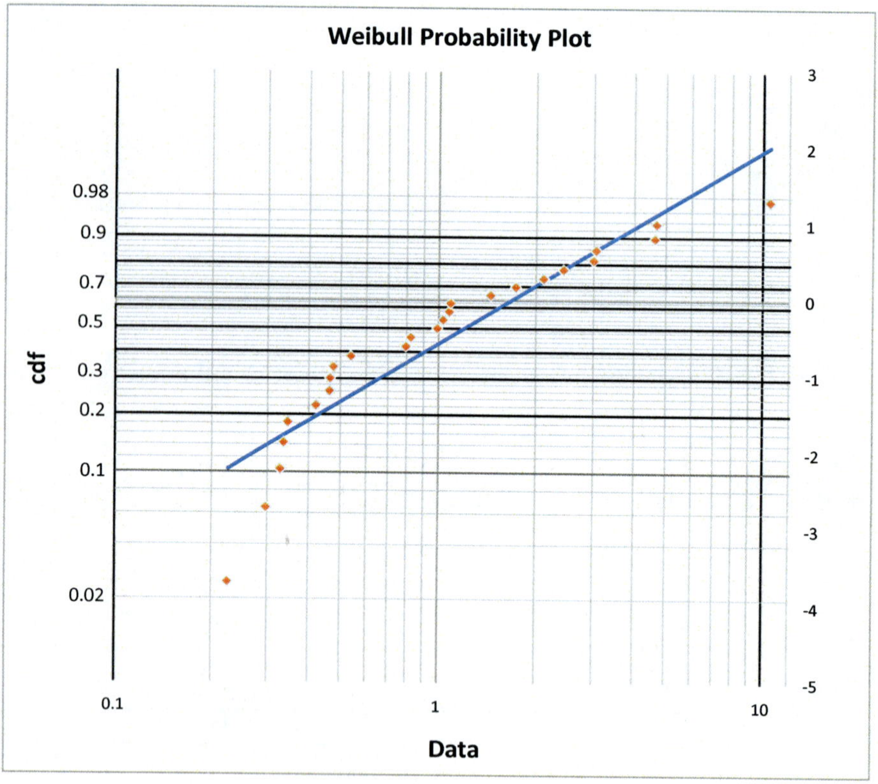

Linear regression analysis:

Slope	2.536491
Y—intercept	− 0.56621
R^2	0.856269

$$\text{shape factor}, \beta = \frac{2.536491}{2.303} = 1.10$$

Annexure C
Yoga and Consciousness, Cognition—Occupational Human Physical and Mental Wellbeing and Health

Based on the evidence found in Indus Valley civilization, modern literature records that rich Yoga tradition and practices dates to more than 4500 years. The Hindu knowledge-based mainly Vedas and Upanishad asserted that consciousness (Chetana) is fundamental to existence, in particular human existence. There is growing interest in modern science, particularly, to develop the neuroscience of yogic theory of consciousness. The modern science is still proposing the model of consciousness through materialistic elements where the mind is a subelement along with internal sense bases and action organism along with intellect, ego and seer (the higher dimension that might include state of self-realization or progress toward enlightenment) [1].

The relevance of Yoga and related spiritual practices form part of risk-conscious operations management (RCOM) system. The risk-conscious culture has two-pronged approach where a) development of human resilience against inadvertent human errors is postulated by spiritual, particularly Yogic practices and b) developing an understanding of the human root causes are envisaged such that human error probability for a given task is either eliminated or reduced through a mechanism that uses input from statistical model of human performance and insights from human root cause analysis. This model complements the conventional approaches that include training/retraining, motivational mechanism, administrative provisions, etc. The major spin-off is that the Yoga, and related practices also form part of RCOM culture to improve the consciousness levels (awareness, alertness, etc.) and cognitive capabilities by improving the agility and memory, etc. Further, the Yoga and meditation being spiritual practices, the ethical or moral aspects can also be addressed as they offer possibility of eliminating or reducing chances deliberate or malicious, actions and thereby address, along with safety, the security aspects of the plant. One of the major features of the human reliability in RCOM is, apart from dealing with the factors responsible for human error, to model methods and techniques where the root causes of human error can be either eliminated or reduced. The traditional science of spirituality that includes Yoga, pranayama and meditation are the proven techniques that not only increases human wellbeing but also help achieve higher goals, like self-realization and enlightenment. In this

P. V. Varde, *Risk-Conscious Operations Management*, Risk, Reliability and Safety Engineering, https://doi.org/10.1007/978-981-19-9334-3

system, consciousness is considered as fundamental which while working at physical level occupies universal dimension. Of course, in RCOM the focus has been to look for the ways where human performance can be enhanced. References of the Meditation (sitting in a posture still and focusing on a single object, like self-breath and part of body, godly figure, on a point, e.g., on a flame of a lamp, etc.), pranayama (breath exercises), Yoga (physical exercises called asanas or postures) can be traced to over 5000–10,000 years to shrutis (Rig Veda) effectively passed on over generations by as oral knowledge in the form of elegant shlokas) includes, Hindu religious book that mainly Vedas and further down the generations as written records like Upanishads, Bhagavad Gita, Ramayana. Some later developments directly relevant to this section are Patanjali School of Yoga is prominent traditional school while in recent times we see many versions of Yoga, pranayama and meditation, e.g., Kundalini Yoga, Sudarshan Kriya, Vipassana (has origin in Buddhism) and Jainism. The modern derivative, being extensively used, for the higher consciousness and cognition and wellbeing that helps develop positive human attitude and attributes is termed as **mindfulness**. The fundamental principles have its root in the traditional systems of wellbeing, like Yoga and associated disciplines. There are efforts in health care areas to investigate the effect of traditional Yoga, mindfulness-based cognitive therapy and the conventional cognitive behavioral therapy on physical wellbeing employing randomized controlled experiments [2].

*"**Yoga** means 'union' is a group of physical, mental and spiritual practices which originated in ancient India and aims to control and still the mind, recognizing detached witness-consciousness untouched by mind and mundane suffering [11]"*

"**Pranayama** is an ancient breath technique that originates from yogic practices in India. It involves controlling your breath in different styles and lengths. It has more recently gained popularity in the western world because of the many health benefits that come from a pranayama practice [12]".

One more element of Hindu system of wellbeing is mudras, often performed as part of Yoga. **Mudras**: Mudra = 'Mud' Dhatu means rejoice and happiness. Mudras help to link the brain to the body, stimulate endorphins, change the mood and increase our vitality. The reason is, as per Indian rishis the body has four major constituting elements, water, earth, Akash, Vayu and the five fingers are associated with each of these elements, viz., little, ring, middle, index and thumb, respectively. Mudra is creating a pattern by folding, touching, bending and pressing these fingers/thumb such that these configurations render mental and physical wellbeing, or helping one to attain higher dimension—self-realization or enlightenment. Also, this mudra also works to cure many diseases. Further various mudras have been discussed in detail in the Table.

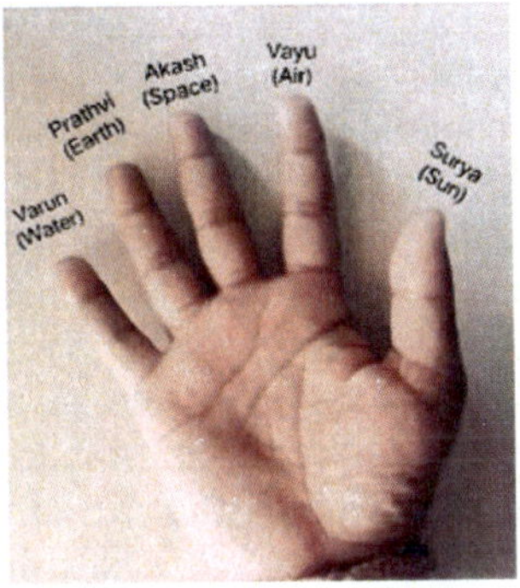

The major aim of these Yoga, meditation and pranayama is human wellbeing toward creating a equanimous society where Dharma, i.e., the welfare and well-being of not only self, as also of social system that includes apart from human and living entities and the environment'. Even though the major objective of these techniques is self-realization and enlightenment, the wellbeing and happiness are the default or spin-off these techniques. The practices are actively being used in physical and mental health improvement. It is asserted that healthy mind-and-body is the precondition to many human attributes, like enhanced focus, attention, alertness, resilience to distraction and wandering mind—which are critical to human performance. This knowledge and its practices provide an effective cultural fabric for organizational and individual aspects. In fact, the increasing application of these techniques is evident in the available open literature.

***Meditation** is an ancient practice that is believed to have originated in India several thousandyears ago. Throughout early history, the practice was adopted by neighbouring countries quickly and formed a part of many religions throughout the world* [13].

Yoga asanas and pranayama are performed regularly in group and alone employed not only for therapeutic treatment but also to enhance wellbeing. These techniques help in developing positive attributes like cooperation, emotional feelings, compassion and sharing in family and has social component of togetherness in a group. Enhanced consciousness and cognitive attributes manifest in the increase in awareness, alertness, focus and concentration. These techniques also support development of resilience against the major root causes against distraction, disinterest in work and activities, manifest as lack of dedication, commitment, etc. Particularly Yoga along with meditation *gently encourage enhanced awareness of body, mind and environment. The 12 basic poses or asanas open the energy channels, chakras and psychic centers of the body while increasing flexibility of the spine, strengthening bones, and stimulating the circulatory and immune systems. Along with proper breathing or pranayama, asanas also calm the mind and reduce stress. With regular practice, one can increase overall physical and mental health and the aid in the prevention of diseases such as diabetes, hypertension, and arthritis. In time, the poses consciously become a mental exercise in concentration and meditation [10].*

Meditation is an ancient practice that is believed to have originated in India several thousand years ago. Throughout early history, the practice was adopted by neighboring countries quickly and formed a part of many religions throughout the world [13]. These ancient techniques developed as part of Sanatana Hindu Darshan was adopted by Buddhist and Jainism. As Shri S.N. Goenka the great teacher of Vipassana Meditation (originated as Part of Buddhist way of life) puts it in his teaching entitled 'The Art of Living' authored by William, 'My interest is in the truth, the teachings of all enlightened ones—that leads from misery to happiness, from defilements to purity, from bondage to liberation from ignorance to enlightenment'.

In the following section, the table shows various mudras, Yoga asanas, pranayama and meditation postures. Essential features of each posture associated benefits have been given in brief. Of course, these asanas and postures are performed under a guide or Guru. The asanas, pranayama and meditation not performed under a supervision might cause harm physically or mentally. For example, the breathing patterns during the Yoga and pranayama are critical otherwise one may invite problem. The postures are indicative and do not contain all the details for further details join a Yoga or Meditation Class.

This annexure has been developed keeping in view the expected benefits that could be obtained; in terms improving performance through enhanced mental and physical attributes, viz. mental agility, awareness, focus, concentration, memory, physical wellbeing in terms of required capability of sense and action organs; through Yoga and related practices with an objective that it offers an approach for occupational staff's performance.

Important Instruction

It may be noted that the Yoga, Meditation, Pranayama and related practices discussed in here should be performed under a trained teacher (Guru). It is important that before taking up these activities one should bring to the notice of teacher, medical and psychological condition, apriori such that teacher can advise on the asanas to be performed, particularly practices to be avoided.

The objective here, is to highlight the potential benefits of Yoga practices in performance improvement, however, the adequacy and completeness of description is not assured.

Reference Sitting Postures for Pranayama, Yoga Asanas, Mudras, and Meditations

A	**Basic Sitting postures for Yoga, Pranayama and Meditation**: The three asanas introduces a practitioner with early advantages of asanas, three benefits include a feeling of calmness, peace, harmony you like everything around you. As one graduates from Sukhasana, to Ardhpadmasana and finally to Padmasana the benefits increases in terms of less time with feeling of more positive attitude	
A1	**Sukhasana (Easy Pose):** Sitting posture, with back erect, head in line with back and straight, legs natural and crossed position without any stress in the body, eyes closed, while practicing natural alternate breathing, is referred as Sukhasana or Easy Posture shown in Figure. Maintain this posture for some time This posture relaxes body and mind a must for productivity and right attitude. Mostly the learners start the sitting postures in Yoga **[19]**	

(continued)

(continued)

A2	**Ardh-Padmasana (Half-lotus Pose)**: Same as Sukhasana except lift one leg and rest it over other. Slight more stress will be felt in the leg that has been lifted and rested over other. The enhanced training makes Yoga learner comfortable in this posture. [19]	
A3	**Padmasana**: Same as Ardh-Padmasana except that the other lower leg is also lifted and rested on the first lifted leg. This posture requires both the legs in locked position. There are three Steps, (a) Sit in Sukhasana, (b) Ardha-Padmasana next take the lower leg to upper leg and rest it, (c) take the lower leg (now) lift and rest it on other to attain Padmasana. This posture generates higher physical stress in the knee joints, however with practice the stress reduces and after some time the Yogi feels comfortable as in Sukhasana. It takes pretty long may be a month or two with regular practice to feel comfortable in this posture	
A4	**Vajrasana (Strong or Thunderbolt Posture)**: Vajrasana is one of the most popular and effective posture. Basically, it is a Dhyana and at times a meditation asana. In fact, this is the only asana that can be performed after meal to accelerate the process of digestion. To perform this asana, one need to sit folded legs with back and head erect and hands resting on the knee as shown in the image opposite. Often the practitioners sit in Vajrasana during Dhyana or meditation with eyes closed and as usual hands resting on the knees. Breathing should be normal in Vajrasana Vajrasana, enhances metabolism, increases the flow of blood in the upper part of the body. It makes leg nerves, bones and muscle stronger. This asana is also effective in making the backbones stronger At mental level, vajrasana is effective in increasing consciousness levels, particularly concentration and focus	

(continued)

(continued)

B	**Pranayama (Breathing Exercises): Regular pranayama practice enhances cognitive performance. Three main pranayama practices, Anuloma-Viloma, Ujjai and Bhramari**	
B1	**Anuloma-Viloma** (breathing from one nostril and breathing out from other): Sit in Sukhasana or Dharmasena, else if you are comfortable then on Padmasana; with back vertical, neck in line with back as far as possible; both hands resting on knees. Start with the right hand as shown in the image. Close your right nostril with thumb and breath-in with left nostril as you can and comfortable, now close the left nostril with second finger and open the right nostril and exhale out through the right nostril. Repeat the same pattern i.e., breathing in from right nostril and breathing out with left nostril. This is one cycle. Repeat this cycle as instructed, at least five times in the beginning [20] Anulom-vilom pranayama reduces the negative tendencies like anger, forgetfulness, anxiety and hopelessness which are critical to having positive attitude and teamwork	
B2	**Kapal Bhati (That help develop light/aura around head) Pranayama**: Sitting posture, as shown in the Image, is same as above, except hand should rest on the knees and they do not have any role other than remaining in this position. Close your eyes with the eyelids softly resting and ensure no physical tension in the body anywhere Breath-in slowly through your stomach deflating in and hold the breath for a second and breath out. This asana is easiest asana. Repeat this cycle for at least 10 times in the beginning and later increase the cycles as per the advice of the Yoga teacher or supervisor Directly works on the cognitive faculties by ways of refreshing the brain, balancing the chemicals and enhances energies levels [18]	
B3	**Ujjayi: (The Victory Pranayama)**—Sit in Sukhasana, Ardh-padmasana, or then on Padmasana whichever asana you are comfortable; with back vertical, neck in line with back as far as possible; both hands resting on knees as shown in the image This pranayama is all about making a peculiar noise by contracting throat region while inhaling till possible and you feel comfortable and bottling up for a while with head lowered	

(continued)

(continued)

	but not touching the chest, e.g., for 5 s. Then in next step exhale slowly. DO NOT HOLD THE BREATH IN THE BEGINNING FOR MORE THAN 5 s. With practice, the holding time can be increased slowly. In fact, this pranayama can be learnt more effectively on video images available in social media. This pranayama expands lungs, refreshes the brain and broaden the chest, which relieves anxiety, depression, and sleep apnea [18] It reduces perceived stress levels; also it increases mindfulness. Ujjayi pranayama is effective to reduce the problem with sleep disorder	
B4	**Bhramari (Humming) Pranayama**: This pranayama is also one of the top five pranayamas. Here also the sitting position is same as earlier two pranayamas discussed. Only difference the sound that is made is humming bees, with closed mouth and creating the humming. Again, here also, since sound is involved, it required to visit a social website and observe how the sound is made. To begin with four to five sessions are adequate and later the frequency can be increased This pranayama involves a calming breathing practice that soothes the nervous system and helps to connect us with our truest inner nature [18]	
C	**Yoga Asanas**: Balancing asanas Like Vrikshasana and Akasaka increases the balancing capacity of the body. It also enhances focus and concentration. It also increases mindfulness and steadiness of mind	
C1	**Sirsasana (Headstand posture)**: This is a headstand posture requires training and guidance under a professional teacher. Should not be attempted on your own. The sirsasana posture is shown in the image. This asana enables enhanced flow of blood toward head and face mainly eyes and other sense organs. Maintaining this posture for say 10 s to couple of minutes ensures the benefits like relieving mental stress, increased concentration and	

(continued)

(continued)

	focus, helps flushing of adrenal glands (responsible for production of stress harmones), improves digestion, increases muscles strength and stimulates the lymphatic system that helps remove body toxins [26]	
C2	**Sarvangasana (Shoulder stand posture)**: Sarvangasana is sthiti or supine posture. Inhale, raise the leg to 90 degree position with respect to upper body, exhale, support the lower portion below buttock and acquire the vertical position of rest of the body with respect to head by bringing pushing up the upper part by hands, inhale and maintain the normal breathing and position, as shown in image, for the time ranging from few seconds to a minute or more, depending on your comfort level but limiting to 5 min. Slowly return to horizontal/sleeping on back position called sthiti This asana simulates and keeps the thyroid health that is responsible for secretion of hormones offers all the benefits of inverted position, makes shoulder stronger, and offer therapeutic uses, e.g., piles and hernia [25]	

(continued)

(continued)

C3	**Halasana (Plough posture)**: This asana also belong to the sthiti posture and initiated from sleeping face up position. Inhale and take the legs up and exhale while raising the buttock up (while keeping the hand supported from ground) and reach the position as shown in the image. Breathe normal and maintain this position depending on your ability and requirement This asana is beneficial when stretching the back muscle, spinal joints, and lumber nerves. Halasana increases the flow of blood to the neck, while activating the thyroid thereby enhanced secretion of hormones [25]	
C4	**Matsyasana (Fish posture)**: Lay on back on floor straight in comfortable position. Next step is, while exhaling, fold both the legs to padmasana position as shown in figure. Ensure that you are not feeling pressure or pain in knees, and the back in horizontal condition. If possible raise the chest and head up (almost in vertical position) and crown of head on the floor. Have normal breathing in this position for time as advised or you feel comfortable but not more than 5 min. Take reverse steps to return back to lay down position	
C5	**Paschimothanasana (Seated forward bend posture)**: This asana starts in sitting position with both the legs stretched forward with tows in upright vertical position. Inhale while raising both the hand upward, keeping in parallel to each other, stretch it up with palm facing each other. Next exhale and bend the trunk forward from lower back stretch the arms horizontally, exhale fully, hold the toe with ring finger and maintain this posture with normal breathing till comfortable or assigned time. Return to sitting position following the reverse steps, i.e., taking hands up with palms parallel and then lowering to acquire the initial position This asana enhances flexibility to the backbones and muscles and bones and agility to spinal nerves [25]. This is an effective exercise for increasing sensitivity to the body systems	 [28]
C6	**Bhujangasana (Cobra posture)**: Lye down with face downward. Keep the arms parallel under the chest with palm resting on the floor. Raise the chest portion by exerting force on the floor with neck backward while inhaling.	

(continued)

(continued)

	Achieve the final position, as shown with normal breathing and not overexerting any part of the body, maintain this position. Return back to normal laying down position and relax [25, 26]	
C7	**Dhanurasana (Bow posture)**: Laydown on the floor on stomach with chin touching the ground and make the body horizontal to come to starting position. Put the hand horizontal on ground. Further, while exhaling raise both the legs simultaneously and slowly and stretch the heels forward as for as possible. Next, hold the ankles with your hands and while inhaling, bring the leg forward and chest backward to come to the position as shown in this figure. Maintain normal breathing This asana has many advantages including improving function of kindly, liver, also helps improving digestion and works wonder in weight loss	
C8	**Pada Hasthasana (Standing forward bending posture)**: Pada and hastha means, leg and hand, respectively. Standing position with hand closed to body and legs at least two inch apart, is the first step in this asana. Next, while inhaling, raise both the hand till it acquires a body in vertical stretching out standing position, with palm parallel to each other. Next, exhale out slowly and bend forward with head down till the palm touches the floor or toe or till you can stretch forward as you can. Stay in this position with breathing normal. Once you feel get back to the standing position by following reverse steps in sequence	

(continued)

(continued)

	This asana makes the backbone flexible. This asana comes under the category of difficult asana, and people with medical condition related back, heart, stomach should check with healthcare staff before trying this asana	
C9	**Virbhadrasana (Warrior posture)**: Handstand pose like Viprit Karni Mudra—By doing inversions, reverses blood flow and increases blood circulation occurs, especially to the brain. As the blood rushes to head, it provides the brain with more oxygen and improves cognitive function The asanas which are a forward bend asanas like Yoga mudra reduce stress, anxiety and fatigue. It also calms the mind	
C10	**Bakasana**: This asana involves resting your head on the floor and lifting the body as shown in the figure. This posture can be maintained, depending on the training, particularly by beginners for nil to 5 s and then time can be increased with training. This is a good posture for starting balancing exercises Bakasana makes wrist, forehand and hand joints as also back stronger	

(continued)

(continued)

C11	**Mayurasana [19] (Peacock Pose)**: This asana begins by putting both the hands on the floor stretching legs back while the body in forward direction. Maintain the balance and normal breathing in this position. Mayurasanas belong to the categories of difficult asanas and beginners should not attempt on their own. It should be done in the supervision of a trained teacher as also follow the preparation rule that this asana should be performed empty stomach preferably in the morning Regular practice of mayurasana renders strength to the hands, leg, back particularly the spinal cord and reenergies like any other physical exercise and reduces health issues	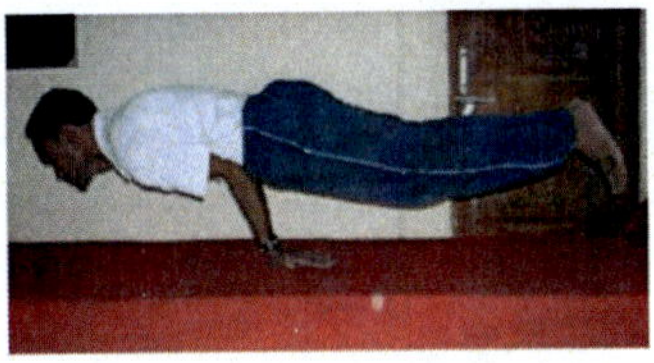
C12	**Natarajasana (Dancer's pose)**: This asana basically is a balancing asana, dancer's need to perfect as part of dancing practices in Indian traditional dance system [19]. Major steps are a) stand upright with back and neck erect in one line, b) hold the right leg with right hand take it back and up slowly by maintaining the balance, c) bend forward by extending the left hand, d) for better balance and position take forward the chest and extend the leg and hand such that they are in one line as shown in figure. Repeat the exercise with left leg and hand following the above four steps Stronger physical balance is a must for performing physical operation, particularly in areas where the site conditions imposing physical constraints	
C13	**Vrikshasana** (Tree Pose): This asana is performed with eye open like preceding, Natarajan posture and helps in increasing and maintaining physical balance. The major steps are a) stand upright with firmly standing on the ground in anticipating that the next step requires balancing of body on one feet, b) lift the right foot with left hand and place it on the inner side of thigh, c) raise both the hands with palm in contact and stretch it above as shown in image. Repeat this above sequence, with left leg on thigh and stretching of hands above head [19]. Keep in this position and maintain the balance for at least 30 s. With time one can increase time for improved benefits, i.e., enhanced balancing of the body	

(continued)

(continued)

	As mentioned above, these balancing exercises makes the physical work easier at site with physical constraints	
C14	**Viparita Karani**: **or the inverted body**: The major steps to practice viparita karni are (a) lie down on ground, (b) raise both the legs, (c) support the back with both the hands, (d) further lift the body with legs upward till legs are in vertical position, and (e) ensure that the head is resting on the ground as shown in the image opposite [25]. Maintain this posture depending on instruction or your capability The major advantages of viparita karni include relief from anxiety, sleep disorder, depressions and breathing difficulties [19]	
C15	**Savasana (Corpse Pose)**: Savasana is part of a Yoga exercise and performed at the end of a Yoga session to relax the mind and body. The teacher narrates beautiful natural aspects, and the performer flows with the thought toward enhancing the feeling of relaxation [19]	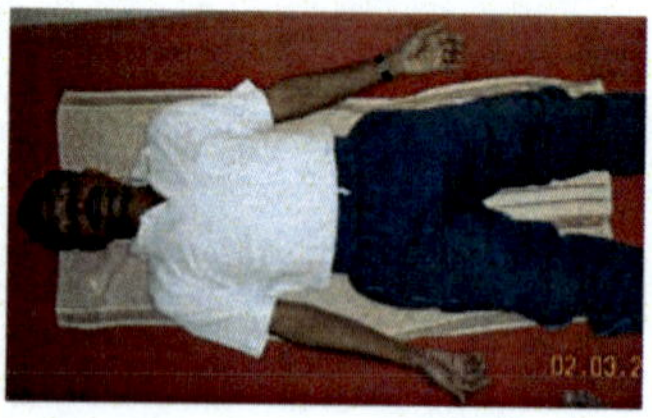
D	**Mudras:** Generally, mudras in Yoga system is referred to standard hand gestures. Mudras generally forms part of Yoga, meditation and pranayama in respect of Hand gestures and when practiced along with these techniques enhances the benefits mind body and soul in numerous ways. For healthy body the balance of five elements corresponding to a finger, viz. earth (ring), fire (thumb), water (little), air (index) and space (middle) is essential. This is achieved by forming a gesture using the five fingers. We will present here the mudras relevant to the subject, i.e., mudras that promote emotional mental and physical wellbeing [21]	
D1	**Gyan (Knowledge Gesture) Mudra**: Gyan mudra is a common hand posture used during sitting Yoga asana, pranayama and meditation and very effecting in calming the mind and promotes conscious thinking capabilities. Works improve the quality of life by enhancing cognitive capabilities by increasing concentration and memory, by reducing sleep disorders, it works to for stress and anger	

(continued)

(continued)

	management. This mudra also helps in management of issues like depression and headache [24]	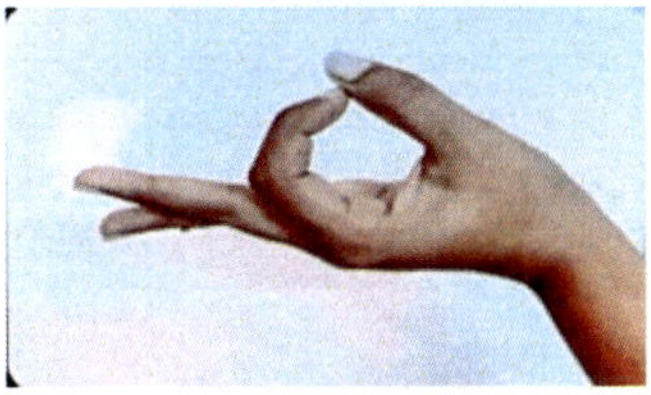
D2	**Vayu (Air Gesture) Mudra**: This gesture is effective in management or correction of deficiencies at physical or mental level due to air element imbalances**.** Works to produce relief from joint pain in general and cervical issues in particular	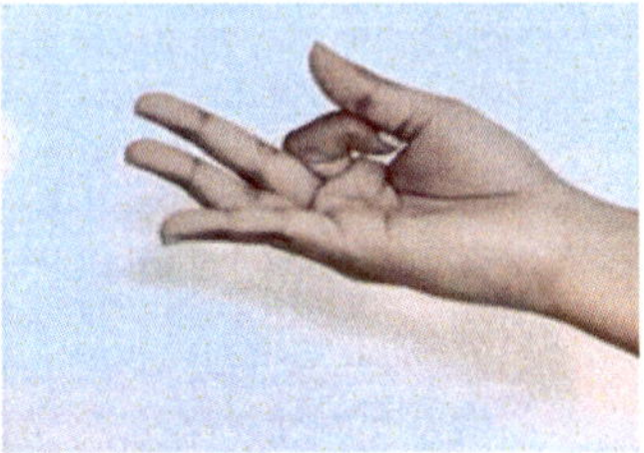
D3	**Akash/Sunya (Zero/Space Gesture) Mudra:** Sunya mudra deals with the element or balancing of space/akasha element in the body. In sunya mudra tip of the middle figure is placed at the base of the thumb and thumb and middle figure mutually pressed in this position. Some systems of meditations resort/prefer sunya mudra. This mudra renders relief from Vertigo, ear disorders and numbness in body [21]	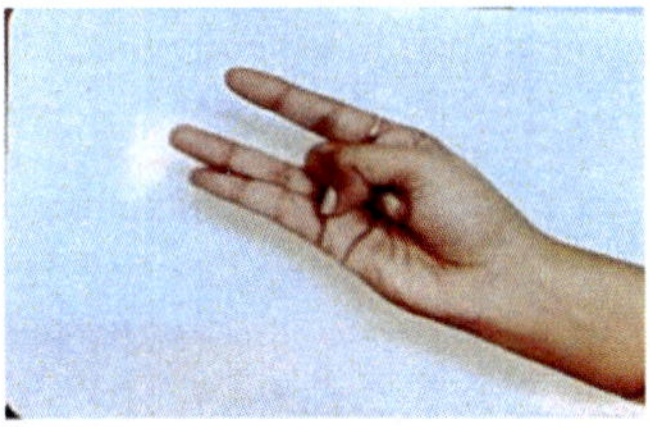
D4	**Apana Mudra (Descending Energy Gesture)**: This mudra, as shown in figure requires the two fingers, ring and middle pressed in contact with each other while keeping the ring index and little figure open and straight. This mudra detoxes the body revitalizes all the physical systems as also work for relief from high blood pressure-related issues, diabetes hemorrhoid [18]	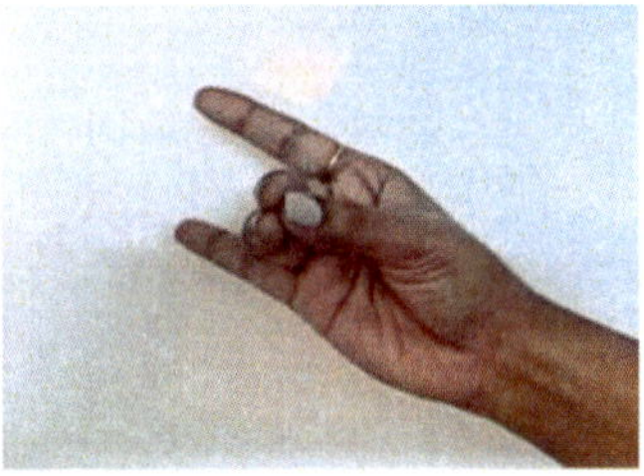
D5	**Prithvi Mudra (Earth Gesture)**: For attaining this mudra during any meditation, or pranayama fold the ring finger and press gently it under thumb while stretching the little,	

(continued)

(continued)

	middle and index finger outwardly. Befitting its name 'Prithvi' this mudra, when performed regularly, relieves one from emotional distress, anxiety and depression. Develop resilience against fatigue and develops mental equanimity [18]	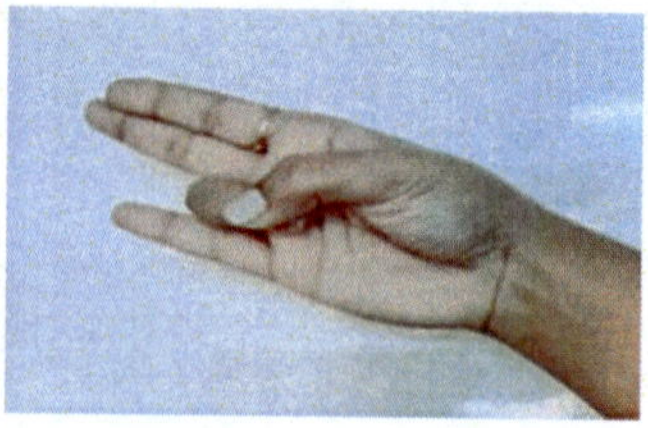
D6	**Surya Mudra (Sun Gesture)**: As can be seen in Figure, this mudra is attained by folding the ring finger and covering it with thumb creating a feeble pressure to maintain the posture. As the name suggest 'Sun' that is the source of life on earth, by catering to heat energy requirements. Heat energy is critical for digestion, fighting out lethargy, management of stress and a balanced attitude and overall wellbeing [18]	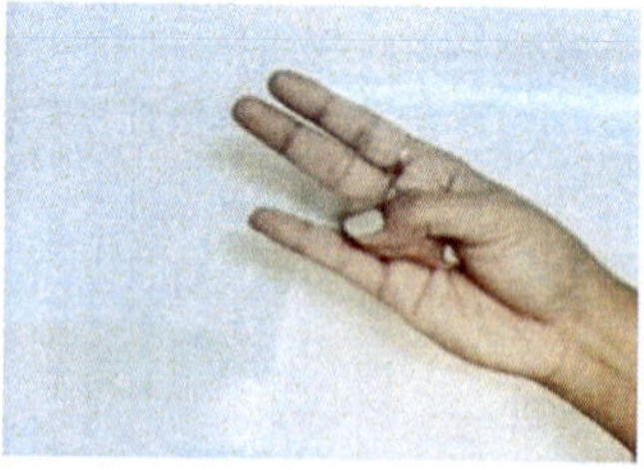
D7	**Prana Mudra (Energy Mudra)**: Prana energy relates to life energy of life quotient of a living being. This energy is a must for well-being, particularly for early recovery or generally useful to recover from the medical conditions. Prana is at the core of vitalizing energy levels [24]	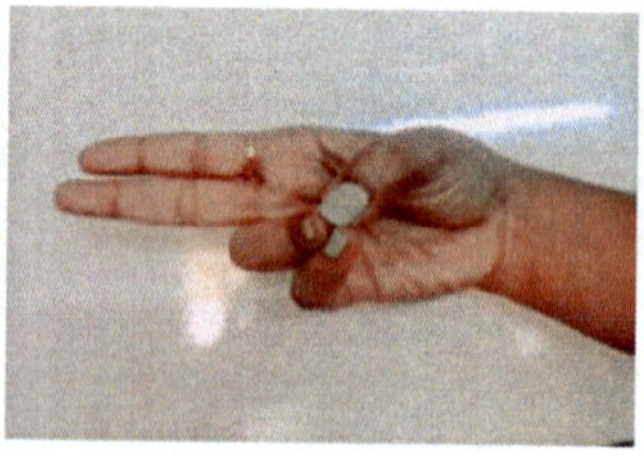
D8	Varun Mudra (Water Gesture): This mudra is created by pressing gently the thumb against the small figure or folding the small figure and pressing down gently under the thumb. 70% of human body weight is due to water, and in this sense, this mudra occupies an important dimension related to fluid balance in the body, and helps upkeep of the organs that maintain and processes the fluid quality and quantity, like kidney, moisturization of skin, wetness in the mouth and rest of the tracks, taste sensation, etc.[18]	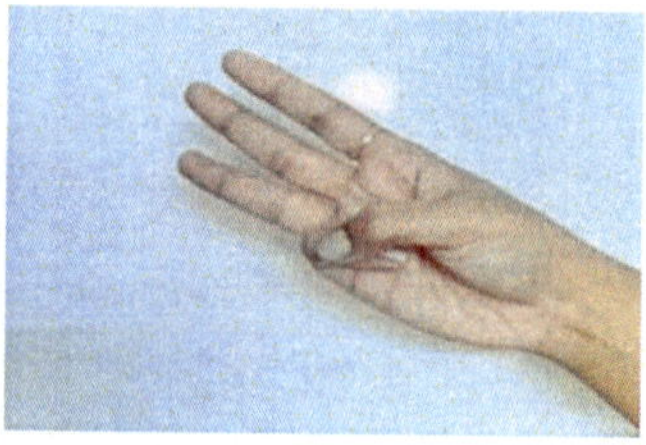

(continued)

(continued)

D9	**Brahma Mudra**: In Yoga Vidhya Niketan (Mumbai) system of Yoga, the Brahma mudra has four major postures [19], and the Brahma mudra is performed as shown in Figure: **Step I**: Sit in Sukhasana/Ardhpadmasana/Padmasana **Step II**: With tongue pulled outward and eyes focusing on the tip of the nose lower the neck as for as possible or till it touches the chest. Remain in this position for a sometime. Retract the tongue and eye to normal to reach the normal posture (as shown in Step I). **Step III**: With tongue touching the upper part of mouth, rolled-in and locked position and eyes focusing on the tip of the nose raise the neck up as shown in image III. Remain in this position for a sometime. Retract the tongue and eye to normal position and attain the normal posture (as shown in Step I) **Step IV:** Turn the head to right and maintain this position for some time with normal breathing while feeling the tension in backside and compression in right side of neck, respectively. After some time (say 1—5 min) return back to Step I posture **Step V**: Turn the head to left and maintain this position for some time with normal breathing while feeling the tension in backside and compression in right side of neck, respectively. After some time (say 1—5 min) return back to Step I posture **Step V:** Maintain the normal posture with normal breathing for say 1—3 min Brahma mudras help to link the brain to the body, stimulate endorphins, change the mood and increases our vitality [19]. Brahma mudras rejuvenate the mind and body through flow of prana/energy. Produces calming effect and improves consciousness attributes like focus and concentration. Other benefits, particularly useful to operations management include stability or balance of mind and body [There are many schools of Yoga where Brahma mudra practices involve Step I, III with head up (without eye and tongue posture) IV (head-right), V (Head left) and VI]	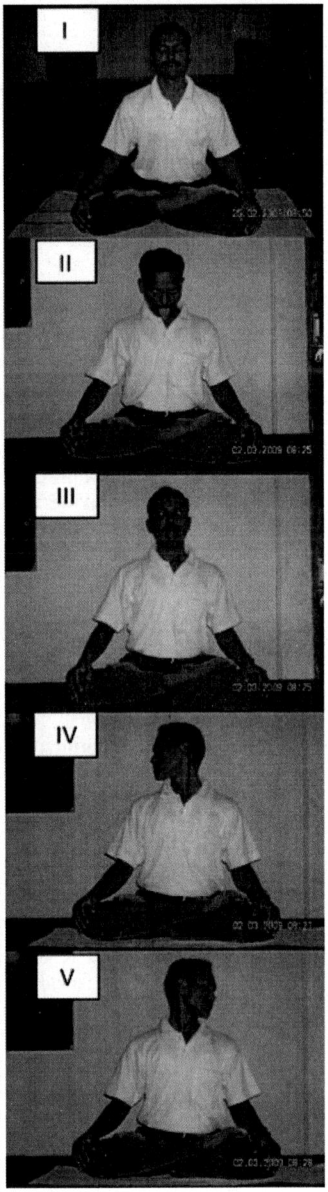

(continued)

(continued)

		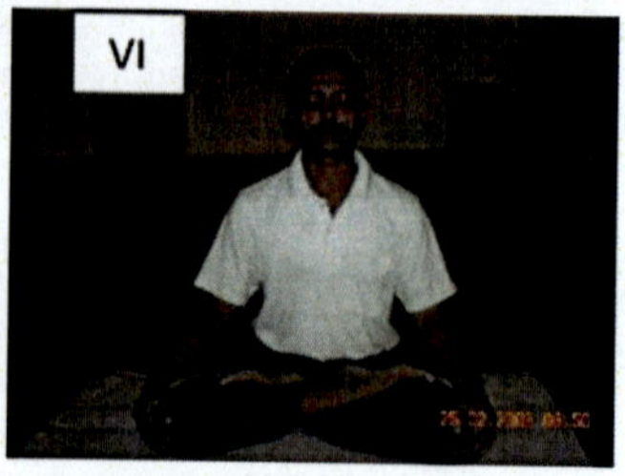
E	**Meditation: Meditation is a powerful, technique as it works not only at mind or physical or cognitive level but further deeper state where enlightenment is also possible. The minimum the practice of meditation assures for a normal occupational worker is defilements from various mental issues and thereby ensures clarity of thinking, enhances analysis and diagnostic capability, reduces susceptibility to mental stresses and thereby conditions or makes the mind resilient to wandering and destruction to support decision making. Experiments in the laboratory employing modern techniques show that meditation ensures instantaneous calming effect recorded through EEG by monitoring brainwave pattern that reaches to alpha or theta level with practice. We will discuss here the salient feature of only Vipassana Meditation discovered by Buddha over 2500 years ago**	
E1	**Vipassan Meditation:** Vipassana means 'insight' [14] or in the context of the meditation 'looking insight', is one of India's most ancient techniques of meditation [22]. This definition forms the central principle of Vipassana Meditation. In fact, it can be argued that this technique is scientific as it produces evidence instantaneously may be one can get the evidence of its claim in just 30 min of meditation and the calming effect it produces at mental level while the body feels energized As per the Vipassana guidelines this technique can be learnt in the recognized Vipassana centers spread out world over. However, based on the available open literature and the hands on experience with Vipassana meditation courses attended by the author, its salient features in terms of major steps can be discussed in brief as follows: • The place should be calm and quite be it outside or in a room or if collectively performed then in a hall or auditorium • The sitting posture could be Sukhasana, Ardhpadmasana, or Padmasana or else if someone has difficulty then on chair also with	

(continued)

(continued)

	hand folded together resting at naval level [23] • Do not force the mind for concentration its wandering can be reduced by focusing on breath with practice • Focus on the breath—the inhalation and exhalation • Feel the air blow through the nostril and on upper lip area (montage region) • Practice it till the wandering of mind is noticeable • Next step is now the body sensation can be observed • Scan the mind from top to bottom and bottom to top and keep observing the sensation without getting involved with it • Maintain the equanimity and continue the practice • Start initially and limit to 30 min increase the time as you feel Vipassana is not limited to as a meditation technique. It brings along with it a lifestyle that takes one to enlightenment or self-realization (Vipassana School of Meditation should learned and performed at the designated Vipassana Center ONLY)	

Acknowledgements This annexure is based on input from Ms Madhurika Desai and Mr Avinash Desai, Yoga experts from Yogavidhya Niketan, Mumbai; Ms Nidhi Mohite Chowhan (Masters in Yogic Science and therapy) Yoga School DAVV, Indore; and open literature that include web-sites and books on Yoga. Since Yoga and related fields, like meditation, pranayama and mudra postures are a specialized field, a care has been taken to maintain authenticity of the input. Even though the references to the source of input have been given in in the following section, if by chance any individual, agency or website/book is found to be missing, kindly let me know, we will update the section based on the new information. The author sincerely thanks and acknowledges the support and advise from the Yoga expert, Ms Desai and Ms Nidhi Mohite and organizations/authors as without their input, this section would not have been possible.

References

1. Tripathi V, Bharadwaj Neuroscience of the yogic theory of consciousness. Neurosci Conscious 7(2):1–15 (2021)
2. Grensman A, Acharya BD, Wändell P, Nilsson GH, Falkenberg T, Sundin Ö, Werner S Effect of traditional yoga, mindfulness—based cognitive therapy, and cognitive behavioral therapy, on health related quality of life: a randomized controlled trial on patients on sick leave because of burnout
3. Manjit K Khalsa (Ed.D.), Greiner-Ferris JM, LICSW, Stefan G. Hofmann, and Sat Bir S. Khalsa (2015) Yoga-enhanced cognitive behavioral therapy (Y-CBT) for anxiety management: a pilot study. Clin Psychol Psychother 22(4):364–371 (2015). https://doi.org/10.1002/cpp.1902.
4. The five phases of emergency management. www.bexar.org/694/five-phases.
5. https://www.emergency-management-degree.org/faq/what-skills-are-important-to-be-an-emergency-management-director/. downloaded on 22 Aug 2022
6. Stlouis-Mo-Gov, Steps of emergency management. https://www.stlouis-mo.goverment/departments/public-safety/emergency-management/about/Steps-ofEmergency-Management.cfm
7. PharmEasy, 11 Simple tips to keep your mental health stable, https://pharmeasy.in/blog/11-simple-tips-to-keep-your-mental-health-stable/. downloaded on 22 Aug 2022
8. 12 Yoga Poses for Beginners that will Give you Countless Benefits (trendymami.com)
9. https://en.wikipedia.org/wiki/Bhujangasana
10. https://sivanandalondon.org/twelve-basic-asanas/
11. https://en.wikipedia.org.Yoga
12. https://www.webmd.com/balance/what-is-pranayama#: ~ :text=Pranayama%20is%20an%20ancient%20breath,come%20from%20a%20pranayama%20practice
13. https://www.news-medical.net/health/Meditation-History.aspx#: ~ :text=Meditation%20is%20an%20ancient%20practice,many%20religions%20throughout%20the%20world
14. Hart W (2019) The art of living—Vipassana Meditation as taught by Shri S.N. Goenka. Embassy Books
15. Raghuram YS, Manasa Brahma Mudra—meaning, method of doing, benefits: ayurveda analysis
16. https://www.dragonfly-yoga.org/blog/ten-yoga-mudras-and-their-benefits-1#: ~ :text=Mudra%20is%20a%2https://www.lionsroar.com/how-to-practice-vipassana-insight-meditation/0Sanskrit%20term,with%20universal%20or%20cosmic%20force
17. https://www.fitsri.com/yoga-mudras#mudras-meaning-benefits
18. Iyer G. "Illuminating lives with YOGA, Publisher Geeta, Iyer.www.geetayoga
19. Dalal D.K., Desai M., Input from—Yoga Vidhya Niketan, Mumbai
20. https://www.amarujula.com/photo-gallery/yoga -and-health
21. https://www.yogapedia.com/definition/6859/shunya-mudra
22. https://www.dhamma.org/en/about/vipassana
23. https://unifycosmos.com/wp-content/uploads/2019/09/vipassana-sitting-position-examples.jpg.png
24. https://www.azulfit.com/hand-mudras-power-and-meaning/#: ~ :text = Mudras%20help%20to%20link%20the,in%20Yoga%2C%20meditation%20and%20dance
25. Nagarathna R, Nagendra HR Yoga for digestive disorders. Swami Vivekananda Yoga Prakashan, Bangalore, India
26. https://healthspaceclinics.com.au/blog/benefits-of-headstand-sirsasana
27. https://www.freepik.com
28. http://www.arhantyoga.com

Annexure D
Relevance of Spiritual Knowledge/Insights to RCOM Culture

D.1 Spiritual

Spiritual knowledge forms one of the source of input, for example, development of a reference human model, moral and ethical aspects, representation of knowledge, etc. in RCOM. This section describes major sources and some examples.

D.1.1 Nirvana Shatakam

Nirvana Shatakam, Author—Adi Shankaracharya (509–477 BC) [1]

Language: Sanskrit

मनोबुध्दयहंकार चित्तानि नाहम्
न च श्रोत्र जिह्वे न च घ्राण नेत्रे
न च व्योम भूमिर्न तेजो न वायुः
चिदान्द रूपः शिवोऽहम् शिवोऽहम् ॥१॥

(Mano Buddhi Ahankara Chittani Naham,

Na Cha Shrotra Jihve Na Cha Ghrana Netre |

P. V. Varde, *Risk-Conscious Operations Management*, Risk, Reliability and Safety Engineering, https://doi.org/10.1007/978-981-19-9334-3

Na Cha Vyoma Bhumir Na Tejo Na Vaayuhu,

Chidananda Rupah Shivoham Shivoham ||1||)

Translation (1): I am neither mind, Intelligence nor Ego, or memory; I am not the ears, the skin, the nose or the eyes;
Not even the five elements: space, fire, water or wind; I am the pure form of consciousness—the blissful and eternal Shiva.

न च प्राण संज्ञो न वै पञ्चवायुः

न वा सप्तधातुर्न वा पञ्चकोशः

न वाक्पाणिपादौ न चोपस्थपायू

चिदान्द रूपः शिवोऽहम् शिवोऽहम् ॥२॥

Na Cha Praana Sanjno Na Vai Panch Vaayuhu,

Na Va Sapta Dhaatur Na Va Pancha Koshah |

Na Vak Pani-Padau Na Chopastha Paayu,

Chidananda Rupah Shivoham Shivoham ||2||

Translation (2): I neither breath, nor five elements, I am also not seven elements neither five sheaths of consciousness; I am also not the speech, the hands, nor the feet; I am the pure form of consciousness—the blissful and eternal Shiva.

न मे द्वेष रागौ न मे लोभ मोहौ

मदो नैव मे नैव मत्सर्य भावः

न धर्मो न चार्थो न कामो ना मोक्षः

चिदान्द रूपः शिवोऽहम् शिवोऽहम् ॥३॥

Na Me Dwesha Raagau Na Me Lobha Mohau,

Mado Naiva Me Naiva Maatsarya Bhaawah

Na Dharmo Na Chartho Na Kaamo Na Moksha

Chidananda Rupah Shivoham Shivoham ||3||

Translation (3): I have no like or dislike, no greed or delusion, I am not pride or jealousy, I neither attached to duty, nor desire for wealth have no lust, or liberation. I am the pure form of consciousness—the blissful and eternal Shiva.

न पुण्यं न पापं न सौख्यं न दुःखम्

न मन्त्रो न तीर्थं न वेदा न यज्ञाः

अहं भोजनं नैव भोज्यं न भोक्ता

चिदान्द रूपः शिवोऽहम् शिवोऽहम् ॥४॥

Na Punyan Na Paapam Na Saukhyum Na Dukham
Na Mantro Na Tirtham Na Vedaah Na Yajnah
Aham Bhojanam Naiva Bhojyam Na Bhoktaa
Chidananda Rupah Shivoham Shivoham ||4||

Translation (4): I am over and above, good and bad deeds, happiness and unhappiness/pain, I do not identify myself with Mantra, religious places, neither scripture like Vedas or any rituals. I am neither source of experience, the experience nor the experienced. I am the pure form of consciousness—the blissful and eternal Shiva.

न मृत्युर्न शङ्का न मे जातिभेदाः
पिता नैव मे नैव माता न जन्मा
न बन्धुर्न न मित्रं न गुरुनैव शिष्यः
चिदान्द रूपः शिवोऽहम् शिवोऽहम् ॥५॥

Na Mritur Na Shanka Na Me Jaati Bhedah
Pitaa Naiva Me Naiva Maataa Na Janma
Ma Bandhur Na Mitram Guru Naiva Shishyah
Chidananda Rupah Shivoham Shivoham ||5||

Translation (5): I have no death or fear of death, nor any cost or creed, I have no father, nor mother, as I have no birth. Have no brother or friend nor teacher or student. I am the pure form of consciousness—the blissful and eternal Shiva.

अहं निर्विकल्पो निराकार रुपो
विभुत्वाच्च सर्वत्र सर्वेन्द्रियाणाम्
न चासंगतं नैव मुक्तिर्न मेय
चिदान्द रूपः शिवोऽहम् शिवोऽहम् ॥६॥

Aham Nirvikalpo Nirakaara Rupo
Vibhutvaatcha Sarvatra Sarvendriyananaam
Na Cha Sangathan Naiva Muktir Na Meyah
Chidananda Rupah Shivoham Shivoham ||6||

Translation (6): I am devoid of duality and formless, I exist everywhere, pervading all senses, I am neither attached nor free. I am the pure form of consciousness—the blissful and eternal Shiva.

Summary and relevance to RCOM

Adi Shankaracharya while wandering in Himalayas met for the first time Guru Bala Shankar. Guru Bala Shankar asked his introduction—'Who are you'. The answer of Adi Shankaracharya to his Guru, in compiled form of six Sanskrit shlokas became the famous 'Nirvana Shatakam' [1]. His answer reflects as: After explain in first six shlokas as to What he is not, while a pure consciousness, further asserts 'What he is', i.e., he is devoid of duality, formless, all pervading, neither attached nor free, while asserting that I am the form of consciousness, the bliss and eternal—The Shiva

These six shlokas, or Shatakams (in Sanskrit) of enlightenment (Nirvana) asserts the fundamental concept of existence, which is all pervading, and in contrasts to the modern science's concept of 'consciousness' where consciousness essentially remains at the level of brain. Of course, from the point of human behavior aspects this concept which is confined to material world is directly relevant to modern science and even RCOM where it forms subset of minds; however, it is also important to understand that Nirwan Shatakam points to highest level of consciousness and consciousness of mind is a subset of this consciousness. This argument or interpretation of consciousness while providing the fundamental form of consciousness enable modelling fundamental of RCOM philosophy that is 'the materialistic or mind-consciousness' which enables development of concept of sense bases required to develop perception of external world as also the physical action conditions required to act in the internal and external world. These shlokas bring out the higher reality—the seer consciousness of existence which is vital and fundamental to future research where the journey starts with the conventional conditional mind to reach a state of self-realization—indeed a long journey for the modern science defined by Spirituality.

Here, we connect the original and fundamental or higher level of consciousness to the lower level or consciousness of mind that is required and relevant to RCOM the postulation that consciousness is fundamental that connects all of us while working at individual level, further at organizational, national and further at universal level and re-affirms another thought of Hindu culture 'Vasudha Iva Kutumbkam' in Sanskrit that comes from Maha Upanishad, which translates to 'The World is one Family'.

D.1.2 Bhagavad Gita, Attributed to Veda-Vyasa (Krishna Dvaipayana) [2]

Ved Vyasa : Author of Bhagvad Gita

भागवद् गीता—A brief background: The available sources indicate that the great epic creation called The Bhagavad Gita authored and credited to Maharishi Veda Vyas, letter many holy souls translated it as per their understanding of the knowledge imparted in battle ground by Lord Shri Krishna to the great worrier Arjun one of the five brothers of Pandava in war to protect Dharma—the laws of nature against a team of Kaurawas. Therefore, Bhagvad Gita is a book that records the dialog between Lord Shri Krisna and Arjuna. This section reproduces the shlokas from one of the translation of verses of Bhagavad Gita by A.C. Bhaktivedanta Swami Prabhu Pada [3] and interpreted by the author for application of these translation to RCOM. This input is mostly related to development of human models and associated attributes, understanding of pure consciousness, role of Dharma (Science of Nature) that can be extended to development of psychology and conscience related aspects in RCOM culture.

D.1.2.1 Chapter 18/Shloka 18

ज्ञानं ज्ञेयं परिज्ञाता त्रिविधा कर्मचोदना ।
करणं कर्म कर्तेति त्रिविधः कर्मसंग्रहः ॥18॥

Translation and relevance to RCOM

This shloka underlines the role of knowledge and associated factors in responsible for actions. Knowledge, the object of knowledge, and the knower are the three constituents that motivate action. There are three kinds of impetus for daily work: knowledge, the object of knowledge, and the knower. The instrument of work, the work itself, and the worker are called the constituents of work. Any work done by any human being has three elements. Before one acts, there is some impetus, which is called inspiration and that forms motivation. Any solution arrived before work is actualized is a subtle form of work. Then work takes the form of action. First one must undergo the psychological process of thinking, feeling and willing, and that is called impetus. The inspiration to work is the same if it comes from the scriptures or from the instruction of the spiritual master. When the inspiration is there and the worker is there, then actual activity takes place by the help of senses, including the mind, which is the Center of all senses. The sum of all the constituents of activity are called the accumulation of work.

In RCOM the human action has two major components: the cognitive part and the action part. The above shlokas bring out how the inspiration precedes motivation and in turn motivation precedes the action be it aspects related to reliability, safety and security. Therefore, conditioning for right knowledge leads to right intentions and finally right action. This aspect is vital for structuring the training program and later monitoring of adequacy of available expertise in the field.

D.1.2.2 Chapter 18/Shloka 19

ज्ञानं कर्म च कर्ता च त्रधैव गुणभेदतः ।
प्रोच्यते गुणसंखयाने यथावच्छृणु तान्यपि ॥

Translation and Relevance to RCOM

According to the three different modes of material nature, there are three kinds of knowledge, action and performer of action. Three modes of material nature are: Mode of Goodness is spiritual, i.e., illuminating; Mode of passion is when one sees that everybody is different type of living entity; this will separate him from others, and the actions will be in mode of passion; and Mode of ignorance is when one is not connected with the reality or true knowledge that might lead to conducive to laziness and indolence.

In RCOM, two modes of Material Nature, Mode of passion and mode of ignorance are more relevant as the spiritual mode may have the relevance to higher level of consciousness, i.e., pure consciousness. The mode of goodness, considering that it requires seer level of consciousness and attainment of self-realization or nirwana, cannot be expected in a normal human being. Since our conditioning is more connected material world; therefore two modes, i.e., passion and ignorance are relevant. In Karma Yoga, passion is when one is attached with the work, for RCOM this is the condition for higher performance. However, certain attributes of goodness, like compassion for others, an individual who is initiated into Karma Yoga, follows principles of Ahimsa, follows ethical and moral value system which are also relevant to RCOM culture. The basic purpose of training and higher administration is to train and ensure trained staff in perpetuity and monitoring to ensure that risk levels are acceptably low.

D.1.2.3 Chapter 2/Shloka 47

कर्मण्येवाधिकारस्ते मा फलेषु कदाचन ।
मा कर्मफलहतुर्भुर् मा ते संगोऽस्त्वकर्माणि ॥४॥

karmanye evadhikaras te
ma phaleshu kadachana
ma karma-phala-hetur bhur
ma te sango ‘stv akarmani

You have a right to perform prescribed duty, but you are not entitled to the fruits of the action. Never consider yourself the cause the results of your activities, and never be attached to not doing your duty.

Though this shloka appears to have idealistic thought that might be difficult to accept. But if one considers over attachment to the fruits of work, one will find that this aspect is possibly root cause of many evils. Further, if one follows the message of this shloka and dedicate to the duty whole heartedness, since he is predominantly in the mode of Karma Yoga success and highest level of bliss can be expected.

D.1.2.4 Chapter 10/Shlokas 4 & 5

बुध्दिर्ज्ञानमसम्मोहः क्षमा सत्यं दमः शमः ।
सुखं दुखं भवोऽभावो भयं चाभयमेज च ॥४॥
अहिंसा समता तुष्टिस्तपो दानं यशोऽयशः ।
भवन्ति भावा भूतानां मत्तप पृथग् विधाः ॥५॥

biddhir jnanam asammohah
kshama satyam damah samah
sukham duhkham bhavo bhavo
bhayam cabhayam eva ca
ahimsa samata tustis
tapo danam yaso yasah
bhavanti bhava bhutanam
matta eva prthag-vidhah

Translation and Relevance to RCOM

Intelligence, knowledge, freedom from doubt and delusion, forgiveness, truthfulness, control of the senses, control of the mind, happiness and distress, birth, death, fear, fearlessness, nonviolence, equanimity, satisfaction, austerity, charity, fame and infamy—all these qualities of living beings are created by Me alone.

As for RCOM culture is concerned, these two shlokas provides invaluable input for developing human attributes as also root causes responsible for human behavioral aspects. Some attributes are tangible, like intelligence, knowledge, freedom from doubt and delusion, resilience to equanimous state through distractions and emotions, that is again governed by control of senses and mind, normal happiness quotient and fear. Others are like charity, fame, austerity ranks high on spiritual scale; they are advantageous for quality of human life but intangible and beneficial in long run.

D.1.2.5 Chapter 6 Shloka 34

चञ्चलं हि मनः कृष्ण प्रमाथि बलवद्दृढम ।
तस्याहं निग्रहं मन्ये वायोरिव सुदुष्करम् ॥३४॥

(Chanchalam hi manah krsna

pramahi balvad drdham

tasyaham nigraham manye

vayor iva su-duskaram)

असंशयं महाबाहो मनो दुर्निग्रहं चलम ।

अभ्यासेन तु कौन्तेय वैराग्येण च गृह्यते ॥३५॥

asamsayam mahabaho

mano durnigraham chalam

abhyasena tu kaunteya

vaireggyena cha grhyate.

Translation and relevance to RCOM

The mind is restless, agitating, turbulent, obstinate and very strong, O krsna and to subdue it, I think, is more difficult than controlling the wind. Lord Sri Krsna said: O mighty-armed son of Kunti, it is undoubtedly very difficult to curb the restless mind, but it is possible by suitable practice and by detachment.

The Chap. 6 on Dhyana (Meditation) Yoga, brings out one of the major human weakness, the 'restless of mind' that keeps wandering. It adversely affects, apart causing the hindrances or distractions to achieve spiritual goal like self-realization, the other qualities like focus, attention, sustaining higher level of awareness and leads to frequent distractions from the job or work in hand that tends to human error or failure.

The Dhyan Yoga or Practice of one of the techniques of Dhyana, e.g. Vipassana Meditation, where one takes a sitting posture and focusses on one of the objects, it could be breath, god or even scanning the body as an independent observer. These practices provides the knowledge and help the goal of self-realization [5].

D.1.2.6 Chapter 18/Shloka 6&7

एतान्यपि तु कर्माणि संगं त्यक्त्वा फलानि च ।

कर्तव्यानीति मे पार्थ निश्चितं मतमुत्तमम् ||6||

etany api tu karmani

sangam tyaktva phalani ca

kartavyaniti me partha

niscitam matam uttamam

Translation and relevance to RCOM

This shloka underlines the role of renunciation of material worldly aspects, like comforts and making sacrifices for purification of mind and body. Renunciation is of three kinds, sacrifice, acts of charity and penance. These activities should be performed without attachment or any expectation of results. Although all sacrifices are purifying, one should not expect any result by such performances. In other words, all sacrifices which are meant for material advancement in life should be given up, a sacrifice that purify one's existence and elevate one to the spiritual plane should not be stopped. Everything that leads to Krsna consciousness must be encouraged. In the Srimad-Bhagavatam also, it is said that any activity which leads to devotional service to the lord should be accepted. That is the highest criterion of religion. A devotee of the Lord should accept any kind of work, sacrifice or charity which will help him in the discharge of devotional service to the Lord [2].

नियतस्य तु संन्यासः कर्मणो नोपपद्यते ।

मोहात्तस्य परित्यागस्तामसः परिकीर्तित ||7||

niyatasya tu sannyasah

karmano nopapadyate

mohat tasya parityagas

tamasah parikirtitah

Translation and relevance to RCOM

In RCOM, the assigned duty occupy one of the major dimension. In fact, duty performed with sincerity, dedication and with utmost attention and awareness following plant procedures and rule books is a hallmark of superior performance. This shloka advises that prescribed duties should never be renounced. If one gives up his prescribed duties because of illusion, such renunciation is said to be in the mode of ignorance.

To further clarify, it is emphasized that work for material satisfaction must be given up, but activities which promote one to perform assigned duty is recommended. As performing assigned duty is part of Karma Yoga—one of the four types of Yoga which is considered directly relevant to material world. A professional should who is performing Karma Yoga is also entitled to the path of enlightenment as Bhakti Yoga, etc. The advantage of Karma Yoga is that apart from performing duty sincerely one has entitlement for access to Gyan Yoga when the purity of profession takes him spiritual and philosophical advancements.

D.1.3 Art of Living—Vipassana School of Meditation

Ignorance has been identified as one of the leading root causes of failure. This is the reason this aspect has been dealt with not only at spiritual, but also at philosophical and psychological level. Modern science though has a systematic academic system to impart knowledge of art, science, language, commerce, etc. through a formal system, the moral aspects appear to be lacking. The moral and philosophical aspects, like ethics, duties, responsibilities, care and compassion, consideration to inner happiness and realization that it is the inner happiness that is important as material approach to life, provide short-term happiness and often long-term misery. For sustained happiness, one has to go inside or look inside (Vipassana in Pali language) to achieve self-realization [Art of Living]. Vipassana School of Meditation asserts that ignorance is the root cause of suffering [Art of Living], and the sequence of events/phenomenon that leads to ignorance is as follows:

If ignorance arises, reaction occurs;

if reaction arises, consciousness occurs;

if consciousness arises, mind-and-matter occur;

if mind-and-matter arise, the six senses occur;

if the six senses arise, contact occurs;

if contact arises, sensation occurs;

if sensation arises, craving and aversion occur;

if craving and aversion arise, attachment occurs;

if attachment arises, the process of becoming occurs;

if the process of becoming arises, birth occurs;

if birth arises, decay and death occur, together with sorrow, lamentation, physical and mental suffering, and tribulations. Thus arises this entire mass of suffering.

Relevance To RCOM: Recognizing that ignorance and its adverse effect have potential for degraded or wrong action and consequences, the RCOM adopts the above cause-effect paradigm to understand the chain of events that provides a cause-effect framework to improve the performance in operational ecosystem, particularly to reduce the chances of human error. The RCOM cause-effect paradigm is as follows:

If right consciousness occurs, right mind-matter occurs,

If right mind-matter occurs, right sense-bases occurs,

If right sense-bases occurs, right contacts and connect occurs,

If right contacts and connects occurs, right cognition occurs right cognition occurs, right analysis and concentration occurs

right analysis and concentration occur, If right perception occurs,

right perception occurs, right physical response occurs

If right physical response occurs, intended outcomes occurs

If intended outcome occurs, successful completion of task is achieved

In this paradigm starts with right consciousness. Right consciousness or a consciousness has been assumed to be devoid ignorance. Here the right consciousness captures knowledge, wisdom, free of distraction and right conscience as we are also addressing moral values). Further, in this paradigm the consciousness is fundamental to individual human or plant and further organization as whole. Therefore, consciousness has two states, nondual form which is fundamental, and the manifestation of this as material consciousness, i.e., dualistic model (in line with the concept of neuroscience). This right conscience gives rise to mind-matter or physical manifestation now we call consciousness that operates at the level of brain and activates right mind—an accepted material entity. The right mind-matter at the level of brain in association with the respective brain processing faculties along with memory develops neural correlates of consciousness (NCC) or for the sake of argument the neural correlates of mind (NCM) gives rise to right six sense bases. These sense bases enable establishing the contact with inside and outside environment. This state creates right cognition that enable operation of mental process to understand the external world (in the context of operational task). This cognition apart from rendering right agility supports the focus and concentration and further a sharp awareness and alertness. This state having right cognition and consciousness factors enable development of perception of external environment as also about the implementation scheme enabled through 4 W + 1H. The right perception drives the five action organs (hands, feet, reproductive organ, skin and anus) supported by the nervous system that facilitate to-and-fro communication and command between brain faculties and the individual action organs. Finally, the right physical response from the external environment based on the task and the feedback obtained after altering the state of the system leads to completion of intended task and thereby results into right action that is vital for successful human action.

D.2 Poems, Poetry, Prose for Risk-Conscious Operations Management (RCOM) Culture (जोखिम-चैतन संचालन प्रबंधन)

Music song and poetry is considered to be an effective way to connect to the object of desire. The poetic or musical expressions are considered to be superior, in terms of expressing emotions, motivation, inspiration, particularly when the same items becomes part of group or systems consciousness. It binds together and proved to effective for team work. In RCOM, the poetic expressions serve two purposes. One for any operational activities it helps evaluate the involvement of sense bases as also action bases and provide an approach to estimation of human success/failure probability-based employing CQB approach as illustrated in Chap. 7. Second, the poetry and songs produce an emotional charge necessary for tasks success. Further, it is easy to remember the expression when it comes as song or jargon or prose then simple text or phrases. Therefore, poetic expressions form part of RCOM culture, to express declarative, procedural, diagnostic and analysis. Following are some compositions in Hindi with translation in English for illustrations purpose.

Author: प्रभाकर वि वर्दे **(P.V. Varde, Author)**

D.2.1 RCOM-Formula for Success of Work

काम करने के तरीके अनेक,
पर, सफलता का सुत्र एक,
"चैतन-मन, चैतन-अहसास,
चैतन-गंध, चैतन-श्वास,
चैतन-श्रुति, चेतन-दृष्टि,
होगी उपर तेरी प्रतिश्ठि" ||

Translation and relevance to RCOM

There are many ways and levels to perform your work or duty in an operational ecosystem but there is only one formula for that enables successful completion of your task. This points to the requirements of sense bases that part of conception of the RCOM. The formula is have Conscious-mind that refers to the highest alertness level, focus and concentration; Conscious-feeling or awareness of the surrounding and related aspects; Conscious-nose or smell power to detect any foul smell; Conscious-breathing or be aware of breathing—that points to your level of mental stability or one of the symptom of stress levels, while facing a challenging task in the field or in control room situations; Conscious listening/or hearing to detect or distinguish the abnormal from normal in the plant surrounding, and a the last but most important, i.e., Conscious-seeing again supported by conscious-mind to develop a perception and inference from what is being observed. With this formula in place, the last line means that either the individual or the organization will achieve higher reputation enabled by higher operational performance.

The context is that the human reliability does not deal with only to 'What causes human error ', but also 'How to understand the human root causes' the concept that finds an exclusive component of human reliability in RCOM, where the root cause analysis does not remain limited to the level limited error of commission, error of omission, mistake, lapse, etc. but enables the root cause analysis what was lacking at physical, physiological or mental level, for example, 'What fundamental factors distract the human that lead to inadvertent action'. Based on this human root cause analysis and understanding, take action at individual and organizational level such that the organization can have an effective mechanism to improve human performance.

D.2.2 अभियांत्रिकी संचालन *(Reactor Operation)*

साथी चल ओ साथी चल,
लेके सपनो का ताजमहल ।
करे पुरा भारत जिस पर है गर्व,
वह हमारा प्यारा, शक्तिपिठ ध्रुव ॥

चलो बनाए नये आयाम, ऊत्कृष्ट पल,
हमारे पास है प्रशिक्षीत संचालन दल ।
गर कोइ आये संयन्त्र मे रुखावट या घाव,
गुणवत्ता हमारी जान लो, करेंगे रखरखाव ॥

देखे वेज्ञानिक तके तेरी ओर,
चले संयंत्र हो नये अविश्कार ।
लहराएँ तिरंगा विश्व पटल पर,
विज्ञान का हो जय जय कार ॥

साथी चल ओ साथी चल,
लेके सपनो का ताजमहल ।
करे पुरा भारत जिस पर है गर्व,
वह हमारा प्यारा शक्तिपिठ ध्रुव॥

Translation and relevance to RCOM

This song is dedicated to reference operational ecosystem of Dhruva—A 100 MWth Indian research reactor Dhruva at BARC, where the author has served since its commissioning phase and first criticality—all these 35 years and now working as Raja Ramanna Fellow. Dhruva operations had a motivated expert operational, maintenance and services team and that enabled smooth and successful operation of Dhruva for over 35 years. The basic emotions are the teamwork walking further with is enhanced motivation, dedication such that every individual takes pride in identifying himself with this great organization.

The translation is as follows: Lets move ahead with higher vigor, dedication, with a dream of highest level of excellence for this reactor Dhruva. Not only us the organization but the whole of India feels a proud about this powerhouse of energy which makes this reactor amongst one of the top (in terms of power and thermal neutron flux) research reactors in the world).

Further, in second stanza the message is, let us work further for new achievements and create beautiful moments for the country. Of course, if confronted with any operational or safety challenge we have a highly trained operational and maintenance staff, they have capability to address the challenge to meet risk and reliability targets.

The research reactors facilitate R&D platform for scientists and development of isotopes and other experiments for societal development, like agriculture, healthcare, forensic, etc. The poem says that the scientists and the end users eagerly wait for the operational time or operational reliability such that they can serve their societal purpose. The achievements and advancements in the area of application of nuclear and science and engineering will further enhance India's recognition as leading nation at global level.

For RCOM, approach has been developed keeping in view the generic experience and insights developed towards working on developing improved models and methods for human reliability. In fact, development and operation of Dhruva Simulator [3] was the first step improve the training of operating staff and experiments on developing the RCOM approach. It is established that the research reactor platform provides research for developing the risk reliability models right from components, structures and human reliability.

D.2.3 *Sanchalan* सुरक्षा

रिऐक्टर प्रचालन एवम् रखरखाव,
लक्ष हो, कम से कम रुखाव, और चलाओ,
देने को सर्विसेस सामाजिक
एवं पलेटफोर्म वैज्ञानिक ।
पर, साथ मे चेतना जगी रहे,
ज्योत सतत ये जलती रहे ।
की मुल मंत्र है,
'अधिभावी कारक है संरक्षा',
और फीर सुनीश्चित हो,
संरक्षा के साथ सुरक्षा,।

अच्छे से यह जानले,
और टाले
अवांछित जोखिम,।
सुरक्षीत और संरक्षीत
तंत्र करे बडा काम
उच्च पैमाने संरक्षा सुरक्षा,
और क्वलिटी सर्विसेस के,
बनाए रिऐक्टर ईको-सिस्टम को
वहनियता प्रामाणित तंत्र ।

Translation and relevance to RCOM
This highlights the role of safety and security. It underlines the role of maintain reliable and safe operation of the research reactor with the rule that safety is an overriding factor and operational decisions should be made keeping in vie the highest importance to safety. As the reliable operation, i.e., operation without interruption for a given mission is the key to ensuring required beam time for experiments and irradiation levels for isotope production and tests and experiments.

The operation team should ensure safety and security provisions while meeting operational goals and services toward ensuring the norms of sustainable operations.

In RCOM, there are considerations of safety and security risk assessment. Apart from this, there is a recognition that unreliability or unavailability of reactor due to frequent tripping and or loss of plant availability due to operational challenges adversely affect the scientific and societal goals. For example, the delay or nonavailability of certain isotopes comprises the delivery to healthcare systems that has an explicit risk component for the end user. Therefore, the reliability risk is one of the major components for generating an integrated statement of risk.

D.2.4 जोखिम-चैतन संचालन प्रबंधन—*Mantra 5W+1H'*

जिवन मे सफल, होना सब चाहे,
काशँ सफलता का, मंत्र मील जाऐं,
तो फीर देर कीस बात की,
सोचो न फीर बार भी,
वैश्विक पता इस प्रकार है

सफल-कैरीयर@सफल-काम.एच

होए पासवर्ड: 4W + 1H'
है सफलता कि यही कुजिं,
या कहलो पासवर्ड या फीर मंत्र
ओर समझलो इस, एक्शन तंत्र का मंत्र

'चार-डब्ल्यु-प्लस-वन-एच',
पहला W फाँर 'व्हाट टु डु',
दुसरा W फाँर 'व्हाय टु डु'
तिसरा W फाँर 'व्हेर टु डु',
चौथा W फाँर 'व्हेन टु डु',
एक H फाँर 'हाउ टु डु',
फीर मीलेगा एक्शन का भान,
जागृत रखे ईन्द्रियाँ महान,
और रहे परिस्थितियो का ध्यान,
जब एक्शन मे होगा भारी एक्सेलेंस,
सफलता आयेगी तोडकर सारी फेन्स ||

Translation and relevance to RCOM

This section proposes a formulation for enhancing the success rate for the tasks in various endeavors of operations. This formulation is a function five Ws and one H or 4W + 1H. Before a task is initiated be at mental level (analysis/diagnosis/logics) or physical level, one should think in terms of a) *What* is to be done: this involves development of an objective and aim of the task as also the end results or expectations from the task, b) the second W refers to *Why* this task is to be performed this will help assess the justification and benefits of the job as also the targets that need to be achieved, c) the third W refers to *'Where'*, i.e., physically in which area the task will be performed and the local conditions and accordingly planning for logistics and other support including the human resource, associated risk or hazards and actions related to having adequate protections, d) the third W deals *When*, i.e., with planning for appropriate time for taking action after giving due considerations to various related factors, including the follow-up actions; e) the last or 5th item in the formulation is *How*, i.e., this requires many factors and it could be range from resources, selection, prioritization, optimization, modelling, expertise, analysis, diagnosis. This step deals with aggregating or revisiting 4 other Ws in the iterative manner to arrive or answer How factors associated with the task.

This formulation in RCOM is one of the approaches to reduce the error during task planning, execution and follow-up after the actions are performed. It is a common practice and very closed to human way of thinking and memorizing, prioritizing the actions using jargons, or musical or rhythmic notations or by abstracting to effectively perform a task without missing any element of the task. The sequences of Ws and H is an indicative one; therefore, the order of these Ws and H are changed depending on the nature of the task. Each W can have involvement of one or more than one sense or action bases. This formulation provides a vital step in arriving at analysis and evaluation of the tasks through CQB model.

Examples for Operational Jargons/Poems/Songs/Pros—Declarative Knowledge

D.2.5 Types of Valve

वाल्व होते टाइप, तीन | रहे यह नित सदैव ध्यान ||

ग्लोब, बटरफलाइ, और गेट | वाल्व बिना चले न प्लाटं ||

Translation and relevance to RCOM

This prose carries declarative knowledge, i.e., it is a response to Declarative Knowledge Check 'What are the major types of Valves in the plant'. The prose provides the answer as: There are three types of Valves: globe valve—used where the flow control is required, butterfly valve again for isolation; however, small leak pastness can be tolerated in the system, and third the gate valve for Isolation to rule out chances of leak pastness.

D.2.6 Valves Function

गेट और बटरफलाइ, करे आइसोलेशन का काम |

जब करना हो फलो कन्ट्रोल, ग्लोब बनाये काम ||

Translation and relevance to RCOM

D.2.7 Role of Reactors

सांई इतना जानिये, चेन रिएक्शन रिएक्टर की जान,

है आधुनिक विज़ान का, यह एक अद्भुत वरदान |

ऊष्मा, न्युट्रान दोनो बने, करे मानव कल्याण।
गहन ऊष्मा शक्ति (se ho), विध्युत की आपुर्ति ।
न्युट्रान flux पुरी करे अनुसंधान विज्ञान की चुनोति ॥

Translation and relevance to RCOM
The nuclear reactors are source of heat energy as well neutron enabled by a precious phenomenon called chain reaction. The poetry says friends please know that the chain reaction in the reactor core is central to reactor operation and this can be considered as one of the wonderful blessings of modern science. The heat energy as well as the neutron flux obtained generated in the reactor core serves the humanity and society at large. The heat energy produced in the power reactor provides for electricity generation while the research reactor is a source of neutron that enables advanced research and experiment in the area of science.

D.2.8 Reactor General Description (By Comparing with Human Body Functioning)

पाइपींग धमनियां प्लांट की, पम्प दिल और जान,
बडे बडे हिट ऐक्सचेंजर करे, कोर-उष्मा का विसर्जन,
कन्ट्रोल-इन्स्ट्रुमेन्टेशन बने कान्शियसनेस, कोगनिशन,
विश्वसनिय विद्युत ऊर्जा मानिये, बने प्लांट का भोजन,
ईन्डीकेटर्स रेकार्डस मिमीक्स, ट्रिप अलार्म कंडिशन,
सुनिश्चित करे स्वास्थ मोनिटरिंग, आटो सेफ्टि ऐक्शन
हजार्ड कन्फाईन्मेंट के लिये कंटेनमेन्ट हे आवरण

Translation and relevance to RCOM
In this short poem for the sake of explanation, the salient features of nuclear plant operation have been explained by making a comparison with human to human body. The poem says the piping system is akin to the vanes, while the pump is like heart in human body. These two together enable circulation of fluid, blood in human body and coolant in nuclear plant through heat exchanger, in the system to provides for process of circulation necessary for operation. The plant Control and Instrumentation (C&I) is like cognitive system in human body which collects data from various systems, subsystems, and faculties in the system performs control and regulation through logic operations, computations and ensures functional and process requirements for the system while ensuring safety functions. Both human body and plant require energy for its reliable operation. What is food and drink for

human body, the electricity for the plant. The instrumentation and control function in the control room of the plant can be compared with various activities in the brain that ensures health monitoring and management while ensuring auto safety function. What is the body for human is containment for the plant that safeguards the plant along with controlled access and exit functions.

D.2.9 Preparing for an Emergency Procedure Execution

साध, पहले मन, फिर आत्मविश्वास ।
जागृत चेतन इन्द्रिया, ले काम का भास ।
ले अब संज्ञान, समय कितना है पास ।
लगा लक्ष पर निशाना, नही होगा तू निराश ॥

Translation and relevance to RCOM
Before attempting any task be it mental or physical, the poem envisages conditions. First, have composed and stable mind, then the required self-confidence, next awareness and alertness through right consciousness levels. Then develop a perception of the task to be done while assessing the time available for completing the task. Once you perform the job with known target, you will not be disappointed, i.e., success is assured.

D.2.10 Guidelines for Physical Action

जहा हाथ, वहा नजर ।
जहा नजर, वहा मन ॥
जहा मन, वही चेतन ।
जहा चेतन, वही भगवन ॥
जहा भगवन, वही प्राण ।
जहा प्राण, वही है मान ॥

Translation and relevance to RCOM
Normally, it is the case that mind precedes the physical task. However, when actually performing the task, the objective is successful completion of the task as per the identified operational procedure. Here, the above poem suggests the sequence for performing the task at locations or in field as: Mind your hand and ensure that your vision is focused on the hand, be its grip, direction of rotation, pressing of

switch or rotation of an encoder, further, ensure that mind is there where you have your vision basically to have right focus, concentration for 'what to do component' of the job. It further asserts that the mind follows consciousness. The next two lines defines the saying 'Work is worship'. Worship is considered here as dedication, sincerity, moral values—the object of dedication: the plant.

D.2.11 Purpose of Centrifugal and Positive Displacement Pump

Flow ल्फो जहा उद्देश्य हो
सेन्ट्रईफ्यूगल पम्प करे काम ||
पोसिटिव डिस्प्लेसमेन्ट लगे वहा |
प्रेशर बनाए जहा काम ||

Translation and relevance to RCOM

The plant operation training system check lists have mainly three categories of knowledge base checks, general system features, declarative knowledge checks that deals mostly with the question 'What', e.g., what are the design features, what is the function of.., that covers normal design and operational information about the SSCs and process functions and associated parameters, second category is, procedural knowledge check, action components of jobs, like altering the state of system, specific surveillance tasks, and third is trouble shooting, that includes fault diagnosis, prognosis, analysis of direct cause of failure, corrective actions and recovery, and decision making specially in emergency scenario, etc.

The above para of a poetry presents an example of declarative knowledge checks. This stanza is an answer to the check point 'What are the major categories of pumps used in the plant ? And the answer is communicated as, there are two common types of pumps found in a plant, centrifugal and positive displacement pump. Centrifugal pumps are used where there required flow or to be precise higher flow is required, whereas the positive displacement pump is used where higher pressure is required.

D.2.12 Ensuring Safety and Security

अलर्ट रहो अवेअर बनो
गलति करने से बचो |
और बडाओ सुरक्षा

अलर्ट रहो अवेअर बनो
संदिग्ध काम स्थितियो
परिवेश पहचानो ।
और सुनिश्चित करो संरक्षा ॥

Translation and relevance to RCOM
This stanza of a poetry highlights the requirements ensuring safety and security at individual or at plant level. It translates to—be alert and aware such that inadvertent or random human failure is avoided while performing general surveillance or performing a task at mental or physical level to ensure safety. Again, to ensure security the poem says, remain alert and aware such that intended malicious or acts or suspicious acts or foul symptoms by individual or group of people inside or outside of the plant can be detected. Even though the plants physical and cyber protection systems are provided as part of design basis threat strategy, a heightened alertness and awareness could be an effective strategy to avert the deliberate actions to harm the plant.

D.2.13 Performance of Oral or Simple Calculations

का करिके पुनी भाग कर,
फीर गुण ले ही सुजान ।
ता पिछे धन रिण कर,
भिन्न रिती सो जान ॥

Translation and relevance to RCOM
It is a common observation that oral or elementary calculations involving only simple four operations, i.e., addition, subtraction, multiplication and division, are performed typically in a control room setting to understand or evaluate plant process system performance. Often it requires the lumping of parameters in groups using bracketed terms. The above poetry is a formulation that says while first step perform multiplication, then division followed by addition and then subtraction. This action falls in the category of rule base tasks at mental level.

D.2.14 Thumb-Rule for Operation of Valve (वाल्व खोलन के वास्ते)

ध्यान रहे AO/CC, यह मन्त्र ।

माने, Anticlockwise फार Open—AO

Clockwise फार Close—CC

फीर जरा instrument reading

देख लिजीये, केसा चल रहा है तंत्र ॥

Translation and relevance to RCOM

The above is a formulation for procedural check point and the question is: Please tell the direction of rotation of hand wheel for closing and opening of a valve? Valve manipulation is a routine activity. The above poem provides an answer that is easy to memorize called AO/CC formulation. Anticlockwise rotation is for opening the valve and clockwise rotation for closing the valve.

D.2.15 Limitation of Slogan

नारो से नाता रखलो

यह है अच्छी बात,

लेकीन, काम तो सिर्फ बने

करने से, रहे हमेशा याद ॥

Translation and relevance to RCOM

The message in the above phrase or prose is that use the slogans are good idea to spread the message and information. However, it might work more effectively if there is a mechanism that helps at working level by playing out at consciousness and cognitive level where the sense and action organs are involved with right perception. The message should go deep down such that it becomes characteristic or defining quality at individual well as organizational level.

D.3 Sage Kabir's Poetry—On Moral and Spiritual Guidance

कबीर

कबीर

सांच बराबर तप नही, झुठ बराबर पाप ।

जाके हिरदे सांच है, ताके हिरदे आप ॥

ऐसी वाणी बोलिये, मन का आपा खोय ।

औरन को शितल करे, आप ही शितल होय ॥

गुरु गोविंद दोउ खड़े, काके लागूं पांय

बलिहारी गुरु आपनो, गोविंद दियो बताय।

अर्थः गुरु और गोविंद (भगवान) दोनों एक साथ खड़े हैं। पहले किसके चरण-स्पर्श करें। कबीरदास जी कहते हैं, पहले गुरु को प्रणाम करूंगा क्योंकि उन्होंने ही गोविंद तक पहुंचने का मार्ग बताया है।

जग में बैरी कोई नहीं, जो मन शीतल होय

यह आपा तो डाल दे, दया करे सब कोय

अर्थः अगर हमारा मन शीतल है, तो इस संसार में हमारा कोई बैरी नहीं हो सकता। अगर अहंकार छोड़ दें तो हर कोई हम पर दया करने को तैयार हो जाता है।

साईं इतना दीजिए, जा में कुटुंब समाय

मैं भी भूखा न रहूं, साधु ना भूखा जाय

अर्थः कबीरदास जी कहते हैं कि परमात्मा तुम मुझे केवल इतना दो कि जिसमें मेरे गुजारा चल जाए। मैं भी भूखा न रहूं और अतिथि भी भूखे वापस न जाए।

References

1. Adi Shankaracharya, Nirwana Shatakam Sanscrit Script. https://shlokam.org/nirvanashatakam/
2. The great epic creation called The Bhagawad Gita authored and credited to Maharishi Ved Vyas. https://en.wikipedia.org/wiki/Bhagavad_Gita
3. A.C. Bhaktivedanta Swami Prabhupada, Bhagavad—Gita as it is, The Bhaktivedanta, Book Trust, 1996, Mumbai
4. A.C. Bhaktivedanta Swami Prabhupada, Bhagavad—Gita as it is, The Bhaktivedanta Book Trust, 1996, Mumbai
5. Hart W. "The art of living—Vipassana Meditation as taught by Shri S.N. Goenka, Vipassana Research Institute, Igatpuri, India, 2019.
6. https://www.jagran.com/spiritual/sant-saadhak-10-inspirational-do-has-by-sant-kabirdas-16165083.html
7. Vivek 'Nishu' (Complier), Kabir ke Dohe – Sant Shiromani Kabirdas Ji ke Arth Sahit Anmol Vachan, Vivek Publication, Mathura, India, 2018
8. Kumar J, Joshi NS, Singh T, Varde PV (2020) Dhruva simulator: core neutronics and thermal hydraulics modules' SRESA Journal of Life Cycle Reliability and Safety Engineering, Vol. 9, Issue 1, 2020.

Annexure E
Evaluation of Precursor Risk Factor for Typical Operations Activities

$$R = J_x^* \epsilon_p^* [t/T]^* (C/n))$$

$$n = 50$$

P. V. Varde, *Risk-Conscious Operations Management*, Risk, Reliability and Safety Engineering, https://doi.org/10.1007/978-981-19-9334-3

Precursor risk factor for typical operations activity

S. No.	Precursor description	Activity type	Human CQB Attributes for Precursors			Human-factors J_c	PE rate ∈/ year	Time Window t/T	Potential consequence of PE	Consequence level (Table 4.2)	Risk factor for PE (R/year)
		D/E/O /M/S	$(Q_1 * w_1)$	$(Q_2\ w_2)$	$(Q_3\ w_3)$		P_f	t/Tw	PE_{PC}		
1	Control room supervisory activities	O	4 * 0.3 = 1.2	4 * 0.6 = 2.4	2 * 0.1 = 0.2	3.8	1.5	0.67	Reactor trip/unavailability risk	L-1 1.45	0.12
2	Management of emergency scenario	O/M	8 * 0.6 = 4.8	1 * 0.3 = 0.3	1 * 0.1 = 0.1	5.2	1.0	0.91	Failure to recovery	L-5 10	0.75
3	Field supervision	O	2 * 0.6 = 1.2	6 * 0.2 = 1.2	2 * 0.2 = 0.4	2.8	0.1	0.5	Judgmental/error	L-1 1.2	$3.4 * 10^{-3}$
4	Procedure execution—Op	O	3 * 0.5 = 1.5	4 * 0.4 = 1.6	3 * 0.1 = 0.3	3.4	1.0	0.59	Individual attribution	L-2 2	$8 * 10^{-2}$
5	Process diagnosis	O	5 * 0.5 = 2.5	5 * 0.5 = 2.5	0 = 0	5.0	1.0	0.77	DG diagnosis failure during Class IV PS failure	L-1 1.7	0.13
6	Correlating the evolving scenario during a significant event	O	6(0.8) = 4.8	4(0.2) = 0.8	0(0) = 0	5.6	1.0	0.84	Loss of team capability to comprehend the scenario	L-3 3.5	0.32
7	System state identification	O	7 * 0.7 = 4.9	3 * 0.3 = 0.9	0(0) = 0	5.8	1.0	0.91	Identification for cause of the event	L-2 2.0	0.22
8	Field activities—Status	O/M	3 * 0.3 = 0.9	2 * 0.5 = 1.0	5 * 0.2 = 1.0	2.9	2.0	0.67	Local or reactor availability risk	L-2 1.5	0.116
9	Training lectures	S	5 * 0.5 = 2.5	3 * 0.4 = 1.2	2 * 0.1 = 0.2	3.9	2.0	0.5	Organizational cultural gap	L-2 3	0.24
10	Training and qualification—checklist	S	5 * 0.4 = 2.0	1 * 0.5 = 0.5	4 * 0.1 = 0.4	2.9	1.2	0.56	Organizational cultural gap	L-2 3	0.12
11	Training—and qualification Simulator experiments for	S	5.0 * 0.6 = 3.0	3 * 0.2 = 0.6	2 * 0.2 = 0.4	4.0	1.5	0.84	Organizational cultural gap	L-2 3	0.30
12	Engineering modifications	S	7 * 0.6	2 * 0.3	1 * 0.1	4.9	1.0	0.56	Planning error	L-3 4.0	0.22

(continued)

(continued)

S. No.	Precursor description	Activity type	Human CQB Attributes for Precursors			Human-factors J_c	PE rate ϵ/ year	Time Window t/T	Potential consequence of PE	Consequence level (Table 4.2)	Risk factor for PE (R/year)
		D/E/O /M/S	$(Q_1 * w_1)$	$(Q_2 w_2)$	$(Q_3 w_3)$		P_f	t/Tw	PE_{PC}		
13	Analysis and development	O/S/M	8 * 0.8	2 * 0.2	0(0.)	6.8	1.0	0.56	Individual and training issues	L-3 4.0	0.3
14	In-service inspection	O/M	3 * 0.5 = 1.5	3 * 0.2 = 0.6	4 * 0.3 = 1.2	3.3	2.0	0.77	Inadequate inspection has potential for loss of structural integrity	L-3 4.5	0.57
15	Maintenance of hardware mechanical and electrical	M	2 * 0.2 = 0.4	2 * 0.4 = 0.8	6 * 0.4 = 2.4	3.6	1.8	0.77	CCF potential	L-3 5.0	0.5
16	Maintenance of digital systems and software	M	5 * 0.4 = 2.0	2 * 0.3 = 0.6	3 * 0.5 = 1.5	4.1	1.5	0.91	CCF potential	L-2 4.5	0.50
17	Instrument calibrations	M	3 * 0.1 = 0.3	4 * 0.3 = 1.2	3 * 0.6 = 1.8	3.3	2.0	0.91	CCF potential	L-3 4.5	0.54
18	Mechanical Maint. pump	M	3 * 0.2 = 0.6	2 * 0.3 = 0.6	5 * 0.5 = 2.5	3.7	1.2	0.84	Potential for fire due to assembly error	L-2 2.5	0.185
19	Logging and reporting	O/M	1.0 * 0.5 = 0.5	1(0.3) = 0.3	8(0.2) = 1.6	2.4	4.0	0.77	Human and organizational error	L-1 1.5	0.22
20	Coordination with agencies—control room	O	5 * 0.6 = 3.0	4 * 0.3 = 1.2	1 * 0.1 = 0.1	4.3	2.0	0.9	Individual/organizational attribute	L-3 4.0	0.61
21	Communications and coordination *during* control room in emergency	O	4 * 0.3 = 1.2	3 * 0.2 = 0.6	3 * 0.5 = 1.5	3.3	2.4	0.9	Organizational cultural gap: Error in communication	L-5 7.4	0.57
22	Emergency management actions in control room	O	7 * 0.5 = 3.5	2 * 0.3 = 0.6	1 * 0.2 = 0.2	4.3	2.0	0.84	Organizational cultural gap	L-4 6.0	0.87

(continued)

(continued)

S. No.	Precursor description	Activity type	Human CQB Attributes for Precursors			Human-factors J_c	PE rate ϵ/year	Time Window t/T	Potential consequence of PE	Consequence level (Table 4.2)	Risk factor for PE (R/year)
		D/E/O/M/S	$(Q_1 * w_1)$	$(Q_2\ w_2)$	$(Q_3\ w_3)$		P_f	t/Tw	PE_{PC}		
23	Human–machine interface/interaction as part of recovery procedure during emergency	O	6 * 0.4 = 2.4	2 * 0.4 = 0.8	2 * 0.2 = 0.4	3.6	1.7	0.9	Individual/organizational	L-5 8.0	0.89
24	Updating documents and engineering drawing	E	3 * 0.8 = 2.4	1 * 0.7 = 0.7	6 * 0.2 = 1.2	4.3	1.8	0.77	Potential for missing as-built information leading to error in O&M	L-2 2.7	0.32
25	Shift change over-charge exchange in shifts	O	5 * 0.3 = 1.5	4 * 0.4 = 1.6	1 * 0.2 = 0.2	3.3	2.0	0.9	Organizational cultural gap/individual gap	L-2 2.5	0.18
26	Interpretation of results of analysis—control room	O	8 * 0.5 = 4.0	6 * 0.3 = 1.8	0	5.8	2.0	0.9	Individual capability limitation	L-1 1.5	0.32
27	Plant areas upkeep	O	1 * 0.5 = 0.5	2 * 0.2 = 0.4	7 * 0.3 = 2.1	3.0	2.1	1.3	Organizational cultural gap	L-2 2.0	0.33
28	Plant status monitoring —control room	O	3 * 0.5 = 1.5	4 * 0.4 = 1.6	3 * 0.1 = 0.3	3.4	2.0	1.1	Events induced due to inadequate awareness and attention	L-2 2.5	0.37
29	Events and significant events analysis and corrective actions	O/S	8 * 0.5 = 4.0	2 * 0.5 = 1.0	0.0	5.0	2	0.84	Organizational cultural gap/individual gap	L-3 3.5	0.58
30	Regulatory compliance	S	3 * 0.2 = 0.6	5 * 0.6 = 3	2 * 0.2 = 0.4	4.0	3.0	1.0	Organizational cultural gap	L-3 3.8	0.92
31	Safety documentation	S	4 * 0.4 = 1.6	4 * 0.5 = 2.0	2 * 0.2 = 0.4	4.0	3.0	0.9	Organizational cultural gap	L-2 2.5	0.54

(continued)

(continued)

S. No.	Precursor description	Activity type	Human CQB Attributes for Precursors			Human-factors J_c	PE rate ϵ/ year	Time Window t/T	Potential consequence of PE	Consequence level (Table 4.2)	Risk factor for PE (R/year)
		D/E/O /M/S	$(Q_1 * w_1)$	$(Q_2\ w_2)$	$(Q_3\ w_3)$		P_f	t/Tw	PE_{PC}		
32	Operations documentation	O	3(0.3) = 0.9	4(0.5) = 2.0	3(0.2) = 0.6	3.5	1.0	0.67	Organizational cultural gap	L-1 1.5	0.07
33	TechSpec compliance	S	6(0.7) = 4.2	4(0.3) = 1.2	–	5.4	1.7	0.9	Individual/organizational cultural gap	L-2 2.5	0.41
34	Procedure execution—maintenance	M	5 * 0.6 = 3.0	4 * 0.3 = 1.2	1 * 0.1 = 0.1	4.3	1.5	0.77	Individual limitation/training adequacy	L-2 2.8	0.28
35	Maintenance of tools and support	M	3 * 0.4 = 1.2	2 * 0.4 = 0.8	5 * 0.2 = 1.0	3.0	1.0	0.9	Organizational cultural gap	L-2 2.5	0.14
36	Inadequate or improper testing standby safety systems	O/M	5 * 0.4 = 2.0	3 * 0.4 = 1.2	2 * 0.2 = 0.4	3.6	1.0	1.0	Potential disabilities that affect equipment/system availability	L-4 7.0	0.51
37	Quality control and quality assurance	S	6 * 0.3 = 1.8	3 * 0.5 = 1.5	1 * 0.2 = 0.2	3.5	1.9	0.9	Compromised organizational culture	L-3 5.0	0.6
38	Fuel management	S/O	7 * 0.6 = 4.2	2 * 0.2 = 0.4	1 * 0.2 = 0.2	4.8	1.2	0.9	Threat to safety of fuel	L-3 5.0	0.52
39	Resilience activities related external events	S	8 * 0.5 = 4.0	2 * 0.5 = 1.0	0	5.0	1.4	0.9	Reduction in safety margins	L-5 8	1.0
40	Staffing quality and positions	O	7 * 0.5 = 3.5	3 * 0.5 = 1.5	0	5.0	1.0	0.9	Organizational attitude	L-3 4	0.36
41	Spare inventory for Safety functions	M	3 * 0.3 = 0.9	3 * 0.4 = 1.2	3 * 0.3 = 0.9	3	2.0	1.0	Equipment maintainability issues	L-3 4	0.48

(continued)

(continued)

S. No.	Precursor description	Activity type	Human CQB Attributes for Precursors			Human-factors J_c	PE rate e/ year	Time Window t/T	Potential consequence of PE	Consequence level (Table 4.2)	Risk factor for PE (R/year)
		D/E/O /M/S	$(Q_1 * w_1)$	$(Q_2\ w_2)$	$(Q_3\ w_3)$		P_f	t/Tw	PE_{PC}		
42	Reactor physics and chemistry services	S	8 * 0.6 = 4.8	2 * 0.4 = 0.8	0	5.6	1.1	0.77	Analysis, trend monitoring and core safety activities compromised	7	0.66
43	Financial and administrative system/support that has potential to impacting plant safety	M	8 * 0.8 = 6.4	2 * 0.2 = 0.4	0	6.8	0.15	0.77	Threat to safety management	8	0.13